McGraw-Hill Illustrated
Telecom Dictionary

Other McGraw-Hill Telecommunications Books of Interest

McGraw-Hill Illustrated Telecom Dictionary

Jade Clayton

Third Edition

McGraw-Hill

New York   Chicago   San Francisco   Lisbon   London   Madrid
Mexico City   Milan   New Delhi   San Juan   Seoul
Singapore   Sydney   Toronto

Cataloging-in-Publication Data is on file with the Library of Congress.

McGraw-Hill

*A Division of The **McGraw·Hill** Companies*

1 2 3 4 5 6 7 8 9 0   DOC/DOC   0 7 6 5 4 3 2 1

P/N 137202-4
Part of ISBN 0-07-137201-6

The sponsoring editor for this book was Stephen S. Chapman and the production supervisor was Sherri Souffrance. It was set in ITC Century Light by Techbooks.

Printed and bound by R. R. Donnelley & Sons Company.

   This book is printed on recycled, acid-free paper containing a minimum of 50% recycled de-inked fiber.

McGraw-Hill books are available at special quantity discounts to use as premiums and sales promotions, or for use in corporate training programs. For more information, please write to the Director of Special Sales, Professional Publishing, McGraw-Hill, Two Penn Plaza, New York, NY 10121-2298, or contact your local bookstore.

This work is dedicated to the memory
of Brian Blackburn and all others
that have lost their life,
or a loved one, to cancer.

Contents

Preface

The purpose of this Dictionary is to assist the general understanding of telecommunications through it's most illusive aspect—the terminology. The CD-ROM included with this book is a complete electronic version of the Dictionary readable on Windows and Macintosh computing platforms via Adobe Acrobat Reader. The CD-ROM version of this Dictionary contains "click-able" links to passages within other McGraw-Hill handbooks and textbooks. It is intended to provide additional information and familiarize readers with some of the best educational materials published for the telecommunications industry. They are written from a practical standpoint by some of the best educational authors in the world, and cover their branch of telecommunications very thoroughly. I strongly recommend the books I have referenced within this CD-ROM for further reading.

Telecommunications terminology can be sometimes intentionally and unintentionally misleading. As the telecommunications industry and network evolves throughout the world, there are two kinds of terms (two-and-a-half if you count slang). There are "*Standard*" terms that have been derived from the technologies and standards that services and equipment are comprised of, and there are the many "*Marketed*" terms that sometimes become better known than the standard term. Marketed terms do not shape the future of telecommunications, Standard terms and the technologies related to them do.

Within this Dictionary, you will be able to reference almost all of the major "standard" terminology. There are also many well-known "marketed" terms. During one's educational pursuits of telecommunications it is more important to understand the "standard" terms, know which sector of telecommunications they pertain to, and then pursue further study within your chosen sector of the industry. The true masters of the telecommunications industry can easily identify and see beyond marketing hype. This skill is becoming more crucial to career minded professionals. It can be developed by studying the past, present and direction of industry sectors. This Dictionary and CD-ROM will help you navigate your direction of study, develop a practical stance in the industry, and serve as a good reference when you need to refer to a term or acronym.

If any of my readers has questions or comments pertaining to the material in this book or CD-ROM, I can be reached by e-mail at jadeclayton@att.net. I will do my best to respond promptly.

Jade Clayton

Introduction

This *McGraw-Hill Illustrated Telecom Dictionary* was written to provide convenient and easy to understand definitions of commonly used terminology and technology in the telecommunications industry. It is meant to be used as a reference for IT professionals, public telecommunications company professionals, telephone and computer equipment/services vendors, telecommunications equipment manufacturers and distributors, as well as instructors and students of all levels that have subject matter relating to computer science, information systems, telecommunications, and electronics.

Because the telecommunications industry will constantly change due to improvements in technology and protocol standards, there will always be new terms and acronyms. As these new terms evolve they will be defined in future editions of this book.

This Dictionary has over 3,300 definition entries, and over 8,000 references including appendices and tables. Of these definitions nearly 500 have a picture, diagram, or chart to assist in the definition. Many of the remaining 2,800 definitions refer to a diagram or picture located somewhere within the book.

Acknowledgments

The CD-ROM provided with this book is a team effort product created by the following professionals:

Steve Chapman of McGraw-Hill, Kevin Fry and Doug Havens of DEMOCARD, Greg Evans for Adobe Acrobat software consulting, and John Clayton, for building the actual file and link structure that makes the CD work as nicely as it does.

The author would like to express his sincere appreciation and gratitude to the following individuals for their support during the creation of this work:

Lucas Clayton (my son), Kathy Thomson (my mom), Don Thomson (my dad), Mary Kirkendall, Johnny Clayton, Zachary Thomson, Robbie Thomson, John and Gayle Kirkendall, Drew Kirkendall, Jim and Monica Kirkendall, Dawn Pensiero, Velma Clayton, John F, Dave, and Cari Clayton, George and Lucille Strelich, Paul and Tracy Anderson, David Anderson, Margurite Anderson, Todd and Holly Fraker, Brian, Claudia, Brianna and Cassidy Blackburn, Brett, Vicki and Tristin Cooper, Chad, Nicole and Jordan Player, Mike and Rose Palmer, Brian and Becky Henrie, Corie Wright, Rydell Mitchell, Mike, Donna, and Ricky King, Carrie Hughes, Tamie Jensen, Ty Mutchler, Dan Cummings, Rod Millar, Chuck Geltz, Steve McLiesh, Scott Varley, Jim Holloway, Dan and Nyla Dick, Chuck Griebe, Tricia Drake, Scott Sargent, Dianne Phillips, Bryan Barr, Dave Hepworth, Tim Taylor, Scott McLean, Randy Hornbeek, Paul Stroh, and Chaz Benson.

A big thank you is also given to the following companies and individuals for their material assistance, experience, training, employment, friendship, and help in gathering information for this work:

Co-Workers of American Express—Technologies

Cary Wood, Gary Callister, Allan Vollmer, Allan Wood, Brice Cox, Carol Burbidge, Dan Dick, David Hull, Davin Stillson, Dell Stillson, Dennis Moore, Don Fell, Doug Klentzman, Evan Jones, Gene Farrell, Janet Roothoff, Jeff Hatch, Merritt May, Norman Sarabia, Peggy Young, Rob Ford, Robert Palfreyman, Rod Kearl, Seth Miner, Shawn Llewelyn, Stan Hong, Steve Trauba, Tim Cassady, Troy Ostmark, and Wayne Leota.

Acme Broadcasting
Jim Facer and Mario Heib.
AT&T
Paul Cabibi, Lee Goettling, Woody Wellons, and Craig Parker.
Co-Workers of Microwave Tower Services
Kelly Keisel, Mike Nelson.
Periphonics IVR
Frank Tatulli and Bob Verda.
Co-Workers of AT&T Local
Phil Tonick, Frank Croan, Dave Young, Fred Kroll, Robert Kellett, Paul Backman, Bill Stieneger, Laurie Guluarte, Dick Forfar, Suzy Mang.
Co-Workers of Qwest-USWest
Lee and Kathy Guthrie, Paul Stroh, Kyle Kalian, Reed Madsen, Gary Chavez, Daryle Starr, Ewe, Ken Olson, Curt Breen, John Dutt, Roland (row your Bayliner) Eysser, Lynn Montgomery, Marty Moss, Dave Smart, Randy Ipsen, Scott Everts, Sam Alvarado, Lonnie Mair, Ken Morley, Kent Forbush, Kris Peterson, Anita Shoblume, Maureen Slusher, Randy Smith, Kevin Smith, Ann and Bob Smith, Jan Hembury, Scott Ross, Dina Eddleman, Julie Barney, Ina Allred, Del Seaton, Mike Nielsen, Steve Pryor, Dan Kidd, Kent Brown, Teresa Price, Linda Lujan, Scott Lutz, Ray (come to Jesus) Spencer, Lonnie Kresser, Drew Broadwater, Juan Martinez, Scott Bullock, Dave Evans, Rob North, Jeff Watters, Mark Giesler, Shane Mozaffari, James Chidester, Monte Shosted, Clint Maxfield, Braden Bollen, Darrell Armstrong, William Liberty, Justin Standing, Marty Montano, Georgia and Mike Harris, and Johnny Clayton.
Co-Workers of SITA
Vic Szalankiewicz, Bob Schutz, Karen Carmack, Dave Johnson, Wayne Matrose, Brian McDonald, Rhonda Butler, Marlene Shaffer, Jose Torra, Sheila Simbeck, Doyle Streeter, Judy Newkirk, Celine Birre, C.C. Chaney, Ramon De La Cruz, Alicia Fields, Amanda Irlbeck, and other co-workers of SITA.
Co-Workers of Equant
Jennifer Baer, Azail Dogans, Steve Moore, Ken Prock, Roz Toliver, Carrie Dunbar, Steve Mushet, Khanh Nguyen, Diane Phillips, Jacqueline Poissant, Paul Rabe, Roger Rader, Beth Bailey, Cherie Randolf, Jill Jurenka, Kannan Rengaraju, Denise Rodriguez, Mark Rosa, Reggie Sangha, Rob Wezwick, Stacie Sharp, Anser Siddiqui, Kathy Sims, Rhonda Stephens, Azhar Zaidi, Susan Sweeney, Chinh Tran, Dianne Trotter, Tina Nguyen, Doug Vaughn, Brian Watts, Melissa Wilson, Andre Wright, Azhar Zaidi, Dave Maynard, Wajid Siddiqui, Erik Mathies, Joan Mason, Felicia Lyles, Quoc Chau, Rhonda Lewis, Sterling Levell, Chris Larkin, John Kurth, Dottie Kossman, Jill Hausner, Debbie Knox, Xavier Avendano, Sergio Henrique, Breann Baldwin, Roshann Black, Mark Abrahim, Earl Barnes, Mike Blue, Marc Langroth, Don Brook, Robi Bratton, Julie Cash, Minh Vu, Brian Clark, Don De Vingo, Mark Dunham, Suzanne Engum, Allan Floyd, Dan Gage, Derrick Harris, Pricilla

Farmer, Freddy Henderson, Pricilla Hernandez, Frank Tabone, Rick Holdge, Pam Ingram, Bob Johnston, Edy Kaufmann, Jim Kellum, Veronica Robin, Deone Kemp, Tom Markou, and Benoit Fertil.

Co-Workers of Cisco Systems

Marc Combs, Paul Reid, Mark Wilhelm, Marshall Smith, Jeff Edwards, Mickey Stewart, Johnny McKeever, Bob Monks, Frank Jimenez, Stacie Murray, Tommy Randle, Dewayne Walker, Curtis Palmer, Terri Wolf, Teresa Barton, Mike Burris, Mike Bolla, Brian Donath, Becky Carlson, Glenn Finch, Ed Hodges, Charles Travis, Reno Madsen, Erin Thompson, Brian Morgan, and John Maxwell.

American Airlines

Karen Manson, Laura Ellis, Sandy Dunlap, Brett Henry, Jim Sawyer, Brenda Arnold, Brian Moreno, Lorinda Crawford, Rachel Runfola, Priscilla Sakmary, Doug Helm, and others the of American Airlines/Sabre organization.

Sprint

Dawn Pensiero

Utah Public Service Commission

Judith Johnson and Peggy Egbert.

Nortel Networks

Roger Harry, Tony Daniels, Dave Kahn, Larry Wang, Kevin McGarrell, Cheryl Heinz, and John Bracken.

Blue Cross Blue Shield IT

Mike Leyva

Nucentrix

Donna Ream

First Security Bank IT

Nolan Bitters

CCS Engineering

Clint Smith

Democard—CDROM Operabilty and Multimedia

Doug Havens and Kevin Fry—kevin@democard.com

Adobe Acrobat Consulting and Artwork

Greg W. Evans—GregWEvans@Yahoo.com

The author would also like to thank Thomas Farley and www.privateline.com, who has been instrumental in promoting this book.

The adobe file structure and links on the CD-ROM were created with operational performance in mind by John T. Clayton

About the CD-ROM

The CD-ROM is completed in Adobe Acrobat and can be viewed with Adobe Acrobat Reader. Adobe Acrobat Reader can be downloaded from the Internet at Adobe.com.

As mentioned in the Preface of this book, the CD-ROM contains additional reference material from other McGraw-Hill telecommunications handbooks and textbooks (approximately 500 pages from 14 different books). The additional material is accessed by clicking on icons that appear in the margins as miniature books. This will allow serious readers to become *conveniently* familiarized with additional reading material while referring to definition entries.

It is recommended that a computer with the following minimum hardware requirements be used to operate the CD-ROM: 586 microprocessor, 8 MB of RAM, and an 8X CD-ROM drive.

When the CD is inserted to a CD-ROM drive in a personal computer running Windows 95/98/00 or Macintosh OS, it runs automatically on all computers we have tested it on. The first viewable computer output after the CD-ROM runs is a Multimedia test created by DEMOCARD. DEMOCARD is the creator of the CD-ROM functionality, and the multimedia inserts within the ebook. After the test is complete, the book opens in Adobe Acrobat Reader, and you may begin browsing. There are shortcuts to the beginning of every letter in the alphabet on each page. When the hand controlled by the mouse changes to a pointing finger while placed over a word or picture, you may click the mouse button to be taken to the reference that is referred to on the page.

Acrobat Reader tips to get you started:
To increase the viewable size of the page, use Ctrl+.

To decrease the viewable size of the page, use Ctrl−.

To go to the next page, or previous page, use the arrows on the toolbar that look like the ones below.

To go to a previous view, as in a previous book, or page in another section of the book, use the arrows on the toolbar that look like the ones below.

◄■ ■►

CD-ROM Installation:
FOR WINDOWS: Insert disk into CD-ROM drive, click START, then click RUN. Where it says "look under" highlight the drive number for the CD-ROM. Click browse, then double click on Start_CD.exe. Click on the open button.

FOR MACINTOSH: Insert disk into CD-ROM drive, open CD-ROM, then double click on the "Dictionary.pdf alias".

About the Author

The definitions have been written by Jade Clayton, a telecommunications/ networking professional with over 13 years of experience in various facets of the telecommunications industry including manufacturing, traffic switching, private line, outside plant construction, inside plant construction, broadband transport, cable TV, LAN networking, WAN networking, broadcast radio, point-to-point microwave radio, call center/PBX management, computer telephony integration, and public telephone services installation and maintenance. Mr. Clayton has been a professional author for over 5 years, published his first amateur work over 10 years ago, and has a degree in electronic engineering. He is currently a Systems Engineer at Cisco Systems in Dallas, Texas.

0-99

0 The "in fact" standard number to dial for reaching a local phone company operator or answering service.

1-Pair Gas Lightning Protector Used in Siecor telephone network interfaces (Fig. 0.1).

Figure 0.1   1-Pair Gas Lightning Protector

1FB A service code that defines a flat-rate business telephone line. A line where a subscriber can make unlimited local calls and not be billed extra, regardless of the number of calls or their duration.

1FR A service code that defines a flat-rate residential telephone line. A line where a subscriber can make unlimited local calls and not be billed extra, regardless of the number of calls or their duration.

1MB A service code that defines a measured-rate business telephone line. A line where the subscriber is billed either for the number of calls made or by the minute.

1MR A service code that defines a measured-rate business telephone line. A line where the subscriber is billed either for the number of calls made or by the minute.

10/100 (Ten/One-Hundred) A reference to the newer family of Ethernet as a whole. 10BaseT is 10 Mbps, 100BaseT is 100 Mbps, and 10BaseF is 100 Mbps over fiber optic. It is also referred to as *802, 10/100*. Because the 10BaseT and 100BaseT can interconnect, the network as a whole is frequently called *Ten-One Hundred network*. For a diagram of the IEEE 802 Ethernet family, see *IEEE 802 Ethernet*.

10Base2 A *Local-Area Network (LAN)* protocol, Standard IEEE 802.3. It is a 10-Mbps Ethernet specification that uses RG-58 50-ohm thin coaxial cable. It also has a distance limit of 606.8 feet (185 meters) per segment. As of 1995, 10Base2 is rarely found in service anymore. See also *Cheapernet, Ethernet, IEEE 802.3*, and *Thinnet.*

10Base5 A *Local-Area Network (LAN)* protocol. It is a 10-Mbps Ethernet specification using standard (thick) 50-ohm coaxial cable. 10Base5, which is part of the IEEE 802.3 baseband physical layer specification, has a distance limit of 1640 feet (500 meters) per segment. The additional distance beyond the 10Base2 standard is because of the thicker coax, which has a lower loss than its RG-58 counterpart (10Base2). 10Base5 became outdated quickly because of speed, cost, and the fact it was cumbersome to work with. Thick coax was also unsightly compared to newer twisted pair, and its flexible management and connectivity. See also *Ethernet* and *IEEE 802.3.*

10BaseF A 10-Mbps Ethernet specification that has three subcategories, or accessories; 10BaseFB, 10BaseFL, and 10BaseFP. These standards are for Ethernet over fiber-optic cabling. See also *10BaseFB, 10BaseFL, 10BaseFP,* and *Ethernet.*

10BaseFB An accessory to the 10-Mbps Ethernet specification 10BaseFP that uses fiber-optic cabling. 10BaseFB is part of the IEEE 10BaseF specification. It is not used to connect user stations, but instead provides a synchronous signaling backbone that allows additional segments and repeaters to be connected to the network. 10BaseFB segments can be up to 1.24 miles (2000 meters) long. See also *10BaseF* and *Ethernet.*

10BaseFL A 10-Mbps Ethernet specification using fiber-optic cabling. 10BaseFL is part of the IEEE 10BaseF specification and, although able to interoperate with FOIRL, it is designed to replace the FOIRL specification. 10BaseFL segments can be up to 3280 feet (1000 meters) long if used in conjunction with FOIRL, and up to 1.24 miles (2000 meters) if 10BaseFL is used exclusively. See also *10BaseF, Ethernet,* and *FOIRL.*

10BaseFP A 10-Mbps fiber-passive baseband (single channel) Ethernet specification using fiber-optic cabling. 10BaseFP is part of the IEEE 10BaseF specification. It organizes a number of computers into a star topology without the use of repeaters. 10BaseFP segments can be up to 1640 feet (500 meters) long. See also *10BaseF* and *Ethernet.*

10BaseT 802.3 Ethernet 10Mb/s LAN standard. See *Ethernet,* and *IEEE 802 Ethernet.*

10Broad36 A 10-Mbps broadband (multichannel) Ethernet specification using coaxial cable. 10Broad36, which is part of the IEEE 802.3 specification, has a distance limit of 2.24 miles (3600 meters) per segment. See also *Ethernet* and *IEEE 802.3.*

12-Pack Coax Cable A bundle of twelve 50-ohm coaxial cables used to transport *STS-1 (Synchronous Transport Signal 1)* signals throughout a central office or node (Fig. 0.2). Commonly, the cables run from a SONET carrier unit to a *DCS (Digital Cross-Connect System).*

100BaseFX A 100-Mbps baseband (single channel) Fast Ethernet specification using two strands of multimode fiber-optic per link. To guarantee proper signal timing, a 100BaseFX link cannot exceed 1312 feet (400 meters) in length. It is based on the IEEE 802.3 standard. See also *100BaseX, Fast Ethernet,* and *IEEE 802.3.*

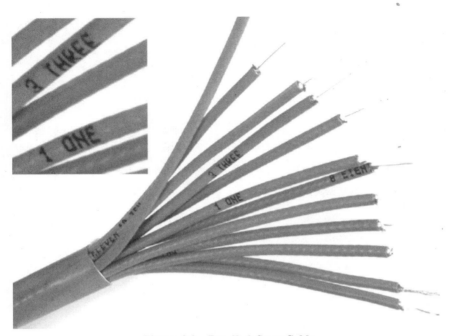

Figure 0.2   Bundled Coax Cable

100BaseT   802.3 Ethernet 100Mb/s LAN standard. See *Ethernet*.

100BaseT2   A physical layer media standard. A twisted-pair segment that uses two pairs of Category 3 voice-grade twisted-pair wires. 100BaseT2 is not meant to be intentionally deployed in networks due to its distance limitations.This standard came about to give Ethernet/802.3 a way to be inexpensively deployed on existing telephone wires.

100BaseT4   A physical layer media standard. A twisted-pair Ethernet segment that uses four pairs of Category 3, 4, or 5 UTP cable. 100BaseT4 uses a standard RJ-45 connector with the same pinout as the 10BaseT specification, plus two bidirectional pairs (transmit on 1 and 2, receive on 3 and 6; bidirectional on 4 and 5; bidirectional on 7 and 8). The distance of wire enabled for Ethernet transmission over 100BaseT4 is dependent on the category of wire. The longest distance for transmission is for Category 5 UTP, which is 328 feet or 100 meters. See also *Appendix G*.

100BaseTX   A physical layer media standard. A twisted-pair segment type based on two pairs of Category 5 twisted-pair wires. The 100-based specification uses two pairs of Category 5 unshielded twisted-pair (UTP), two pairs of 100-ohm shielded twisted-pair (STP), or Type 1 STP cable.

100BaseTX uses a Category 5 certified RJ-45 connector and the same pinout used in 10BaseT (transmit on 1 and 2, receive on 3 and 6). 100BaseTX supports full-duplex connection for switches, NICs, and routers. See also *Appendix G.*

100BaseX   100-Mbps baseband Fast Ethernet specification that refers to the 100BaseFX and 100BaseTX standards for Fast Ethernet over fiber-optic cabling. Based on the IEEE 802.3 standard. See also *100BaseFX, 100BaseTX, Fast Ethernet,* and *IEEE 802.3.*

100VG (AnyLAN)   100-Mbps Fast Ethernet and token-ring media technology using four pairs of Category 3, 4, or 5 UTP cabling. This transport technology, developed by Hewlett-Packard, can operate on existing 10BaseT Ethernet networks. Based on the IEEE 802.12 standard. See also *IEEE 802.12.*

100-Pair Cable UTP   Telephone twisted copper pair commonly used in building horizontal distribution or in riser systems, (connectivity between floors). The cable illustrated in Fig. 0.3 is *100 UTP (Unshielded Twisted Pair) plenum.*

Figure 0.3   100-Pair Plenum UTP Cable

101B Closure   A closure/housing used to protect service wire splices and inside wiring splices (Fig. 0.4).

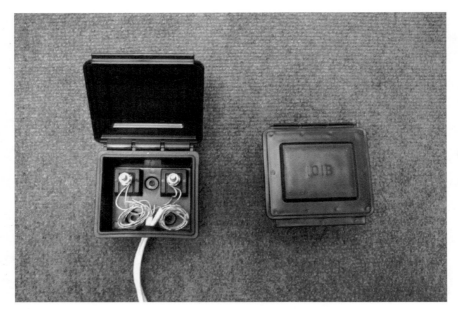

Figure 0.4   101B Closure

110 Punch Tool   A tool used to terminate solid twisted-pair copper wire on AT&T 100 termination blocks (Fig. 0.5).

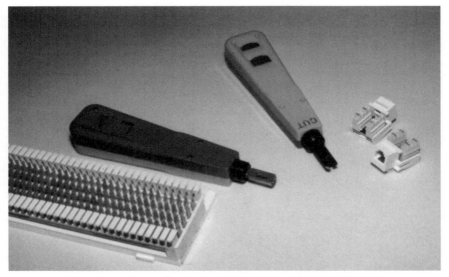

Figure 0.5   Punch Tool with 110 Blade and CAT5 RJ45 Jacks

110 Termination Block   Also called *AT&T 110 ("one-ten") blocks.* Devices used to mount twisted-pair wire so that different devices in a network can be cross connected easily (Fig. 0.6).

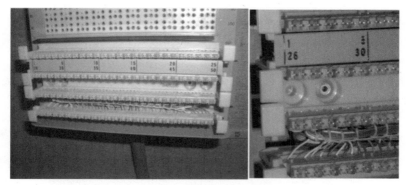

Figure 0.6   AT&T 110 Termination Blocks

145A Test Set   An analog telephone cable test set that measures the length of twisted pairs, and tests for grounds and shorts. This test set can also send a tone (Fig. 0.7).

Figure 0.7   145A Test Set

2B1Q (Two Binary One Quarternary)   A type of *Pulse Amplitude Modulation (PAM)*, where two bits presented at different possible voltage levels represent four bits at one voltage level. This line line code is a mainstay for ISDN, and is also used in some ADSL and IDSL implementations.

2-Line Network Interface   Old style with interchangeable lightning protectors. The white paint on the tops of the protectors indicates "gas type," rather than the carbon type (Fig. 0.8).

Figure 0.8   2-Line Network Interface

2FR   A service code for a flat-rate party line with two subscribers. For more info, see *Selective Ringing Module* and *Party Line.*

25-Pair Modular Splice   Used in a modular splice tool to splice PIC telephone cable (Fig. 0.9).

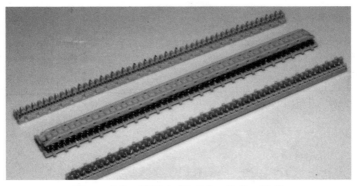

Figure 0.9   25-Pair Modular Splice Unit

25PR Connector   Also called an *Amphenol, Amp connector, P connector (male),* or *C connector (female)* (Fig. 0.10).

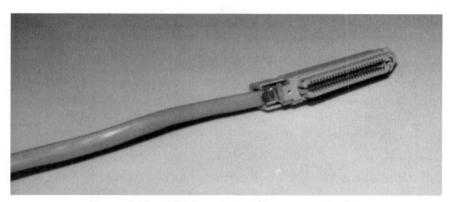

Figure 0.10   25PR Female Amp Connector (50 pin)

25PR PVC   Common telephone cabling used for horizontal and vertical wiring in buildings (Fig. 0.11).

Figure 0.11   25PR PVC UTP Unshielded Twisted Pair

258A Adapter   Adapter used to connect 25-pair Amphenol cables to RJ45 patch cords (Fig. 0.12).

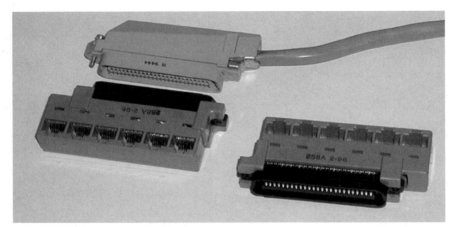

Figure 0.12   258A Adapter—Harmonica Adapter (50 pin amp to RJ45)

267A Adapter   Also called a *one-line splitter.* This is a simple "T" adapter that splits a single jack into two (Fig. 0.13).

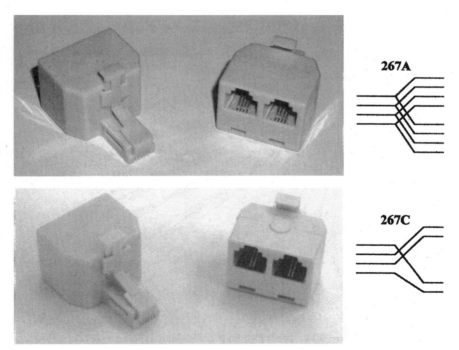

Figure 0.13   267A Adapter (top) and 267C Adapter (bottom)

267C Adapter   Also called a *two-line splitter*. This adapter is frequently confused with a 267A adapter, which is a one-line splitter. The 267C unit is designed to split two separate lines (with two separate phone numbers) out of one RJ-11 jack. See also *267A* (Fig. 0.13).

2500 Set   A frequently referred to telephone because it is known to be "noncordless," to have a switch-hook that is not built into the handset, and to have a mechanical bell ringer equivalence between 0.8 and 1. The 2500 was the Western Electric model number of this once most widely used telephone set. A "2500 telephone" is also a reference to a traditional analog telephone (Fig. 0.14).

Figure 0.14   Desktop Telephone

3 Command Set   A reference to Cisco System's method of interacting with a router or switch, particularly the 5000 series line of switches and many router models. The three commands used to manipulate and view the IOS settings are SET, CLEAR, and SHOW.

3DES (Triple DES)   Data Encryption Standard. A 168-bit encryption method that incorporates an algorithm developed by the United States National Bureau of Standards.

3FR   A service code for a flat-rate party line with three subscribers. For more information, see *Selective Ringing Module* and *Party Line.*

3G (Third Generation Network)   In wireless communications, a convergence of voice, data, and multimedia services at initial bandwidths of 144 Kbps, with a future bandwidth maturity to 1 Mbps and beyond. A simple identification of wireless communications technology evolvement is defined in generations. The first generation was AMPS, which utilized FDM technology to carry one call on each analog channel. The second generation is referred to as *CDMA/TDMA/GSM,* and placed multiple digital calls within PCS bandwidths as well as provided enhanced services. The third generation, 3G, is intended to unify not only voice, data, and multimedia, but also application formats. The standard application interface for whatever radio is used will be IP. In the United States and Japan, among other countries that have deployed both GSM and CDMA for wireless technology, CDMA2000 will be the (OSI layer 2) G3 migration path. For countries that use the European TDMA and GSM formats, GPRS (General Packet Radio Service, or W-CDMA) will be the migration path. Regardless of the wireless link between end users, the applications that are accessed will be done through IP. Therefore, end users will be able to exchange application information via open standards. For multinational users, handset manufacturers are planning to produce devices that are compatible with both technologies by incorporating both types of radio technology.

3720   A common reference for an IBM 3720 communications controller. For more information and a diagram, see *Communications Controller.*

3725   A common reference for an IBM 3725 communications controller. For more information and a diagram, see *Communications Controller.*

3745   A common reference for an IBM 3745 communications controller. For more information and a diagram, see *Communications Controller.*

3746   A common reference for an IBM 3746 communications controller. For more information and a diagram, see *Communications Controller.*

4FR   A service code for a flat-rate party line with four subscribers. For more information, see *Selective Ringing Module* and *Party Line.*

4 Pair   Shown in Fig. 0.15 is 4-pair, *PVC (Polyvinyl Chloride Jacketed) UTP (Unshielded Twisted Pair).*

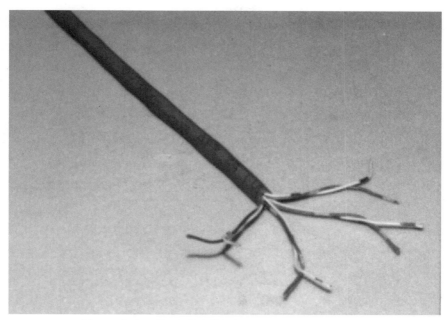

Figure 0.15   4PR PVC UTP (Unshielded Twisted Pair)

4B/5B Coding (4-Bit/5-Bit Coding)   A physical layer coding/compression method used by the *FDDI (Fiber-Distributed Data Interface)* for 100-Mbps physical layer applications. In applications where the ATM cell format is transmitted over 4B/5B FDDI, an additional byte of overhead results from the encoding and cell delimiting method. However, the net transmission rate remains the same because of the compression. The mechanics of this compression method are the same (yet a smaller version) of 8B/10B coding. See also *8B/10B Coding.*

49A Ready Access Terminal   A common terminal found in an outdoor aerial copper telephone plant (Fig. 0.16). These terminals will soon be very uncommon because better splice closures and weatherproof access devices have come to market. The 49A is a ready-access pic terminal, which means that the copper pairs are not pre-spliced to binding posts within the cover. To install service from one of these terminals, a technician splices directly into the pair. These older terminals have been a mainstay for telephone companies for decades because they are inexpensive and flexible in making service changes.

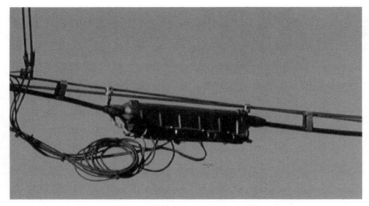

Figure 0.16   49A Ready Access Terminal

6-Pair Can   A termination or splicing enclosure designed especially for 6-pair aerial or buried service wire (Fig. 0.17). 6-pair cans are available with lightning protectors (protected 6-pair can).

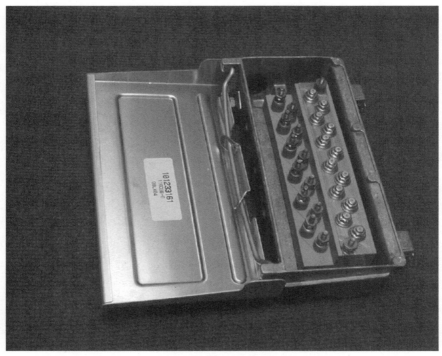

Figure 0.17   6-Pair Can

66 Block   The 66M150 termination block is used to terminate twisted-pair wire on distribution frames and any other solid 22 to 24 wiring application (Fig. 0.18).

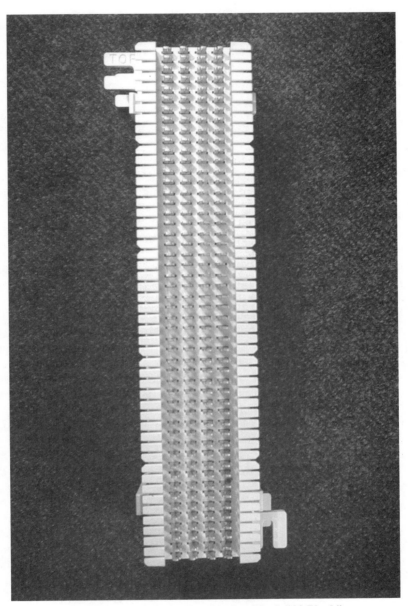

Figure 0.18   66M150 Termination Block "66 Block"

8 Click Rule/8 Second Rule   A guideline that Internet website creators use. The objective is to get the user the information they need within 8 clicks or 8 seconds or less. It is believed that if an end user must wait longer than 8 seconds or make more than 8 clicks, they will move on to an alternative website.

8B/10B Coding (8-Bit/10-Bit Coding)   A physical-layer compression method developed by Fiber Channel that is used to transfer ATM transmissions from OC-3 SONET to STS-3 twisted pair. This is an ATM LAN application that is recommended up to 100 meters. The 8B/10B coding technique combines overhead with data. With the 10 bit/baud, 1024 symbols can be transmitted (0000000000 to 1111111111). Because payload data is based on 8 bits, which, in turn, enables 256 different symbols, 768 spare symbols remain that can represent a combination data-character/overhead-information. If a user is transporting ATM over SONET OC-3, that transmission can be directly transferred to a twisted pair with no buffering and no delay, so long as 8B/10B coding is used.

8FR   A service code for a flat-rate party line with eight subscribers. For more information, see *Selective Ringing Module* and *Party Line*.

80/20 (Eighty/Twenty)   1. A rule of thumb used by telephone companies whereby switch and transport facilities would be increased in a certain area when the utilization reached 80%. This rule worked in the former monopolistic business model well. By the time the remaining 20% was utilized, there would be additional network facilities installed to accommodate growth. In the newer competitive business environment, additional network facilities are built based only on marketing forecasts and the revenue potential of the market that the network serves. This change in business structure and methods has had good and bad effects on the vast majority of telecommunications service customers. The good part is that economically rich areas get an abundance of telecommunications services; the bad part is that areas that are remote or poor do not.   2. A design consideration in enterprise networks where servers are *distributed* or spread out throughout a network. The purpose of this design method is to keep 80% of the traffic on the same autonomous system, subnet, or broadcast domain (possibly a building or floor). This design was used when switching equipment for the backhaul of traffic to or through a core was very expensive. Since the drop in price for LAN switching equipment, the cost of managing an 80/20 network has proved to exceed the cost of the switching equipment that would provide one central core. The core design is called a *hierarchical network design.* See also *Hierarchical Network Architecture* and note the placement of servers (Fig. 0.19).

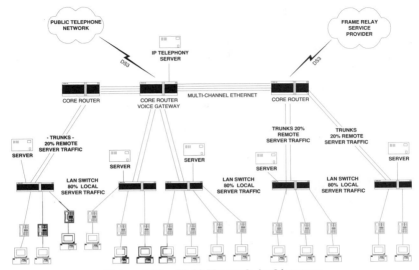

Figure 0.19   80-20 Network Architecture

89B Bracket   The bracket that is used to attach 66M150 blocks to back boards in telephone closets or distribution frames (Fig. 0.20).

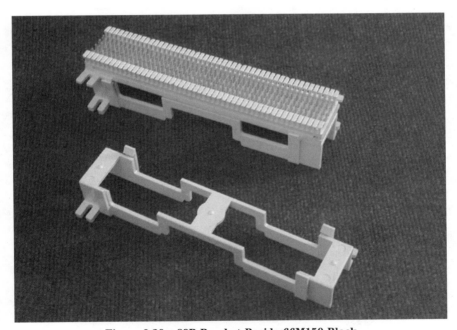

Figure 0.20   89B Bracket Beside 66M150 Block

802.1d   The IEEE standard for spanning tree algorithm that prevents loops in redundantly connected LAN switches. Spantree is automatically enabled when redundant bridges are connected. If redundant bridges were connected to a network without Spantree enabled, the dual connected bridges would forward the same frames to each other in an endless loop. This condition saturates bandwidth immediately, and renders all devices associated with the loop useless. The way that Spantree works is that when bridges are initialized (powered on), they send a signal to other networked devices called a *Bridge Protocol Data Unit* (BDPU). When bridges/switches receive these BDPUs from other devices, they become "aware" that other bridges are connected to the network and whether any are connected in redundancy to them. Using BDPU information, bridges on the network elect a "root bridge" and a "designated bridge." Depending on the way the bridges are physically connected, all ports are "blocked" or partially disabled except for "root ports" and "designated ports," which are bridge ports closest (by number of hops) to a designated or root bridge. If a link is lost, an alternate port then becomes the root port. New BDPU messages are sent to notify other bridges of the status change. Most makers of bridging hardware set the default to automatically send BDPUs and enable Spantree to on. This is so that if a network is unknowingly connected with bridges in parallel, it will not bring the network down. The 802.1d standard evolved from Digital Equipment Corporation's (DEC) Spantree algorithm. 802.1d and the original *DEC Spantree* are not interoperable. Further, when incorporated with 802.1Q (VLANs), one instance of spanning tree must be set up for *each and every* VLAN.

802.1p   The IEEE standard for prioritization of LAN traffic among Ethernet switches based on either the switch port, MAC address, or IP address associated with the communicating end appliance (whether it is an IP phone, video monitor, host PC, printer, or server). Packets are tagged as belonging to a queue, which determines the priority of the packet. By the 802.1p standard, queue 0–3 is normal, and 4–7 are high priority. 802.1p functions hand-in-hand with 802.1Q or VLANs.

802.1Q   The IEEE standard that evolved from Cisco Systems' ISL (Inter-Switch Link) protocol. ISL and 802.1Q are not interoperable. The reference 802.1Q is better known as the VLAN or tag switching standard. It is a feature on post 1998 LAN Switches that makes selected ports behave as if they were attached to the same segment, or hub. Another good name for this feature would be *V-Segment,* or *Virtual Segment.* Devices/users that exchange a large amount of information are usually placed within the same Virtual LAN segment. This helps make the operation of the LAN switch more efficient, keeping traffic contained within specified ports. This allows other ports on separate VLANs to carry other nonrelated traffic simultaneously. VLANs are configured by a network engineer, network analyst, or network administrator. When IP telephony is implemented over an Ethernet switched network, the telephone devices connected to the network are best placed into their own VLAN. Most switches that are 802.1Q compatible can recognize more than 1,000 VLANs. Further, there are two kinds of VLANs: static, and dynamic. Static VLANs are associated with switch ports, and dynamic VLANs are associated with the MAC addresses of devices attached to the switch. Dynamic VLANs allow users to move to another office that could have a switch port connection preinstalled. The switch would recognize the MAC address of the device and automatically include its traffic in the same VLAN as the previously connected switch port. See also *Frame Tagging*.

802.3ab   Ten gigabit Ethernet over copper UTP and fiber standard (10,000BaseT). Still in the process of standardization as of this writing. It is expected that the technology being implemented in the 802.3ab standard and/or its revisions will extend Ethernet more than 40 km.

802.3u   The IEEE Ethernet 100BaseT feature specification that provides for flow control (pause frames) and full-duplex operation. Full duplex allows for 100 Mbps send and 100 Mbps receive, for a total 200 Mbps Ethernet connection over Cat5 twisted pair. The auto negotiation is an enhancement of the link integrity signaling method used in 10BaseT networks, and is backward-compatible with link integrity. Auto negotiation allows the NIC or the network device to adjust its speed to the highest speed that both ends are capable of supporting. To be able to use this feature, both the network device (switch port) and the NIC must contain the auto negotiation logic. This specification also allows for DTE to DTE links of 400 meters, or a one repeater network of approximately 300 meters.

802.3x Full-Duplex Transmission   The Full-duplex portion of this standard provides the means of transmitting and receiving simultaneously on a single wire. Full duplex is typically implemented between two endpoints, such as between switches, between switches and servers, or between switches and routers. Full-duplex transmission is not used for desktop workstation PCs because the PCs work at a maximum 90 Mbps, which makes 200 Mbps impractical. Full duplex allows bandwidth Fast Ethernet (802.3) networks to be easily and cost-effectively doubled from 100 Mbps to 200 Mbps.

802.3z   1000BaseX specification. Also called the *Gigabit Ethernet specification*. Gigabit Ethernet is defined for fiber optic multimode and single mode. There is also a copper version that runs distances of 25 meters. Gigabit Ethernet is mostly interfaced with SC fiber optic connectors, and there is a wide variety of laser-diode adapters available from manufacturers (i.e., the Cisco Systems GBIC connector) in the LX, LS, and LH range. Gigabit Ethernet standards use 8B/10B encoding and decoding schemes(Fig. 0.21).

		GIGABIT ETHERNET				
		802.3z and 802.3ab distances				
STANDARD	SPECIFICATION	Wavelength L nanometers	Fiber Type	Modal Bandwidth MHz-km	Recommended Maximum Distance	
802.3z	1000BaseLH*	1300nm	9/10 Single Mode	n/a	10km	32,810ft
802.3z	1000baseLX	1300nm	5um Single Mode	n/a	3km	9,843ft
802.3z	1000BaseLX	1300nm	62.5/125um Multimode	500	550m	1804ft
802.3z	1000BaseLX	1300nm	9um Single Mode	500	5km	16,405ft
802.3z	1000BaseSX	850nm	62.5/125um Multimode	160	220m	722ft
802.3z	1000BaseSX	850nm	62.5/125um Multimode	200	275m	902ft
802.3z	1000BaseSX	850nm	50/125um Multimode	400	500m	1640ft
802.3z	1000BaseSX	850nm	50/125um Multimode	500	550m	1804ft
802.3ab	1000BaseT	n/a	Cat5 UTP	n/a	100m	328ft
802.3ab	1000BaseCX	n/a	balanced copper	n/a	25m	82ft
802.3u	100BaseFX	850nm	62.5/125um Multimode	400	400m	1,312ft
	*proposed 802.3z Cisco Systems Proprietary as of 10/2000					

Figure 0.21   Gigabit Ethernet Distances

802.11b   Wireless LAN standard update to 802.11DS for increased speed to 11 Mbps at an operating frequency of 2.4 GHz. The modulation technique used in 802.11b is DSSS (Direct Sequence Spread Spectrum). WEP (Wired Equivalent Privacy) is also an addition in the 802.11b standard, which allows manufacturers to implement security up to and including 128-bit key encryption.

802.11DS   The 802.11 Wireless LAN standard for the Direct Sequence method of line coding. In the standard, there are 11 22-MHz wide stationary channels. This allows for an 11 Mbps throughput, with up to three nonoverlapping radio channels operating in the same area, which means three separate radio units (called *Access Points* in some wireless circles) can operate in the same area without interfering with each other.

802.14   The IEEE standard for the operation of cable telephony modems that enables cable TV networks that are coax and hybrid fiber-coax in composition to carry Ethernet 802 traffic as well as ATM based traffic. There are multiple MAC layer interfaces defined in 802.14 to make cable telephony services equally as flexible to the end user as traditional services enabled by DSL or ATM.

900   The "in fact" standard for services billed through telephone companies. Some 900 services include weather information, stock exchange information, and "erotic" information. Phone companies have their own individual criteria for providing 900 services to companies that wish to sell services over the telephone network. Individuals that call 900 numbers are billed a pre-determined amount for the call on a per-minute basis. Most telephone companies require that anyone selling information services on their network to inform callers of the charges in advance and to allow them a certain amount of time to hang-up before any billing begins.

911   The standard emergency service telephone number. 911 calls are not answered by the telephone company; they are answered by an emergency dispatch service. This is why 911 service has a separate charge on telephone bills.

965TD   3M Dynatel 965TD loop analyzer is shown in Fig. 0.22. Used for testing twisted-pair telephone cable. The 965TD is also a data terminal capable of accessing a database via an internal modem. The TDR (*Time Domain Reflectometer*) is another great feature of the 965TD.

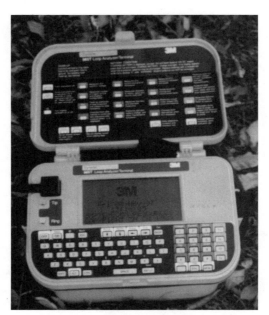

Figure 0.22   3M 965T Loop Analyzer

A (Amp, Ampere)   A unit of electrical current flow that is equal to one volt applied to one ohm of resistance. The Ohm's Law definition of amperage is:

$$I = \frac{E}{R} \text{ or } Amps = \frac{Voltage}{Resistance}$$

The ampere can also be defined as one coulomb of charge flowing past a point in one second. One coulomb of charge is equal to 6,300,000,000,000,000,000 electrons.

A Law   An ITU-T standard companding method used to convert analog voice to compressed digital in the majority of the world for cellular radio networks. In the United States, the mu-Law standard is used. See *Companding.*

AA (Automated Attendant)   Most voice-mail systems come with an automated attendant built in. An automated attendant is an answering machine that asks the caller to push 1 for sales, 2 for service, etc. They are also capable of routing callers to a dial by name directory. See also *Directory Tree.*

AAL (ATM Adaptation Layer)   A transfer format, cell header format, and functional section of the ATM transport method. There are five variations of the ATM Adaptation layer. Each is intended to be used with a specific type of data. For a diagram of ATM layers, see *Fig. A.1.*

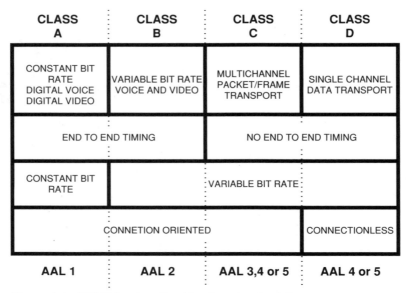

CLASS A	CLASS B	CLASS C	CLASS D
CONSTANT BIT RATE DIGITAL VOICE DIGITAL VIDEO	VARIABLE BIT RATE VOICE AND VIDEO	MULTICHANNEL PACKET/FRAME TRANSPORT	SINGLE CHANNEL DATA TRANSPORT
END TO END TIMING		NO END TO END TIMING	
CONSTANT BIT RATE	VARIABLE BIT RATE		
CONNETION ORIENTED			CONNECTIONLESS
AAL 1	AAL 2	AAL 3,4 or 5	AAL 4 or 5

Figure A.1   ITU-T Service Classifications for the ATM Adaptation Layers

AAL1 (ATM Adaptation Layer One)   The part of the ATM protocol that enables the transfer of time-sensitive data, such as voice or video. AAL1 uses an adaptive clock method, where the devices at each end of the link negotiate a clock agreement, then incorporate a small buffer to monitor the rate at which cells are being transferred across the link. AAL1 is used for DS0, DS1 emulation, and other voice and video.

AAL2 (ATM Adaptation Layer Two)   For class-B traffic (see diagram under AAL), packet technologies, and the transport thereof. It is similar to voice over frame, video over frame, etc.

AAL3/4 (ATM Adaptation Layer Three and Four)   For class C and D (see the *AAL* diagram) layers that are designed to handle nontime-sensitive data transfer. This layer class adds header information that incorporates error-checking functions before and after the original data. Also, a Message ID function allows multiplexed or interleaved transmissions to be sent directly over the single ATM virtual channel. This layer would be used as a backbone to carry many X.25 or Frame-Relay logical links, or could be used in a campus application to carry Ethernet from one building to another.

AAL5 (ATM Adaptation Layer Five)   The layer created for class C and D types of traffic (see the *AAL* Figure A.1 diagram). The cell header remains the same except larger buffers are used and a *CRC (Cyclic Redundancy Check)* is appended to the end of the last cell of the packet's cell stream. No Message ID function is available to directly transfer multiplexed data. Cell payloads are 48 bytes, and a PTI bit is used to indicate the last cell of a packet.

AAR (Automatic Alternate Routing)   A feature of some networks and protocols to reroute traffic on the fly without interrupting or corrupting traffic.

AB Switch   A mechanical/manual switch used to switch a signal between two source or destination devices. For example, if you have two computers and one monitor, you could implement an AB switch to control which computer the monitor is connected to. The monitor would connect to the "C" port on the switch, and the two computers would connect to the "A" and the "B" ports. AB switches suit many applications of connectivity from computer to audio/video (Fig. A.2).

Figure A.2   Common AB Switch

Abandoned Call   When you make a call, are put on hold, then hang up before someone answers, you have abandoned the call. Customer-service call centers like to know the number of abandoned calls they have so that they know how many people to employ answering calls, etc. Believe it or not, if you hang up when calling a call center, someone that cares eventually finds out!

Ablation   To burn holes into metal film with a laser. The holes represent ones and zeros for optical storage on disks.

ABR (Available Bit Rate)   *Quality of Service (QOS)* defined by the ATM Forum for ATM networks that is used for connections that are not time or delay sensitive. A connection would be rightfully commissioned as an ABR connection if it carried only spontaneous or bursty data. Other QOSs defined by the ATM forum for ATM connections include *CBR (Constant Bit Rate), UBR (Unspecified Bit Rate),* and *VBR (Variable Bit Rate).*

Absorption Loss   The weakening of light intensity as it travels a length of optical fiber. The unit for absorption loss is dB/Km (decibels per kilometer).

AC (Alternating Current)   Alternating current is electricity that changes/ alternates its direction of flow in a steady cycle or period. The line voltage in most American homes is somewhere between 110 V and 120 V AC RMS, which makes the actual peak to peak voltage about 325 V.

AC-to-DC Converter   This is an electronic device that defines itself. Large-scale AC/DC converters are mostly referred to as *rectifiers.* They convert alternating current to direct current (or voltage) by incorporating a large capacitor and two or four diode rectifiers for a half wave or full wave, respectively. Almost all rectifiers have regulated output, meaning the output DC voltage is kept at a steady level, regardless of the electronic device it is providing power to. Rectifiers are also available with battery backup and redundant circuits so if a component fails, the output voltage won't be disturbed. AC to DC converters are rated by input voltage requirement output voltage/current ability.

Acceptable Angle   The maximum angle that a fiber optic accepts light and doesn't reflect it away.

Access Charge (Carrier Common-Line Charge)   What local phone companies charge long-distance companies to connect the far-end local portion of a call. A fee that everyone pays for every phone line to make up for subsidies that long-distance services paid to help the less-profitable local services before the divestiture of AT&T and the RBOCs (Regional Bell Operating Companies).

Access Layer   One of the three LAN network design layers. In LAN network design, the three switch layers are Core Layer, Distribution Layer, and Access Layer (Fig. A.3). The Core Layer provides redundancy for distributing traffic across multiple Access Layers. The Access Layer provides high-speed switching and routing among a group of switches networked by trunks. The Access Layer provides switches where users connect, so high port quantity is desired in this layer.

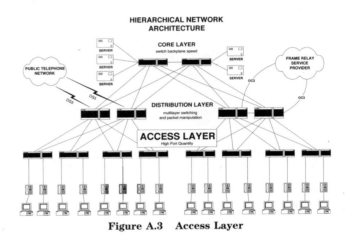

Figure A.3   Access Layer

Access Line   A connection provided by a telephone company that runs from a customer's premises to a central office or to a co-location within the central office.

Access Link   The local phone line that connects you to a central office switch and gives you access to a long-distance carrier. It's an access line with all the electronics that give you dial tone or private line communications capability.

Access List   A list kept by routers to control access to or from the router for a number of services. A good example is the prevention of packets within a specific IP address leaving a particular interface on the router.

Access Network   Regarding xDSL, the portion of a public switched network that connects access nodes to individual subscribers. It is also called a *local network*. Today's access network is predominantly passive twisted-pair copper wiring and fiber optic, and is owned by such telephone companies as SBC, Ameritech, Qwest, and GTE.

Access Node   A connecting point for a data transport or data-packet network. Access nodes usually reside in a central office environment, or are a part of a leased space agreement. Connections to access nodes are provided by local carrier loops. Access devices at the end of the customer loop are generally provided by the data-network service provider or by the customer.

Access Point (AP)   Another name for a cross-box where telephone cables are cross connected. See also *Aerial Cross Box* (Fig. A.4).

Figure A.4   Access Point (AP) Cross-Box

Access Server   A communications processor that connects asynchronous devices to a LAN or WAN through network-emulation software that resides in its memory. It performs synchronous and asynchronous routing of such supported protocols as Ethernet, token ring, frame relay, and

X.25. Access servers are sometimes referred to as *network-access servers* or *communications servers*.

Access Service Request   When a special-service provider (frame relay or long distance private line) needs wire facilities from their point of presence in the city to your location, they call the local telephone company and make an access service request to provide a line that runs from your network interface to them. Many special service providers have their equipment located in the local phone company's central office as a part of a co-location agreement. When a CLEC (competitive local exchange carrier) needs to provide service where they don't have facilities, this is how they do it by using the RBOC's (Regional Bell Operating Company) wire facilities.

Access Switch   A point in a network where multiple services/protocols are differentiated and routed for enterprise networking services.

Access Tandem   A telephone company central office or node that contains a switch in which all inter and outer area code traffic is handled. The main *LEC (Local Exchange Carrier)* central office in an area code where the hand-off for long-distance service happens (Fig. A.5).

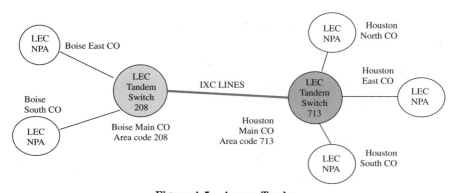

Figure A.5   Access Tandem

Account Code   In accounting for communications costs, an account code is used. If you have multiple employees in your office using multiple phone lines to make long-distance calls, figuring out who made those calls can be impossible. Some long-distance companies offer a service where employees enter an account code before their call is completed. When the phone bill comes, it is easy to see who is making long-distance calls, to where, and for how long. Most importantly, it is easy to distribute the costs among a group. Most *PBX (Private Branch Exchange)* switches have call accounting systems available that allow a telephone extension (or group of extensions) to be attached to an account code that shows how much each phone is utilized, and what calls are made and received.

ACD (Automatic Call Distributor/Distribution System)   A separate footprint or built-in feature of a *PBX (Private Branch Exchange)* that equally distributes incoming calls to agents. As calls come in, they are placed into a queue (or a waiting line) for the next available agent. ACD systems are very versatile and relatively easy to program as some incorporate their own script programming feature. For incoming calls, the waiting times, pre-recorded announcements and other call treatments can be set up by the users/companies to their discretion. Some well-known ACD systems are made by ACCENT, Lucent, and Northern Telecom. See also *ACS*.

ACIS (Automatic Customer/Caller Identification Service)   This feature comes with many *ACD (Automatic Call Distribution)* systems and enables them to make useful the DNIS signal attached to an inbound call. The *DNIS (Dialed Number Identification Service)* can be used to identify the caller or call type. If you have a call center where more than one business is served, then the Automatic Customer/Caller Identification feature can forward the call to a certain group of agents that are associated with the dialed number. It can even tell them what kind of call they are answering on the display of their phone. If a customer calls from Mexico, then the ACD ACIS feature will relate the dialed number with an ACD group that speaks Spanish. Of course, the ACD system needs to be set up or programmed by an administrator to do this.

ACK   ASCII control code abbreviation for acknowledgement. Binary code is 0110000 Hex is 60.

ACM (Advanced Communications Function)   Also known as an *advanced communications function control program.* The ACM program resides within a communications controller and interfaces with the *SNA (System Network Architecture)* access method in the host computer/mainframe to control network communications.

Acoustic   This term refers to the natural sound vibrations of an object or space. In telecommunications, acoustics are a concern when using hands-free or speaker-phone devices. If the acoustics of the electronic hands-free device are poor, it will vibrate or resonate when the volume level is increased. If the acoustics of the room that the device is in are bad, the device will cut in and out as it "hears" its own echo. Cloth cubicles have a good acoustic vibration-dampening effect. Wide open rooms with no ceiling tile and sheet rock walls have a poor acoustic-dampening effect.

Acquisition   The process of a terrestrial-based device locking on a satellite's *GPS (Global Positioning System)* signal. Included in the process of acquiring the signal is *AGC (Automatic Gain Control)* for optimum signal level, synchronization, and processing of the data signal.

ACR (Allowed Cell Rate)   A parameter defined by the ATM Forum for ATM Traffic Management. The ACR varies between the *Minimum Cell Rate (MCR)* and the *Peak Cell Rate (PCR)*. It is managed by the protocol congestion control mechanisms.

ACS (Automatic Call Sequencer)   If you can't afford an ACD system, this could be the answer you have been looking for. An ACS answers the call, plays a recorded announcement, and puts the caller on hold. The calls coming in (all the calls) appear as lights on a telephone (or multiple telephones). The calls that have been on hold the longest blink the fastest (or have other signaling methods). ACS systems are designed primarily for or as a part of key systems.

Activated Return Capacity   The ability of your cable TV box to send information back to the cable-TV office head end and the ability of the head end to receive the data. This information can include the ID number of the cable TV box and what station you are watching.

Active Device   An active device is an electronic component that requires external power to manipulate or react to an electronic input for a desired output. Examples of active devices are: transistors, op amps, diodes, cathode ray tubes, and ICs. If it's not active, it is a passive device. Included in the passive-device category are capacitors (condensers, if you want to use a really old term) resistors, and inductors (or coils), which include transformers.

Active Matrix Display   Also called *TFT (Thin Film Transistor)* displays. A type of laptop computer display technology. In an active matrix display, each picture element has its own control transistor. The display performance of active matrix displays is significantly sharper, faster, and are less stressful on human eyes, yet they consume more battery power than passive matrix displays. See also *Passive Matrix Display*.

Active Vocabulary   A list of words that a voice-recognition system has been programmed to recognize. Each voice-recognition system has its own set of words that are selected to fit its application. This is done so that when a voice says "pair," the voice recognizes a word that means two, not a fruit ("pear").

ACU (Automatic Calling Unit)   A device that IBM computers use to access outside dial tone for communications. It does the job of a modem, but uses its own protocols to communicate with the computer.

AD (Analog to Digital Converter, ADC)   A part of a channel bank that encodes analog voice signals into a stream of binary digits. The digital to analog converter or analog to digital converter samples a caller's voice at a rate of 8000 times per second. (The sample rate for a T1 channel is 8000 times per second.) Each sample's voltage level is measured and converted to one of 256 possible sample levels. These levels are from the lowest, 0000000, to the highest, 11111111. The reason for 256 levels is because if you count in binary from 00000000 to 11111111, you end up with 256, the highest number possible with 8 bits. The bits are then transmitted one after another at a high rate of speed to their destination, where the same process happens in reverse. For a diagram, see *Analog to Digital Conversion.*

Ad Insertion Module   In cable TV or broadcast radio networks, a device that broadcasts commercial advertising during pre-determined time segments. Ad-insertion equipment enables local cable TV companies to sell and insert advertising space to local businesses. The same is applicable for broadcast TV/AM/FM radio stations.

Adaptable Digital Filtering   A method of conditioning twisted-pair telephone lines to carry data more efficiently up to 12,000 feet before regeneration. The adaptive digital filter can be customized to the characteristics of any given pair (that is in good condition). This method of line conditioning is not compatible with XDSL.

Adaptive Differential Pulse Code Modulation (ADPCM)   A family of techniques used to compress audio information for data storage and wireless telephone networks. This is accomplished by using algorithms that code the difference between sound samples and dynamically switching the coding scale to compensate for audio signal amplitude and frequency. This technique requires less disk storage space and less bandwidth to transmit, as opposed to PCM, which encodes absolute quantitative values.

ADCCP (Advanced Data Communications Control Protocol)   A bit-oriented data link protocol developed by ANSI. ADCCP was similar in make-up to that of HDLC and SDLC. See also *HDLC.*

Add On   A PBX, Centron, or Central Office feature (also known as *three-way calling*). Some telephone stations have a button that is designated "add on." To add a third caller, you push the add-on key and dial the number of the third party, then push add-on again to bridge the calls together.

Address Mapping   A method of making dissimilar protocols work together. It is done by translating the address of one protocol to another protocol. For example, when routing IP over X.25, the IP address must be routed or forwarded to the X.25 address, and vice versa on the opposite end of the transmission.

Address Mask   A bit combination used to describe which portion of an address refers to the network or subnet, and which part refers to the host. The address mask is usually referred to as *the mask*. See also *Subnet Mask.*

Address Resolution   A general term that refers to overcoming the differences between computer addressing schemes. Address resolution is most commonly a method of mapping or transferring layer 3 (network layer) addresses to layer 2 (datalink layer) addresses.

Address Resolution Protocol (ARP)   The ARP function is to match higher-level network IP addresses with the physical hardware address of a piece of hardware, such as a *NIC (Network Interface Card),* which is the subsequent router, PC, server, or terminal that it belongs to. ARP provides the link of the physical address (NIC) and the appropriate IP address. For example, an IBM mainframe attempting a connection to a server on an IP network would send an ARP broadcast (it contains the IP address of the target server, as well as its own IP address) to the network. If the mainframe receives the ARP, it will recognize its own IP address in the packet, and respond by sending its physical address (Fig. A.6).

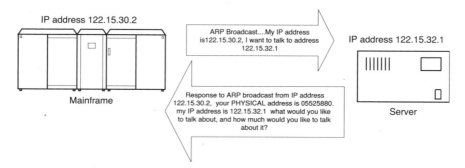

Figure A.6   ARP Broadcast

Address Signals   The digits you dial on your phone pad, the phone number is actually an address signal to the local central office that you are connected to.

Address Translation Gateway   A reference to a router software function. This software provides address format conversions as data packets are routed from one network to another. Most address translation gateways enable a router to do specific address format conversions for transmissions between specific nodes of separate networks.

Addressable Programming   For Pay Per View, cable TV companies use an addressable programming system. When you call the phone number to activate the Pay-Per-View movie or event, an IVR system receives your ANI signal or asks for your phone number. Then, it uses your phone number as your customer ID code for billing; in some cases, it identifies which cable-TV converter box to enable. Your cable TV box has an ID code or address code in its memory. When it receives its own address signal from the cable-TV office head end, it enables the horizontal sync or digital channel for the Pay-Per-View program. It's called *addressable programming* because the converter box is programmed after it receives its address, which acts like a password.

Addressed Call Mode   Another way of saying "dial-up mode." Sometimes the term *phone number* is substituted with address by some of the standards committees. This happened in the V.25bis standard. Addressed call mode permits DTE and DCE to establish and terminate calls by dialing user-determined phone numbers, based on the V.25bis modem standard.

Adjacency   A relationship formed between selected neighboring routers and end nodes for the purpose of exchanging routing information. Devices using the same media segment are considered to be adjacent.

Adjacent Channel Interference   In terrestrial microwave radio as well as other types of radio communication, adjacent channel interference results when another RF link is using an adjacent channel frequency. In selecting a site, a spectrum analyzer can be used to determine if any strong signals are present at the site, and if they are, to determine how close they are to the desired frequency. The further away from your proposed frequency, the less likely they are to cause a problem. Antenna placement and polarization as well as the use of high-gain, focused antennas is the most effective method of reducing this type of interference.

Administrative Distance   A rating of the trustworthiness of a routing information source. In Cisco routers, administrative distance is expressed as a numerical value between 0 and 255. The higher the value, the lower the trustworthiness of the routing information. For example, if routing data has traveled through 1 router and 1 media segment, the chances that it has been corrupted are low, so the administrative distance is considered to be about 2. If the routing data has traveled through 100 routers and 100 media segments, its administrative distance rating could be about 200.

ADPCM (Adaptive Differential Pulse Code Modulation)   A family of techniques used to compress audio information for data storage and wireless telephone networks. This is accomplished by using algorithms that code the difference between sound samples and dynamically switching the coding scale to compensate for audio signal amplitude and frequency. This technique requires less disk storage space and less bandwidth to transmit, as opposed to PCM, which encodes absolute quantitative values.

ADSL (Asymmetric Digital Subscriber Line)   Also referred to as *ADSL Full Rate* or *G.992.1*. ADSL is a physical-layer protocol that supports up to 8 Mbps bandwidth downstream and up to 1 Mbps upstream (Fig. A.7). The asymmetrical aspect of ADSL technology makes it ideal for Internet browsing, video on demand, and remote *Local-Area Network (LAN)* access. Users of these applications typically download more information than they send. ADSL also allows simultaneous voice communication by transmitting data signals outside of the voice frequency

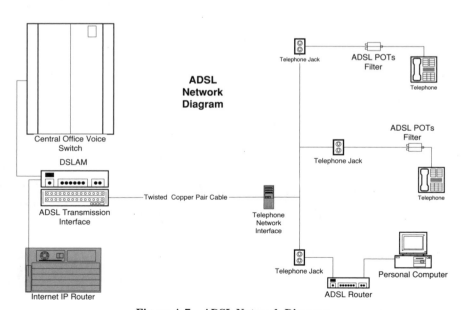

Figure A.7   ADSL Network Diagram

range. Sometimes a faint hiss can be heard on the line. To eliminate the hiss, a voice/data splitter, commonly called a *POTS splitter (Plain Old Telephone Service)* is installed at the jack of each phone. The base transmission range for an ADSL line is 18,000 feet. ADSL can be extended to remote communities by using ADSL repeaters (to 48,000 feet) or fiber optic. For more information on the DSL family of protocols, see *xDSL*.

ADSL Router   The device that interfaces a customer's personal computer with the ADSL telephone line. The ADSL router needs its own *NIC (Network Interface Card)* installed in the personal computer. See also *Fig. A.8.*

Figure A.8   ADSL Router

ADSU (ATM DSU, Asynchronous Transmission Mode Digital Service Unit)   A terminal adapter used as a demarcation point or interface point to an ATM network.

Advance Replacement   The process of getting a replacement component (card, phone, power supply, software, etc.) by calling the distributor or manufacturer and obtaining an advance-replacement reference number. When you receive your advance replacement item, you replace it in the box with the bad item, mark the box with the advance-replacement reference number and send it back. Hopefully, the replacement item doesn't go bad so that you don't have to go through all that again.

Advanced Data Communications Control Protocol (ADCCP)   A bit-oriented data link protocol developed by ANSI. ADCCP was similar in make-up to that of HDLC and SDLC. See also *HDLC*.

Advanced Peer-to-Peer Networking (APPN)   An enhancement to the original IBM *SNA (System Network Architecture)*. APPN handles the following: session establishment between peer nodes, dynamic transparent route calculation, and traffic prioritization for Advanced Program-to-Program Communication.

Advanced Program-to-Program Communication (APPC)   The IBM *SNA (System Network Architecture)* software that allows high-speed communication between programs on different computers in a distributed/mainframe computing environment. APPC establishes and tears down virtual connections between programs that require communication. It has two software interfaces; the first interface is the programming interface, which replies to programs requiring communications. The second interface is the data-exchange interface, which establishes the sessions or "connections" between the programs.

Advertising (Router)   A process in which routers send routing table updates and/or service updates at specified intervals. This is done so that all routers maintain accurate information about their network surroundings, which assists in the efficient and accurate passing of data packets. See also *Router Protocol*.

Aerial Cable   Twisted copper pair, coax, or fiberoptic cable that is attached to power or telephone poles strung through the air. Electrical (power), telephone (fiber optic and twisted pair), and cable TV (coax) are frequently aerial. Aerial cable is attached to a steel strand with lashing wire in most cases. It is sometimes attached during manufacturing as a part of the jacket or sheath (this kind is called *figure-8 cable*). The steel strand is attached to pole with strand clamps and other pole attachments. This is all done with pole-attachment agreements with the owner of the pole, which is the power company, in most cases.

Aerial Cross Box   A cross box that is mounted on a pole away from the ground. Aerial cross boxes (also called *tree stands*) are installed in areas where easement rights are narrow or in areas where vandalism is a high risk (Fig. A.9).

Figure A.9   Aerial Cross Box (AP) "Tree Stand"

Aerial Service Wire Splice   A common device used to splice aerial service wire (also called a *football* or *potato*). See *Fig. A.10.*

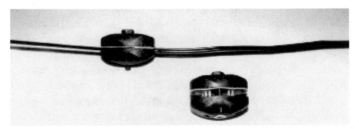

Figure A.10   Aerial Service Wire Splice

AGC (Automatic Gain Control)   Built-in to every radio is an AGC circuit that compensates for the strength of the signal you are receiving (Fig. A.11). If your radio had no AGC and was tuned to a distant station, tuning to a local station would cause the volume to blare. The way it works is as follows: After the tuner has selected the frequency to be processed, the signal goes to an AGC circuit, which is very similar to a regular intermediate-frequency amplifier (single transistor), except that the gain (amplification) of the circuit is controlled by a level detector. The level detector samples the output voltage of the first pre-amp and converts it to a DC voltage that is applied to the base configuration of the AGC transistor. The DC voltage controls the bias (amplification configuration) of the AGC circuit, which directly controls how large of a signal is output to the first preamp. The entire system is designed for an optimum signal into the second preamp. When the signal is optimal, then the level of the AGC control signal is zero in most AGC circuits.

AGC is an important part of digital microwave. Some microwave links are often miles apart. When the path of the two dish antennas are aligned, the technician connects a volt meter to the AGC control signal. As the dish is rotated on its axis (azimuth), the technician watches as the AGC control signal changes. When the signal peaks, it is pointed directly at the other antenna. Even though terrestrial microwave links do not move or switch stations, they still need AGC to compensate for weather changes.

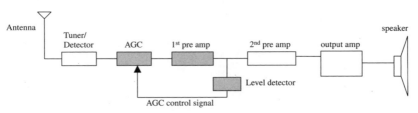

Figure A.11   AGC

Agent   1. In data administration, software that processes queries then returns replies on behalf of an application program. 2. In network-management systems, such as Novell RIP or Cisco IGRP, an agent is a subprocess that resides in all managed devices (such as routers and servers) and reports the values of specified network variables to management stations. 3. An extension in a PBX or call center environment that has variable or conditional availability.

Aggressive Back-off Algorithm   A reference to autonegotiation features within Ethernet. It is the IEEE 802.3u 100BaseT feature specification that provides for flow control (pause frames) and full-duplex operation. Full duplex allows for 100 Mbps send and 100 Mbps receive, for a total

200 Mbps Ethernet connection over Cat5 twisted pair. The autonegotiation is an enhancement of the link integrity signaling method used in 10BaseT networks and is backward compatible with link integrity. Autonegotiation allows the NIC or the network device to adjust its speed to the highest speed that both ends are capable of supporting. To be able to use this feature, both the network device (switch port) and the NIC must contain the autonegotiation logic.

AGP (Accelerated Graphics Port)   A newer bus/interface architecture developed by Intel Corporation to be used as an interface for computer monitors. AGP is based on the PCI bus architecture, except that AGP provides a dedicated connection to main memory for a video card. This enables throughput that exceeds 1 Gbps. The push for a faster monitor interface has come from the demands of 3-D graphics. AGP slots/sockets can easily be distinguished by their offset pin configuration (Fig. A.12).

Figure A.12   An AGP Video Card

Aggregate Bandwidth   The total bandwidth of a broadband circuit and all of its tributaries, including the payload and overhead. A T1 has an aggregate bandwidth of 1.544 Mb/s.

Aggregation Device   An ISDN terminal adapter that can combine two B channels (64 Kb/s each) together for a single channel that has twice the bandwidth (128 Kb/s). These adapters can switch back and forth from aggregated to non-aggregated while the circuit is in use.

Aggregator   A long-distance reseller. They sign up with a long-distance company as a reseller and all their customers are "aggregated" together for a bulk discount. The long-distance company provides the service and does the billing. The advantage to the long-distance company is that they have more people selling their long distance. The advantage to the customer is the value-added service (consultation/expertise) that the aggregator offers.

AH (Amp Hour)   A battery rating for UPS system and other batteries. The amp-hour rating is derived by multiplying the amount of current that a battery can supply by the time it can supply it. It works out to be a ratio so that you can calculate how long your battery back-up system will last if the power goes out.

For example: If a battery has an amp-hour rating of 100, then it can supply 100 amps for 1 hour. Or it can supply 50 amps for 2 hours, 25 amps for 4 hours, 1 amp for 100 hours, etc.

AIN (Advanced Intelligent Network)   The ability of a communications network to determine the routing or handling of a call based on the way the caller desires. AIN is used by local and long-distance companies to give customers a choice as to how they would like their calls routed. A particular trunk can be specifically programmed to route a specific path through switching centers across a geographical area. AIN is ultimately an upgrade to SS7. Some AIN trunks can be made to route to an IVR (Interactive Voice Response) system that gives the customer options for their call handling.

AIOD (Automatic Identification of Outward Dialing)   This is a call-accounting system feature of PBX and some key systems that captures every number dialed by a specific telephone extension and prints it out on a report for accounting and cost-tracking purposes.

AIR (Allowed Information Rate)   The maximum data transfer rate that a frame-relay *DLC (Data Link Connection)* will allow. The AIR is equal to the *CIR (Committed Information Rate)* plus the *EIR (Excess Information Rate)* of the particular DLC.

Air-Pressure Cable   Telephone cable that is equipped with air-pressure equipment. In many cables nitrogen is used instead of air because it is noncorrosive (air contains humidity and oxygen that corrodes copper pairs). Nitrogen is pumped into the cable and the pressure is monitored. If the cable is cut, the pressure drop notifies the telephone company of a cable problem and the nitrogen rushing out of the cable helps prevent any water from entering the cable (Fig. A.13).

Figure A.13   Air-Pressure Cable

Airline Mileage   The mileage between two cities that long-distance private-line pricing is based on. AT&T developed a grid coordinate system (coordinates shown in V&H table) that gives every telephone central office in the United States a vertical and horizontal grid number. To calculate the mileage between two cities, the Pythagorean theorem is used.

To calculate mileage between two cities, follow these steps:

1. Take the difference of the V coordinates and square it.
2. Take the difference of the H coordinates and square it.
3. Add the two squared numbers together.
4. Divide by 10.
5. Take the square root of that number. This is the mileage.

Example: What is the airline mileage from Los Angeles, CA to New York, NY?

- The V coordinate of Los Angeles is 9213. The V coordinate of New York is 4977. The difference is 4236.
- H coordinate of Los Angeles is 7878. The H coordinate of New York is 1406. The difference is 6472.
- Next, square both numbers: $4236^2 = 17,940,000.$ $6472^2 = 41,890,000.$
- Now, add these numbers: $17,940,000 + 41,890,000 = 59,830,000.$
- Now, divide these numbers: $59,830,000 \div 10 = 5,983,000.$
- Take the square root: $\sqrt{5,983,000} = 2446.$
- 2446 miles is the airline mileage between Los Angeles and New York.

Algorithm   A well-defined rule or process for arriving at a solution to a problem. In networking, they are commonly used to determine the best route for traffic from particular source to a particular destination, and used to create error detection and correction processes.

Alignment Error   In most Ethernet networks, an error that occurs when the total number of bits of a received frame is not divisible (a factor of) by eight. Alignment errors are usually caused by collisions.

All-Routes Explorer Packet   In a source route bridging network, a signal sent by a user device that hunts for another end device across an entire network.

All Trunks Busy   You might try to make a call and get a fast busy signal or an intercept message that says "I'm sorry, all circuits are busy now. Please try your call again later." This situation can happen for a number of reasons. If you are dialing long distance, you get this message because all of the trunks that your long-distance company has between their interlata central office *POPs (Points of Presence)* are busy. If you are making a local call that terminates to a different local CO and you get this message, it is because all the inter-office trunks are busy. If you are calling your neighbor and you get this message, then the inter-grouping trunks within the local CO switch are all busy. Inter-grouping trunks are used in large switches to interconnect "smaller CO switch groups" within the CO switch.

Alligator Clips   Most analog test equipment comes equipped with alligator clips. For a photo, see *Bed of Nails Clips.*

Allowed Cell Rate (ACR)   A parameter defined by the ATM Forum for ATM Traffic Management. The ACR varies between the *Minimum Cell Rate (MCR)* and the *Peak Cell Rate (PCR)*. It is managed by the protocol congestion control mechanisms.

Allowed Information Rate (AIR) The maximum data transfer rate that a frame-relay *DLC (Data Link Connection)* will allow. The AIR is equal to the *CIR (Committed Information Rate)* plus the *EIR (Excess Information Rate)* of the particular DLC.

ALPETH (Aluminum/Polyethylene) The sheath or jacket of an outside plant telephone cable that is used mostly for aerial applications. It is basically an aluminum wrap around the conductors, which resembles a serrated tin can, coated with ⅛″ of black plastic (Fig. A.14).

Figure A.14 ALPETH

Alternate Answering Position A second attendant console where the first console can forward calls if the first console attendant is absent. An alternate answering position can also be used for overflow of calls that the first attendant can't keep up with.

Alternate Routing A switch feature that enables all trunks to have alternate outgoing assignments. If the primary routing or least-cost routing is all busy or out of service, then the switch will route the call to an alternate trunk to connect the call. It is a good idea to use multiple long-distance and local services in conjunction with alternate routing in case of a service outage.

AM (Amplitude Modulation)   AM is a technique of making a voice or other signal ride on (or modulate with) another frequency (the carrier frequency). For a diagram, see *Amplitude Modulation.*

AMA (Automatic Message Accounting)   What RBOCS call their call tracking system for billing.

Ambient Current   The result of the voltages created by random movement of electrons in a circuit when the power is off. There is always ambient voltage, which is why oscillator circuits start oscillating when the power is turned on. The natural oscillations of the electrons become filtered and amplified when the power is applied to the circuit.

Ambient Noise   Noise caused by the random movement of electrons in an electronic circuit when the power is off or by the random movement of air.

Ambient Voltage   Electromotive force created by the random movement/vibration of electrons in a circuit when the power is off. There is always ambient voltage, which is why oscillator circuits start oscillating when the power is turned on. The natural oscillations of the electrons become filtered and amplified when the power is applied to the circuit.

American National Standards Institute (ANSI)   A nongovernmental nonprofit standards setting institute that publishes standards that industries voluntarily follow. ANSI works very hard to bring together the interests of the private and public sector. ANSI is the official U.S. member body to the world's leading standards bodies.

American National Standards Institute Communications Standards
Some examples of ANSI communications standards are:

ANSI T1.110-1987 SS7	General information
ANSI T1.111-1988 SS7	Message Transfer Part (MTP)
ANSI T1.112-1988 SS7	Signaling Connection Control Part (SCCP)
ANSI T1.113-1988 SS7	ISDN user part
ANSI T1.114-1988 SS7	Transaction Capability Application Part (TCAP)
ANSI T1.206	Digital Exchanges and PBX loop-back test lines
ANSI T1.301	ANSI ADPCM standard
ANSI T1.401-1988	Interface between carriers and customer installations for voice-grade switched analog lines Loop Start and Ground Start
ANSI T1.501-1988	Network performance/network encoding limits for 32Kb/s ADPCM

ANSI T1.601-1988	Basic access interface/electrical loops for the network side
ANSI T1.T1.Q1	Network performance standards for switched exchange and IXC
ANSI TIX9.4	SONET
ANSI X3T9.5 TPDDI	FDDI on UTP/STP
ANSI character set	ANSI 256 character set code

AMI (Alternate Mark Inversion)   AMI is a line format that serves two advantages to sending a digital signal directly over a twisted pair. The feature of AMI that makes it unique is that each bit is inverted. This makes the first bit +5 V, the second −5 V, the third =5 V, etc. Alternating the bit polarity also makes the signal look like it is half the frequency to the twisted pair (Fig. A.15).

Ref +5 ▶   Unformatted digital signal   AMI formatted signal
Ref
Ref −5 ▶

Reference voltage for T1 carrier is typically +135 Volts over the public network, so AMI actually switches from +140V to +130V

Figure A.15   AMI

Amp   Abbreviation for Ampere, see *Ampere*.

Amp Hour (AH)   A battery rating for UPS system and other batteries, the Amp Hour rating is derived by multiplying the amount of current that a battery can supply by the time it can supply it. It works out to be a ratio so that you can calculate how long your battery back-up system will last if the power goes out.

For example: If a battery has an amp hour rating of 100, then it can supply 100 amps for 1 hour, 50 Amps for 2 hours, 25 Amps for 4 hours, 1 amp for 100 hours, etc.

Ampere   6,300,000,000,000,000,000 electrons moving past a point in one second (a coulomb) is equal to one ampere (also known as an *amp*) of electrical current. The shortcut/alternative to counting all the electrons as they run by is to use the Ohm's law formula and calculate the amperage instead. If you know two of the following about your circuit, voltage, resistance, or watts you can perform the calculation. The formulas are:

$$Current \text{ (in amps)} = \frac{Voltage \text{ (in volts)}}{Resistance \text{ (in ohms)}}$$

or

$$Current \text{ (in amps)} = \frac{Power \text{ (in watts)}}{Voltage \text{ (in volts)}}$$

or

$$Current \text{ (in amps)} = \sqrt{\frac{Power \text{ (in watts)}}{Resistance \text{ (in ohms)}}}$$

Amplified Handset   A handset with a built-in amplifier for the hearing impaired. Amplified handsets can be purchased for virtually every kind of PBX telephone. Walker Electronics is a well-known manufacturer of these devices.

Amplifier   An electronic circuit designed to increase an input level characteristic to a desired output level characteristic. Some amplifiers are designed to amplify the voltage level of a signal and others are designed to amplify the current of a signal flowing through a load. A typical stereo system has both of these types of amplifiers. If you are listening to a CD player, the signal (after digital to analog processing) is fed to a voltage amplifier to increase its ability to drive a current amplifier, which amplifies the current that is driven through the loudspeaker. Both of these amplifiers (voltage and current) combined make a power amplifier, hence power (in watts) is a function of voltage and current. Amplifiers are usually rated by the amount of power that they are capable of producing in a loudspeaker. Peak power is calculated by using the peak value of a sinusoidal waveform. RMS (root mean square) power is calculated by using the RMS value of the sinusoid waveform, which mathematically works out to be 70.7% of the peak value. Many amplifier manufacturers use the peak-power rating because it looks better. If you compare a JVC amplifier that is rated at 71 watts RMS output, it is the same output as the 100-watt peak-power "other brand" amplifier. High-quality audio amplifiers are usually rated in RMS power.

Amplitude   The peak or peak-to-peak amplitude of a signal measured in volts. The AC signal below has a peak amplitude of 10 volts or a peak-to-peak amplitude of 20 volts (Fig. A.16).

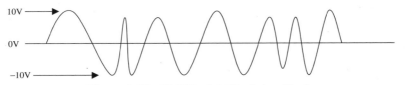

Figure A.16   20 V Peak-to-Peak Amplitude

Amplitude Modulation   AM is a technique of making a voice or other signal ride on (or modulate with) another frequency (the carrier frequency). See *Fig. A.17.*

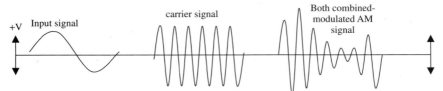

Figure A.17   Amplitude Modulation

AMPS (Advanced Mobile Phone System)   This is the cellular/PCS network as we know it today. The first mobile phone system was called *MTS (Mobile Telephone System)* and was developed during World War II. To make a call on an MTS system, a human operator was needed to hand-off/connect the call. In the early 60's *IMTS (Improved Mobile Telephone System)* evolved. IMTS did not require a human operator to connect a call from one mobile phone to another, but calls could only be made within one cell. In 1983, the implementation of the *AMPS (Advanced Mobile Phone System)* began. AMPS allows callers to call from one mobile phone to another, from one cell to another, and connect calls between the land-based network and the mobile network without the need for an operator.

Analog   A signal having an infinite number of levels per cycle, in contrast to digital, which has only two possible levels per cycle (i.e., on or off). See *Fig. A.18.*

Figure A.18   Analog

Analog-to-Digital Conversion (Digital-to-Analog Converter/Analog-to-Digital Converter)   A part of a channel bank that performs the function of encoding analog voice signals into a stream of binary digits. The analog-to-digital converter samples a caller's voice at a rate of 8000 times per second. (The sample rate for a T1 channel is 8000 times per second.) Each sample's voltage level is measured and converted to one of 256 possible sample levels. These levels are from the lowest, 0000000, to the highest 11111111. It has 256 levels because if you count in binary from 00000000 to 11111111, you end up with 256, that is the highest number possible with 8 bits. The bits are then transmitted one after another at a high rate of speed to their destination, where the same process happens in reverse (Fig. A.19).

A callers analog voice pattern

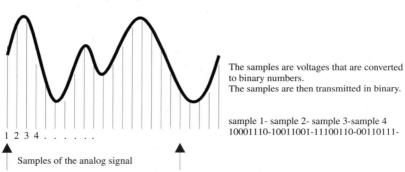

The samples are voltages that are converted to binary numbers.
The samples are then transmitted in binary.

sample 1- sample 2- sample 3-sample 4
10001110-10011001-11100110-00110111-

1 2 3 4 . . . .

Samples of the analog signal

The binary numbers are actually voltages, the binary bit stream of four samples is shown as voltages below.

1 000 1 1 1   0 1 00 1111 00 1 1 1 1 00 11   000   11 0 1 1 1

At the other end of the transmission line, the binary code is converted back into its original analog form using the same process in reverse.

Figure A.19   Analog-to-Digital Conversion

ANI (Automatic Number Identification)   ANI is also called Caller Identification or Caller ID. This feature, offered by local phone companies, sends the phone number (and often the name of the caller) down the phone line in a digital data packet between the first and second ring. To receive the data, a subscriber that has signed up for the service needs to have a caller-ID unit (also called a *caller-ID box*) plugged into the phone line. The caller-ID unit displays the name and the number of the calling party for each incoming call. Caller ID only works if the caller and the called party's phone service is fed out of a central office that has caller-ID capability. If the central office does not have caller-ID capability, then the display will read "out of area" to the called party. If the called party does not have caller service they will get a display that says "no data sent."

Annex A   A frame relay standard extension that outlines the provisioning of a *Local Management Interface (LMI)* that goes between the customer's equipment and the frame relay network. The LMI will provide network monitoring and status through the transmission overhead.

Annex D   The second frame relay standard extension that outlines the provisioning of a Local Management Interface (LMI) that goes between the customer's equipment and the frame relay network. The LMI will provide network monitoring and status through the transmission overhead.

Annular Ring   A marking around a cable to show length—some are in meters and some are in feet.

Anonymous Call Rejection   A feature that can be provided by the local phone company that will not complete anonymous calls to your line. You can also reject anonymous calls by installing a caller-ID unit that has anonymous call rejection built-in to it.

ANSI (American National Standards Institute)   A nongovernmental nonprofit standards-setting institute that publishes standards that industries voluntarily follow. ANSI works very hard to bring together the interests of the private and public sector. ANSI is the official U.S. member body to the world's leading standards bodies.

ANSI Standards   See *American National Standards Institute Communications Standards* for some examples of their communications standards.

Answer Back   A command that a local data terminal sends to a computer or device across a network asking it to send its address so that the local terminal can verify that it has connected to the correct computer.

Answer Supervision   The ability of a central office to know when a long-distance call has been answered so that the call can be billed. This feature is a combination of the SS7 network and CO switch software that integrates with the telephone company's call accounting and billing system. Anyone who bills people for phone calls should have this system, but some do not. Hotels are famous for not having answer supervision. If you make a call from your hotel room and the end you are calling rings for more than 30 seconds (eight rings) you will probably be billed for the call. The hotel's PBX has no way to know if anyone picked up the phone on the other end, so it starts billing for the call after a certain time. The hotel's PBX is supervising and billing according to you being off-hook and the digits you dialed, not according to when the "answerer" picked up the line. To this day, some long-distance companies don't have answer-supervision capability.

Antenna   A device for receiving and transmitting electromagnetic signals. The optimal antenna for a given transmission or reception of a frequency has a length equal to the wavelength (or a usable fraction) of that frequency. CB radio antennas are very long in comparison to the antenna on your PCS cellular phone. CB radio is transmitted at frequencies that are low, which have a long wavelength and PCS cellular is transmitted at high frequencies, which have shorter wavelengths. See also *Yagi Antenna, Parabolic Dish Antenna,* and *Monopole Antenna.*

Antennas are available in many varieties that are designed to manipulate the incoming or outgoing signal.

- *Single pole, full wavelength*   For picking up weak signals or transmitting over long distances.
- *Single-pole half or quarter wavelength*   Smaller size and for receiving nearby stations.
- *Dipole*   For directional applications. Rabbit ears, for example, are directional.
- *Phased array*   For picking very distant signals.
- *Parabolic dish*   For focusing a signal from or to one direction.
- *Vertical Loop*   For low-noise, directional reception. Common for UHF TV.
- *Horizontal Loop*   For low-noise, directional reception.
- *Multielement*   For low-noise, directional reception or transmission.
- *Monopole*   Used in cellular and PCS applications.

Antenna Beamwidth   The dimensional characteristics of a radio field projected by an antenna element. In general, different antennas focus a radio field to achieve a stronger signal over a particular direction or area. The areas are described as a ratio of front-to-side, front-to-back, and other comparative methods. This is done under the assumption that a transmission power loss in one area constitutes a gain in another and vice versa. When a radio link is subject to interference from other nearby radios, an antenna beamwidth with high dimensional ratios is preferred and often required in licensed applications. Keeping your radio beam focused increases efficiency and reduces the likeliness of interference from other radio transmissions. The only drawback is that these types of antennas are more expensive.

Antenna Farm   A collection of satellite dish antennas located at a cable-TV head end or satellite telecommunications center.

Antenna Gain   An indicator of how well an antenna focuses RF energy in a preferred direction. Antenna gain is expressed in dBi (the ratio of the power radiated by the antenna in a specific direction to the power radiated in that direction by a nondirectional antenna fed by the same transmitter). Antenna manufacturers normally specify the antenna gain for each antenna they manufacture.

Anti-Static   Materials coated or manufactured with semiconductive materials makes them anti-static, which is good for *CMOS (Complementary Metal-Oxide Semiconductor)* components. CMOS components are highly sensitive to static discharges and static fields (ESD). Your body can easily hold a static charge of 40,000 V on a dry day. About 25,000 V is required to get a static shock from a door knob. Exposure to static electricity can ruin a CMOS component instantly or just weaken it, which would cause it to fail unpredictably. CMOS components include microprocessor chips, transistors, RAM and ROM chips, and many others. If a CMOS device must be used in an ESD hazard area, TTL components are used as an alternative. TTL components are not as static sensitive, but, they are not as fast, not as small, use more electricity, and produce more heat.

AOSP (Alternate Operator Service Provider)   A long-distance company that works like the old days, when a live operator would assist you with your call. Some calling-card companies incorporate this in their service. AOSP service is great, and it's a good thing because we pay for each use.

AP (Access Point)   Another name for a cross-box where telephone cables are cross connected. See *Fig. A.20*.

Figure A.20   Access Point (AP)

APD (Avalanche Photo Diode)   A device used as a light-to-electricity converter and signal amplifier at the same time. They are incorporated in optoelectronic circuits used in fiber-optic terminating applications.

Application Layer   The seventh and highest layer of the OSI communications model. The applications layer is the function of connecting an application file or program to a communications protocol. The latest model or guideline for communications protocols is the OSI (open systems interconnect). It is the best model so far because all of the layers or functions work independently of each other. Older proprietary communications models are shown below alongside the OSI. For a diagram of the OSI, SNA, and DNA layers, see *Open Systems Interconnection*.

API (Application Program Interface)   A set of routines, protocols and other tools that are used by software programmers to create applications. Microsoft Windows has an API that enables programmers to create applications that are consistent with the Windows user environment.

APON (ATM Passive Optical Network)   A passive network consists of devices that do not require external power. The physical characteristics of the carrier (light, in this case) are used to route and distribute signals. In xDSL networks, it is possible to extend and route the xDSL signal using fiber optic. Passive optical networks are less expensive to implement and operate. An APON is a passive optical network that is carrying ATM. ATM is a practical and frequently used method to transport xDSL. For more details on the physical characteristics of light transmissions, see *Fiber Optic, WDM,* and *Refraction*.

APPC (Advanced Program-to-Program Communication)   The IBM *SNA (System Network Architecture)* software that allows high-speed communication between programs on different computers in a distributed/mainframe computing environment. APPC establishes and tears down virtual connections between programs that require communication. It has two software interfaces; the first interface is the programming interface, which replies to programs requiring communications. The second interface is the data-exchange interface, which establishes the sessions or "connections" between the programs.

Applications Processor   An add-on to a PBX (Private Branch Exchange) system or CO (Central Office) switch that expands its ability to provide extended services or process additional protocols. An example of an applications processor is a voice-mail system, ACD, frame relay or ISDN interface. Physically, the applications processor is often an additional shelf, module, or card that interfaces into the PBX system's bus architecture.

APPN (Advanced Peer-to-Peer Networking) An enhancement to the original IBM *SNA (System Network Architecture).* APPN handles the following: session establishment between peer nodes, dynamic transparent route calculation, and traffic prioritization for Advanced Program-to-Program Communication.

ARCnet (Attached Resource Computer Network) A token-bus local-area network protocol/package developed by Datapoint Corporation. ARCnet was a popular coax media network solution after its initial inception in the late 1970s. Newer versions have evolved for use on other physical media.

Area In OSPF (Open Shortest Path First) routing, a group of routers that share identical link state databases provided by a designated router. It is recommended that no more than 40 routers exist within one OSPF area.

Area Code An area code is a three-digit code that designates a toll center in the North American Numbering Plan. To call outside of your toll center, you first dial 1, then the area code for the toll center or "area" you wish to call. See Appendix C for a listing of area codes by area. See Appendix D for a listing of area codes by number.

ARP (Address Resolution Protocol) A set of communications rules and instructions used by computers and communications interface equipment to map or forward *IP (Internet Protocol)* addressed data packets to a hardware address. A hardware address is also known as a *MAC (Media Access Control)* address. ARP can only be used in networks that have broadcast capabilities.

ASCII (American Standard Code for Information Interchange) ASCII is a code developed by ANSI (Fig. A.21). It is a seven-bit character and command code. More than one variation of ASCII exists, such as extended ASCII, which is eight bit and has many more characters. ASCII is the standard code that PCs use and is the code that is transmitted into your computer every time you push a key on your keyboard.

Least significant bits (hexadecimal)	Most significant bits (hexadecimal) 000 (0)	001 (1)	010 (2)	011 (3)	100 (4)	101 (5)	110 (6)	111 (7)
0000 (0)	NUL	DLE	SP	0	@	P	`	p
0001 (1)	SOH	DC1	!	1	A	Q	a	q
0010 (2)	STX	DC2	"	2	B	R	b	r
0011 (3)	ETX	DC3	#	3	C	S	c	s
0100 (4)	EOT	DC4	$	4	D	T	d	t
0101 (5)	ENQ	NAK	%	5	E	U	e	u
0110 (6)	ACK	SYN	&	6	F	V	f	v
0111 (7)	BEL	ETB	'	7	G	W	g	w
1000 (8)	BS	CAN	(	8	H	X	h	x
1001 (9)	HT	EM	)	9	I	Y	I	y
1010 (A)	LF	SUB	*	:	J	Z	j	z
1011 (B)	VT	ESC	+	;	K	[	k	{
1100 (C)	FF	FS	,	<	L	\	l	
1101 (D)	CR	GS	-	=	M	]	m	}
1110 (E)	SOH	RS	.	>	N		n	~
1111 (F)	SI	US	/	?	O		o	DEL

DEFINITIONS OF ASCII CONTROL CODE ABBREVIATIONS

ACK - ACKNOWLEDGE
BEL - BELL
BS - BACKSPACE
CAN - CANCEL
CR - CARRIAGE RETURN
DC - DIRECT CONTROL
DEL - DELETE IDLE
DLE - DATA LINK ESCAPE
EM - END OF MEDIUM
ENQ - ENQUIRY
EOT - END OF TRANSMISSION
ESC - ESCAPE
ETB - END OF TRANSMISSION BLOCK
ETX - END OF TEXT
FF - FORM FEED

FS - FORM SEPARATOR
GS - GROUP SEPARATOR
HT - HORIZONTAL TAB
LF - LINE FEED
NAK - NEGATIVE ACKNOWLEDGE
NUL - NULL
RS - RECORD SEPARATOR
SI - SHIFT IN
SO - SHIFT OUT
SOH - START OF HEADING
STX - START OF TEXT
SUB - SUBSTITUTE
SYN - SYNCHRONOUS IDLE
US - UNIT SEPARATOR
VT - VERTICAL TAB

Figure A.21   ASCII

ASCU (Agent Set Control Unit) An address identifier for a set of terminals that handle type-A messages in an IBM network environment. An ASCU identifies a workstation to a concentrator and can use P1024B or P1024C protocol.

ASIC (Application Specific Integrated Circuit) A proprietary integrated circuit created to do a specific job. An example is Motorola manufacturing a chip to do multiplexing specifically for a MUX that is made by Nortel. Motorola manufactures the chips for Nortel and stamps Nortel's name on them. Then, only Nortel uses the technology and no one else.

ASR (Access Service Request) If a special service provider (frame relay or long-distance private line) needs wire facilities from their point of presence in the city to customer's location, they call the local telephone company and make an access service request to provide a line that runs from your network interface to them. Many special service providers have their equipment located in the local phone company's central office as a part of a co-location agreement. When a CLEC needs to provide service where they don't have facilities, they do it by using the RBOCS wire facilities.

Asymmetric   Communications transmission that is full or half duplex, where one direction is very fast, compared to the other. Cable TV is an example of asymmetrical communication. The cable TV head end sends massive amounts of video and audio information down a coax one way, and the cable TV set-top decoder boxes send small amounts of ID and status information the other way back to the head end over the same coaxial connection. Sometimes asymmetrical channels are referred to as "upstream" for slow and "down stream" for fast. ADSL is another example of asymmetric communications.

Asymmetric Digital Subscriber Line (ADSL)   Also referred to as *ADSL Full Rate* or *G.992.1*. ADSL is a physical-layer protocol that supports up to 8 Mbps bandwidth downstream and up to 1 Mbps upstream. The asymmetrical aspect of ADSL technology makes it ideal for Internet browsing, video on demand, and remote *Local-Area Network (LAN)* access. Users of these applications typically download more information than they send. ADSL also allows simultaneous voice communication by transmitting data signals outside of the voice frequency range. Sometimes a faint hiss can be heard on the line. To eliminate the hiss, a voice/data splitter, commonly called a *POTS splitter (Plain Old Telephone Service)* is installed at the jack of each phone. The base transmission range for an ADSL line is 18,000 feet. ADSL can be extended to remote communities by using ADSL repeaters (to 48,000 feet) or fiber optic. For more information on the DSL family of protocols, see *xDSL*. For a diagram, see ADSL.

Asynchronous   To communicate without external timing and to have each communicating device work at its own speed. People talk asynchronously. Even though one person talks very fast and another very slowly, their brains still receive the conveyed messages and respond. Modems, FAX machines, and TCP/IP communications are asynchronous.

Asynchronous Transfer Mode (ATM)   An ANSI and CCITT standard communications protocol (Fig. A.22). ATM is a frame-format communications protocol whereby data is transmitted and received 53 bytes or octets at a time. There are 48 customer bytes (Payload) and 5 bytes for control and addressing. Three things make ATM special.

1. It is capable of carrying delay-sensitive transmissions without delay (such as speech, music, or video).
2. Many ATM channels can be concatenated to provide more bandwidth (carrying capacity).
3. It has the ability to carry many different types of data at the same time: LAN, WAN, video, voice, and anything else that is capable of being digitized.

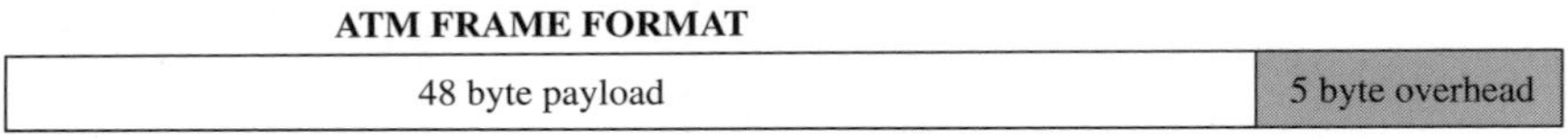

ATM FRAME FORMAT

48 byte payload	5 byte overhead

Figure A.22   Asynchronous Transfer Mode

Another good thing about ATM is that the overhead is a bit more than 10%, which is an improvement over other transport methods. To put ATM into a simple picture, imagine that you have a computer network LAN signal and a video signal. You want to send the signals that your computer and video are generating to other computers and TVs.

Now, take the scenario a step further. Imagine that your computer LAN signals are motorcycles and the video signals are cars. ATM would then be large trucks that carry the motorcycles and cars to their destination and back, linking your communications gap with two great things in mind. Your cost is lower in contrast to many small circuits and you only buy one piece of gear to connect everything (in contrast to many different terminal adapters, CSU/DSU, etc.). See also *ATM*.

AT (Access Tandem)   A telephone company central office or node that contains a switch in which all inter and outer area-code traffic is handled. The main LEC central office in an area code, where the hand-off for long-distance service happens. For a diagram, see *Access Tandem*.

AT (Advanced Technology)   A reference to the IBM clone motherboard standard baby AT form factor. A generation of personal computers that was introduced by IBM in 1984. The first AT computers had an Intel 80286 microprocessor and hard disk drives that ranged from 10 MB to 30 MB. AT computer architecture is able to be upgraded easily by replacing RAM, hard disk drives, plug-in CPUs, and even the motherboard.

AT architecture has enabled PCs to evolve with new technologies, spawning machines that incorporate CPUs with speeds in excess of 300 MHz and hard disk drives that exceed 8 GB in memory. Illustrated is an AT socket-7 motherboard (Fig. A.23). The newer version of AT architecture is ATX. See also *ATX.*

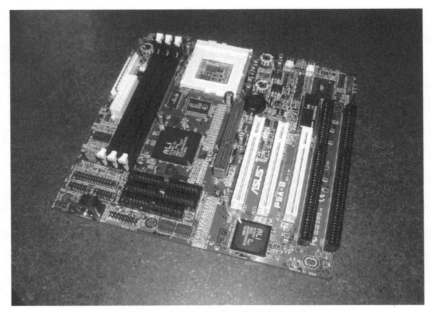

Figure A.23   An AT Motherboard

ATA (AT Attachment)   Also referred to as *IDE (Intelligent Drive Electronics* and *Integrated Drive Electronics). ATA* refers to a drive-interface standard for CD-ROM drives and hard disk drives. It was called *ATA* because of the popularity of the *AT (Advanced Technology)* motherboard at the time of its development by the Small Form Factor Committee. It is a relatively less-expensive alternative to *SCSI (Small Computer System Interface)* and has retained popularity in the PC industry because it manages to continuously improve while remaining price competitive. Newer versions of this interface standard are called *Fast ATA, ATA-66,* and *EIDE (Enhanced IDE).* See also *IDE, SCSI,* and *Hard Disk Drive.*

ATM (Asynchronous Transfer Mode)   An ANSI and CCITT standard communications protocol. ATM is a frame-format communications protocol whereby data is transmitted and received 53 bytes or octets at a time. There are 48 customer bytes (Payload) and 5 bytes for control and addressing. For more information, see *Asynchronous Transfer Mode.*

ATM25   The ATM Forum-defined 25.6-Mbps cell-based user interface that is based on the IBM token-ring network.

ATM Adaptation Layer (AAL)   A transfer format, cell header format, and functional section of the ATM transport method. There are five variations of the ATM adaptation layer. Each is intended to be used with a specific type of data. For a diagram of ATM layers, see *ATM.* For a diagram of AAL classifications, see *AAL.*

ATM Adaptation Layer One (AAL1)   The part of the ATM protocol that enables the transfer of time-sensitive data, such as voice or video. AAL1 uses an adaptive clock method, where the devices at each end of the link negotiate a clock agreement, then incorporate a small buffer to monitor the rate at which cells are being transferred across the link. AAL1 is used for DS0, DS1 emulation, and other voice and video. For a diagram, see *AAL.*

ATM Adaptation Layer Two (AAL2)   For class-B traffic (see diagram under AAL), packet technologies, and the transport thereof. It is similar to voice over frame, video over frame, etc.

ATM Adaptation Layer Three and Four (AAL3/4)   For class-C and -D (see the *AAL* diagram) layers that are designed to handle non-time-sensitive data transfer. This layer class adds header information that incorporates error-checking functions before and after the original data. Also, a Message ID function allows multiplexed or interleaved transmissions to be sent directly over the single ATM virtual channel. This layer would be used as a backbone to carry many X.25 or frame-relay logical links, or could be used in a campus application to carry Ethernet from one building to another.

ATM Adaptation Layer Five (AAL5)   The layer created for class-C and -D types of traffic (see the *AAL* diagram). The cell header remains the same except larger buffers are used and a *CRC (Cyclic Redundancy Check)* is appended to the end of the last cell of the packet's cell stream. No Message ID function is available to directly transfer multiplexed data. Cell payloads are 48 bytes, and a PTI bit is used to indicate the last cell of a packet.

ATM DXI (ATM Data Exchange Interface)   A connectivity and conversion method that allows an existing LAN network element (i.e., router) to access an ATM network via a private line that is equipped with a CSU/DSU (V.35 or HSSI).

ATM DSU (Asynchronous Transmission Mode Digital Service Unit)
A terminal adapter used as a demarcation point or interface point to an ATM network.

ATM DXI (ATM Data Exchange Interface)   A connectivity and conversion method that allows an existing LAN network element (i.e., router) to access an ATM network via a private line that is equipped with a CSU/DSU (V.35 or HSSI).

ATM Forum   An international organization jointly founded in 1991 by Northern Telecom, Cisco Systems, Net/Adaptive, and Sprint. The organization develops and promotes standards based implementation agreements for ATM technology. The ATM Forum expands on official standards developed by ANSI and ITU-T. The organization also develops implementation agreements in advance of official standards.

Attached Resource Computer Network (ARCnet)   A token-bus local-area network protocol/package developed by Datapoint Corporation.

Attendant   Another name for a PBX operator. The person who connects outgoing calls and/or answers, screens and directs incoming calls in a polite mannerly way. If they don't, they would soon be replaced by an auto-attendant.

Attenuation   Reduction of a signal's voltage level as it travels down a line, measured in decibels. Attenuation is also called *loss,* because some signal is always lost through resistance and reactance. Optical light-wave signals are also attenuated when they traverse through a fiber-optic because of impurities in the fiber optic, and the fact that light intensity decreases with distance.

Attenuator   An attenuator is also called a *pad, T pad,* or *H pad.* It is a device that reduces the voltage level of a signal without changing its impedance. Attenuators are frequently used on telephone lines that terminate to customers close to the central office so that the volume in the handset does not hurt their ears. Attenuators are also made for fiber-optic applications. A fiber-optic attenuator works like your sunglasses, it reduces the level of light entering your eyes so that you can see more effectively. For a photo, see *Fiber optic Attenuator.*

ATX A trademark of Intel that is not an abbreviation, but a model number. Many refer to it as *Advanced Technology Expanded* because it is an improvement of the AT *(Advanced Technology)* generation of computer hardware architecture. The ATX generation of PC architecture is based on an improved motherboard design, which is rotated 90 degrees within the cabinet. This places devices within the PC closer to their respective plug-in connectors and places the CPU closer to the power-supply cooling fan. ATX motherboards also have on-board I/O interfaces. The most noticeable esthetic difference between an AT and ATX architecture is that the interface connectors are laid out differently. On the AT, the connectors are set along the edge of the motherboard, on the ATX the connectors are blocked together on a metal mounting plate. Illustrated is an ATX slot 1 motherboard. For a visual comparison of AT and ATX, see *PC* (Fig. A.24).

Figure A.24   ATX Motherboard

Audio Sound. Signal frequencies that if amplified and applied to a loudspeaker can be heard. These frequencies range from 20 Hz to 17 kHz.

Audio Frequency Signal frequencies that, if amplified and applied to a loudspeaker, can be heard. These frequencies range from 20 Hz to 17 kHz.

Auger A device that looks like a giant drill bit, which is used for boring holes into the ground for telephone or power poles. Some utility construction vehicles are equipped with augers.

AUI (Autonomous Unit Interface) A 15-pin connector used in *CTI (Computer Telephony Integration)* applications. For a picture, see *DB15*.

Auto Baud A term that refers to a modem or other communicating device's ability to match or adapt to the bit transmission rate of the device at the far end.

Auto Dialer A device that automatically dials preprogrammed digits when the line it is attached to breaks a dial tone. Auto dialers are used to program long-distance carrier access codes so that people don't have to deal with the confusion of which long-distance company to access and when. The person making a call pushes a key on their phone that accesses a long-distance trunk equipped with an autodialer that dials the access code, then the person making the call dials in the number they want to call. Auto dialers are also used to make a phone dedicated to one phone number or directory extension. In the city where I live there is a restaurant with no waiters or waitresses. When the patrons to the restaurant are seated, they find a phone at their table. The phone has no dial pad. When the phone is picked up, an auto dialer instantly dials the order taker's extension and the patrons place their order.

Automated Attendant The machine that answers the line and plays a message that says: "Thank you for calling company X. To speak to a person in sales press one, to speak to a repair person press two." Advantages of auto attendants are that you can have one number advertised for multiple departments and not have to have a full-time person directing calls.

Automated Voice Response Not to be confused with *integrated* voice response, an *automated* voice-response system or network is a way of guiding callers to a department, agent or pre-recorded information. Automated-attendant/voice-mail systems that have directory trees programmed into them are becoming the most popular ways of accomplishing this. Directory trees are set up by the voice-mail administrator. Directory trees are capable of being as long and complicated as the caller can tolerate. For example, the caller hears a pre-recorded message that prompts them to dial "1" for information about skin ailments, dial "2" for flu-like symptoms or dial "3" for head pain. Options 1, 2, and 3 then branch out into a tree (Fig. A.25):

Sample Information Mailbox/Directory Tree

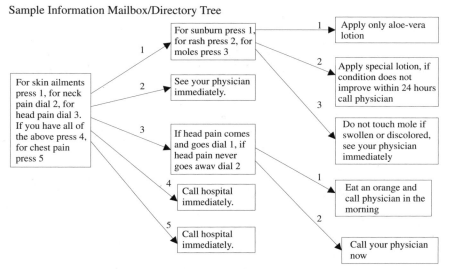

Figure A.25   Automated Voice Response

Automatic Call Distributor (ACD system)   A separate system or built-in feature of a PBX that equally distributes incoming calls to agents. As calls come in, they are placed into a queue (or a waiting line) for the next available agent. ACD systems are very versatile and relatively easy to program because some incorporate their own script programming language. For incoming calls, the waiting times, pre-recorded announcements and other call treatments can be set up by the users/companies to their discretion.

Automatic Gain Control   See *AGC*.

Automatic Number Identification (ANI)   A service provided by local and long-distance telephone companies that sends the name and number of the calling telephone to a display attached to the called phone line. In an in-band signaled phone line or residential telephone line, the ANI signal packet comes as a miniature data burst between the first and the second ring.

Autonegotiation   1. A function or feature of communication devices and/or communication protocols that provides a set of rules where communication methods are set on a per-session basis. The methods agreed on include speed of transmission, packet size, and error correction. In LAN switching, some switches have a feature built into the hardware of the ports that automatically adjust to 10BaseT or 100BaseTX, depending on which type of NIC the host/computer is equipped with. 2. Autonegotiation, Ethernet—The IEEE 802.3u 100 Base T feature specification that provides for flow control (pause frames) and full duplex operation. Full duplex allows for 100 Mbps send and 100 Mbps receive, for a total 200 Mbps Ethernet connection over Cat5 twisted pair. The autonegotiation is an enhancement of the link integrity signaling method used in 10BaseT networks, and is backward-compatible with link integrity. Autonegotiation allows the NIC or the network device to adjust its speed to the highest speed that both ends are capable of supporting. To be able to use this feature, both the network device (switch port) and the NIC must contain the autonegotiation logic.

Autonomous System   In IP/Routing a group of routers that share the same policies. Usually a single corporate network is configured as one autonomous system. In the OSPF routing protocol, all routers of the same autonomous system share routing table updates as well as policies.

Autonomous System Boundary Router   In OSPF routing environments, a router that connects to the outside world, or another separate network. More specifically, it is a router that connects one "network of policies, or routing features" to another. Autonomous System Boundary Routers nearly always have dual Routing Protocols enabled, which enable packet transfer between two networks of unlike policies.

Available Bit Rate (ABR)   *Quality of Service (QOS)* defined by the ATM Forum for ATM Networks that is used for connections that are not time or delay sensitive. A connection would be rightfully commissioned as an ABR connection if it carried only spontaneous or bursty data. Other QOSs defined by the ATM Forum for ATM Networks include *CBR (Constant Bit Rate), UBR (Unspecified Bit Rate),* and *VBR (Variable Bit Rate).*

Avalanche Photodiode   A device used as a light-to-electricity converter and signal amplifier at the same time. They are incorporated in optoelectronic circuits used in fiber-optic terminating applications.

Avalanching   Avalanching occurs when a PN diode or transistor junction is reverse biased (reverse positive and negative) with enough voltage to force it to conduct in the wrong direction. Diodes and transistors have an avalanche voltage rating. When this voltage is exceeded, the device avalanches. The avalanche is actually a sudden steady rush of current that causes lots of heat. This usually damages the device (severely). Some devices (avalanche photodiode, SCR, etc.) are designed to use the avalanche effect in a useful way. They switch "on" or conduct when the reverse or gate voltage applied to them reaches a certain level.

AWG (American Wire Gauge)   A measurement standard for copper wire. The gauge rating is the thickness of a solid copper wire. The larger the gauge, the smaller the wire. Most telephone wire is 19 AWG at the largest and 26 AWG at the smallest. Cat3 is commonly 24 AWG. The electrical wire in your home is probably 12 AWG.

AX.25   A protocol that is based on X.25 recommendations. AX.25 is a connection-oriented version of X.25. Rather than sending separate individual packets, as is done in X.25, the AX.25 protocol sets up a layer-3 *Permanent Virtual Circuit (PVC)* before the transmission begins. When the transmission is complete, the PVC is disconnected. AX.25 is generally useful in older mainframe WAN applications.

Azimuth   In directional radio transmission, the azimuth is the direction in degrees (bearing) that an antenna is transmitting to its far-end counterpart. If a directional radio signal seems weak after a strong windstorm, the azimuth might need to be realigned.

B 911 This is also known as *Basic 911 emergency service*. It is the older version of the 911 system and it is what public service commissions want phone companies to phase out. B 911 does not provide automatic location information. In many cases (depending on the CO switch/911 software serving the customer that makes the 911 call), it does not provide automatic number identification.

B Channel The "bearer" channel of an ISDN circuit. It carries 64 Kbp/s of end-user data. The other ISDN channel is referred to as the *D* or *data* channel, which is 16Kbp/s and carries phone company signaling along with the other stuff that makes the ISDN circuit work. The two categories of ISDN are the *BRI* (*Basic Rate Interface* 2B channels and 1D channel) and the *PRI* (*Primary Rate Interface* 23B channels and 1D channel). For a relational diagram of the two types of ISDN lines, see *Integrated Services Digital Network*.

B Connector A wire-splicing connector for splicing twisted pair wire, also called *beans*. Beans are shaped like a plastic tube that is about as big around as a drinking straw, but only an inch long. They have metal teeth inside them so that when two wires are crimped inside, they make a good connection and don't slide out. Beans can also have a water-retardant jelly inside them as well.

B-ICI (Broadband InterCarrier Interface) This ATM backbone solution uses ATM to multiplex multiple services, such as cell relay, voice

DS1, frame relay (PVC) over one ATM link (i.e., DS3 44.736 Mbps, STS-3c 155.52 Mbps, STS-and 12c 622.08 Mbps).

B Washer, Curved   The washer used between a telephone pole and a strand clamp when installing pole attachments (Fig. B.1).

Figure B.1   Curved B Washer

B3ZS (Bipolar 3-Zero Substitution)   A line-coding/data-transmission format. Transmission formats are used to prevent too many consecutive zeros from being transmitted. If too many zeros go down the line in a row, the transmission line effectively becomes a flat line, with no timing.

B8ZS (Binary 8-zero Substitution)   A line-coding/data-transmission format. Transmission formats are used to prevent too many consecutive zeros from being transmitted. If too many zeros go down the line in a row, the transmission line effectively becomes a flat line, with no timing. If a sequence of eight bits are detected prior to being transmitted, they are replaced with a different pre-determined byte that is not all zeros.

Backbone   The part of a communications network that connects main nodes, central offices, or LANs. The backbone usually has its own high-speed protocol, such as switched token ring or FDDI for LAN interconnections, and SONET for central-office and main-node interconnections.

Back-Feed Pull   When it is difficult to get large cable pulling equipment into end locations of a cable installation, outside plant construction personnel use a technique called a *back-feed pull.* If there is a vault or hand hole that the cable passes through between the end points, the cable will be fed in two parts. The first section of cable will be fed one direction from the mid-point. After the first half is fed, the remaining cable is unreeled and fed through the opposite direction to other end point (Fig. B.2).

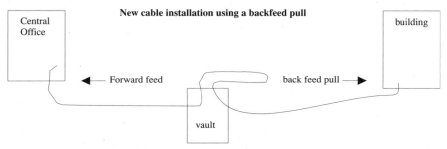

New cable installation using a backfeed pull

Central Office — Forward feed — back feed pull → building — vault

Figure B.2   New Cable Installation Using a Back-Feed Pull

Back Haul   A long-distance service term. Sometimes you can save money by creatively routing long-distance calls. *Back haul* is a routing term that means routing a call past its destination and then back. Many new long-distance companies are providing less expensive long-distance services to specific cities and conglomerate long-distance companies take advantage of this by back hauling phone calls. Airline passengers are often back hauled because it is less expensive to have a lay-over in another nearby larger city—even though it is beyond your destination.

Back Hoe Fade (slang)   The cutting of a buried fiber optic that connects two communications nodes. It is called a *fade* because in some equipment architectures, not all communications are cut off—they are either divided or rerouted. To some, this is a "fade" in communications service, not a complete outage.

Back Plane   1. The main electronic PC board in a communications equipment cabinet that has slots or connectors for circuit cards to be plugged into. The back plane in almost all key service units and PBX card cages is where the power bus, CPU bus, and the control bus are located (Fig. B.3).   2. The matrix of circuitry that performs packet forwarding in the Ethernet switch.

Figure B.3   Backplane

Back Pressure   A method used by receiving end Ethernet devices to stop the transmission until they can process the data they have received. This is similar to the DSR (Data Set Ready) on a connection oriented link such as RS232 being low, or off. See also *Autonegotiation.*

Backward Channel   The channel that flows upstream in an asymmetrical transmission. An asymmetrical communications transmission that is characterized by one direction being very fast compared to the other. Cable TV is an example of asymmetrical communication. The cable TV head end sends massive amounts of video and audio information down a coax one way and the cable TV set-top decoder boxes send small amounts of ID and status information the other way back to the head end over the same coaxial connection. Sometimes asymmetrical channels are referred to as "upstream" for the slow channel and "downstream" for the fast channel, or "forward" for the fast channel and "backward" for the slow channel.

Backward Explicit Congestion Notification (BECN)   A bit in the overhead of a data packet travelling in a frame-relay network. This bit is set to 1 if the packet travels through an area of network congestion that is opposite (or backward) of its flow. This bit is a signal from the frame-relay network to higher-level protocols within DTE and DCE to take flow-control action, as appropriate. An example of flow-control action would be to not attempt to exceed the *Committed Information Rate (CIR)* for the connection. This would prevent the unnecessary transmission of low-priority frames that would surely be discarded.

Bad Line Button   A button on an attendant console that enables an attendant to busy out a trunk that is in trouble (such as noise or static). Some bad line buttons don't busy out the trunk; they just mark it so that it can be identified from the rest when the local phone company is called to fix the problem. Sometimes the trouble is not in the trunk, but at least when you call the phone company, you can tell them which line you are having trouble with.

Baffle   An enclosure for the back of a loudspeaker that improves its sound quality by improving its acoustic profile. Baffles are made of thin, flexible rubber foam and are frequently used on ceiling intercom speakers and automobile speakers. They usually cost about $6 to $10 each, depending on the size of the speaker.

Balance, Circuit   An electronic circuit that can be active (using external power) or passive (using only capacitors, resistors, and inductors) that is attached to a twisted copper pair to even out the electronic characteristics of both wires. Balance is very important in a transmission line. If the two wires or pair is not balanced, noise is created on the line, which interferes with the transmission signal.

Balancing Network   See *Balance, Circuit.*

Balun (Balanced/Unbalanced)   A device used for matching impedances or transmission characteristics between different media so that electronic signals can pass from one to the other without being severely attenuated. Baluns that match twisted pair and coaxial cable are very common in local-area network environments.

Band, Citizens   A low-power two-way transmission radio band in the United States. There are actually two types or bands of CB radio. They are 26.965 MHz to 27.225 MHz and 462.55 MHz and 469.95 MHz.

Band-Elimination Filter   A band-elimination filter is a circuit used to pass a certain range of frequencies away from specific equipment or devices to ground, or somewhere they are wanted. Radio-frequency in-terference is easily eliminated with the correct band-elimination filter. Band-elimination filters are also called *RFI (Radio Frequency Inter-ference)* filters, RFI suppressors and *EMI (Electromagnetic Interfer-ence)* suppressors. They work by providing an easier path for noise to go through, rather than your electronic device (such as your telephone or modem). Many RFI filters are on the market, and some are adjustable to different frequencies. Some are modular instead of hard wired into the NI or jack so that you can plug them right into your phone line. These filters are not XDSL compatible because XDSL has its own built-in line conditioning. Further, filters are not recommended for any digital serv-ice line.

Band, Frequency   The following are the defined boundaries of radio fre-quency bands:

ELF	Extremely Low Frequency	below 300 Hz
ILF	Infra Low Frequency	300 to 3000 Hz
VLF	Very Low Frequency	3 kHz to 30 kHz
LF	Low Frequency	30 kHz to 300 kHz
MF	Medium Frequency	300 kHz to 3000 kHz
HF	High Frequency	3 MHz to 30 MHz
VHF	Very High Frequency	30 MHz to 300 MHz
UHF	Ultra High Frequency	300 MHz to 3000 MHz
SHF	Super High Frequency	3 GHz to 30 GHz
EHF	Extremely High Frequency	30 GHz to 300 GHz
THF	Tremendously High Frequency	300 GHz to 3000 GHz

Band, Marking   A label placed around an insulated wire or fiber optic for identification. Some are printed on during manufacture and some are at-tached during installation.

Band-Pass Filter   A band-pass filter is used in frequency-division multi-plexing, as well as the equalizer in your stereo. It is usually a capaci-tor/resistor/inductor network that has a resonant frequency and a rating of how well it passes a band of frequencies and blocks out others, called the *Q* (quality) of the circuit. The resonant frequency of the circuit is the frequency that the circuit will pass. Band-pass filters are used in ADSL applications. See *DSL Inline Filter.*

Band-Stop Filter   See *band-elimination filter.*

Bandwidth   The difference in frequency between the top end of a chan-nel and the bottom end. A good example of a bandwidth is sound. If you are listening to a sound, such as music, you notice the different pitches. All of these pitches or tones of sound are audio information that your ear can process. The sounds are actually vibrations. Bass tones vibrate at a slow rate, about 20 to 700 vibrations per second. Treble tones vibrate faster, from 3000 vibrations per second to 17,000 vibrations per second. The total bandwidth (vibration range) that you are listening to is about $17,000 - 20 = 16,980$ vibrations per second. This is the range or bandwidth of human hearing.

Bandwidth Allocation   The process of dynamically assigning communications resources to users and software programs within a network. The process incorporates predetermined priority levels for data based on its time sensitivity. Generally, when a network becomes congested, lower-priority traffic is dropped. The dropped traffic is retransmitted at a later time. For an example of a priority level parameter, see *ABR* and *MCR*.

Bandwidth Control   The process of allowing customer traffic to run through a frame-relay network at a rate faster than the *CIR (Committed Information Rate)* setting for the particular *DLC (Data Link Channel)* when traffic is low. This is also referred to as *throttling* or *rate adaptation*.

Bandwidth-Control Elements (BCE)   The parameters of a frame-relay *Data Link Connection (DLC)* that determine the amount of data that will be accepted into the network over a period of time, and what delivery priority they have. Bandwidth-control elements are:

CIR   Committed Information Rate
AIR   Allowed Information Rate
EIR   Excess Information Rate
Be   Excess Burst
Bc   Committed Burst
Tc   Time in seconds

Their relationships are as follows: $CIR = Bc/Tc \; AIR + CIR + EIR$

Bandwidth Reservation   Another term synonymous with bandwidth allocation. See *Bandwidth Allocation*.

Banjo   This is also called a *beaver tail* or *break-out block*. It is used by technicians to connect other devices to modular jack wiring for testing purposes. *Banjo* is a trademark of Harris Dracon Division (Fig. B.4).

**Figure B.4   Harris Dracon "Banjo" RJ14 6 Conductor (Left)
RJ45 8 Conductor (Right)**

Bantam Connector/Plug   A standard plug and jack that is used to interface test equipment with digital circuits (DS1, DS3, STS-1) that are wired to DSX patch panels (Fig. B.5).

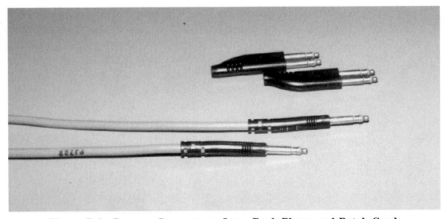

Figure B.5   Bantam Connectors, Loop Back Plugs, and Patch Cords

Barge In   When an attendant or operator adds themselves onto a line that is already in use. See also *Busy Override.*

Barrel Connector   A gender-changing device that connects two male coaxial F connectors together.

Base Station   A device that connects wireless communications to the land-line phone network. Base stations can be integrated into a BTS. See also *Cellular.*

Baseband   The opposite of broadband. Baseband is the transmission of one signal over a media or carrier. Telephone conversations in themselves are baseband. Applying 24 telephone conversations to a T1 carrier is a broadband application. Because T1 has more than one channel within one transmission, it is broadband. See also *Broadband.*

Baseband Modem   A modem that modulates and demodulates a single digital data transmission to and from another modem. Baseband modems can be used over plain twisted-pair wire with no telephone service. In this application, they are better known as *short-haul modems,* which extend peripheral devices (such as printers) 50 to 1000 feet from their host.

BASIC (Beginners All-Purpose Symbolic Instruction Code)   A type of programming language that has many variations, depending on the developer. Microsoft Visual Basic is a newer version of this style of programming.

Basic Rate Interface   The small size ISDN line (the other size is a primary rate interface). It consists of two bearer or "B" channels and one data or "D" channel. The B channels are 64 Kbp/s each. With the appropriate service package from the phone company and correct terminal adapter, you can talk on one B channel while using your computer modem on the other B channel. When your phone conversation ends and you hang up, the terminal adapter will send a message back to the phone company through the D channel that connects both B channels together for a total transmission bandwidth of 128 Kbp/s for your computer automatically.

Battery   A device that converts chemical energy to electrical energy. Batteries of over 1.5 V (nominally) are composed of cells, each cell being a smaller battery that is equal to 1.5 V in electrical potential. A 12-V lead-acid automotive-type battery is comprised of eight 1.5-V cells in series that add up to 12 V. Some batteries are re-chargeable, depending on the two chemically interacting materials that the battery is made of.

Battery Back Up   There are two different types of battery back-up systems. There are rectified power sources, which continuously charge batteries that power a system. This system is used by telephone companies for their central offices. When the power goes out, the charging on the batteries stops, but the system still runs because it's running on the batteries. The other type of battery back up is a UPS (uninterruptable power supply), which is always on standby. When the power goes out, the UPS converts the DC battery power to AC power to run the system. Which system you use depends on the type of power that your phone system requires. If you have the option of running on DC, the rectified battery back-up is far superior in reliability, and they are available in many different sizes. Reltec is a well-known manufacturer of these systems.

Baud Rate   The actual bit rate on a communications line. Not to be confused with bit rate, which includes data compression.

Baum   An impedance-matching transformer used in RF applications. Fifty to 75 ohms is typical.

Bay   A place in a computer cabinet where a peripheral device, such as a disk drive, can be installed.

BBN (Bolt, Baranek and Newman, Inc.)   The company that developed and maintained the ARPANET (later the Internet) core gateway system.

BBS (Bulletin Board System)   A Website that can be accessed by users that acts as a central source of information. BBS Websites are usually set up by particular interest category.

Bc (Committed Burst)   An amount of data that is permitted onto and over a frame-relay network *DLC (Data Link Connection)* over a specific amount of time. See also *Bandwidth-Control Element.*

BCD (Binary Coded Decimal)   A four-bit code that represents the numbers zero through nine in binary. It is basically implemented as a short cut for entering many binary numbers into a machine-language program. Logic circuitry decodes the BCD to binary for the microprocessor. The code is

Decimal	BCD
0	0000
1	0001
2	0010
3	0011
4	0100
5	0101
6	0110
7	0111
8	1000
9	1001

BCM (Bit-Compression Multiplexer)   A multiplexer that increases bandwidth by encoding data bits into a special format.

BDFB (Breaker Distribution Fuse Bay)   The point in a central-office power system where each DC feed to all rows of equipment are equipped with a fuse or a breaker. The BDFB is a central location for power distribution. Generally, each rack of electronic equipment is also equipped with its own *FAP (Fuse Alarm Panel)*.

Be (Excess Burst)   One of the frame-relay bandwidth-control elements. It is sometimes a negotiated service between customers and service providers. An excess burst is an amount of data in bytes (not a bandwidth, like Bc is). Excess burst data over a frame-relay network is marked *Discard Eligible (De)* by the network. If the network is not congested, there is a very good probability that the data marked *De* will be delivered. If the frame relay network is congested, it will discard data marked De. It is the amount of data that exceeds the *Committed Burst (Bc)* setting of a frame-relay *DLC (Data Link Connection)*. See also *Bandwidth-Control Elements*.

Beacon   A frame from a token-ring or FDDI device that indicates a serious problem within the ring's network, such as a broken fiber optic. The beacon frame contains the address of the station assumed to be down.

Beamwidth, Antenna   In radio environments, including wireless LAN, the dimensional characteristics of a radio field projected by an antenna element. In general, different antennas focus a radio field to achieve a stronger signal over a particular direction or area. The areas are described as a ratio of front-to-side, front-to-back, and other comparative methods. This is done under the assumption that a transmission power loss in one area constitutes a gain in another and vice versa. In the case when a radio link is subject to interference from other nearby radios, an antenna with a beamwidth of high dimensional ratios is preferred and often required in licensed applications. Keeping your radio beam focused increases efficiency and reduces the likeliness of interference from other radio transmissions. The only drawback is that these types of antennas are more expensive.

Bean   A wire-splicing connector for splicing twisted-pair wire. Beans look like a plastic tube that is about as big around as a drinking straw, but only an inch long. They have metal teeth inside them so that when two wires are crimped inside, they make a good connection and don't slide out. Beans can also have a water-retardant jelly inside them as well. For a photo, see *Plain B Wire Connectors*.

Beaver Tail Another name for a Harris Dracon Division Banjo or a similar break-out device. The Banjo is a device for connecting test equipment to the wiring in modular jacks. For a photo, see *Banjo*.

BECN (Backward Explicit Congestion Notification) A bit in the overhead of a data packet travelling in a frame-relay network. This bit is set to 1 if the packet travels through an area of network congestion that is opposite (or backward) of its flow. This bit is a signal from the frame-relay network to higher-level protocols within DTE and DCE to take flow-control action, as appropriate. An example of flow-control action would be to not attempt to exceed the *Committed Information* Rate *(CIR)* for the connection. This would prevent the unnecessary transmission of low-priority frames that would surely be discarded.

Bed-of-Nails Clip A test clip similar to an alligator clip, except that it has a section of very sharp needle-like objects bunched together that poke through a wire's insulation when the clip is applied. These clips achieve a good connection for testing without stripping the insulation off the wire (Fig. B.6).

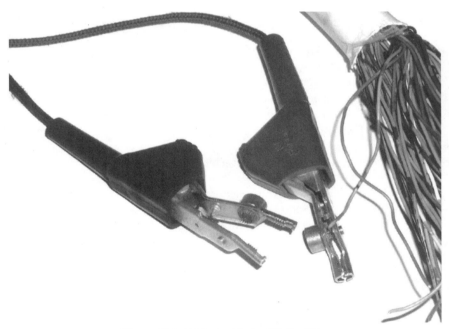

Figure B.6   Alligator/Bed-of-Nails Clips

BEL   This is an ASCII control-code abbreviation for bell. The Binary code is 0111000 Hex is 70.

Bell Operating Company (BOC) or Regional Holding Company   At the time of the 1984 divestiture, there were 22 BOCs, grouped into seven Regional Bell Operating Companies (RBOCs) in the United States.

BOCs:

Bell Telephone Company of Nevada
Illinois Bell Telephone Company
Indiana Bell Telephone Company
Michigan Bell Telephone Company
New England Telephone and Telegraph Company
US West Communications Company
South Central Bell Telephone Company
Southern Bell Telephone and Telegraph Company
Cincinnati Bell Company
Mountain Bell Telephone Company
Mountain States Telephone and Telegraph Company
Southwestern Bell Telephone Company
The Chesapeake and Potomac Telephone Company of Maryland
The Bell Telephone Company of Pennsylvania
The Chesapeake and Potomac Telephone Company of Virginia
The Chesapeake and Potomac Telephone Company of West Virginia
The Diamond State Telephone Company
The Ohio Bell Telephone Company
The Pacific Telephone and Telegraph Company
New Jersey Bell Telephone Company
Wisconsin Telephone Company

RBOCs:

Ameritech
Bell Atlantic
Bell South
NYNEX
Pacific Telesis
Southwestern Bell
US West

Bell System Practices (BSPs)   A volume of standards that explain how to do everything from terminate an RJ11 jack to install a central office. They even had a standard on how to collect a past-due phone bill. The BSPs were a pre-1984 (divestiture) tool for operating phone companies.

They are no longer widely embraced by the RBOCS or AT&T. New equipment manufacturers have their own instructions for operating and installing their products, and each RBOC has its own way of operating a communications company.

Bend Loss   The loss of transmission in a fiber-optic or twisted-pair cable because of a bend. Bending fiber-optic cable causes the light traversing through it to reflect outward, instead of down the core of the fiber. A bend in a twisted-pair wire causes the dielectric or insulation to change its electrical properties, which results in a loss of signal.

Bending Radius   The smallest or tightest bend that a fiber can withstand under a tensile-pulling force without damaging its transmission characteristics.

BER (Bit Error Rate)   A way to measure data-transmission integrity. The bit error rate is a ratio of bad bits to good bits.

BERT (Bit Error Rate Test)   A way to measure data transmission integrity. The test gives a result as a ratio of bad bits to good bits.

Beta   A way of referring to a test site or test product. If a new revision is released by a manufacturer, the first sites that it is installed at is referred to a *beta site.*

BGP (Border Gateway Protocol)   An interdomain routing protocol that has been used in Internet core router applications to exchange reachability information with other "same system type" routers.

BGP4 (Border Gateway Protocol version 4)   The predominant interdomain routing protocol used on the Internet. It is capable of aggregating routes listed within a router's memory, which reduces the size of routing tables.

BICSI (Building Industry Consulting Service International)   The provider of the *RCDD (Registered Communications Distribution Designer)* certification. The RCDD certification is often referred to as a *BICSI* (pronounced "bik-see") *certification.* The RCDD certification is designed to educate professionals in the area of physical network distribution, including twisted pair and optical media. More information can be found regarding BICSI certifications at *http://www.bicsi.org.*

Bidirectional Bus   A bus that connects devices that clock bits in as well as out of their shift registers. The devices that a bus is connected to make it unidirectional or bidirectional, not the bus itself. All buses are merely a group of parallel conductors that connect the shift registers of components, as well as the power for the devices.

Big Endian   A method of storing or transmitting data in which the most-significant bit or byte is presented first. The opposite is referred to as *little-endian,* which presents the least-significant bit first.

Billed Telephone Number   The number that is the regarded as the billing account number on a phone bill. Sometimes when a customer calls a phone company for service, the customer-service representative will ask the customer for the billed telephone number because that is the number that all the other customer's phone numbers and charges are referenced to. This method is used so that a customer doesn't get a phone bill for every individual phone line they have.

Binary   A number system that counts with only two digits, 0 and 1. We are all more familiar with the arabic base ten, which counts with ten digits: 0, 1, 2, 3, 4, 5, 6, 7, 8, and 9. The following is a list of numbers and their binary equivalent. For a larger table of binary numbers, see *Appendix E.*

0	0000
1	0001
2	0010
3	0011
4	0100
5	0101
6	0110
7	0111
8	1000
9	1001
10	1010
11	1011
12	1100
13	1101
14	1110
15	1111

Binary Coded Decimal (BCD)   A four-bit code for representing the numbers zero through nine in binary. It is basically implemented as a short cut for entering many binary numbers into a machine-language program. Logic circuitry decodes the BCD to binary for the micropro-cessor. The code is (Fig. B.7):

decimal	BCD
0	0000
1	0001
2	0010
3	0011
4	0100
5	0101
6	0110
7	0111
8	1000
9	1001

Figure B.7   Binary Coded Decimal (BCD)

Binary-to-Decimal Conversion   For a conversion table of Binary to Decimal and Hexadecimal, see *Appendix E.*

Binary-to-Hexadecimal Conversion   For a conversion table for binary to decimal and hexadecimal, see *Appendix E.*

Binder   A method of separating groups of 25 pairs in a twisted-pair cable with counts of more than 25 (Fig. B.8). Colored plastic ribbon binds, designates and separates each group of 25 pairs. The first binder group is white/blue pairs 1 to 25, the second is white/orange pairs 26 to 50, the third is white/green pairs 51 to 75, the fourth is white/brown pairs 76 to 100, etc. If you would like to see the entire

Figure B.8   Binder

list of binder groups and their associated pairs, see *Color Code, Twisted Pair.*

Binder Group   A method of separating groups of 25 pairs in a twisted-pair cable with counts of more than 25. See *Binder.*

Binding Post   A reference used to identify where twisted copper pairs are terminated in access points, cross boxes, and terminals. Physically a binding post is a pair of teeth on a 66M150 block or a pair of $\frac{9}{16}''$ lugs. Each binding post has a number. When a technician looks for a specific pair in a cable (called a *cable pair*), they refer to documents that list the pairs and which binding posts they are spliced to.

BIOS   Basic Input Output System residing in a PC. It contains the shift registers (dynamic RAM) used as buffers for sending bits to the specific hardware that they are intended for.

Biphase Coding   A bipolar (positive negative alternating) coding scheme originally developed for use in Ethernet. The clocking/timing signal is embedded into the data stream as the positive and negative switching cycle itself. This encoding scheme eliminated the need for separate clocking leads.

Bipolar   1. A copper twisted-pair transmission method (or line format), where bits that are transmitted are alternated positive and negative. This transmission technique increases the distance a transmission can travel on a twisted pair.   2. A transistor.

Bis   French for *encore.*

BISDN (Broadband Integrated Services Digital Network)   A conceptual telecommunications service. When the idea of BISDN was conceived, it would have the ability to provide "on-demand" bandwidth to customers for various services, such as video, data transfer, etc. BISDN types of services have evolved through ATM and the Internet. Newer data-compression techniques and high-speed local telephone lines have made bandwidth on demand available through services other than ISDN. Frame-relay and ATM services provide excess information rates across networks and are available to customers at a relatively low cost. The standardization and widespread use of Internet protocols, ATM feature flexibility, and the implementation of xDSL in local networks are the enablers of BISDN.

Bisync   A nickname for a tradition-breaking data-communications protocol developed by IBM. It was one of the first and perhaps the "original" of the non-character-code oriented protocols. Hence, the name came from *bit-synchronous.* The advantage of a noncharacter-oriented or bit-synchronous protocol is that they are flexible in the types and sizes of data characters they can send (e.g., 7-bit or 8-bit, ASCII or EBCDIC). The official name for *Bisync* is *SDLC (Synchronous Data Link Control)* and eventually evolved to *HDLC (High-Level Data-Link Control).* HDLC was the basis for the *LAP (Link-Access Procedure)* developed by the ITU (then the CCITT), which ultimately became the basis for the X.25 standard in 1976.

Bit (Binary Digit)   A unit of data that is represented as a one or a zero. Inside most data devices, a bit is physically a positive 5 volts or 0 volts.

Bit Error Rate (BER)   A way to measure data transmission integrity. The bit error rate is a ratio of bad bits to good bits.

Bit Interleaving   A simple way to time-division multiplex by interleaving individual bits, instead of bytes or packets. Timing of the two ends is not as complex with this method. Used in X.25 and HDLC, not in T1 or T3.

Bit Oriented   Communications protocols that use bits to represent control information in contrast to bytes. A byte can mean different things in the different varieties of character sets that are transmitted. Bit-oriented protocols are not character code sensitive.

Bit Parity   A way to check that transmitted data is not corrupted or distorted during the transmission. The way parity works is as follows. Take a bit stream that will be transmitted, add all the bits as binary numbers mathematically, and the resulting number is odd or even. Add a 1 at the end of the stream if the number is even and a 0 if the number is odd. When the bits are received at the other end, they are added up and compared to the last bit. If they add up to be an even number, then the last bit should be a 1. If they add up to an odd number, then the last bit should be a 0. If the case for either does not hold true, then the receiving end sends a request to retransmit the stream of bits. They are retransmitted, with the parity bit attached all over again.

For example, a computer sends a bit stream of 10101011. Simply adding the bits gives a sum of $1 + 0 + 1 + 0 + 1 + 0 + 1 + 1 = 5$. This is an odd number, so add a 0 to the end of the stream to make it 101010110. The bits are received at the other end, added together, and compared to the parity bit the same way. There are new and more sophisticated ways of checking for errors in data transmission, such as cyclic redundancy checking.

Bit Rate   The average net number of bits being transmitted over a communications line in a second, including compression and encoding techniques, as well as retransmission of corrupted data.

Bit Robbing   Bit robbing is often known as *in-band signaling*. The practice of taking a bit here and there in the beginning and end of a digital transmission for use in the overhead of the transmission equipment. Bit robbing is bad when the signals being multiplexed into the transmission are data. Robbing a bit from a data stream severely corrupts data. Bit robbing is a technique reserved for multiplexing multiple voice circuits onto a T1. Circuits intended to transmit data use out of band signaling or clear-channel signaling.

Bit Stream   A series of voltage pulses that represent a binary code. A serial data transport. A bit stream can exist on a transmission line or within the electronics of a data device.

Bit Stuffing   The temporary modification of user data during transmission so that it does not interfere with lower-layer signaling functions. For instance, in X.25, a flag byte is put before and after the frame. This byte is a 01111110. If the data being sent contains a 01111110, such as a lower case "w" (0111111), followed by a lower case "s" or lower case "a" (which both start with a 0), the data would be mistaken for a flag. This would cause random characters and bits to be lost, which would ultimately make transmission impossible beyond the error-correction stage. To fix the scenario, a "0" is inserted (or "stuffed") any time that five consecutive ones appear. The data is framed, transmitted, deframed, and then after every five ones, a zero is deleted (or "unstuffed"). The data is then handed to the packet layer.

Bits Clock   A device that provides a timing pulse in the form of a 1-0-1-0-1-0-1-0 bit stream. Bits clocks are used extensively in SONET networks. A bits clock provides the timing pulse that nodes in a network synchronize to (Fig. B.9).

Figure B.9   Bits Clock

Bits Per Second   The average net number of bits being transmitted over a communications line in a second, including compression and encoding techniques, as well as retransmission of corrupted data.

Black Box   Usually a device that converts or routes one type of data or signal applied to the input to a desired useful output for a specific application. One company called "Black Box Corporation" specializes in the manufacture of these specialized devices.

Black Hole   A name for an area (or router) of an Internetwork where packets enter, but do not emerge. Thus, data is lost. Black holes can be caused by any of the following: system (router table) configuration, bad hardware, or an adverse operating environment (noncompatible equipment).

Blended Agent   An agent in a call center that receives calls from outside customers. When the incoming call rate slows down, it makes outgoing calls.

Blended Call Center   A call center that receives calls from customers and also calls customers. Sometimes agents are dedicated to either inbound or outbound calls.

BLF (Busy Lamp Field)   A part of or an add-on module to a phone or console that allows the user to see multiple extensions and if that extension is in use (or busy).

Blocked Call   A call that cannot be completed because the Central Office or PBX switching capacity is full at the time the call was attempted. Blocking can occur at any point in a network where a call is switched (from CO to CO or from local to long distance). The caller with a blocked call either hears a fast busy or an intercept message that says "I'm sorry, all circuits are busy now. Please try your call again later."

Blocking   When a central office or PBX has fully utilized its capacity to connect calls, it blocks them. Callers trying to call in or out of a switch that is blocking calls will get a fast busy signal.

BNC (British Naval Connector)   A type of connector used on all different types of coax. It is keyed so that it locks into place and it has better transmission characteristics than an "F" connector (Fig. B.10).

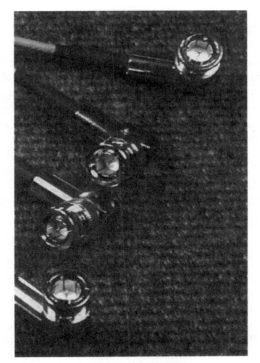

Figure B.10   BNC (British Naval Connector)

BOC (Bell Operating Company or Regional Holding Company)   At the time of divestiture, there were 22 BOCs, grouped into seven *Regional Bell Operating Companies (RBOCs)*. For a listing of the BOCs and RBOCs, see *Bell Operating Company*.

Body Belt   Used by communications/power/construction personnel to harness themselves to telephone/power poles or tower structures. This is also called a *safety belt* or a *climbing belt*. For a photo, see *safety belt*.

Bolt, Baranek and Newman, Inc. (BBN)   The company that developed and maintained the ARPANET (later, called the *Internet*) core gateway system.

Bond   1. What telephone company construction personnel call the connection between the sheath of a telephone cable and an electrical ground. 2. An electrical connection.

Bonding   In ISDN the joining of two 64Kbp/s B channels together for one 128 Kbp/s channel.

Boomerang Server   A method of layer 7 networking. In WAN networking or Internet applications, a boomerang server is reference to a router that is equipped with special software that enables it to measure the network distance between two network servers that provide the same application. To provide an end user accessing a redundant application with the quickest response time, it makes sense to have them connect with the closest server. The boomerang server receives the initial request for connection, and then creates a "race" request among multiple servers. The server that responds the quickest is the closest by factual network performance, and gets the request packet forwarded to it. The transaction between end user and appropriate server then takes place. This is also called a *race conditional routing scheme.*

Boot   To restart a computer or CPU-based system by physically turning it off, then back on, which resets the CPU. This is also called *bootstrap.*

Boot PROM (Boot-Programmable Read-Only Memory)   A memory chip with permanent programmed instructions burned into it. It is used to provide executable instructions to a computer device when it is initially turned on or restarted.

BootP   A feature added to networking that uses the User Datagram Protocol (UDP) to formulate network requests to allow a diskless device to obtain and configure its own IP information, such as IP address and subnet mask. The workstation looks to the BootP server for files during power-on/start-up (the BootP server is a PC or server that is running the BootP program and database). To incorporate and use BootP, a BootP server must be configured on the network. Before BootP was defined by RFC951, IP addresses had to be configured manually by an administrator in order to operate diskless workstations.

Border Gateway Protocol (BGP)   An interdomain routing protocol that is used in Internet core router applications to exchange reachability information with other "same system type" routers.

Bounce   A common term used by technicians in place of *reset,* with regard to digital communication channels, such as T1s.

Boundary Router   In OSPF routing environments, a router that connects to the outside world or another separate network. More specifically, it is a router that connects one *network of policies* or *routing features* to another. Autonomous system boundary routers nearly always have dual routing protocols enabled, which enable packet transfer between two networks of unlike policies.

BPAD (Bisynchronous Packet Assembler Dissasembler)   A hardware-based device that inserts bytes into packet frames and vice-versa in packet multiplexing/transmission equipment.

BPS (Bits Per Second)   The average net number of bits being transmitted over a communications line in a second including compression and encoding techniques, as well as retransmission of corrupted data.

BPSK (Binary Phase-Shift Keying)   A method of transmitting binary bits in a form of frequency shift, or FM, that is the same concept as frequency-shift keying. The difference is that with BPSK, you change the phase of the frequency, instead of the frequency itself. The two are shown in the diagram. If you look closely, you can see the changes in the waveforms that represent the switch from a one to a zero value.

Breakdown Voltage   The voltage at which insulation in a cable or an electronic device fails.

Break-Out Box   A test device that plugs into a data cable (i.e., RS232) and provides easy test access for each wire in the cable.

Breaker Distribution Fuse Bay (BDFB)   The point in a central-office power system, where each DC feed to all rows of equipment are equipped with a fuse or a breaker. The BDFB is a central location for power distribution. Generally, each rack of electronic equipment is also equipped with its own FAP *(Fuse Alarm Panel)*. See Fig. B.11.

BRI (Basic Rate Interface)   The small-size *ISDN (Integrated Services Digital Network)* line (the other size is a Primary Rate Interface). It is made up of two bearer or "B" channels and one data or "D" channel. The B channels are 64 Kbp/s each. With the appropriate service package from the phone company and correct terminal adapter, you can talk on one B channel while using your computer modem on the other B channel. When your phone conversation ends and you hang up, the terminal adapter will send a message back to the phone company through the D channel that connects both B channels together for a total transmission bandwidth of 128 Kbp/s for your computer automatically. For a diagram, see *ISDN*.

Bridge   A bridge is a device that connects two networks at the data-link level. The data-link level is the simplest protocol layer in the OSI to make

Figure B.11   Breaker Distribution Fuse Bay (BDFB)

a connection between two networks because no data interpretation is needed at this level. As a result, data can be transmitted with virtually no delay. In other words, a bridge simply decides whether the datagram/packet should pass or not pass. They operate over segments that use the same protocol, such as Ethernet or token ring. Bridges can be used to extend a network over similar or dissimilar media, such as twisted pair to fiber optic. See also *Source Routing Bridge* and *Transparent Bridge.*

Bridge Clip   A metal clip (sometimes plastic insulated) used to electronically connect or bridge across the left side of a 66M150 block with the right side (Fig. B.12).

Figure B.12   Bridge Clip

Bridge/Source Routing   A type of network bridge that relies on routing information provided by an external sending system.

Bridge Tap   A Y splice in a copper twisted-pair communications cable. It gets its name because the first splice is a straight through splice, and the second splice connects the wires by cutting or "tapping" them into the first splice. A bridge tap adds flexibility to telephone plant. A telephone company never knows which customer is going to use lots of pairs for their service and who will use only a few. To remedy the situation without installing 100 pairs of copper to each individual building, the telephone company installs the same 100 pairs into three or four buildings. This is done by bridge tapping the cable splices. The bad thing about bridge tapping is that it lengthens the pair. Length is bad for digital services (additional loss, and reactance), such as T1 and ISDN, so the bridge taps must be disconnected before these services are installed. On a cable drawing, bridge taps are shown as arrows and telephone cable is shown as a line (Fig. B.13).

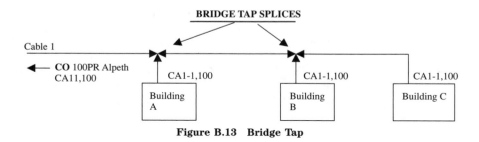

Figure B.13   Bridge Tap

Bridge/Transparent   A type of network bridge that learns which systems are on each network by listening to traffic and building its own reference tables.

Broadband   Incorporating more than one channel into a communications transmission. T1 is a broadband communications protocol because it carries 24 conversations over four wires. Cable TV is also broadband because it carries many TV channels over one coax.

Broadband Integrated Services Digital Network (BISDN)   A conceptual telecommunications service. When the idea of BISDN was conceived, it would have the ability to provide "on-demand" bandwidth to customers for various services, such as video, data transfer, etc. BISDN types of services have evolved through ATM and the Internet. Newer data-compression techniques and high-speed local telephone lines have made bandwidth on demand available through services other than ISDN. Frame-relay and ATM services provide excess information rates across networks and are available to customers at a relatively low cost. The standardization and widespread use of Internet protocols, ATM feature flexibility, and the implementation of xDSL in local networks are the enablers of BISDN.

Broadband Inter-Carrier Interface (B-ICI)   A reference to an ATM backbone solution. It uses ATM to multiplex multiple services, such as cell relay, voice DS1, frame relay (PVC) over one ATM link (i.e., DS3 44.736 Mbps, STS-3c 155.52 Mbps, and STS-12c 622.08 Mbps).

Broadcast   To send information in any form to more than one place.

Broadcast Address   A special address reserved for sending a message to all stations. A broadcast address is a destination address of all ones (in decimal, 255.255.255.255). See also *IP Broadcast* and *Multicast.*

Brouter   A marketing term for a multiport router, which inherently performs layer 2 (bridging or switching) functions as well as layer 3 (routing) functions (Fig. B.14).

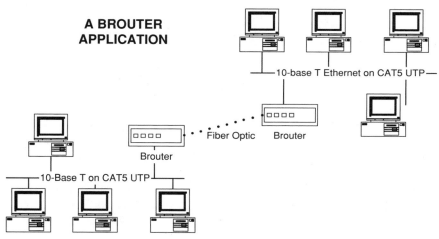

**A BROUTER
APPLICATION**

10-base T Ethernet on CAT5 UTP

Fiber Optic   Brouter

Brouter

10-Base T on CAT5 UTP

Figure B.14   Brouter

Browser   A computer program that allows users to download World Wide Web pages for viewing on their computers. Two popular browser programs are Netscape Navigator and Microsoft Internet Explorer. The first browser program was called *Mosaic,* which was a text browser, as opposed to the newer graphical browsers.

BS   An ASCII control code abbreviation for backspace. The binary code is 1000000 and the hex is 80.

BSP (Bell System Practice)   A volume of standards that explain how to do everything from terminate an RJ11 jack to install a central office. They even had a standard on how to collect a past-due phone bill. The BSPs were a pre-1984 (divestiture) tool for operating phone companies. They are no longer widely embraced by the RBOCS or AT&T. New equipment manufacturers have their own instructions for operating and installing their products, and each RBOC has its own way of operating a communications company.

BSS (Base Station System) A wireless communications device that manages radio traffic and bandwidth between a group of base transceiver stations.

BTA (Basic Trading Area) Geographical boundaries defined within a cellular radio license.

BTN (Billed Telephone Number) The number that is regarded as the billing account number on a phone bill. Sometimes when a customer calls a phone company for service, the customer-service representative will ask the customer for the billed telephone number because that is the number that all the other customer's phone numbers and charges are referenced to. This method is used so that a customer doesn't get a phone bill for every individual phone line they have.

BTS (Base Transceiver Station) A station that transmits mobile radio signals.

Buffer A temporary storage (memory) device for data. A buffer is basically a box with RAM inside it. A common application for buffers is to collect a stream of data and temporarily store it until another device, such as a PC or server asks the buffer to download it. This is useful when the PC, server or LAN could be out of service for a period of time. When the server or PC is returned to service it just asks for the data from the buffer and it is downloaded. The buffer is then empty and ready to receive more data.

Building Entrance Agreement A building entrance agreement gives a telephone company or other utility the privilege to construct communications facilities into a building and to occupy their own space for equipment, power for the equipment, and access to said equipment. Even though having a CLEC present in a building is a great advantage for tenants, smart building management companies use their position to an advantage. They only allow the CLEC to construct the facilities into the building the way they want, when they want. The facilities (fiber, cable, conduit, not electronics) when completed belong to the building. In some cases, the building management then charges rent back to the CLEC for the use of the facilities that they paid to have designed and constructed. The CLECs must agree to all of the building management's terms because without building entrance agreements the CLECs can only exist by using another phone companies facilities (usually an RBOCs) to provide service.

Building Industry Consulting Service International (BICSI) The provider of the *RCDD (Registered Communications Distribution*

Designer) certification. The RCDD certification is often referred to as a *BICSI* (pronounced "bik-see") *certification.* The RCDD certification is designed to educate professionals in the area of physical network distribution, including twisted pair and optical media. More information can be found regarding BICSI certifications at *http://www. bicsi.org.*

Bulletin Board System (BBS)   A Website accessed by users that acts as a central source of information. BBS Websites are usually set up by particular interest category.

Buried Cable Terminal   Where buried service wires are fed from between the feeder cable and the standard network interface. For a photo of a *Buried Terminal, see Pedestal.*

Buried Service Wire Splice   A special watertight splice that is filled with an encapsulant. Common types of these splices are made by Keptel and Communications Technology Corporation (Fig. B.15).

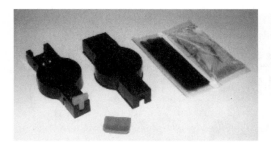 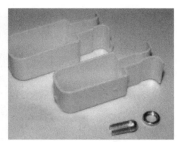

Figure B.15   Buried Service Wire Splice Kit

Bus Topology   A LAN physical topology, which means the way that the individual devices within the LAN are physically connected (Fig. B.16). The method at which the devices connected to the LAN access the media (coax is common for this topology) is called the *logical topology.* The bus topology is a wire (UTP) that behaves as a street that connects a number of PCs. When a PC wants to access the network or wire to a server or another PC, it looks at the street to see if there is no traffic. If there is no traffic, then the PC (or server) sends data down the wire (or street). The data has an address attached to it and all the other devices connected to the network see the address. If the address belongs to a certain device, that device reads the data attached to the address. This scheme of sending data is called *Ethernet.* One of the inefficiencies of Ethernet is something called a *collision.* A collision

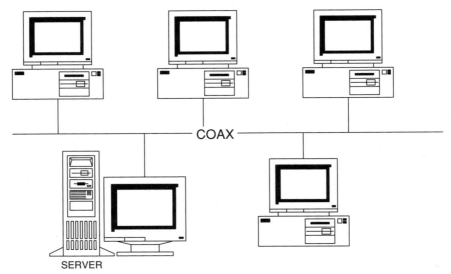

Figure B.16   Bus Topology

happens when two or more devices look at the network (the wire) and see that it is clear at the same time, then attempt to send data at the same time. The data that is transmitted by the multiple devices collides and becomes corrupted. The devices on the network sense the collision and try to send their data again when the network is clear. This method of control is classified as a contention-based protocol because all the devices on the network contend for its use. All of the Ethernet protocols are contention based. See also *CSMA/CD, CSMA/CA,* and *Ring Topology.*

Busy Hour   The hour in the day or month when a central office or PBX connects the most calls. The busy hour is an important factor in designing a switch for blocking.

Busy Lamp Field   The lights on an attendant console that indicate who is on their line (what lines are busy) and who is not. Some busy lamp fields are an add-on module that can be attached to a phone.

Busy Out   A temporary fix or condition of a phone service. To "busy a line out of a hunt sequence." If a business phone line becomes defective and it is in a hunt or roll over sequence, calls will not hunt or roll past this line. Say that you have four lines coming into your business. The first line is the main number and if that first line is busy, then calls come in on the second line, etc. If line one goes bad then it can't be called, so it can't be busy. Because it is not busy, then calls will not hunt or rotate to the next three lines. When you call the phone company repair service they busy out the bad line, which makes it look busy to the network. Your calls then start coming in on the other three lines. When a repair technician finishes with repairing the problem on the bad line, he has it unbusied.

Busy Override   When an attendant or operator adds themselves on to a line that is already in use. This is also called a "barge in." If a local phone company operator barges in on a phone conversation the people on the line will hear a beep tone, then subsequent short beep tones as long as the operator is connected. The operator can converse with the two parties on the line after barging in and pass on urgent information. This is a common feature of PBX systems. A PBX system can be programmed to not warn the people in the middle of their call that another person is listening. This feature is often used to monitor the quality of customer service in call centers.

Busy Signal   There are two types of busy signals. The most common type is a *slow busy*, which means that the number you are trying to call is being used. The other type of busy signal is a *fast busy*, which means the phone company central office could not understand the digits that you dialed or the actual phone network is too busy to take your call. Many fast busy signals are being replaced with "intercept messages," which tell the dialer what the problem is. Examples of intercept messages are: "I'm sorry, all circuits are busy now. Please try your call again later." And "The number dialed is out of service. Please check the number and try your call again."

Busy Verification   The local telephone company test to see if a particular phone number (or circuit, or loop as they call them) is busy. The test is run by a *DATU (Direct Access Test Unit)* in the central office. When the test comes back, it says the line is "in use busy speech" or "ROH *(Ringer Off Hook),"* which means that the handset has been taken off the hook and left there.

Butt Set (slang)   A test telephone set used by telephone installation and repair personnel. Instead of a plug on the end of the cord, it has a pair of alligator/bed-of-nails clips. For a photo, see *Craft Test Set.*

Bypass Trunk Group   A method of connecting one central office to another without going through a tandem. This method reduces tandem traffic and reduces blockage between the two offices that have bypass trunk groups installed. A bypass trunk group is shown in Fig. B.17 on page 98.

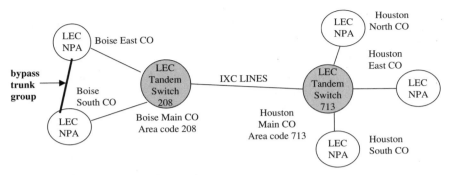

Figure B.17   Bypass Trunk Group

Byte   Eight bits, also known as an octet (Fig. B.18).

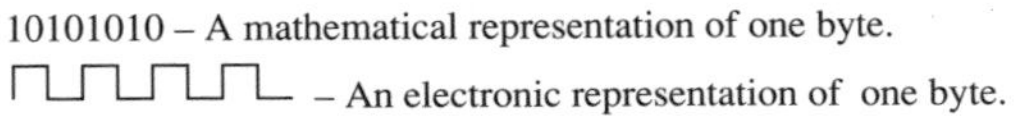

10101010 – A mathematical representation of one byte.

⎍⎍⎍⎍⎍ – An electronic representation of one byte.

Figure B.18   Byte

Byte-Oriented Protocol   An older class and "era" of data-link communications transmission rules that use a specific character from a character set code (ASCII or EBCDIC) to delimit or separate transmitted
frames. These protocols have been almost completely replaced by bit-
oriented protocols. Byte-oriented methods are avoided when possible.
Even less technologically advanced parts of the world have migrated to
bit-oriented protocols, such as X.25.

Byte Reversal   An electronic logic process used in Intel microprocessors.
It stores least-significant integer and address bits first. See also *LSB
(Least Significant Bit)*.

Byte Stuffing   Byte stuffing is what some communications protocols do
to make data more suitable for transmission. If a customer wants to send
30 bytes, but one transmission frame carries 48 bytes, then the transmission equipment adds 18 bytes so that the frame is full. Think of byte
stuffing as the styrofoam peanuts or wadded-up paper you put in a box
when you ship something through the mail.

C

C A high-level programming language that is somewhat similar in application to the Basic programming language.

C+ A high-level programming language.

C++ A high-level programming language.

C3Po (Cisco 3 Port Switch) A design for Cisco Systems desktop IP phones, where the phone has three switch ports. One switch port is an Ethernet VLAN trunk for interfacing to the switch network, the second is a switch port for an interface to the PC workstation, and the third is to the internal telephone instrument. (Incidentally, there is nothing in telecom for R2D2.)

C7 The European version of SS7. SS7 and C7 are not the same protocol even though they perform the same functions. Gateway switches (class 1 central offices) convert the two different international standards when calls are handed through.

C Band The band of frequencies designated by the IEEE between 4 GHz and 8 GHz (7.5 cm to 3.75 cm). For a table, see *IEEE Radar Band Designation.*

C Connector Also called a *female amp connector* or *25-pair female connector.* For a photo of a C connector, see *25-Pair Connector.* The male version is called a *P connector.*

C Drop Clamp   A clamp used to fasten service wire to homes/buildings (Fig. C.1).

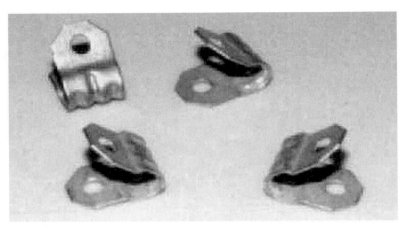

Figure C.1   C Drop Clamp

C Plane   One of the three entities of frame-relay network management. The three planes are: The *User Plane* (the U Plane defines the transfer of information), the *Management Plane* (the M Plane defines the LMI, Local Management Interface), and the *Control plane* (the plane is delegated for signaling and switched virtual circuits).

C Shell   One of the UNIX program operator access levels. The other levels are V Shell and Root. To access C Shell or V Shell, a password must be entered. To work in a UNIX program under the root comman set, an additional password must be used. The different shells permit different operations, which have different command sets. This allows a root user to allow limited access to C-Shell and V-Shell users.

C Wire   Wire that is strengthened with steel for "long-span" aerial plant applications. Some C wire is noninsulated, so it is also called *open wire.* Open wire fits the application better as it is used in wide open or very rural areas. The old telegraph system was an open-wire system.

Cable   The general name for copper-based media to transport electrical voice, data, and video signals. It can be twisted pair or coax, and indoors or outdoors. In general, two types of cable are installed indoors, PVC and Plenum. Many types of cable are used outdoors. The basic difference in outdoor cables is how many PVC and aluminum sheaths are around the conductors and if the cable is filled with a water-proofing jelly.

Cable Act of 1984   An act passed by The United States Congress in 1984 that deregulated almost all of the Cable TV industry and was eventually superseded by another act in 1992. After the 1984 Act was passed, the FCC had control of cable TV in only the following six areas: 1. Each CATV system had to be registered with the FCC prior to any operations.   2. Enforcement of all subscribers having access to an AB switch so that they could easily switch from CATV to regular TV programming.   3. Rebroadcast of local television stations without any alteration or deletion. 4. Non-duplication of local broadcast regular TV programs.   5. Fines and/or imprisonment for broadcasting indecent material.   6. Licensing for receive-only earth stations for satellite delivered via pay cable.

Cable Head End   The signal-processing point for telecommunications services provided by companies that are traditionally known as *cable-TV companies*. Since these companies have began offering traditional two-way telecommunication services, such as Internet connectivity and basic telephone service, the equipment housed in the cable-TV head end been expanded (Fig. C.2).

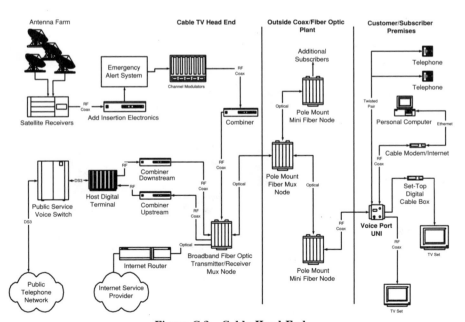

Figure C.2   Cable Head End

Cable Knife   A knife used by communications cable installers to strip ALPETH (aluminum/polyethylene) jacketed cable (Fig. C.3).

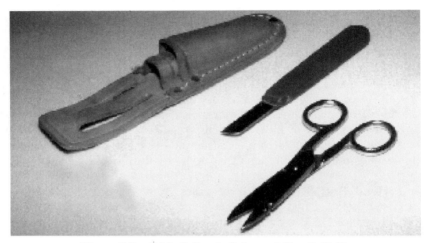

Figure C.3   Cable Splicer's Snips and Sheath Knife

Cable Mapping   The tracking of installed cable and pairs in a network. Sometimes an actual map is used and sometimes a tracking system of cable, pair, address and binding post is used. The RBOCs use both.

Cable Modem Internet   A service provided by cable TV companies that allows for up to 5 Mbps downstream and 512 Kbps upstream connectivity to Internet-service providers. With various compression methods, cable-modem data-transfer rates extend beyond 10 Mbps downstream (Fig. C.4).

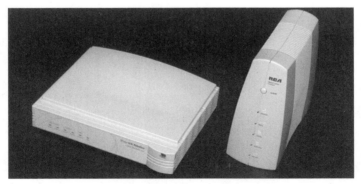

Figure C.4   Cable Modems for Cable Internet Access, Manufactured by 3Com (Left) and RCA (Right)

Cable Plant   A term that refers to a communications utility's twisted pair and/or coax network that winds through towns and neighborhoods. It includes terminals, pedestals, cross boxes, and vaults.

Cable, Riser   A twisted-pair cable (usually several hundred pairs) distribution system that progresses from the telephone company Demarc or point of entrance in a building to each floor of that building.

Cable Router   A router located at a cable-TV head end that manages Internet traffic among cable modems located at subscriber locations. Cable routers work the same way that Internet-service provider routers do, except that the IP-based transmission between the router and various cable modems are compressed and modulated into an RF signal. The analog RF signal is then converted to digital within the cable modem at the customer premises.

Cable Span   The cable suspended in the air between two telephone or power poles.

Cable Stripper   Several tools are used to strip the jackets off of ALPETH and lead-jacketed telecommunications cable. The most popular is a cable knife and snips (for a photo, see *Cable Knife*). An alternative tool is shown in Fig. C.5.

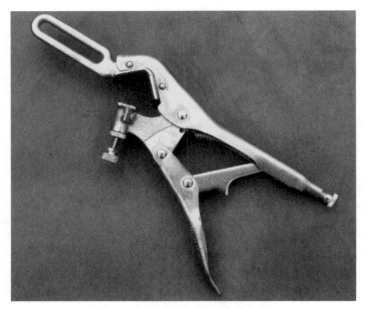

Figure C.5   Cable Sheath Cutter

Cable Telephony   A reference to telecommunications services that are provided over the coaxial outside plant that is owned and operated by cable-TV companies. Services provided include Internet access, telephone, digital cable-TV, and analog cable-TV (Fig. C.6). See also *HDT (Host Digital Terminal)* and *Cable Modem/Internet.*

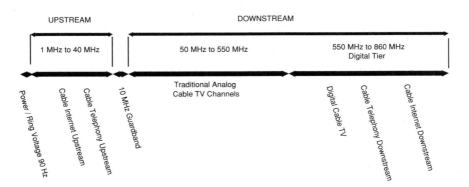

Figure C.6   Cable Telephony Bandwidth Allocation

Cable TV (Community Antenna Television)   A cable-TV company has satellite dishes in a central location that pick up TV signals from around the country. They retransmit those channels down a coax that branches out through a geographical area where subscribers to the antenna service can be connected. Some cable-TV companies add their own information channels to their broadcast and local advertising between programming on selected channels.

Cable Vault   Formerly known as a *manhole.* A concrete enclosure that is usually underground and varies in size from a small crate to a large room. They are specifically designed for the housing and easy access to telephone and cable-TV splices. Cable vaults are extremely dangerous places because of lack of oxygen and accumulated carbon monoxide. Natural gas, and other dangerous gasses have a tendency to build up inside them (Fig. C.7).

Cable Voice Port (Also called a *UNI, User Network Interface,* or *Voice Port*)   In cable-TV networks, a device located at the telephone

Figure C.7   Cable Vault (Inside View). Street Access is at Top

subscriber's premises that modulates and demodulates the DS0 voice upstream and downstream channels. The modulated DS0 voice signal is
sent by the *HDT (Host Digital Terminal)* located at the cable-TV head
end. The voiced port provides connection to the customer-premises' telephone wiring. Channel control between the head end and the customer
site voice port can be maintained remotely, from the head end or other
office. Because the voice service in cable telephony is provided via radio channels, the channels can be switched when a subscriber is experiencing static or radio interference on line. This would be equivalent to
a pair change in a twisted-pair-based telephone network. A pair change
cannot be done by remote control and requires a technician on site. For
a photo, see *Voice Port.*

Cable Weight   A specification of outside plant cable used by outside plant engineers when figuring how long a span can be, and how big of a messenger to use.

Cache Memory   Pronounced "cash memory." A type of RAM incorporated into personal-computer CPUs. Cache memory retains frequently used instructions, and the locations of those instructions in main memory. When a CPU operates, it "looks" for addresses to retrieve instructions. When the cache memory is holding the address the CPU is needing to access, it retrieves the information from that cache memory bank. This is called a *cache hit.* When the address needed is not in the cache memory bank, it is called a *cache miss,* and the CPU accesses the instructions from the main memory address. See also *L1 Cache.*

Call-Accounting System   A computer (usually a dedicated PC) that connects to a switch via a serial data port and monitors the details of every phone call made through that switch. The call details are stored as call records. With the appropriate software, the records can be retrieved, sorted, processed, and queried to almost any specific nature that the call-accounting system administrator desires. These systems are used by hotels to track all the calls you make from your room so that they can bill you appropriately. They are also used by companies to do bill-back reports for individual departments within the company (Fig. C.8).

Figure C.8   Call Accounting System

Call Admission Control   A traffic management feature of ATM networks that ensures that virtual channel connections are not offered unless enough bandwidth is available on its network.

Call Announcement   A way of transferring a call to another extension. A feature of PBX and key systems where you have a chance to talk to the person you are transferring the call to before you make the transfer. The way it works on most systems is: You are talking to a person that you want to transfer, you push the transfer button on your phone, dial the extension you would like to transfer the call to, and you hear it ring. When the person at the extension you have called answers the phone, you can notify them that you are transferring a very important client to them. You then push the transfer button again to complete the transfer. If the person you want to transfer to does not answer their phone, you can switch back to the call you wanted to transfer and speak to that person again. The other way of transferring calls is called a *blind transfer,* where you simply transfer the caller to another extension with no intervention or control of the call.

Call Attempt   An uncompleted call because of blocking, where callers cannot get through because all lines are busy. This is a report statistic of central office switches as well as PBX systems.

Call Blending   The mix of typical incoming calls in a call center with outgoing calls made by an auto dialer with call-blending capability. Call centers blend calls to fully utilize their employees working at the time. If inbound calls are slow, the auto dialer increases the number of outgoing call attempts so it can connect people that answer with agents. Auto dialers are usually programmed with a predetermined list of numbers to dial, usually for customers that are expecting follow-up on their services from the company making the call.

Call Detail Recording (CDR)   The initial function of a call-accounting system is to receive detailed information on telephone calls connected through a PBX switch and store them in memory. This function is called *call detail recording.* Call details include number dialed and duration of the call for outbound calls, and the trunk ID (or phone number) and call duration for inbound calls. Each call event (transfer, connect, disconnect, etc.) gets a time stamp.

Call Duration   The length of time a phone call lasts from the time both ends are off-hook, until the time that one end hangs up.

Call Forwarding   A service offered by local phone companies to their subscribers and a feature of PBX systems that allows a user to make calls dialed to their phone ring to a different phone or phone number.

Call Hand Off   When your call is transferred from one cellular site (or cellular transmitter) to another.

Call Letters   The station identification letters assigned to broadcast radio and TV stations by the FCC. Some examples are: WKND, WLIS, WPPR, WKRP, KXRK, and CKLW.

Call Menu   A recorded message that gives callers options to choose from by using their dial pad. See also *Auto Attendant.*

Call Mix   A statistical account of calls in a system. This information is useful for determining architecture upgrades and troubleshooting. An example of a call mix would be 40% DID calls, 30% autoattendant-routed calls, and 30% zero out to human-attendant calls. From this call mix, you can see exactly how each percentage of your calls are handled.

Call Packet   A packet of data that contains X.25 *Switched Virtual Circuit (SVC)* addressing information and other overhead.

Call Park   A PBX feature that allows a call to be placed on hold in a way that it can be picked up from any extension in the office. The way it works is: The attendant or anyone else who wishes to park the call presses a park key on their telephone. The display on the phone then shows the extension that the call is parked in (park extensions are imaginary, just reference numbers). In this example, the display said "parked in 60." The person that parked the call then pages the person that they wish to pick up or receive the call over a loudspeaker "Johnny please pick up park 60." Johnny hears the page, goes to the nearest telephone set, and presses the park key and then 60. He is immediately connected with the calling party that was parked by the attendant that answered the call.

Call Pick Up   The ability to answer a ringing phone that is not yours. You hear a phone in the next office ringing, pick up the receiver and press the pick-up key on your telephone, then enter the extension number of the phone you wish to intercept the call from. Your phone is immediately connected with the call and you say "Hello, John Doe's (or name of the person's extension) line."

Call Pick-Up Group   A group of telephones that receive or "ring" when a certain number is called. An example of a pick-up group would be all of the phones on all the desks that ring simultaneously when a 1-800 hotline is called. When someone picks up the handset of any one of those ringing phones, they answer the call.

Call Priority   1. In an *Automatic Call Distribution (ACD)* system, incoming calls can be given a preference as to which are answered first. The preference is based on which trunk the call comes in on, or on the *DNIS (Dialed Number Identification Service)* for the call. As of this writing, the Nortel Meridian 1 PBX (Private Branch Exchange) voice switch has a capability of four different call priority levels.   2. In data communications, call priority is a preference assigned to each origination port in a circuit-switched system (i.e., switched Ethernet). This preference determines which ports are connected first (and which ones wait) if the network is congested.

Call Queuing   The function of placing incoming calls on hold and in line for the next available call-answering agent. Call queuing is a function of *Automatic Call Distribution (ACD)* systems. The queue is an extension number within the ACD system that calls are transferred to. An ACD system integrates with a PBX system.

Call Record   One of the many reports generated by a call-accounting system. A call record details a telephone call made or received by a telephone or extension by the number dialed, incoming trunk ID, duration of the call, and the time of connection and disconnection.

Call-Request Packet   A packet that is sent by the originating DTE equipment in a frame-type data transmission that requests a network terminal number, network facilities, and call user data (or X.29 control information).

Call Restrictor   A device that can be attached to a phone line or trunk that prevents certain numbers from being dialed through or that only allows a certain group of numbers to be dialed. When installing or purchasing a call restictor, it is a good idea to be sure that 911 is dialable under any application.

Call Return   Also called *last call return.* A service offered by local telephone companies that enables telephone customers to return a call they missed by dialing *NN (the two numbers after the * depend on the local company). This service is handy for those times when you are running to the phone and the caller hangs up right when you pick up the receiver. All you have to do is dial the *NN code, you hear a recording that tells you the phone number of the last caller, and gives you the option of ringing them back by pushing 1 or just hanging up. Using this service can cost 50 to 75 cents each time you use it up to a maximum amount per month, usually about $6.00.

Call Second   One phone call for one second. This is the smallest unit of telephone switch traffic. One hundred call seconds is equal to a *Centum Call Second (CCS)*. A one-hour call is equal to 36 CCS, which is one Erlang. The Erlang is the standard unit of measure for telephone-switch traffic.

Call Setup Time   The time from when you go off-hook, dial a number, the phone network checks to see if the number you are calling is busy, a path is established between central offices, the other end rings and is picked up. Even though call setup costs money, customers making long-distance calls don't pay for the call setup time.

Call Trace   A service offered by local phone companies. If you receive a malicious or obscene telephone call, you can have the call traced by immediately dialing *57 after the call. This only works for the last call received. After the *57 is pressed, a recorded message is played that gives further instructions. The call is tagged in the local telephone company's call detail-recording log. The caller is usually warned for a first offense. The person that made the report never finds out who made the obscene call, but is not surprised to see how quickly the obscene calls cease. *57 is the North American Standard for last-call trace service.

Call Transfer   A feature of PBX systems that allows users to transfer a conversation or call connection to another extension. The feature is usually executed by pressing a transfer key while on the line with the person or party they wish to transfer, dialing the digits of the extension they wish to transfer to, then pressing the transfer key again.

Call Waiting   A feature offered by local phone companies that allows someone that is talking on their phone line to receive another incoming call by briefly pressing the switch hook. The person knows they are getting another call because they hear a short beep or click on the line. If a person does not have call waiting, the caller that is trying to reach them while they are on the phone will receive a busy signal, as opposed to a ring when they have this service.

Caller ID (Caller Identification)   Also known as *ANI (Automatic Number Identification)*. A feature offered by local phone companies that sends the phone number (and often the name of the caller) down the phone line in a digital data packet between the first and second ring. To receive the data, a subscriber that has signed up for the service needs to have a caller-ID unit (also called a caller-ID box) plugged into the phone line. The caller-ID unit displays the name and number of the calling party for each incoming call. Caller ID only works if the caller and the called party's phone service is fed out of a central office that has caller ID capability. If the central office does not have caller-ID capability, the display will read "out of area" to the called party. If the called party does not have caller service, they will get a display that says "no data sent."

Caller Identification (Caller ID)   See *Caller ID*.

Caller-Independent Voice Recognition   A voice recognition system that recognizes a certain number of words, rather than a specific voice.

Camp (Camp On)   A way of placing incoming callers on hold. A feature of PBX systems. If you are trying to call someone and they are on their phone, you or a PBX attendant can put your call on hold in a way that when the person you are calling hangs up, your call rings through to their phone instantly.

CAN   The ASCII control-code abbreviation for cancel. The binary code is 100001 and the hex is 81.

Canonical   Also known as *MSB (Most Significant Bit)* first or *Big Endian*. FDDI and token ring transmit source addresses and destination addresses canonically, with the most significant digit being put on the wire first. Non-canonical protocols such as Ethernet place the LSB (Least Significant Bit) on the wire first.

CAP (Competitive-Access Provider)   A company that offers private line services in competition with the *RBOC's (Regional Bell Operating Companies)* in providing private line access to long-distance carriers. The private line service offered by a CAP can carry a long-distance company's dial tone. A CAP shouldn't be confused with a *CLEC (Competitive Local Exchange Carrier)*, which not only provides private-line service, but also provides their own switched dial-tone services with their own switches. Some CAP companies are Electric Light Wave, Teleport Communications Group (the first CAP founded in NY City), Teligent, and Metropolitan Fiber Systems. Some of these companies operate as CLECs in certain cities.

CAP (Carrierless Amplitude Phase Modulation)   A version of *QAM (Quadrature Amplitude Modulation)* used in xDSL transmissions that suppresses the carrier frequency at the central office. The carrier is essentially recreated in the electronics of the end DSL device for decoding purposes. CAP requires more electronic circuitry at the end premises, and does not adjust to noise conditions as well as QAM. However, CAP has better latency characteristics (does not delay bit throughput as much). Vendors that implemented single-carrier design (nonDMT) in their xDSL products often used CAP. The multiple carrier *DMT (Discrete Multi-Tone)* has been selected by ANSI as the design forward line code technique for xDSL.

Capacitance A measure in farads of a capacitor. See *Capacitor.*

Capacitive Coupling In audio amplifiers, the different stages of amplification are linked by capacitors. The capacitors prevent the DC transistor bias voltage from passing onto and interfering with the next amplifier, yet it allows the AC audio signal to pass through and be further amplified. Some very high end and very expensive audio amplifiers are direct coupled, which means all the different amplification stages are connected by only a conductor and are integrated with each other. The transistors in direct-coupled amplifiers are usually biased with other active devices instead of resistors. They are very complicated and expensive, in contrast to capacitive-coupled amplifiers, but the low-end frequency-response approaches 0 Hz.

Capacitor A capacitor is an electronic device that has two special properties. It only allows alternating current to pass through it (blocks DC current) and it can store an electric charge (Figs. C.9 and C.10). One of

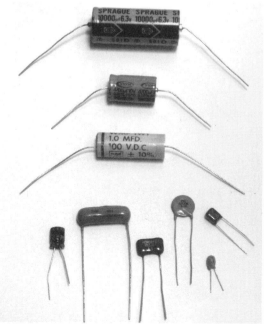

Figure C.9   Various Capacitors: Electrolytic (Top), Mylar and Film (Bottom)

non-electrolytic   electrolytic
Figure C.10   Schematic Symbols for Capacitors

the many applications of capacitors is to filter alternating current (AC) out of DC power supplies and rectifiers. This is done by placing a capacitor from the DC output to ground. The capacitor appears as an easier path to voltage fluctuations and RFI, and an impossible path to direct current (DC). Physically, a capacitor is two plates of metal separated by an insulator (mylar is common). The physical size of a 1-F capacitor would be two sheets of tin foil the size of a football field insulated (or separated) by a thin sheet of mylar. The farad is a huge unit of capacitance. This is why most capacitors are microfarads (μF) in value.

Cap Code   An ID code for a pager. The code is usually labeled on the outside of the pager and is the actual address ID code for that pager. When the ID code is broadcast, your pager receives the information that follows, then it beeps (or vibrates) and displays the transmitted information.

Carbon   1. A semiconductor that is used to make electronic components, such as resistors and microphones.   2. A name for a lightning protector. Carbon-type protectors are being replaced with gas-type protectors.

Carbon Transmitter   A microphone in a handset that is made from small grains of carbon packed into the shape of a diaphragm. The way these microphones work is that as the sound waves from your mouth strike the carbon in the receiver (microphone) they vibrate the carbon. These vibrations in the carbon change its electronic resistance in conjunction with your voice, converting your voice into electronic fluctuations, which traverse down the phone line. This is an older type of microphone. The most common problem with carbon microphones is that after a period of time, the fine carbon grains stick to themselves or become settled. A few good whacks usually fixes the problem. When a telephone customer with this problem saw the telephone man pound the handset with the handle of screwdriver and say "there you go, all fixed," they probably did not know that this was a standard procedure for repairing defective carbon receivers.

Card Cage   A box frame that has one open side for inserting electronic circuit cards and a back plane on the opposite side of the opening that the circuit cards plug into. For a photo of a card cage, see *Backplane.*

Carrier Band   A range of frequency that is used by a specific transmission system. The carrier band for a T1 is 1.544 Mb/s. The carrier band for a DS0 is 64 Kb/s. See also *Cable Telephony*.

Carrier Common Line Charge   What local phone companies charge long-distance companies to connect the far-end local portion of a call. Also called an *access charge*. A fee that everyone pays for every phone line to make up for subsidies that long-distance services paid to help the less profitable local services before the divestiture of AT&T and the RBOC's (Regional Bell Operating Companies).

Carrier Detect (CD)   Most modems have a little red LED with "CD" next to it. When that light is on, your modem is connected to another modem or communications device that it can communicate with.

Carrier Failure Alarm (CFA)   A notification that timing has been lost in a digital transmission because of excessive zeros in the transmission. When a carrier failure alarm occurs, all of the calls and data on that transmission are dropped until the carrier equipment regains timing.

Carrier Frequency   In radio, cable-TV, and television communications, the frequency that carries the audio or TV signal. A radio carrier frequency is specified by its location on the dial. For instance, the radio station at 1590 on your AM radio dial rides on a 1590-kHz carrier.

Carrier ID Code   The code that is entered by a long-distance caller that wishes to bypass the preselected long-distance company. Each local telephone line has a preselected long-distance company that the subscriber chooses when the phone line is ordered and initially put into service. If you would like to bypass this long-distance company and use another, you dial 1, then 0, then the carrier ID code, which is three digits long. After you enter the carrier ID code, you dial the area code and the seven-digit number. Carrier-ID codes are available in almost every exchange for long-distance equal access.

Carrier, Long Distance   A company that is licensed by the FCC and local *Public Utility Commissions (PUCs)* to provide intrastate and/or interstate communications services. Many companies provide long-distance and local telecommunications services, including voice, wireless, Internet, and virtual private network service (i.e., frame relay, ATM, and X.25).

Carrier Loss   In T1 transmissions, a carrier loss occurs when too many consecutive zeros are transmitted or when a component of the T1 circuit fails. In other transmissions, carrier loss is simply an unintentional loss of signal, regardless of the reason.

Carrier-Provided Loop   A carrier-provided loop is a local phone line that is bought by a long-distance company and re-sold as a part of a WAN service. In most WAN services, the long-distance portion of the service is billed separately from the local portion, just like your residential (phone in your home) service.

Carrier Sense Multiple Access (CSMA)   An Ethernet LAN protocol. In *local-area networks (LANs)* with CSMA, PCs check the network to see if it is clear before transmitting. They do this because if more than one PC sends data at the same time, the data gets garbled and is meaningless to the other PCs. This simultaneous data transmission is called a *collision*. The two other types of LAN protocols that are advanced versions of this one are: *Carrier Sense Multiple Access/Collision Avoidance (CSMA/CA)* and *Carrier Sense Multiple Access/Collision Detection*. In Ethernet networks, PCs sense and transmit hundreds of times per second. If the network looks clear for a tiny fraction of a second, the PC will try to transmit.

Carrier Sense Multiple Access/Collision Avoidance (CSMA/CA)   An Ethernet LAN protocol. In *Local-Area Networks (LANs)* with CSMA/CD, PCs check the network to see if it is clear before transmitting. If the network is clear, it sends a jam signal, then waits a specified time to allow all the other PCs to receive it. It transmits its data and sends a clear signal. They do this because if more than one PC sends data at the same time, the data gets garbled and is meaningless to the other PCs. This simultaneous data transmission is called a *collision*. In Ethernet networks, PCs sense and transmit hundreds of times per second. If the network looks clear for a tiny fraction of a second, the PC will try to transmit.

Carrier Sense Multiple Access/Collision Detection (CSMA/CD)   An Ethernet LAN protocol. In *Local-Area Networks (LANs)* with CSMA/CD, PCs check the network to see if it is clear before transmitting. If it is clear, it transmits its data. They do this because if more than one PC sends data at the same time, the data gets garbled and is meaningless to the other PCs. This simultaneous data transmission is called a *collision*. CSMA/CD senses these collisions and attempts to retransmit the same data again when the network is clear again. In Ethernet networks, PCs sense and transmit hundreds of times per second. If the network looks clear for a tiny fraction of a second, the PC will try to transmit.

Carrier Serving Area   The geographical boundaries of a telephone service provider. Also referred to as the *Local Serving Area*.

Carrier Shift   A change in frequency. Carrier shift is also a way of transmitting binary ones and zeros over a phone line or radio carrier, which is called *frequency-shift keying*.

Carrier Signal   A signal that carries another signal. In telecommunications, the word *carrier* has a broader meaning than in the broadcast radio or TV industry. In telecommunications, a carrier can simply be a digital signal connecting two modems or a T1 circuit. The data in itself is the carrier. In radio, TV, or cable-TV, a carrier is a continuous unchanging waveform (ac sine wave) of a specific frequency. The sound or video sent over the carrier signal changes or alters the once unchanging carrier. In simple terms, the sound or video "rides" within or on the carrier. See *AM* for more radio carrier information.

Carrierless Amplitude Phase Modulation (CAP)   A version of *QAM (Quadrature Amplitude Modulation)* used in xDSL transmissions that suppresses the carrier frequency at the central office. The carrier is essentially recreated in the electronics of the end DSL device for decoding purposes. CAP requires more electronic circuitry at the end premises, and does not adjust to noise conditions as well as QAM. However, CAP has better latency characteristics (does not delay bit throughput as much). Vendors that implemented single-carrier design (nonDMT) in their xDSL products often used CAP. The multiple carrier *DMT (Discrete Multi-Tone)* has been selected by ANSI as the design forward line code technique for xDSL.

CAS (Channel Associated Signaling)   In IP telephony circles, another name for in-band T1/E1 signaling. Within the 64 Kbps DSO channel, 8 Kbps are robbed from the data stream to provide on-hook, off-hook, and dialed digit information. This is true for all 24 channels of the T1 and all 32 channels of the E1. For more information on T1 Signaling, see *T1, In-Band Signaling*, and *Out-of-Band Signaling*. For more information about the interaction of IP telephony with these legacy signaling methods, see *H.323, RTP, Skinny Protocol*, and *MGCP*.

Cascaded Amplifier   An amplifier that consists of two or more amplifiers coupled together. Almost all consumer audio electronics products have more than one stage of amplification or cascaded amplifier circuits inside them. This design is very common in all kinds of amplifiers.

Cascaded Stars   A LAN physical topology where star networks are connected together (Fig. C.11).

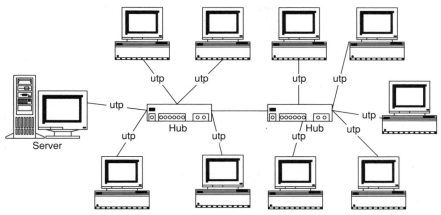

Figure C.11   Cascaded Stars Topology

Case Sensitive   When a computer's software recognizes a difference between upper and lower case alphabet symbols. Sometimes passwords are case sensitive.

CAT 1   Category 1. Any wire other than phone wire used for transmission, except coax. CAT 1 is nontwisted and can be any AWG. Applications include audio speaker wire, telephone quad conductor (four wire, red-green-black-yellow), electrical, doorbell, thermostat, and other control wire.

CAT 2   Category 2. Twisted-pair wire of 22 to 26 AWG. UTP or STP digital or data-transport media good to speeds of up to 1.5 MHz at 300 feet. The nominal impedance is 100 Ω ($\pm$10%). Typical applications include analog telephone and lesser analog transmission and control.

CAT 3   Category 3. Twisted-pair wire of 22 to 24 AWG. UTP or STP digital or data-transport media that is good to speeds of up to 16 MHz at 300 feet. Nominal impedance is 100 Ω ($\pm$10%). Typical applications include analog telephone, 10 base-T, and T1 (on conditioned pairs).

CAT 4   Category 4. Twisted-pair wire of 22 to 24 AWG. UTP or STP digital or data-transport media that is good to speeds of up to 20 MHz at 300 feet. Nominal impedance is 100 Ω ($\pm$10%). Typical applications include analog voice, 10 base-T, token ring, and T1.

CAT 5   Category 5. Twisted-pair wire of 22 to 24 AWG UTP or STP, where each pair of wire within the sheath has a different number of twists per foot (Fig. C.12). Digital or data-transport media that is good to speeds of up to 100 MHz at 300 feet. Nominal impedance is 100 Ω (±610%). Typical applications include 10 base-T, 100 base-T, token ring, switched token ring, ATM, and T1.

Figure C.12   CAT 5 UTP Plenum-Jacketed Twisted-Pair Cable

CAT 7   Category 7. Twisted-pair wire 22 to 24 AWG. Each pair of wire placed side by side within a sheath has a different number of twists per foot (Fig. C.13). The unusual flat shape of CAT 7 makes it very distinguishable from other twisted-pair wire types. *UTP* or *STP (Unshielded Twisted Pair* or *Shielded Twisted Pair)* digital or data-transport media that are good to speeds of up to 250 MHz at 300 feet. Nominal impedance

Figure C.13   CAT 7 UTP Plenum-Jacketed Twisted-Pair Cable

is 100 Ω (±10%). Typical applications include 10 base-T, 100 base-T, token ring, switched token ring, ATM, T1, T3, and STS-1.

Catalyst   The Cisco trademark for their family of network switches. Included is the Cisco Catalyst 5000, which is a modular switching system that allows connection between Ethernet, CDDI, FDDI, ATM, frame relay, and other LAN segments, or host/servers.

Category 3 Twisted Pair (CAT 3)   See *CAT 3*.

Category 4 Twisted Pair (CAT 4)   See *CAT 4*.

Category 5 Twisted Pair (CAT 5)   See *CAT 5*.

Category 7 Twisted Pair (CAT 7)   See *CAT 7*.

Cathode   The more negative end of a diode or other electronic device, such as a vacuum tube. The screen of your TV set and monitor are actually the cathodes of large vacuum tubes. Figure C.14 shows rectifier symbols and the cathode locations.

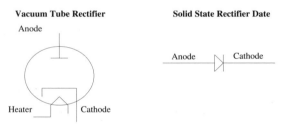

Figure C.14   Cathode

Cathode-Ray Tube (CRT)   The real name for a TV or monitor screen.

CATV (Community Antenna Television)   Better known as *cable TV*. A cable-TV company receives satellite and terrestrial TV signals. They retransmit those channels down a coax that branches out through a geographical area, where subscribers to the antenna service can be connected. Some cable-TV companies add their own information channels to their broadcast and local advertising between programming on selected channels.

CB (Citizens Band)   A frequency band from 26.965 MHz to 27.225 MHz and 462.55 MHz to 469.95 MHz that are set aside for unlicensed two-way communications. CB radios are limited to a transmission power of 4 Watts.

CBR (Constant Bit Rate) A *Quality of Service (QOS)* defined by the ATM Forum for ATM Networks that provides precise clocking to ensure undistorted delivery. A connection would be rightfully commissioned as a CBR connection if it carried such time-sensitive data as voice, video, or real-time computing information. The CBR quality of service guarantees freedom from cell loss and delay. Other QOSs defined by the ATM forum for ATM connections include *UBR (Unspecified Bit Rate), ABR (Available Bit Rate),* and *VBR (Variable Bit Rate).* See also *AAL.*

CBS (Certified Banyan Specialist) This well-known industry certification is offered by Banyan Systems, Inc. There is also a Microsoft Windows NT extension to this program. It is designed for network professionals who utilize Banyan Vines and other Banyan products. More information regarding Banyan training programs can be found through *http://www.banyan.com.*

CCC (Clear-Channel Coding, Clear-Channel Capability) A reference to a type of T1 service. A clear-channel T1 is formatted for out-of-band signaling, which means there is no bit robbing and the dial tone, hook flashes, and DTMF digits are sent over the 24th (the last) channel in the T1 circuit. Clear-channel signaling is usually best for data circuits and in-band signaling is best for voice circuits. In a clear-channel circuit, all 64 Kb/s in each channel of the T1 instead of 56 Kb/s (except for the 24th, which is 100% dedicated to signaling for the other 23 channels) are available to the end user.

CCDA (Cisco Certified Design Associate) This well-known industry certification/training program is offered by Cisco Systems and Cisco Systems training partners. The CCDA program is provided to train individuals for simple-routed LAN, WAN, and switched-LAN networks. More information can be found regarding Cisco Systems training programs through *http://www.cisco.com.*

(CCDP) Cisco-Certified Design Professional This is a well-known industry certification/training program offered by Cisco Systems and Cisco Systems training partners. The CCDP program is provided to train individuals for complex-routed LAN, WAN, and switched-LAN networks. A *CCNA (Cisco-Certified Network Associate)* certification must be acquired as a prerequisite for the CCDP program. More information can be found regarding Cisco Systems training programs through *http://www.cisco.com.*

CCFL (Cold-Cathode Fluorescent Lamp)   A technology used to light LCD screens. The Dynatel 965T has a CCFL built in so that technicians can see the display in the dark.

CCH (Connections Per Circuit Hour)   The number of connections or calls completed at a switching point per hour.

CCIE (Cisco-Certified Internetwork Expert)   This well-known industry certification/training program is offered by Cisco Systems and Cisco Systems training partners. The CCIE program is the most advanced level of certification for Cisco Systems associates, and is regarded as a very challenging certification that only the experienced can obtain. The program prerequisites are CCDP, CCNA, and CCNP certifications. This program is offered to provide training and certification to individuals who work in the most-complex networking environments. More information can be found regarding Cisco Systems training programs at *http://www.cisco.com.*

CCITT (Consultative Committee on International Telegraphy and Telephony)   The CCITT is one of the four parts of the ITU (International Telecommunications Union), which is based in Switzerland. The CCITT makes recommendations for the manufacture and interoperability of telecommunications equipment. The recommendations are not enforced by anything other than the peer pressure of the industry and the fact that following standards greatly improves the chances for a product's success.

CCK (Complementary Code Keying)   In Wireless LAN radio, a modulation technique that utilizes a complex set of mathematical functions known as *complementary codes* to transfer more data over a link. CCK is less affected by Multipath distortion than other modulation methods such as QPSK (Quadrature Phase Shift Keying) and BPSK (Binary Phase Shift Keying).

CCNA (Cisco-Certified Network Associate)   This is a well-known industry certification/training program offered by Cisco Systems and Cisco Systems training partners. The CCDA program is provided to train individuals for installation and maintenance of routed-LAN, WAN, and switched-LAN networks using Cisco Systems products. The course also lightly covers general LAN and WAN telecommunications. More information can be found regarding Cisco Systems training programs through *http://www.cisco.com.*

(CCNP) Cisco-Certified Network Professional   This well-known industry certification/training program is offered by Cisco Systems and

Cisco Systems training partners. The CCNP is an advanced certification and the prerequisite for the course track is the *CCNA (Cisco-Certified Network Associate)*. The CCNP program is provided to train and certify individuals for complex ISP environments using Cisco Systems networking products. More information can be found regarding Cisco Systems training programs through *http://www.cisco.com.*

CCS (Centum Call Second)   1. A centum call second is 100 seconds of telephone conversation. 36 centum call seconds is one Erlang, which is one call hour (one hour of phone conversation). Erlangs are measurements of telephone switch traffic.   2. Common Channel Signaling: Another term for out-of-band signaling on a T1/E1 circuit. In IP telephony circles, T1 signaling is often referred to as *CCS (Common Channel Signaling)* or *CAS (Channel Associated Signaling)*. CCS uses one channel in a T1 to carry signaling information such as on-hook or off-hook and touch tones for the remaining 23 channels (also called *DS0s*). E1 CCS uses 2 of the 32 channels for carrying the same type of signaling information. See also *CAS, T1,* and *H.323* for IP telephony signaling information.

CCS7 (Common-Channel Signaling No. 7)   ISDN version of SS7. An out-of-band signaling system between central offices throughout the telephone network that carries information and signaling for each phone call (such as billing, ANI, and ringing), as well as information about each central office (such as trunks busy or blocking and routing information). CCS7 is uncommonly used in North America, Malaysia, and Japan.

CCTV (Closed-Circuit Television)   Usually CCTV is a network of security cameras that terminate into a video processor, which displays the camera images on one or more video monitors.

CD (Carrier Detect)   Most modems have a little red LED. When that light is on, your modem is connected to another modem or communications device that it can understand.

CDDI (Copper Distributed-Data Interface)   The twisted-pair version of *Fiber Distributed-Data Interface (FDDI)*. Pronounce them the way they look, "fiddy" and "siddy." These two token-passing systems are intended to be backbone applications for LAN environments. CDDI is capable of transmission speeds of 100 Mb/s. For more information on the way it works, see the original version, *FDDI.*

CD-E (Compact Disc Erasable)   The original name for *CD-RW (Compact Disc Re-Writeable)* technology when it was developed in 1995. By the time the product was offered to consumers in 1997, its name had evolved to *CD-RW.* CD-RWs are capable of being erased and rerecorded 10,000 times.

CD-I (Compact-Disc Interactive)   A multimedia education and entertainment system, developed by Philips and Sony, consisting of a proprietary compact disc player that connects to a television set. The CD-I system incorporates *CD-RTOS (Compact-Disc Real-Time Operating System),* which synchronizes interactive video, audio, and text with real time.

CDMA (Code Division Multiple Access)   A radio transmission format used in North America for wireless telephone or cellular telephone service over PCS allocated bandwidths. A company called QUALCOMM developed the primary CDMA method. CDMA is a breed of spread spectrum radio that uses O-QPSK (Offset—Quadrature Phase Shift Keying) as the modulation technique. The CDMA method creates multiple logical channels within a single bandwidth via Walsh coding, which is a random code key generated by random electronic noise. In the Walsh coding process, 64 separate logical channels can be defined in one physical bandwidth, which is 1.25 MHz wide. Each logical channel is preceded by a code key during transmission that identifies data packets to the end user.   Each logical channel can be used for a voice conversation, and four of the channels are used for signaling synchronization, pilot, and paging.   Mobile phones in the transmission control of a specific base station all use these four channels for signaling, sync, key, and timing. The remaining channels carry voice traffic.

CDP (Cisco Discovery Protocol)   In Ethernet, a device *autosensing* method. CDP is currently a proprietary method developed by Cisco Systems to enable its Ethernet switch ports to discover what kind of device they are connected to.   After discovering what type of device, CDP determines whether the end device requires –48V DC power, as in the case of an IP telephone. As of this writing (early 2001), CDP is being reviewed by standards committees and will most certainly become a standard.

CDPD (Cellular Digital Packet Data System)   A data packet transfer standard for sending data over cellular that was developed by the *CTIA (Cellular Telephone Industry Association)* in 1993. It is offered by some cellular companies. With CDPD-compatible equipment, you can send data over cellular the same way you send data over a land-based data-packet service (such as frame relay). For the cellular telephone company (operator), the CDPD equipment is physically and functionally separate from the cellular switching equipment, but it shares the cell site and radio spectrum.

CD-R (Compact Disc Recordable) A compact-disc standard format that, when used with a CD-R or CD-RW drive, can record up to 650 MB of unerasable data one time. The advantage of CD-R discs is that they record slightly faster, and cannot be erased or modified during subsequent uses (although they can be copied to CD-RW discs). CD-Rs can also be played in regular CD-ROM drives, and audio compact-disc drives.

CDR (Call Detail Recording) See *Call Detail Recording.*

CD-ROM (Compact-Disc Read-Only Memory) A read-only nonmagnetic data storage device in the form of a reflective disk that is 4.7 inches in diameter. Data stored can be audio, video, data, or a combination of these (Fig. C.15). Maximum storage is typically 650 MB. The CD-ROM is read by a laser diode. When the CD-ROM is manufactured, tiny reflective spots are burned into the surface of the disk.

Here is a simple description of how a CD-ROM works: Imagine that you are driving a car through a tunnel with the headlights pointing straight up. On the ceiling of the tunnel are mirrors. Every time you pass a mirror, you see a flash of light. The flashes of light and periods of darkness would be ones and zeros. The tunnel is a track on the CD-ROM disk. Your headlights would be the laser diode and your eyes would be the optical receiver.

Figure C.15   CD-ROM Disk Drive (Internal)

CD-ROM XA (CD-ROM Extended Architecture) A version of the CD-ROM that was released to consumers in 1991 that enabled software, audio, and video to be interleaved on the recorded tracks of the CD-ROM disc. To utilize this format, a regular CD-ROM can be used, but an XA controller card needs to be installed in the PC.

CD-RW (Compact Disc Re-Writeable) A compact-disc drive that can write, read, or erase discs made for this purpose (CD-RW discs). CD-RW has a storage capacity of 650 MB, and can be recorded and erased 10,000 times. Discs that are recorded with the MultiRead standard can also be used in CD-ROM drives and compact-disc audio drives.

CD-V (Compact-Disc Video) A disk that is about 3 inches in diameter that is capable of storing about five minutes of audio and video.

CDVT (Cell Delay Variation Tolerance) A parameter defined by the ATM Forum for ATM traffic management. In constant bit-rate transmissions, this parameter determines the level of jitter that is tolerable for the data samples received by the *Peak Cell Rate (PCR)*. See also *PCR* and *ABR (Available Bit Rate)*.

CED (Called Equipment Identification Tone) A 2100-Hz tone with which a fax machine answers a call.

Ceiling Distribution System Also called a *ceiling rack*. It consists of rows of ladder-shaped iron (usually painted gray), supported above electronic equipment by more iron posts bolted to the floor. It is used as a safe, out-of-the-way place to mount the cables that connect the electronic equipment below. They are used in telecommunications central offices and large computing environments.

Cell A geographical area in cellular communications. Each cell consists of a cell site. A cell site consists of an antenna, a hut, and a doghouse (the doghouse contains the transmitting electronics and is in the hut).

Cell The basic unit for ATM switching, better referred to as an *ATM frame,* which consists of 48 bytes of payload and 5 bytes of overhead.

Cell Delay Variation Tolerance (CDVT) A parameter defined by the ATM Forum for ATM Traffic Management. In constant bit-rate transmissions, this parameter determines the level of jitter that is tolerable for the data samples received by the *Peak Cell Rate (PCR)*. See also *PCR* and *ABR (Available Bit Rate)*.

Cell Loss Priority (CLP) An *ATM (Asynchronous Transfer Mode)* cell header bit that determines the probability of that cell being dropped if the network becomes congested. Cells with CLP = 0 are insured traffic, which are unlikely to be dropped. Cells with CLP = 1 are best-effort traffic, which will be dropped in congested conditions in order to free up resources for insured traffic. Whether cells are dropped interdepends on many congestion-handling parameters within the ATM network. See also *Constant Bit Rate*.

Cell Loss Ratio (CLR)   An *ATM (Asynchronous Transfer Mode)* quality-of-service parameter that might be implemented on a subscribed connection. The CLR is equal to the ratio of discarded data cells to successfully transported data cells. See also *Cell Loss Priority* and *Insured Rate*.

Cell Relay   A type of network-level communications protocol based on the use of small, fixed-size packets. The packets are called *cells* because they carry "micro amounts" of data. Because cells are small in size (a fixed length of 16 to 128 bytes, plus an overhead of 4 to 8 bytes), they can be processed and switched in hardware at high speed. The reason for fixed size is that the overhead required to decipher the length of the data contained in each chunk of the transmission is not needed, and errors do not require a large portion of data to be retransmitted. The cell relay is the basis for many high-speed network protocols, including *ATM (Asynchronous Transfer Mode), IEEE 802.6,* and *SMDS (Switched Multimegabit Data Service).* When the ATM standard was developed, 48 bytes was chosen for the payload size because it suited neither the North American desires for a 64-bit cell nor the Eurasian desires for a 32-byte cell. The 5-byte overhead was agreed upon by all. The ATM standard was created in a way that it would not give any country an economic advantage by having the standards meet their existing needs perfectly. This is one of the reasons that ATM has been implemented slowly over the years.

Cell Site   A cell site consists of an antenna, a hut, and a doghouse (the doghouse contains the transmitting electronics and is in the hut). A cell site is the transmit and receive center for a geographical area, called a *cell.*

Cell Switching   The process of handing a call from one cellular broadcast site or antenna to another without interrupting the call. This process is controlled by a *Mobile Telephone Switching Office (MTSO)*, to which all the cell sites within a region are connected.

Cellular   A wireless design method where multiple transmitters are strategically placed throughout a geographical area to provide two functions. The first is to radiate the area with an adequate signal, the second

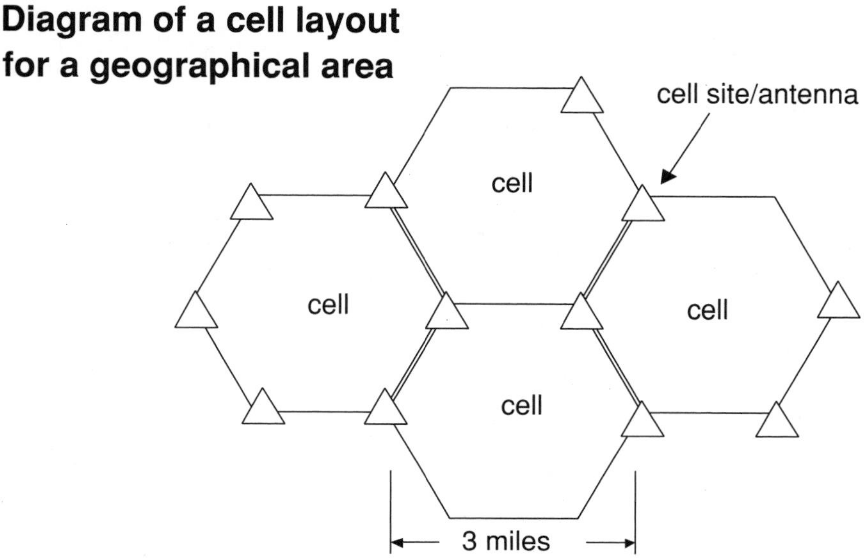
Diagram of a cell layout
for a geographical area
cell site/antenna
cell
cell
cell
cell
3 miles

Figure C.16 Cellular

is to make the size of the radiated areas adequate to accomodate the number of users. Cellular designs are used in the local PCS service as well as indoor wireless LAN (such as 802.11ab) designs. See also, *PCS*.

Cellular Data-Link Control (CDLC) A protocol for sending data over cellular. The equipment is not integrated with the cellular-service provider. It works the same way a modem does for a regular phone line. It error checks and retransmits corrupted data.

Cellular Digital Packet Service (CDPD) A service offered by some cellular companies. With CDPD-compatible equipment, you can send data over cellular the same way you send data over a land-based data-packet service (such as frame relay).

CEMH (Controlled Environment Man Hole) The new nondiscriminatory name for this is *CEV (Controlled-Environment Vault)*. It contains heating and cooling equipment and communications electronics, unlike plain old vaults that only contain splices.

Centel A company that was bought by Sprint in 1992.

Central Office A building that houses a telecommunications switching or trafficking system (Fig. C.17). Typical switching systems installed in central offices in North America are Lucent Technologies' 5ESS and Northern Telecom's DMS family of switches. There are five classes of central offices and five major parts to a central office. As a whole these parts are referred to as *inside plant.*

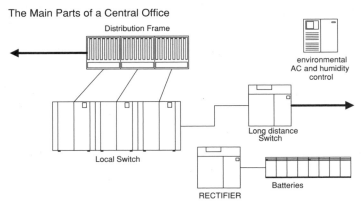

The Main Parts of a Central Office

Distribution Frame

environmental
AC and humidity
control

Long distance
Switch

Local Switch

RECTIFIER

Batteries

Figure C.17   Central Office

Central Office Battery   In theory, "central office battery" is –48 Volts. In reality, it is –52 Volts. The deviation is because of the difference between rectifier output voltage and the true battery voltage. The batteries that the central offices are powered from are arrays of 12-V batteries in series and parallel (Fig. C.18). These 12-V batteries actually output 12.7 V.

Figure C.18   Central Office Batteries: Lead Acid (Left) and Gel Cells (Right)

If four 12.7-V batteries are placed in series, they add up to 50.8 V. The rectifiers in the central office that charge the batteries have an output voltage of –52 V. This is a difference of 1.2 Volts, which is the trickle charge for the batteries. Ultimately, if the power is on at the central office, you are getting –52 V, which is the output of the rectifiers that power the switching system and charge the batteries at the same time. If the street power is out and the back-up generator is not running, the real central office battery voltage is –50.8 V.

Central Office Code   The address of a central office (Fig. C.19). The second three digits of your phone number (including the area code). It is also referred to as the *NXX*. It defines an exchange area, which is the boundary area of a certain central office.

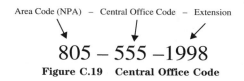

$$805 - 555 - 1998$$
Figure C.19   Central Office Code

Central Office Trunk   A communications path between central offices. A central office trunk is usually multiplexed into T1 formats.

Central Processing Unit (CPU)   The device within a computer (or switch or other machine that performs complex tasks) that controls the transfer of the individual instructions from one device connected to its bus (the data or I/O bus) to another, such as ROM, RAM, subcontrollers, decoders, and I/O ports. Some communications equipment manufacturers actually call a certain card or portion of the system the *CPU*. That is because they include all of the RAM, sub processors, buffers, clocking circuitry, and ROM as a part of the CPU. This is OK because we know that a real CPU is actually a small integrated circuit.

Centrex   A service provided by local telephone companies that mimics an on-premises PBX. The customer purchases a block of telephone numbers (e.g., 555-1000 to 555-1999), then every telephone on the customer's premise is connected to the telephone company as an individual phone line. Each line is associated with one of the numbers in the customer's block. The telephone company then programs those specific lines to route calls as desired by the customer. Voice mail can also be incorporated into Centrex.

Centrex LAN   A service that uses your modem, the phone company's wire, and the phone company's switch to connect equipment in an office or campus environment. The signal from a peripheral in your office goes all the way to the phone company's central office, then back, just to connect to a computer in the next room. Ethernet or token ring is a much less expensive and better-performing way to connect your LAN in the long run.

Centronics Connector   A connector developed by Centronics that is widely used in many applications. One standard is a 30-pin version for personal-computer parallel-printer connections. Another standard is the Centronics 50-pin connector, which is used widely in telecommunications applications. This connector is often called a *C connector* for a female socket and a *P connector* for a male plug (Fig. C.20).

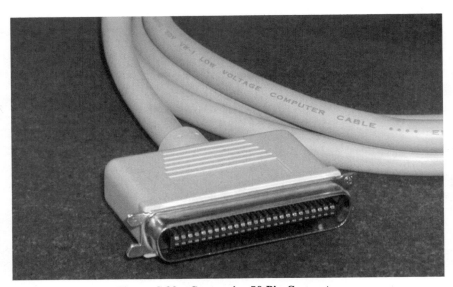

Figure C.20   Centronics 50-Pin Connector

Centum Call Second   See *CCS*.

CEPT (Confrence Europenne des Postes et des Telecommunications)   The association of 26 European *PTTs (Post, Telephone, and Telegraph)* that resolves interconnect issues between countries and recommends communication specifications to the ITU-T standards committee.

Certified Banyan Specialist (CBS)   This well-known industry certification is offered by Banyan Systems, Inc. There is also a Microsoft Windows NT extension to this program. It is designed for network professionals who utilize Banyan Vines and other Banyan products. More information regarding Banyan training programs can be found through *http://www.banyan.com.*

Certified Internet Professional (CIP)   A well-known industry certification/training program that is offered by Novell and its training partners. The CIP is an advanced certification. The prerequisite for the CIP study track is the possession of a CNE rating. This training program is designed to train individuals who will utilize Novell products in network/Internet integration. Information regarding Novell training can be found at *http://www.education.novell.com.*

Certified Network Expert (CNX)   A well-known industry certification that is offered by the CNX Consortium. It provides an intense training track for LAN and MAN networking in multivendor environments. More information can be found regarding the CNX certification program at *http://www.cnx.org.*

Certified Novell Administrator (CNA)   A well-known industry certification/training program offered by Novell and its training partners. The CNA program is designed to educate those who will maintain and administer installed Novell networking products.

Certified Novell Engineer (CNE)   A well-known industry certification/training program that is offered by Novell and its training partners. The CNE study track is designed to train individuals who will provide complex technical support and installation of Novell products. The CNE program is an in-depth study program that covers Novell specific products and data networking as a whole. An additional level to this certification is called *Master CNE*. Information regarding Novell training can be found at *http://www.education.novell.com.*

Cesium Clock   A clock that is used to synchronize communications equipment (i.e., SONET transport) by providing a perfectly steady output pulse (a very fast one), the same way that a metronome provides steady timing for a musical band. Its timing base is a factor of the atomic vibrations of the element Cesium.

CEV (Controlled Environment Vault)   A vault that is designed to have electronics in it. The environment inside the vault must be kept at a certain temperature and humidity.

CFA (Carrier Failure Alarm)   A notification that timing has been lost in a digital transmission because of excessive zeros in its transmission. When a carrier failure alarm occurs, all the calls and all the data on that transmission are dropped until the carrier equipment re-gains timing.

CGSA (Cellular Geographic Service Area) The geographical area that a cellular company provides service, which means their cellular radio waves can be received within this area.

Channel One segment or time-slot in a broadband communications transmission.

Channel Bank A hardware device used in TDM networks to code/decode a T1 into 24 separate channels (Fig. C.21). The individual channels can be used for TDM data transmission or voice. Channel banks and the T1 connections that they service can be configured to accommodate 23 CCS (Common Channel Signaling or out-of-band) 64 Kbps channels, or 24 CAS (Channel Associated Signaling or in-band) 56 Kbps channels. These are intended for data and voice, respectively. See also *T1*.

Figure C.21   Channel Bank

Channel Capacity The maximum number of bits per second that can be carried by a channel. The channel capacity of a DS0 within band signaling is 56 Kb/s, the channel capacity of a DS0 with out-of-band signaling is 64 Kb/s.

Channel Loop Back A method of testing a digital service line, such as a T1, where the receive channel is connected into the transmit channel (sometimes with only a pair of wires) at the far end (Fig. C.22). The signal can then be tested at the originating location and analyzed for er-rors. Equipment is made where the loop-back can be performed via re-mote control. One example of this equipment is a smart jack.

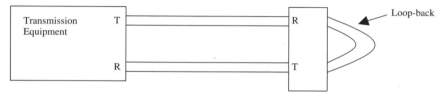

Figure C.22   Channel Loop Back

Channel Modulator   In cable-TV networks, a device that receives a program signal, such as CNN or MTV, and mixes it with an RF carrier (Fig. C.23). The program signal includes video and audio. The cable-TV head-end personnel determine the carrier frequency for which the modulator will be set. The frequency of this carrier determines the station number (i.e., 2 to 99) that the program will be received on by subscribers.

Figure C.23   Channel Modulator

Channel Service Unit (CSU)   This is also called a *CSU/DSU (Channel Service Unit/Data Service Unit)*. A CSU is a hardware device that can come in many shapes and sizes. Rack-mount, shelf-mount, and stand-alone CSUs are available. A CSU/DSU has three main functions. The first function is to act as a demarcation point for a T1 (DS1) service from a local communications company. The second function is to provide line-format and line-code conversion (B8ZS to AMI, SF or D4 to ESF, 135 to 0 V) between the public network and the customer-premises equipment, if necessary. The third function is to provide maintenance or alarm services and loop-back for isolating problems with the T1 line or customer's equipment. For a photo, see *CSU/DSU*.

Channel Termination   Also called a *chanterm*, is a cross connect that links the transmit and receive of two devices. Channel terminations are used to connect private-line services through a central office. Many local phone companies charge an additional amount of money for each channel termination that a private line has. If a private line goes from one building to another building across town, it probably passes through two or three central offices to get there. Each connection through a central office requires a channel termination.

Chanterm   See *Channel Termination*.

Character   A number, letter, or symbol that is represented by a binary code. See *Character Code* for more information.

Character Code   A code in binary numbers that represents the alphabet and other symbols. Figure C.24 shows ASCII character codes.

Least significant bits (hexadecimal)	Most significant bits	(hexadecimal)						
	000 (0)	001 (1)	010 (2)	011 (3)	100 (4)	101 (5)	110 (6)	111 (7)
0000 (0)	NUL	DLE	SP	0	@	P	`	p
0001 (1)	SOH	DC1	!	1	A	Q	a	q
0010 (2)	STX	DC2	"	2	B	R	b	r
0011 (3)	ETX	DC3	#	3	C	S	c	s
0100 (4)	EOT	DC4	$	4	D	T	d	t
0101 (5)	ENQ	NAK	%	5	E	U	e	u
0110 (6)	ACK	SYN	&	6	F	V	f	v
0111 (7)	BEL	ETB	'	7	G	W	g	w
1000 (8)	BS	CAN	(	8	H	X	h	x
1001 (9)	HT	EM	)	9	I	Y	I	y
1010 (A)	LF	SUB	*	:	J	Z	j	z
1011 (B)	VT	ESC	+	;	K	[	k	{
1100 (C)	FF	FS	,	<	L	\	l	
1101 (D)	CR	GS	-	=	M	]	m	}
1110 (E)	SOH	RS	.	>	N		n	~
1111 (F)	SI	US	/	?	O		o	DEL

Figure C.24   Character Code (ASCII)

Characteristic Impedance   The impedance or AC resistance of a transmission media such as CAT 5 twisted pair. The characteristic impedance of CAT 5 twisted pair is 100 ohms. This means that when a CAT 5 twisted pair is terminated (or connected to) a device that also has an impedance of 100 ohms, the twisted pair, regardless of its physical length, "looks" infinitely long to the circuit. The usefulness of this is that the voltage-to-current ratio is the same all the way down the line. So, if you have 2 V of signal at 20 mA at the beginning of the twisted pair, that ratio equals $2 \div 0.02 = 100$—for a photo of different types of coax, see *Coax*. Here are some values of characteristic impedance for common physical media:

X-mission Media	Characteristic Impedance (Z)
RG-6 coax	75 ohms
RG-8 coax	50 ohms
RG-58 coax	50 ohms
RG-59 coax	75 ohms
RG-62 coax	93 ohms
Cat 5 UTP	100 ohms
Cat 7 UTP	100 ohms
3IBM type 1 Data	150 ohms

Character-Oriented Protocol (Another name for Byte-Oriented Protocol)   A pre-X.25 standard protocol used by mainframes that prevented data of unlike character sets to be transmitted over communications lines. Newer communications standards are bit-oriented protocols, and are not character or application sensitive. See also *Byte-Oriented Protocol*.

Character Set   The letters and numbers on computer keyboards. Different standards apply to how the letters and numbers are converted to binary code. The most widely embraced standard for PCs is ASCII.

Chip   An integrated electronic component. *Integration* refers to many circuits integrated into one small device.

Chipping Sequence   In wireless data transmissions, specifically DSSS (Direct Sequence Spread Spectrum), a code of binary digits that are transmitted simultaneously and equate to one bit of data. Chipping is a functional part of CCK (Complimentary Code Keying), which is the modulation method used in DSSS. See also *DSSS* and *CCK*.

Chipset   A group of microchips in a computer that allows hardware peripheral devices (such as modems and network interface cards) in PCI and ISA buses to communicate without using the main CPU. This reduces the CPU processing load, thus helping the computer accomplish tasks faster.

Choke   Also called a *choke coil* or *RF choke*. A coil of wire manufactured with the intent of being a filter that reduces the passing of high frequencies through it. For a photo, see *RF Choke*.

Choke Coil   See *Choke.*

CI (Clear Indication/Clear Request)   The packet that is sent when a device on an X.25 network would like to end a call. The other device confirms the end of the call by sending a *CC (Clear Confirm)* packet. For a diagram of X.25 packet-level header structure, see *X.25 Call-Request Packet.*

CIDR (Classless Interdomain Routing)   A method of reducing the size of routing tables, thus, the computational strain required to route packets using them. The method used to accomplish this is known as *route aggregation.* Route aggregation combines complex route information and reduces it to a simplified port. It is also viewed as combing many networks into a few. Addresses are grouped and then associated with a port, similar to the way an area code is specific to a geographic location containing a large amount of phone numbers. A telephone switch processor utilizes only the area code information to route the call, yet passes the entire dialed digit sequence to be used by the network processor that is next in the call connect sequence. In CIDR, it is not quite as simple because IP addresses are not as easily grouped into routes, but the fact that the absolute value of the most significant address digits are the ones used remains the same.

CIP (Certified Internet Professional)   A well-known industry certification/training program that is offered by Novell and its training partners. The CIP is an advanced certification. The prerequisite for the CIP study track is the possession of a CNE rating. This training program is designed to train individuals who will utilize Novell products in network/Internet integration. Information regarding Novell training can be found at *http://www.education.novell.com.*

Circuit   1. Another name for a phone line. There are many types of circuits: digital, analog, T1, and ISDN circuits.   2. An electronic device that receives a given input and converts it into a desirable output. For instance, a TV converts a transmission input into a picture and sound. A TV can be regarded as one giant circuit or many small circuits.

Circuit Board   Any form of electronics parts placement that is on a flat surface.

Circuit Breaker   An electronic or electrical device installed in a power-distribution system that disconnects (or turns off) the power when the specified current rating of the circuit breaker is exceeded (Fig. C.25). In most cases, when the specified current in a circuit is exceeded, there is a problem. Circuit breakers disconnect the power before the "problem" causes other problems (such as fire or power-supply overloading).

Figure C.25   Circuit Breaker. Three 200-Amp Breakers are Shown on a −48-V Distribution Bus

Circuit ID Code   The part of a CCS or SS7 signaling message that identifies (gives a name to) the circuit that is being established between two points.

Circuit Switching   Also referred to as *line switching.* In early voice telecommunications, the line switching was performed by operators sitting in front of a "cord board." When a caller wanted to make a call, they would pick up the receiver and an operator would ask "what number, please?" The caller would then tell the operator what number to connect them to, and the operator would plug a cord into the line and connect it to the line (or loop) associated with the number they wanted to call. Today, line switching is performed by "switches." *Switches* (also called PBX and *key systems*) are electronic machines that work similar to the operator. Instead of speaking with the operator, the switch and caller "signal" each other to accomplish the switching function. When the caller goes off hook, the central office switch sends a dial tone. The dial tone is a signal to the caller to "enter number please." The caller enters digits that signal the number to which they would like to be connected. The central office then sends a "ring" signal to the party being called and a "ring simulation" signal to the calling party, which lets them know that the central-office switch received the digits ok and is signaling the party being called. The party being called then picks up the handset, which activates the switch hook and places a 1000-ohm short on the line. The 1000-ohm short signals the central office that the party is ready to receive the call. The central office switch then stops the ring signaling and "switches" the talk paths into place for the two callers.

CISC (Complex Instruction Set Computing)   Pronounced "Sisk." A software and hardware architecture method that utilizes a computer's CPU for the majority of the work load, rather than the software. CISC processors support up to 200 instructions. Most personal computers are made with CISC processors. See also *RISC.*

Cisco-Certified Design Associate (CCDA)   This well-known industry certification/training program is offered by Cisco Systems and Cisco Systems training partners. The CCDA program is provided to train individuals for simple-routed LAN, WAN, and switched-LAN networks. More information can be found regarding Cisco Systems training programs through *http://www.cisco.com.*

Cisco-Certified Design Professional (CCDP)   This is a well-known industry certification/training program offered by Cisco Systems and Cisco Systems training partners. The CCDP program is provided to train individuals for complex-routed LAN, WAN, and switched-LAN networks. A *CCNA (Cisco-Certified Network Associate)* certification must be acquired as a prerequisite for the CCDP program. More information can be found regarding Cisco Systems training programs through *http://www.cisco.com*.

Cisco-Certified Internetwork Expert (CCIE) This well-known industry certification/training program is offered by Cisco Systems and Cisco Systems training partners. The CCIE program is the most advanced level of certification for Cisco Systems associates, and is regarded as a very challenging certification that only the experienced can obtain. The program prerequisites are CCDP, CCNA, and CCNP certifications. This program is offered to provide training and certification to individuals who work in the most-complex networking environments. More information can be found regarding Cisco Systems training programs at *http:// www.cisco.com*.

Cisco-Certified Network Associate (CCNA) This well-known industry certification/training program is offered by Cisco Systems and Cisco Systems training partners. The CCDA program is provided to train individuals for installation and maintenance of routed-LAN, WAN, and switched-LAN networks using Cisco Systems products. The course also lightly covers general LAN and WAN telecommunications. More information can be found regarding Cisco Systems training programs through *http://www.cisco.com*.

Cisco-Certified Network Professional (CCNP) This well-known industry certification/training program is offered by Cisco Systems and Cisco Systems training partners. The CCNP is an advanced certification and the prerequisite for the course track is the *CCNA (Cisco-Certified Network Associate)*. The CCNP program is provided to train and certify individuals for complex ISP environments using Cisco Systems networking products. More information can be found regarding Cisco Systems training programs through *http://www.cisco.com*.

Cisco Discovery Protocol (CDP) In Ethernet, a device *autosensing* method. CDP is currently a proprietary method developed by Cisco Systems to enable its Ethernet switch ports to discover what kind of device they are connected to. After discovering what type of device, CDP determines whether the end device requires –48V DC power, as in the case of an IP Telephone. As of this writing (early 2001), CDP is being reviewed by standards committees and will most certainly become a standard.

Citizens Band (CB) See *CB*.

Cladding   One of the two glass sections of a fiber optic (Fig. C.25).

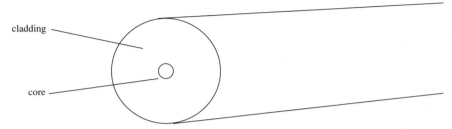

Figure C.26   Cladding

CLAS (Centrex Line-Assignment Service)   A feature offered with centrex service that allows customers to dial into the telephone company's line-assignment computer system and make changes to where extensions are located, ringing and hunt groups, and which extension numbers or phone numbers are in service.

Class-1 Central Office   A centermost point in a large logical network hierarchy of long-distance central office switches.

Class-2 Central Office   An older term for an network hierarchy of switch points. Central offices are now becoming so well connected in so many ways that they are rapidly becoming either a class-5,4 or a class-1 central office.

Class-3 Central Office   An older term for an older network hierarchy of switch points. Central offices are now becoming so well connected in so many ways that they are rapidly becoming either a class-5,4 or a class-1 central office.

Class-4 Central Office   A tandem central office or main switch center for an area code. That might also perform class-5 end-connection functions.

Class-5 Central Office   A local telephone company central office that connects to end customers.

Class-A IP Address   An IP address within the range of 1.0.0.0 and 126.255.255.255. There are a total of 126 Class-A IP addresses. Each Class-A address can support 16,777,216 hosts. Address 127.x.x.x was used during the development of IP, and is not a valid address for network use. Class-A addresses are mostly allocated to and used by very large corporations, governments, and universities to identify their Internet domain. Internet addresses are assigned by InterNIC.

Class-B IP Address   An IP address with the range of 128.0.0.0 to 191. 255.255.255. There are a total of 16,384 Class-B IP addresses. Each Class-B address can support 65,534 hosts. Class-B addresses are used by medium to large corporations to identify their Internet domain. Internet addresses are assigned by InterNIC.

Class-C IP Address   An IP address with the range of 192.0.0.0 to 223. 255.255.255. There are a total of 2,097,152 possible Class-C networks. Each Class-C address can support up to 254 hosts. Class-C addresses are used by small corporations to identify their Internet domain. Internet addresses are assigned by InterNIC.

Class-D IP Address   An IP address with the range of 224.0.0.0 to 239.255.255.255 are used for Multicast addressing. Such modern routing protocols as *ICMP (Internet Control Message Protocol)* and Cisco's *IGRP (Interior Gateway Routing Protocol)* are capable of recognizing multicast addresses. See also *IP Multicast.* Class-D addresses are not assigned to specific hosts, but to groups of hosts on a network. Multicast addresses are assigned by the administrator of the Internetwork.

Class-E IP Address   An IP address with the range of 240.0.0.0 to 255. 255.255.254. These addresses are reserved for special-purpose and future addressing modes. The address 255.255.255.255 is the dejure standard broadcast address. Class-E addresses are not assigned in a network.

Class of Service (COS)   A type of telephone service or telephone line purchased from a telephone company. Some class-of-service examples follow:

- *1FR*   One flat rate, residential. What most residential customers have.
- *1MR*   One measured rate, residential. Where the line has a low monthly fee, but each call beyond a certain number costs up to an additional 10 cents.
- *1FB*   One flat rate, business. What most small-business customers have.
- *1F4*   A four-user party line.

Classless Interdomain Routing (CIDR)   A method of reducing the size of routing tables thus, the computational strain required to route packets using them. The method used to accomplish this is known as *route aggregation.* Route aggregation combines complex route information and reduces it to a simplified port. It is also viewed as combing many networks into a few. Addresses are grouped and then associated with a port, similar to the way an area code is specific to a geographic location containing a large amount of phone numbers. A telephone switch processor utilizes only the area code information to route the call, yet passes the entire dialed digit sequence to be used by the network processor that is next in the call connect sequence. In CIDR, it is not quite as simple because IP addresses are not as easily grouped into routes, but the fact that the absolute value of the most significant address digits are the ones used remains the same.

Clear Channel   See *CCC.*

Clear To Send (CTS)   See *CTS.*

Cleaving   A term used in the splicing fiber optic that means to cut the end of the fiber clean at 90 degrees, with minimal rough edges, in preparation for a fusion splice. Where a mechanical splice is involved, the end of the fiber optic is hand-smoothed with a polishing puck before splicing.

CLI (Command Line Interface)   Another term for "prompt-response" type interfaces used by administrators to configure network devices and systems. CLI systems are somewhat cryptic and syntax oriented, making them frustrating for some people to use. Many manufacturers are integrating web-based (HTML/GUI) management interfaces into their systems that alleviate the need to remember command functions. For most systems, the option to use the CLI is still there.

Client-Server Environment   A type of network environment with requesters (clients) and providers (servers). A service requested could be for processing, a file, or an application.

Climbers   What telecommunications and power company personnel wear to climb wooden telephone and power poles. The official name for these devices are *lineman's climbers.* They are also called *spurs, hooks,* and *gaffs.* They consist of a steel shank with straps so that it can be strapped to a person's leg. On the inside of the shank is a spike that is used to stab into the pole. The climbers in Fig. C.27 were manufactured by Buckingham Mfg. Inc.

Figure C.27   Lineman's Climbers

Climbing Belt   Used by communications/power/construction personnel to harness themselves to telephone/power poles or tower structures. Also called a *safety belt* and *body belt*. For a photo, see *Safety Belt*.

Clipper   A circuit that takes a sine-wave input, like that of a timing oscillator and converts it into a square-wave form that can be used as a clock signal for a digital device. A clipper circuit is simply an amplifier that is overdriven into saturation, which causes a form of distortion (in this case, useful) called *clipping*.

Clipping   You can experience clipping by turning up the volume on an inexpensive stereo and noticing that at a certain point, the sound reproduction is unclear (Fig. C.28). First, the bass tones are affected because

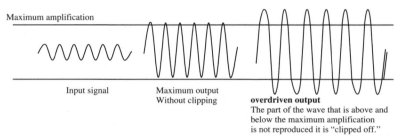

Maximum amplification

Input signal   Maximum output
                      Without clipping   **overdriven output**
                      The part of the wave that is above and
                      below the maximum amplification
                      is not reproduced it is "clipped off."

Figure C.28   Clipping

they take the most energy to reproduce. At this point in increasing the volume, the sound level is not getting any louder, the signal is only becoming more corrupted. Technically, *clipping* is a form of signal distortion where the amplification of a signal in volts exceeds the saturation bias voltage of the transistor in an amplifier. In other words, the peak and negative peaks of the output waveform are not included because the amplifier is not capable of amplifying the input anymore.

CLLI Code   An 11-digit alphanumeric code that identifies physical locations in the phone network. Almost every significant building in the United States has a CLLI code. It is pronounced "silly code."

CLNP (Connectionless Network Protocol)   An *OSI (Open Systems Interconnect)* network layer protocol that does not require a circuit to be established before data is transmitted. The basis of a connectionless protocol is that there is no direct *RTS (Request to Send)* or *ACK (Acknowledge)* type of signaling, thus, no direct connection. See also *CLNS.*

CLNS (Connectionless Network Service)   An *OSI (Open Systems Interconnect)* network layer protocol that does not require a circuit to be established before data is transmitted. CLNS works as a part of or with other protocols that reside in the network layer and other layers. All packets are sent independently of each other. See also *CLNP.*

Clock   A device that provides timing pulses for communications equipment or devices within a computer the same way that a metronome provides a steady time for a musical band (Fig. C.29).

Clock Bias   The difference between a clock's output and true universal time. Simply put, how far off a piece of equipment's timing is. It is usually measured in positive or negative microseconds.

Clocking   See *Clock.*

Clone Fraud   The crime of finding cellular ID codes by monitoring cellular transmissions (with expensive cellular equipment), then copying the code to another cellular phone and making calls with it. The airtime for the copied phone (the clone) is billed to the original phone.

Closure   A casing, pedestal, or cabinet used to house open ends or splices in outside cable plant. Closures mostly refer to splices. Different names for buried splice closures or enclosures are Xaga and Cold-N-Close. A popular aerial splice closure is the TRAC closure.

Figure C.29   Clock Source (Distributed)

CLP (Cell Loss Priority)   An *ATM (Asynchronous Transfer Mode)* cell header bit that determines the probability of that cell being dropped if the network becomes congested. Cells with CLP = 0 are insured traffic, which are unlikely to be dropped. Cells with CLP = 1 are best-effort traffic, which will be dropped in congested conditions in order to free up resources for insured traffic. Whether cells are dropped interdepends on many congestion-handling parameters within the ATM network. See also *Constant Bit Rate.*

CLR (Cell Loss Ratio)   An *ATM (Asynchronous Transfer Mode)* quality-of-service parameter that might be implemented on a subscribed connection. The CLR is equal to the ratio of discarded data cells to successfully transported data cells. See also *Cell Loss Priority* and *Insured Rate.*

Cluster Controller   1. Generally, an intelligent device that provides the data communications management and control for a cluster of terminals to a data link. 2. In *IBM SNA (System Network Architecture),* a programmable device that controls the input/output operations of attached devices. Typically, it is an IBM 3174 or 3274 or newer type of device.

CMOS (Complementary Metal-Oxide Semiconductor)   The reason why many computer and other high-speed components are static sensitive. Complementary Metal-Oxide Semiconductor's largest advantage over TTL (Transistor-Transistor Logic) is their low power consumption (less than $\frac{1}{10}$ of TTL), they switch on without drawing very much current, in contrast to TTL. Since very little current is drawn, very little power is consumed and very little heat is given off. This allows the devices to be much smaller.

CNA (Certified Novell Administrator)   A well-known industry certification/training program offered by Novell and its training partners. The CNA program is designed to educate those who will maintain and administer installed Novell networking products.

CNE (Certified Novell Engineer)   A well-known industry certification/training program that is offered by Novell and its training partners. The CNE study track is designed to train individuals who will provide complex technical support and installation of Novell products. The CNE program is an in-depth study program that covers Novell specific products and data networking as a whole. An additional level to this certification called *Master CNE.* Information regarding Novell training can be found at *http://www.education.novell.com.*

CNX (Certified Network Expert)   A well-known industry certification that is offered by the CNX Consortium. It provides an intense training track for LAN and MAN networking in multivendor environments. More information can be found regarding the CNX certification program at *http://www.cnx.org.*

CO   An abbreviation for an RBOC Central Office. CLEC's often refer to their central switching offices as *Type One Nodes.*

Coax (coaxial cable)   A shielded copper transmission media that has one central conductor surrounded by a dielectric (Fig. C.30). It comes in several varieties. For a listing of the different characteristic impedances of coax types, see *Characteristic Impedance.*

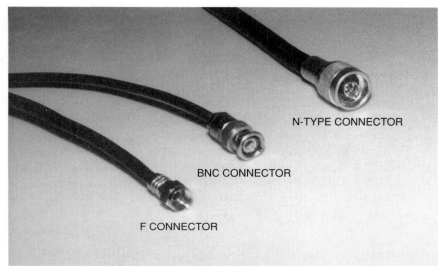

Figure C.30   Coaxial Connector Types: F (Left), BNC (Center), and N Type (Right)

COBOL (Common Business Oriented Language)   A computer programming language.

Co-Channel Interference   In radio such as terrestrial microwave or satellite, co-channel interference is when two radios operating separate links are set at the same frequency. In selecting a site, a spectrum analyzer can be used to determine if any strong signals are present at the site, and if they are, to determine how close they are to the desired frequency. The further away from your proposed frequency, the less likely they are to cause a problem. Antenna placement and polarization as well as the use of high-gain, focused antennas is the most effective method of reducing this type of interference.

Code Blocking   A feature of a telecommunications switch that enables it to restrict specific extensions from dialing long distance or just a specific group of area codes. Local telephone companies use this feature to restrict a customer's long-distance calling ability from their phone if they have a poor credit rating and were not able to make a deposit for their phone service. This feature is frequently used in PBX applications to prevent employees from making any long-distance calls.

Code Division Multiple Access (CDMA)   A radio transmission format used in North America for wireless telephone or cellular telephone service over PCS allocated bandwidths. A company called QUALCOMM developed the primary CDMA method. CDMA is a breed of spread spectrum radio that uses O-QPSK (Offset—Quadrature Phase Shift Keying) as the modulation technique. The CDMA method creates multiple logical channels within a single bandwidth via Walsh coding, which is a random code key generated by random electronic noise. In the Walsh coding process, 64 separate logical channels can be defined in one physical bandwidth, which is 1.25 MHz wide. Each logical channel is preceded by a code key during transmission that identifies data packets to the end user. Each logical channel can be used for a voice conversation, and four of the channels are used for signaling synchronization, pilot, and paging. Mobile phones in the transmission control of a specific base station all use these four channels for signaling, sync, key, and timing. The remaining channels carry voice traffic.

CODEC (Coder Decoder)   Another name for an analog-to-digital (and digital-to-analog) converter.

Codec (Coder/Decoder)   Codecs are widely used in IP telephony applications (Fig. C.31). The standards used are the G series, and can compress a voice that would be 64 Kbps wide in bandwidth to 8 Kbps (in the case of G.729). The Codec standard for traditional circuit based telephony is G.711, which is the coding method for a DS0. When transporting voice over WAN links, coded schemes are implemented to conserve bandwidth. The diagram for H.323 illustrates the G.7XX standards as being a part of the presentation layer.

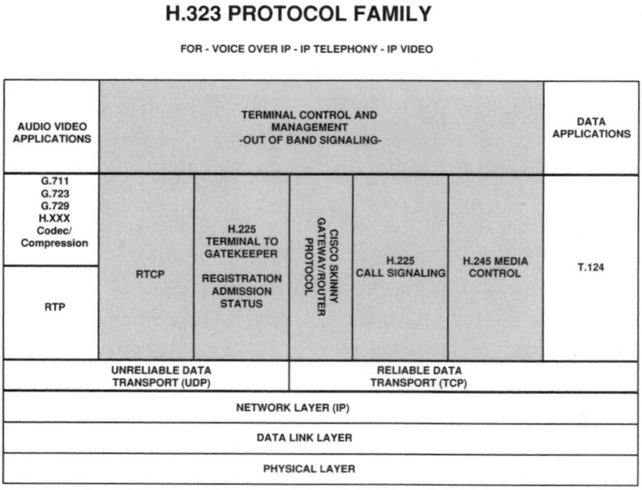

H.323 PROTOCOL FAMILY

FOR - VOICE OVER IP - IP TELEPHONY - IP VIDEO

AUDIO VIDEO APPLICATIONS				TERMINAL CONTROL AND MANAGEMENT -OUT OF BAND SIGNALING-				DATA APPLICATIONS
G.711 G.723 G.729 H.XXX Codec/ Compression	RTCP	H.225 TERMINAL TO GATEKEEPER	CISCO SKINNY GATEWAY/ROUTER PROTOCOL	H.225 CALL SIGNALING	H.245 MEDIA CONTROL	T.124
RTP		REGISTRATION ADMISSION STATUS				
UNRELIABLE DATA TRANSPORT (UDP)			RELIABLE DATA TRANSPORT (TCP)			
NETWORK LAYER (IP)						
DATA LINK LAYER						
PHYSICAL LAYER						

Figure C.31   Codec G.7xx Standards

Coherent Light   Light that consists of only one frequency or very close to one frequency. Coherent light looks to the human eye as a very pure color. Lasers and *LEDs (Light-Emitting Diodes,* like the one that lights when the hard drive in your computer is running) emit light that is very close to being coherent. A light bulb emits noncoherent light, which consists of many colors and wavelengths.

Coil   In telecommunications, a *coil* refers to a *load coil,* which is a voice-amplifying device for twisted-pair wire. A load coil is usually placed on each twisted pair used for a voice line every 3000 feet past a central office. Coils are usually located in vaults, with twisted-pair splices. Other coils, used for other electronic applications are usually referred to as *choke coils.*

Coin-Operated Telephone   A telephone that can be installed, operated, and maintained by a local phone company or purchased from a telephone-equipment distributor, connected to a local phone line and maintained by a private individual or company other than the local phone company. Many different kinds of coin-operated phones are manufactured by different companies, just the same as the different single-line phones you can purchase for your own home.

Collision Detection   A part of *CSMA/CD (Carrier Sense Multiple Access/Collision Detection),* An Ethernet LAN protocol. In LANs with CSMA/CD, PCs check the network to see if it is clear before transmitting. If it is clear, it transmits its data. They do this because if more than one PC sends data at the same time, the data gets garbled and is meaningless to the other PCs. This simultaneous data transmission is called a *collision.* CSMA/CD senses these collisions and attempts to retransmit the same data again when the network is clear again. In Ethernet networks, PCs are sensing and transmitting hundreds of times per second. If the network looks clear for a tiny fraction of a second, the PC will try to transmit.

Collapsed Core   A reference to a LAN hierarchical network design where there is no distribution layer (Fig. C.32). In *centralized* or *hierarchical* LAN network design, the three switch layers are core layer, distribution layer, and access layer. The core layer provides redundancy for distributing traffic across multiple access layers. The access layer provides high-speed switching and routing among a networked group of switches through trunks. It also provides switches where users connect, so high port quantity is desired in this layer.

Collision   In X25, *Logical Channel Numbers (LCNs)* are constantly assigned and reassigned to DTE devices on the network. Outgoing-data LCNs are assigned by the highest unused LCN, and incoming-data LCNs

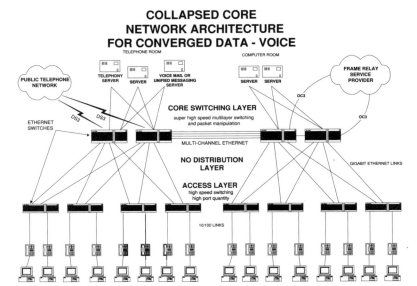

Figure C.32   Collapsed Core

are assigned by the lowest unused LCN. If an X.25 device tries to assign both an incoming link and an outgoing link the same LCN, then the device becomes confused as to where to send the data. This scenario is a collision in X.25. The X.25 protocol has a maximum growth capacity of 1023 LCNs, currently the protocol only utilizes 63 (0–63 and 0 is reserved), because having more than 63 DTE devices trying to communicate on the same X.25 link is extremely impractical (it would be very slow).

Collision Domain   In Ethernet, the network area within which frames that have collided are propagated. This area of the affected network will experience increased latency or delay in relation to the rest of the network. Repeaters and hubs propagate collisions; LAN Ethernet switches, token-ring switches, bridges, and routers do not. See also *Collision, CSMA,* and *Ethernet.*

Colocation, Physical   A colocation is an interconnection agreement and a physical place where telephone companies hand-off calls and services to each other (Fig. C.33). This is usually done between a CLEC and an RBOC. The CLEC installs and maintains interconnection equipment usually consisting of optical carrier (SONET) equipment and a digital cross-connect system. There are other types of colocations. Alarm companies like to have their alarm signaling equipment located in the local central office for the security and convenience of connecting alarm circuits. Long-distance companies colocate with local telephone companies, as well.

Figure C.33   Colocation. A Leased-Space Agreement (LSA)

Colocation, Virtual   A colocation is an interconnection agreement and a physical place where telephone companies hand-off calls and services to each other. This is usually done between a CLEC and an RBOC. A virtual colocation is when telephone company A (the CLEC) requests that their phone company's network be connected to telephone company B's (the RBOC's) network. Telephone company B charges company A lots of money. Company B owns, installs, and maintains the equipment. To company A, the interconnection is virtual because they never physically do anything to it when and after it is installed.

Color Code   The three communications color codes are the twisted-pair cable color code *(PIC),* the fiber-optic color code, and the resistor color code. For a listing of these color codes, see *Appendix F.*

Combination Trunk   A *DID (Direct Inward Dial)* and *DOD (Direct Outward Dial)* trunk all in one. These trunks are basically the same as the phone service in your home. They can be dialed out, and they can ring in.

Combiner   In cable-TV networks, a device that receives multiple RF transmissions (via coax) on separate inputs and places them on one output (Fig. C.34). Combiners are also referred to as *reverse spitters.* They are comprised of RF-matching transformers, and can have attenuation or signal-boost capabilities. Combiners receive the program signals from many modulators and place them on one coax, which distributes the cable-TV signal to subscribers.

Figure C.34   A Passive RF Combiner

Command Set   Parts of a program within a switch or PBX's software. A command is a set of instructions written in a program and attached to a command. A command is a simple entry that a user makes to instruct the PBX to perform a function, such as "Enable Trunk Port 8." When this command is entered, a command set within the software is activated and trunk port 8 looks for calls to come in.

Commercial Building Telecom Wiring Standard   A wiring standard for buildings. The standard states that all wire on each floor of a building is run and terminated to a single location. All wire run between floors is fed by a riser that terminates at each floor where the wire runs terminate. All the riser cables must originate at a single point where the local telephone company point of presence (NI) is located.

Command Line Interface (CLI)   Another term for "prompt-response" type interfaces used by administrators to configure network devices and systems. CLI systems are somewhat cryptic and syntax oriented, making them frustrating for some people to use. Many manufacturers are integrating web-based (HTML/GUI) management interfaces into their systems that alleviate the need to remember command functions. For most systems, the option to use the CLI is still there.

Committed Burst (Bc)   An amount of data that is permitted onto and over a frame-relay network *DLC (Data Link Connection)* over a specific amount of time. See also *Bandwidth Control Element.*

Committed Information Rate (CIR)   The rate in Kb/s or Mb/s that a communications company will guarantee over a frame relay circuit that they provide to their customer. If you purchase a frame relay circuit, there is a place on the order agreement that you state the rate of information that you want to transmit. The choices usually range from 56 Kb/s, to 1.5 Mb/s. If you enter 1.5 Mb/s for your committed information rate, you will pay more for your service than the 56 Kb/s Committed Information Rate. You have the capability to transmit and are permitted to transmit at rates up to the full CIR, unless the frame relay network becomes congested. Then, only your committed rate will get through. This is similar to purchasing the use of lanes on the freeway. If the freeway has very few cars travelling on it, then there is no sense in buying the right for multiple lanes because no one is using them anyway. However, if the freeway is congested, then you are getting your money's worth with your lane rights. You never know how congested a frame relay network will be, or when and where bottlenecks will occur in the network.

Common Audible Ringer   A loud ringer. A bell connected to a telephone line that is in a noisy or wide-open area. When the phone rings, the loud bell also rings.

Common Carrier   A licensed private utility company that supplies communication services to the public at prices that are regulated by the *FCC (Federal Communications Commission)* and/or local government organizations, depending on the state or country.

Common Channel Signaling (CCS)   Another term for out-of-band signaling on a T1/E1 circuit. In IP telephony circles, T1 signaling is often referred to as *CCS (Common Channel Signaling)* or *CAS (Channel Associated Signaling)*. CCS uses one channel in a T1 to carry signaling information such as on-hook or off-hook and touch tones for the remaining 23 channels (also called DS0s). E1 CCS uses 2 of the 32 channels for carrying the same type of signaling information. See also *CAS, T1,* and *H.323* for IP telephony signaling information.

Common, Electrical   Not to be confused with ground. *Common* is a reference point, and is ungrounded. It is usually a signal return or DC reference coupling for transmission circuits.

Common, Return   See *Common, Electrical.*

Communications Controller   1. Also called a *front-end processor* or *proxy server.* It has the capability of receiving multiple data communications transmissions of different protocols and converting all of those protocols to one common protocol, then routing the data to its destination. You could think of a communications controller as a language translator. Many newer routers incorporate communications controller functions.   2. A node in an *SNA (IBM System Network Architecture)* network that coordinates the flow of data payload and overhead to the mainframe, which controls all communications in an SNA network (Fig. C.35).

Communications Protocol   The method that a communications circuit or link exchanges information. If you are a customer, there isn't a whole lot to worry about with protocol. Just imagine that protocol is just another word for "dialect." If one piece of gear "speaks Chinese," don't try to hook it up to a piece of gear that "speaks English." Some examples of communications protocols are ISDN, frame relay, V.32, and Ethernet. They are all very different.

Communications Satellite   Usually a geostationary piece of electronic equipment that stays in the same location, relative to the earth's

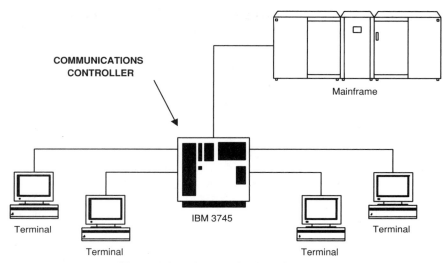

Figure C.35   Communications Controller

surface, but at a distance or "altitude" of 22,000 miles. They are used to overcome the curvature of the earth for radio-transmission applications.

Communications Server   Also called an *access server,* it is a communications-management and protocol-conversion device. A communications processor that connects asynchronous (independently timed) and synchronous (common timing source) devices to a LAN or WAN through network/terminal emulation software that resides in its memory. Older communications servers used to only handle asynchronous protocols, such as IP or Novell IPX. Modern communications servers perform the same functions as access servers, supporting synchronous (common timing source) and asynchronous protocol routing, and can even provide certain levels of security.

Communications Workers of America   The organized labor union of the Regional Bell Operating companies. This organization negotiates the wage scales listed in CWA contract handbook.

Community-Antenna Television (CATV, Cable TV)   See *Cable TV.*

Compact Disc (CD)   See *CD.*

Compact Disc Erasable (CD-E)   The original name for *CD-RW (Compact Disc Re-Writeable)* technology when it was developed in 1995. By the time the product was offered to consumers in 1997, its name had evolved to *CD-RW.* CD-RWs are capable of being erased and rerecorded 10,000 times.

Compact Disc Interactive (CD-I)   A multimedia education and entertainment system, developed by Philips and Sony, consisting of a proprietary compact disc player that connects to a television set. The CD-I system incorporates *CD-RTOS (Compact-Disc Real-Time Operating System),* which synchronizes interactive video, audio, and text with real time.

Compact Disc Recordable (CD-R)   A compact-disc standard format that, when used with a CD-R or CD-RW drive, can record up to 650 MB of unerasable data one time. The advantage of CD-R discs is that they record slightly faster, and cannot be erased or modified during subsequent uses (although they can be copied to CD-RW discs). CD-Rs can also be played in regular CD-ROM drives, and audio compact-disc drives.

Compact Disc Rewriteable (CD-RW)   A compact-disc drive that can write, read, or erase discs made for this purpose (CD-RW discs). CD-RW has a storage capacity of 650 MB, and can be recorded and erased 10,000 times. Discs that are recorded with the MultiRead standard can also be used in CD-ROM drives and compact-disc audio drives.

Compact-Disc ROM Extended Architecture (CD-ROM XA)   A version of the CD-ROM that was released to consumers in 1991 that enabled software, audio, and video to be interleaved on the recorded tracks of the CD-ROM disc. To utilize this format, a regular CD-ROM can be used, but an XA controller card needs to be installed in the PC.

Companding (Compression-Expansion)   A pulse-code modulation technique that takes small samples of an analog signal the same way that is done in *PAM (Pulse Amplitude Modulation),* except the resultant PAM signal is converted to binary. Companding compresses the binary signal using a mathematical algorithm. This allows more analog channels within the same network bandwidth. Companding is used over *PCS (Personal Communications Service)* cellular radio networks. See also *Mu Law* and *A Law.*

Complementary Code Keying (CCK)   In wireless LAN radio, a modulation technique that utilizes a complex set of mathematical functions known as *complementary codes* to transfer more data over a link. CCK is less affected by multipath distortion than other modulation methods such as QPSK (Quadrature Phase Shift Keying) and BPSK (Binary Phase Shift Keying).

Complementary Metal-Oxide Semiconductor (CMOS)   See *Complementary Metal-Oxide Semiconductor.*

Complementary Network Services   An additional service that can be added to your telephone line. Voice mail is one of these. The service can be provided by the local exchange carrier providing the dial tone or it can be provided by a complimentary network services provider.

Completed Call   A call that is connected to its destination. When someone calls a number and the other end is picked up by someone, the call is completed. You would think that a call would be completed when the people were finished talking, but in regard to call routing and switching, that is not the case.

Complex Instruction Set Computing (CISC)   Pronounced "Sisk." A software and hardware architecture method that utilizes a computer's CPU, rather than the software, for the majority of the workload. CISC processors support up to 200 instructions. Most personal computers are made with CISC processors. See also *RISC.*

Component Video   A video signal transmission and/or reception that consists of separate units for each function. In broadcast television, the video is AM, the audio is FM, and color is PM. They are all transmitted together on a single carrier, but processed and decoded separately.

Composite Signal   The whole signal, overhead and payload included. The composite speed of a T1 signal is 1.544 Mb/s.

Compressed Serial Link Internet Protocol (CSLIP)   An extension of serial link Internet protocol that, when appropriate, allows only header information to be sent across a serial Internet connection, reducing overhead and increasing packet throughput. CSLIP and SLIP are early Internet dial-up transport methods for IP. They were made obsolete by *PPP (Point-to-Point Protocol),* which allows for encryption and the ability to support Microsoft Net Beui and Novell IPX in a dial-up environment.

Computer Telephony Integration (CTI)   A wave of products that have been coming out since 1994. CTI applications interface computers with telephone systems, *IVR (Interactive Voice Response)* systems, voice-mail systems, call-accounting systems, and anything else that is telecommunications oriented. A good example of CTI is an OCTEL product that allows users to click and drag voice-mail messages from their phone to any other telephone they would like to get the message. It also provides diagnostic functions, traffic analysis functions, and administration functions through a *GUI (Graphical User Interface,* such as Windows 95).

Concatenation   To join two or more ATM channels, ISDN B channels, or T1 DS0 channels together to make a larger single channel that can carry a broader signal. In the T1's case, concatenation is usually called *fractional T1*.

Concentrator   A device that performs multiplexing and routing functions.

Condenser   An obsolete term for *capacitor.* See *Capacitor.*

Condenser Microphone   A microphone that incorporates a capacitor that changes its capacitance as sound waves strike it.

Conditional Transfer   A call-forwarding feature offered by telecommunications companies. Incoming traffic is call forwarded to different target parties, depending on the time of day or which parties are available to receive calls.

Conditioned Circuit   A twisted copper pair within an outside plant network that is modified to carry a digital data signal instead of an analog voice signal. Conditioned circuits have noise-filtering electronics components attached to them instead of load coils.

Conditioning   A term that refers to modifying a twisted copper pair in an outside plant network so that it can carry a digital data signal instead of an analog voice signal. Twisted-pair circuits (a circuit is a local loop, which is the pair that connects the central office to the customer) are conditioned by adding noise-filtering electronic components to them.

Conductance   The mathematical inverse of resistance. The unit of conductance is Siemons or mhos, which is ohm (the measure of resistance) spelled backwards. To calculate the conductance of a circuit or device, just figure $1/R$ (in ohms).

For example: Convert 500 ohms to a value of conductance.

1/500 ohms = 0.002 siemons.

Conductor   An element or chemical that allows electrical current to flow through it easily under the influence of electric forces. The following list of conductors is in order of conductive properties. Sometimes the quality of a conductor is not only viewed by its ability to conduct electricity but its ability to resist corrosion as well (Fig. C.36).

Most conductive

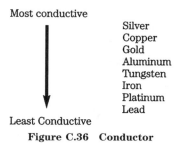

Silver
Copper
Gold
Aluminum
Tungsten
Iron
Platinum
Lead

Least Conductive

Figure C.36   Conductor

Conduit   Conduit is another name for tubing, of which there are many different kinds.

For outside plant applications, 4" PVC conduit is usually used for linking cable vaults with an easy means to add, remove, and upgrade cable. If fiber-optic cable is installed in a 4" PVC conduit, four innerducts of different color are pulled into the PVC conduit first. If conduit is being buried specifically for fiber-optic cable, then a quad-lock is usually used. Quad-lock is four 1" conduits braced together.

For inside plant applications, 4" EMT conduit is normally used. EMT is the standard metal conduit that electricians use. PVC conduit is not allowed in modern buildings because when it burns, it emits very toxic chlorine gas.

Conference Bridge   A service where the number of calls or people connected to a conference call are controlled by a single source. Most conference bridge services have everyone that wants to be on a conference call dial a toll-free number, where a conference attendant will answer their call, then connect them through (or bridge them) to the other conference callers.

Conference Call   A communications connection where three or more different telephone lines (with different phone numbers or extensions) are connected together.

Conference Europeenne des Postes et des Telecommunications (CEPT)   The association of 26 European *PTTs (Post, Telephone, and Telegraph)* that resolves interconnect issues between countries and recommends communication specifications to the ITU-T standards committee.

Confidencer   A device that connects to a telephone handset to block out background noise. Some confidencers are in the form of an interchangeable mouthpiece that replaces the existing mouthpiece on your handset. Walker Electronics Company is a well-known manufacturer of confidencers.

Congestion Control   In LAN switching, ports can be overloaded with devices and/or traffic. Methods used to manage high traffic conditions are called *back-pressure* and *aggressive back-off algorithm*.

Connect Time   The duration that a call path through a switch or network is set up. Simply put, how long your phone call lasts.

Connection Oriented   A protocol model of interconnection that has three phases: connection, transfer of data, and disconnect. Some connection-oriented protocols are X.25, TCP, and a regular telephone call. Many protocols are a mixture of connection/connectionless, such as '
ATM, TCP/IP, and frame relay.

Connection-Oriented Network Protocol (CONP)   A method of communications where the sender and receiver of data are communicating directly via signals sent, such as *RTS (Request to Send)* and *ACK (Acknowledge)*. The contrast to this communications method is the connectionless oriented protocol, which is a method of communications where the sender and receiver do not signal each other directly. Connectionless operation forwards data across a network using an address as routing information for the network. Both provide lower-layer software connectivity to upper-layer protocols.

Connectionless   A reference to packet or cell-based communications protocols that do not require a connection and direct signaling to exchange messages (for example, Ethernet and frame relay). Connectionless communications enables multiple hosts/users to simultaneously share a communications channel (or many channels) through a network. A nonconnectionless method or connection-oriented method is a modem that establishes a connection through a handshake. Connectionless protocols are also referred to as *best effort, packet,* or *cell protocols.*

Connectionless Network Protocol (CLNP)   An *OSI (Open Systems Interconnect)* network layer protocol that does not require a circuit to be established before data is transmitted. The basis of a connectionless protocol is that there is no direct *RTS (Request to Send)* or *ACK (Acknowledge)* type of signaling, thus, no direct connection. Some protocol stacks are a mixture of connection oriented and connectionless protocols, such as TCP/IP, where TCP is connection oriented, and IP is connectionless. See also *CLNS.*

Connectionless Network Service (CLNS) An *OSI (Open Systems Interconnect)* network layer protocol that does not require a circuit to be established before data is transmitted. CLNS works as a part of or with other protocols that reside in the network layer and other layers. All packets are sent independently of each other. See also *CLNP.*

Connections per Circuit Hour (CCH) The number of connections or calls completed at a switching point per hour.

Connector, Genderless Sometimes called a *hermaphroditic connector.* A genderless connector developed by IBM that is usually called a *Data Connector.* The Data Connector does not need complementary plugs (male and female) to make a connection, like all other known communications modular connecting systems. The Data Connector is specifically designed and used for switched token-ring backbone applications. For a photo, see *Data Connector.*

CONP (Connection-Oriented Network Protocol) A method of communications where the sender and receiver of data are communicating directly via signals sent, such as *RTS (Request to Send)* and *ACK (Acknowledge).* The contrast to this communications method is the connectionless oriented protocol, which is a method of communications where the sender and receiver do not signal each other directly. Connectionless operation forwards data across a network using an address as routing information for the network. Both provide lower-layer software connectivity to upper-layer protocols.

Conservation of Radiance A scientific law that basically says that you cannot amplify or increase light without a light-creating source. So, optical fiber does not make the light brighter as the light travels through it and neither do your sunglasses. It would be cool to have glasses that actually made the night brighter, with no electronics, just the lenses. Conservation of radiance simply states that this is impossible.

Console The large telephone with all the keys and/or buttons on it. It is the traffic-control center for a PBX system. The PBX operator or attendant usually has a console. The two types of consoles designed for two different applications are the hotel PBX operator console and the business PBX operator console.

Constant Bit Rate (CBR) A *Quality of Service (QOS)* defined by the ATM Forum for ATM Networks that provides precise clocking to ensure undistorted delivery. A connection would be rightfully commissioned as a CBR connection if it carried such time-sensitive data as voice, video, or real-time computing information. The CBR quality of service guarantees freedom from cell loss and delay. Other QOSs defined by the ATM forum for ATM connections include *UBR (Unspecified Bit Rate), ABR (Available Bit Rate),* and *VBR (Variable Bit Rate).* See also *AAL.*

Content Networking   A strategy of making network resources more efficient by providing them with the ability to transfer data/files from the core of the network to the edge (other side of the WAN) before it is needed by mass users. The strategy behind content networking implementations is to reduce the need for additional bandwidth by better utilizing the existing bandwidth. This is useful in both internet providers as well as WAN environments. Content networking also provides packet direction according to layer seven sessions (also called *server load balancing* or *server synchronization*). There are two primary add-ons to a content networked system. The first is a content cache engine on the edge of the network. Being closer to the end user, it consumes less network resources as well as provides a better response time for the end user. The second device is a content network server, which controls what data files are pushed out to the content cache engine. Content networking will be a crucial part of networking when network service providers begin offering bandwidth intensive services such as on-demand video. See also *Flash Crowd* and *Boomerang Server.*

Contention   A type of LAN control scheme used by Ethernet, where all the users (PCs) on the network fight for use of the network. The PCs check to see if the network is clear, then transmit. Often more than one PC tries to transmit at once, which causes the data to be garbled (this is called a *collision*). This network-control scheme is called *contention.* I guess you could say that the PCs contend to see which one gets their data on the network first. The best alternative to this type of network is token ring—a completely different method of network control.

Context Keys   Buttons on a phone or other device that have a display adjacent to them. They perform different functions, depending on what the display is showing at the time you push that button.

Contiguous Slotting   The banding together of two or more adjacent channels in a T1 to get one larger channel. Also called *fractional T1* and *concatenation.*

Control Plane   one of the three entities of frame-relay network management. The three planes are: The *User Plane* (the U Plane defines the transfer of information), the *Management Plane* (the M Plane defines the LMI, Local Management Interface), and the *Control Plane* (the C Plane is delegated for signaling and switched virtual circuits).

Control Point   In IBM *SNA (System Network Architecture)* networks, an element that identifies the APPN networking components of a PU 2.1 node, manages device resources, and provides services to other devices. In APPN, CPs are able to communicate with logically adjacent CPs by way of CP-to-CP sessions. See also *EN* and *NN*.

Control Signal   A signal sent as a bit, byte, or tone that prompts a communicating device to do something. When you pick up the handset on your telephone, the switch-hook pops up and makes a 1000-ohm connection across the two telephone wires that go to the phone company. This causes a current flow out of the central office switch. This control signal shows that you are off hook and would like to dial a number. The central office responds to your "off-hook" or current flow signal by sending a dial tone, which prompts you to dial digits. The digits are a control signal that tells the central office switch where to route your call. These are all control signals, and all control signals (regardless of the protocol) are equally as systematic and organized.

Controlled-Environment Vault   A vault that is designed to have electronics in it. The environment inside the vault must be kept at a certain temperature and humidity.

Controller   In telephone equipment, *controller* is another name for CPU. In data communications, such as LANs, a controller controls data transfer between two devices.

Convection Cooling   A method that newer telephone electronics uses to cool itself. Rather than having a cooling fan attached to a device to cool it, it is equipped with metal deflectors. The deflectors channel warmer air that is rising out of the top of the equipment, which, in turn, pulls cooler air in through the bottom. With convection-cooled equipment, you still need to have a system that cools and controls the humidity in the room that the equipment is in. This cooling option is available with the Tellabs Titan 5500 Digital cross-connect system, for example.

Convergence   The point in time when all routers within an internetwork have completed the process of sharing and updating all of their routing tables so that they all match. Different routing protocols take different amounts of time to converge.

Cookie Sticky   (Slang) A reference to Internet applications or servers that use cookies (which are ways to track where users have been on the Internet) from a user's computer and remembers what the users did. Some network applications enable users that have "purchase cookies" or "golden cookies" to get priority bandwidth and processor time because they are considered preferred surfers.

Coprocessor   An additional microprocessor chip that is added to the bus architecture of a personal computer. The coprocessor is an extension of the main CPU's instruction set and is generally used for special-purpose operations, such as advanced mathematical calculations.

Cord, Base   The cord that goes between your telephone and the wall. It has RJ6x-type plugs on the ends. Base cords are available in 2-conductor (RJ6x2c plugs on the ends), 4-conductor (RJ6x4c plugs on the ends), 6-conductor (RJ6x6c plugs on the ends), and 8-conductor (RJ8x8c plugs on the ends). RJ4x4c means an RJ modular-type plug that is 4-conductor positions wide with four conductors installed.

Cord, Handset   The cord that goes between your phone and the handset. Also known as a *curly cord.* It has RJ4x4c (an RJ modular-type plug that is four conductor positions wide, with four conductors installed).

Core Layer   One of the three LAN network design layers. In LAN network design, the three switch layers are core layer, distribution layer, and access layer (Fig. C.37). The core layer provides redundancy for

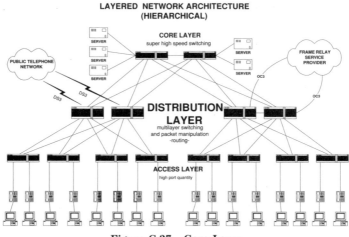

Figure C.37   Core Layer

distributing traffic across multiple access layers (it is almost always desirable to have two switches within a core). It also provides central access for servers (and/or mainframe and AS400 computers). The access layer provides high-speed switching and routing among a group of switches networked by trunks. The access layer provides switches where users connect, so high port quantity is desired in this layer.

Core Network   Another reference to the local telephone network. It is a combination of switching offices and transmission facilities connecting local central switching offices together.

Core Processing Unit   Some communications equipment manufacturers call the card or shelf that controls a communications system (e.g., PBX) or portion of the system the *CPU* (Fig. C.38). This is because they include all of the RAM, subprocessors, buffers, clocking circuitry, and ROM in this part of the system.

Figure C.38   Northern Telecom Option 81 PBX CPU (Core) Shelf

Core Router   In a packet-switched star topology, a device that is a part of the backbone communications link and serves as the gateway for any communications that come from its local peripheral network to other peripheral networks.

Core Switch   An Ethernet LAN switch located within the core of a network design (Fig. C.39). Core switches are generally high speed, with multilayer and quality of service (802.1p and 802.1Q) capabilities. See also *Core Layer.*

Figure C.39   Core Switch

COS (Class of Service)   A type of telephone service or telephone line purchased from a telephone company. See *Class of Service.*

Co-set   A number used to prevent an all zero result from an ATM Header Error Check calculation. The co-set is crucial in locating where cells begin and end in ATM. The co-set is a predefined eight-bit pattern, consisting of both ones and zeros, that is compared as an exclusive OR function to all ATM cyclic redundancy check header values. The co-set prevents a zero-multiplied zero event when identifying ATM cell boundaries and error corrections.

Coulomb   A unit of electrostatic charge equal to 6,300,000,000,000, 000,000 electrons or protons (electrons would be a $-1C$ and protons would be a $+1C$). This unit of electric charge was established by Charles Augustin Coulomb (1736–1806) and is useful because with it we can determine a standard of measurement for electric force, which led to the measurement of electrical charge flow (current), known as the *ampere.* One ampere is equal to one coulomb of charge (the number of electrons) flowing past a point in one second.

Counter Rotating Ring   A backbone network architecture that is common to *FDDI (Fiber-Distributed Data Interface), SONET (Synchron-ous Optical Network), DQDB (Distributed Queue Dual Bus),* switched token ring, and *CDDI (Copper-Distributed Data Interface).*

Coupling   A means of connecting one adjacent circuit to the next. Different types of coupling include: capacitive, inductive, electromagnetic (radio), optical, and direct (hard wire).

Country Code   A code used in international dialing for countries that are not a part of the *North American Number Plan* (*NANP*). To dial international long distance from the United States, dial:

011 + county code + city code + number.

For a listing of country codes, see *Appendix B*.

To dial the United States from another country that is a part of the NANP, simply dial the area code the same way you would call long distance to another state. To call the United States from another country that is not a part of the NANP, consult your long-distance company. The United States has a different country code/access code for almost every country that is not a part of the NANP.

Coverage Area   The geographical area that is serviced by a cellular or PCS telephone system. Within this area, subscribers can access a cellular or PCS radio signal link and make calls. If the subscriber travels outside of this area, the "no service" or "roam" indicator appears on the phone's display. If the roam indicator is on, the subscriber still has a signal and can make calls, they are just within another cellular company's coverage area and the call will be more expensive. If the "no service" indicator is on, no signal is present and no phone call can be made.

CP (Control Point)   In SNA (IBM) networks, an element that identifies the APPN networking components of a PU 2.1 node, manages device resources, and provides services to other devices. In APPN, CPs are able to communicate with logically adjacent CPs by way of CP-to-CP sessions. See also *EN* and *NN*.

CPE (Customer-Premises Equipment)   The equipment that is connected to a phone line. The exact definition is anything beyond the Standard Network Interface, which includes wire, jacks, telephones, answering machines, and any other devices connected to the telephone line.

CPS (Cycles or Characters Per Second)   See *Cycle*.

CPU (Central Processing Unit)   1. The device within a computer (or switch or other machine that performs complex tasks) that controls the transfer of the individual instructions from one device connected to its bus (the data or I/O bus) to another, such as ROM, RAM, subcontrollers, decoders, and I/O ports.   2. *Core Processing Unit.* Some communications equipment manufacturers call the card that controls a communications system (e.g., PBX) or portion of the system the *CPU.* This is because they include all of the RAM, subprocessors, buffers, clocking circuitry, and ROM in this part of the system (Fig. C.40).

Figure C.40   Slot 1 CPU (Central Processing Unit) for a Personal Computer

CR   1. The ASCII control code abbreviation for carriage return. The binary code is 1101000 and the hex is D0.   2. *Call Request.* A packet-level (data link layer) control signal sent from a DTE into an X.25 DCE device to initiate a call. The DCE then creates a Call Confirm packet and sends it in response. This initiates a call through a multiplexed channel. For a diagram of the call-request packet structure, see *X.25 Call Request Packet.*

Craft   A reference to nonmanagement RBOC personnel. Many craft personnel are members of the Communications Workers of America (CWA) labor union.

Craft Test Set   Also called a *Goat* or *Butt-Set* (Fig. C.41). A test telephone that is used by technicians to test analog telephone lines (ringing, dial tone, monitor, etc.).

CRC (Cyclic Redundancy Check)   A method used by modems and other transmission devices to verify the accurate transfer of bits.   CRC is used in connection-oriented transmissions and functions with a retransmit feature when bits received are corrupted.

Crimp Tool   Tools used to place connectors on the ends of different types of coax and twisted pair (Fig. C.42).

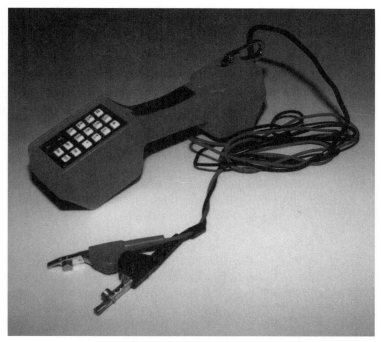

Figure C.41   Harris Dracon Craft Test Set (Also Known as a "Butt Set" or "Goat")

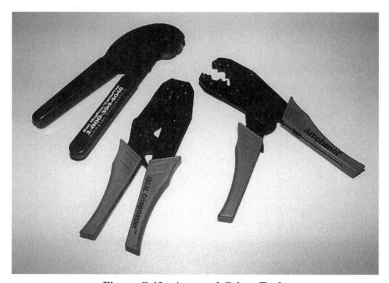

Figure C.42   Assorted Crimp Tools

Cross Bar   An obsolete telephone switch that was analog and had mechanical relays that connected telephone calls. This is where the term *switch* comes from. Old central office switches contained literally thousands of mechanical switches.

Cross Connect   A cross connect is the connection of one circuit path to another via a physical wire. Telephone cable pairs are terminated or "punched down" onto a termination block (usually a 66m150 or an AT&T 110 block) that has extra connections available for each pair so that jumper wires can be easily connected and rearranged between them.

Cross-Over Cable   A connector cable where two or more of the conductors reverse themselves from one end to the other. A null modem cable is a type of cross-over cable.

Cross Talk   The two forms of cross talk are inductive cross talk, because of wires touching (or conducting to) each other that shouldn't be and channel seepage because of inaccuracies in multiplexing equipment timing or components. Inductive cross talk is caused by radio or the use of nontwisted-pair wire. Inductive cross talk travels from one device to another via electromagnetic fields generated by different transmissions. A hard cross or physical cross is most commonly caused by water seeping into a telephone cable and conducting the electric signals on pairs in the cable to each other. Channel seepage is usually a situation where the people on one phone call can distantly hear one side (only one person's voice, not both) of another phone call.

CRT (Cathode-Ray Tube)   The real name for a TV or monitor screen.

Crystal Oscillator   An electronic device that is made from a thin piece of polished quartz crystal (Fig. C.43). When a periodic voltage is applied to a crystal, it has a piezoelectric reaction. This means that the voltage applied to the crystal distorts it. When the voltage is removed, the crystal physically vibrates. With each vibration of the crystal, a very small AC voltage cycle is produced. The physical size of the crystal determines its oscillating frequency. Crystal oscillators are used because of their reliable timing.

CSDC (Circuit Switched Digital Capability)   A 56 Kb/s phone line that can carry voice or data. The conditioning equipment is equipped with a digital-to-analog converter and vice versa.

CSLIP (Compressed Serial Link Internet Protocol)   An extension of serial link Internet protocol that, when appropriate, allows only header

Figure C.43   Crystal Oscillator

information to be sent across a serial Internet connection, reducing over-head and increasing packet throughput. CSLIP and SLIP are early Internet dial-up transport methods for IP. They were made obsolete by *PPP (Point-to-Point Protocol)*, which allows for encryption and the ability to support Microsoft Net Beui and Novell IPX in a dial-up environment.

CSMA (Carrier Sense Multiple Access)   See *Carrier Sense Multiple Access.*

CSMA/CA (Carrier Sense Multiple Access/Collision Avoidance) See *Carrier Sense Multiple Access/Collision Avoidance.*

CSMA/CD (Carrier Sense Multiple Access/Collision Detection) See *Carrier Sense Multiple Access/Collision Detection.*

CSU (Channel Service Unit)   Also called a *CSU/DSU (Channel Service Unit/Data Service Unit).* A CSU is a hardware device that can come in many shapes and sizes. Rack-mount, shelf-mount, and stand-alone CSUs are available (Fig. C.44). A CSU/DSU has three main functions. The first function is to act as a demarcation point for a T1 (DS1) service from a local communications company. The second function is to provide line format and line-code conversion (B8ZS to AMI, SF or D4 to ESF, 135 V to 0 V) between the public network and the customer's equipment, if necessary. The third function is to provide maintenance or alarm services and loop-back for isolating problems with the T1 line or customer's equipment. Some loop backs can be done remotely, depending on the model CSU/DSU. Some are done with bantam loop plugs.

Figure C.44   CSU (Channel Service Unit)

CTI (Computer Telephony Integration)   A wave of products that have been available since 1994. CTI applications interface computers with telephone systems, IVR systems, voice-mail systems, call-accounting systems, and anything else that is telecommunications oriented. A good example of CTI is an OCTEL product that allows users to click and drag voice-mail messages from their phone to any other telephone they would like to get the message. It also provides diagnostic functions, traffic analysis functions, and administration functions through a *GUI (Graphical User Interface,* such as Windows 95).

CTS (Clear To Send)   A frame-layer DTE request in X.25. The signal that a modem sends after it receives an *RTS (Request To Send)* signal. Upon sending the CTS, the modem is ready to receive data. This process happens after the handshake and communication parameters, such as transfer rate and protocol have been negotiated. For more details on X.25 frame-layer communication, see *X.25 Control Field.*

Current   The flow of electricity measured in amperes.

Current, Line   The average off-hook current of a telephone line is about 35 mA (0.035 A).

Customer-Premises Equipment (CPE)   Equipment that is connected to a phone line. The exact definition is anything beyond the standard network interface, which includes wire, jacks, telephones, answering machines, and any other devices connected to the telephone line.

Cut Over   The actual changing use of one type of equipment or system to another. If you install a new PBX in your office, the cut over is when you disconnect your old system and begin using a new one.

Cut-Through Packet Switching   There are three ways that frames/packets transverse through a LAN switch, bridge, or router. The first is *store and forward*. In Store and Forward, the entire frame and its contents are accepted and stored in the switch. Error detection is calculated (CRC) and if the frame is good, the address is looked up in the routing table. When the associated destination port/segment is found, the frame is sent on to its destination. This is a good method for routing traffic because damaged frames, runt frames, and giant frames are discarded before they are transmitted. This method is used where the network infrastructure or media is prone to damaging frames, such as RFI environments or a poor WAN network service. The disadvantage of the Store and Forward method is *latency*. Storing the entire frame while the destination port is retrieved causes a delay, and in multiple hop networks, this can cause slow network performance even when there is very little traffic.

The second way that frames/packets transverse through a LAN switch/router/bridge is the *cut-through switching* method. In cut-through switching, only the address of an incoming frame is processed by memory. The address is associated with its destination port/segment in the routing table, and the entire frame sent directly through. This process happens if the frame is good or not, as long as there is a non-damaged address in the frame. Cut-through switching greatly reduces latency delays through a network, but still transmits bad frames. If an NIC card or host device begins sending lots of bad erroneous frames, the network performance could be slowed greatly. Some LAN switches have safeguards in place to detect and suppress error storms from defective equipment.

The third method of forwarding frames is *modified cut-through*. This method works similar to store and forward, except that it uses a limited number of bytes to check for errors rather than the entire frame. This method helps prevent the retransmission of defective frames and also provides an acceptable level of latency delay through the network.

CVSD (Continuously Variable Slope Delta Modulation)   A method of converting analog voice into digital and vice versa with an on-the-fly variable sampling rate that ranges from 16 Kb/s to 64 Kb/s, depending on how much bandwidth is available.

CWA (Communications Workers of America) See *Communications Workers of America.*

Cycle One cycle of a waveform's pattern (Fig. C.45). Cycles are used as a reference to measure the frequency of a waveform or signal. In the following diagram, two waveforms and one cycle of each are singled out. *Cycles* are usually referred to as a number of cycles per unit of time. *Cycles per second* and *hertz* are measurements of the number of cycles you get per second in an analog transmission. *Bits per second* is a measurement of how many "square-wave" clock sample sequences are being read from a digital transmission. *Frequency* is a measure of the number of cycles per second (in Hz). One Hz (hertz) is equal to one cycle in one second. Two Hz is two cycles in one second, 1000 Hz is 1000 cycles in one second.

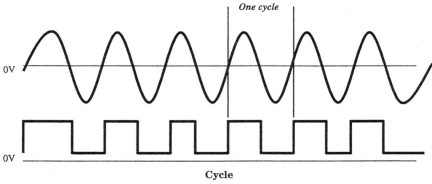

Figure C.45 Cycle

Cyclic Redundancy Check (CRC) A method used by modems and other transmission devices to verify the accurate transfer of bits. CRC is used in connection-oriented transmissions and functions with a re-transmit feature when bits received are corrupted.

D Amps   *Digital Advanced Mobile Phone Service.* See also *Amps.*

D Bank   Another term for a *Channel Bank.* A device that demultiplexes (breaks down) a T1 circuit to its 24 channels.

D Bit   A control bit in an X.25 data packet that requests an acknowledgement from the far-end device on an X.25 network. Acknowledgements (called *RR Receive Ready* in X.25) originate from the nearest DCE device. Because most X.25 networks have multiple handoffs or "hops," it is comforting to be able to get a true acknowledgement from the very far end (technically called the *end-point DTE*). The D bit is the signal in X.25 that requests this type of acknowledgement. It is usually requested when "collect calls" are made. For more information about the data packet header structure in X.25, see *X.25 Data Packet.*

D Channel (Data Channel)   The name of an ISDN out-of-band signaling channel. The two kinds of D channels depend on which ISDN circuit you have. If you have a *BRI (Basic Rate Interface)*, the D channel is 16Kb/s and controls two B (Bearer) channels. If you have a *PRI (Primary Rate Interface)*, then the D channel is 64Kb/s and controls 23 B channels. Both D channels carry the same information and perform the same function. A BRI circuit requires one pair (two wires for transmission) and a PRI requires two pairs for transmission. For a diagram that compares the different types of ISDN circuits, see *Integrated Services Digital Network.*

D Connector   A 25-pin miniversion of the DB25 Connector. See also *mini-connector.*

D4 Channel Bank   A Lucent digital transmission/transport product that is T1 based (Fig. D.1). The D4 channel bank, which is useful for central office intertrunking, is being replaced with SONET transport equipment throughout North America.

Figure D.1   D4 Channel Bank

D4 Framing   More frequently called the *super frame format* for T1 transmissions. The super frame format consists of 12 standard 193-bit T1 frames. The D4 framing format also incorporates in-band signaling.

DAC (Digital to Analog Converter)   A part of a channel bank that encodes analog voice signals into a stream of binary digits. The digital to analog converter or analog to digital converter samples a caller's voice at a rate of 8000 times per second. (The sample rate for a T1 channel is 8000 times per second.) Each sample's voltage level is measured and converted to one of 256 possible sample levels. These levels are from the lowest, 0000000, to the highest, 11111111. The reason for 256 levels is because if you count in binary from 00000000 to 11111111, you end up with 256, the highest number possible with 8 bits. The bits are then transmitted one after another at a high rate of speed to their destination, where the same process happens in reverse. For a diagram, see *Digital to Analog Converter.*

DACS (Digital-Access Cross-Connect System)   A DACS is also called a *DCS (Digital Cross-Connect System),* depending on the manufacturer. A digital cross-connect system is a fundamental part of a local and long-distance company's network. The DACS or DCS is a rack-mountable system that enables any circuit that interfaces with it to be electronically cross-connected from one path to another within the network it is connected to. Circuits that can interface with a digital cross-connect system include DS0, DS1, DS3 (or T3), STS-1, and SONET OC-1. An incoming circuit can be rerouted by simply making path changes in DACS administrative software. In the following, a T1 circuit coming into the input side (left side) could be cross-connected to exit as one of the channels in a DS3 on the right side. For a diagram of a DACS/DCS, see *Digita Cross-Connect System.*

Daisy Chain   A method of connecting devices in a string. The bus topology for Ethernet is an example of a daisy chain. For a diagram of a daisy chain, better known as a *bus topology,* see *Bus Topology.*

DAL (Dedicated Access Line)   A private circuit that provides a direct connection (or access) to a long-distance carrier or other communications service, like frame relay or an Internet service provider. Some DALs are a "full-service circuit," which means that if you have a circuit that connects you directly to your Internet service provider, then the only bill you see for that service is from the Internet service provider. The local circuit is in the Internet service provider's name and they pay the phone bill for that service. You, of course, pay a single bill for the entire service. If you are going to get a direct Internet connection, this is the way to go. If it ever stops working, you just call the Internet service provider and they determine where the problem is and fix it.

Dark Current   In a photodiode, the dark current is the flow of electricity through the diode when no light is present. Photodiodes are used as light-sensitive switches. When they are exposed to light, they act like a switch and turn on. However, even if no light is present, a small amount of electricity still flows through. This is the dark current (Fig. D.2).

Figure D.2   Dark Current: The Schematic Symbol for a Photodiode

Dark Fiber   A fiber-optic cable that is installed in a telephone company's outside plant network, but has no electronics connected to it. Sometimes customers like to lease or buy rights to use dark fiber, and connect their own electronics to it in a point-to-point or ring application.

DASD (Direct-Access Storage Device, "Dazzdee")   A technology that incorporates a memory-retrieval method where a disk drive or RAM can retrieve or save data directly to a specific address without having to scan through addresses in the medium to find it. CD-ROM is an example of a storage device that incorporates DASD technology.

DAT (Digital Audio Tape)   A high-fidelity digital storage media for audio applications. DAT tapes are only usable on DAT tape recorder/players. DAT machines record music in the same manner as Compact Disks. For each channel, (left and right) they convert analog sound or music to a 16-bit sample for every 20 microseconds of sound (a rate of about 48,000 samples per second, which is 12 times the accuracy of a T1).

Data   In the communications industry, data is anything that is transmitted or processed digitally. The only thing that is not data anymore is a *POTS (Plain Old Telephone Service)* or analog line. A T1 is a digital circuit. The channels carry voices, computer transmissions, and video in a digitized data format.

Data Burst   A short transmission of data that is not timing sensitive, such as a transmission for a credit-card authorization or an hourly data download. Audio or video is timing-sensitive data.

Data Bus   A two-way connecting scheme of 8, 16, 32, or 64 wires or conductors that connect a microprocessor (CPU) to RAM, ROM, and I/O devices (Fig. D.3). A computing or controlling device needs to have at least three bus systems, a data bus, address bus, and control bus. The data bus provides for transfer of data between the microprocessor, the ROM and RAM, and the I/O devices. The address bus works in only one direction (from the microprocessor to the other devices) and allows the microprocessor to control memory addressing and retrieval. The control bus allows the microprocessor to control data flow and timing for all the various components.

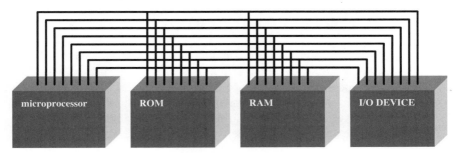

Figure D.3   8-Bit Data Bus

Data Circuit Terminating Equipment (DTE, Data Terminating Equipment)   DTE equipment receives a communications signal. For a data connection to work between I/O (input output) devices, one needs to be designated the "communications sending" equipment and one the "communications terminating" or DTE. A computer's printer port is a DCE port; a printer is a DTE device. A practical way to classify the two is: DCE is the sender of data and the DTE is the receiver of data.

Data Circuit Transparency   A circuit's ability to carry data without any apparent change or restructuring of protocol. In reality, when data is transmitted through the public network, it gets loaded into other protocols, but on the other end of the transmission, the data is received exactly the way it was sent. A clear-channel T1 has data-circuit transparency.

Data Communications   To transmit information encoded in binary. Data communications as a whole has many technologies to accomplish this. These techniques are called *protocols*. Some protocols include ADSL, ISDN, Ethernet, token ring, SONET, switched 56Kb/s, and frame relay.

Data Communications Equipment (DCE)   DCE is the equipment that provides the source communications signal. For a data connection to work between I/O (input output) devices, one needs to be designated the "communications sending" equipment and one the "communications terminating" or *DTE (Data Terminating Equipment)*. A computer's printer port is a DCE port; a printer is a DTE device. A practical way to classify the two is: DCE is the sender of data and the DTE is the receiver of data.

Data Compression   A method of obtaining higher speeds of data transfer with the same number of bits being transmitted per second. Before the data is transmitted, it is encoded one or more steps beyond the original bit stream.

Data Concentrator   Another name for a *multiplexer*. A multiplexer encodes data of many channels to be transmitted on one channel.

Data Connector   A genderless connector developed by IBM (Fig. D.4). This connector is also called a *hemaphroditic connector*. The Data Connector does not need complementary plugs (male and female) to make a connection, like all other known communications modular connecting systems. The Data Connector is specifically designed for switched token-ring backbone networks.

Figure D.4   Data Connector

Data Conversion   The process of converting data from one protocol to another. Many devices are made to accomplish the transformation of data in one form to another.

Data Encapsulation The method used to transfer application/user data from one layer of the OSI/SNA/DNA protocol stack to the next. At the upper three layers, units of encapsulated data are referred to as *packet data units.* At the transport layer, units of encapsulated data are called *segments;* at the network layer, the units are called *frames,* and at the data-link layer, the units are called *frames.* Each layer adds its own header information.

Data-Encryption Standard (DES) A cryptographic algorithm developed in the 1970s that is endorsed by the National Institute of Science and Technology. The standard is still in use today, being encapsulated within routing (and other) protocols. During the encryption process, a 56-bit header (called a *key*) is added to the data stream.

Data Extender A device that extends a data signal such as an RS232 or PBX telephone line (or both simultaneously) to another location via a dial-up public telephone line. Data Extender's incorporate their own proprietary data compression protocols, therefore the data extenders on each side of the line must be of the same manufacturer. The advantage of Data Extenders is that they compress two sizeable digital signals (56 kbps/ and higher each) and modulate them to be transported over an inexpensive POTS (Plain Old Telephone Service) line.

Data-Flow Control Layer Layer 5 of the *SNA (System Network Architecture)* model. This layer determines and manages interactions between session partners, particularly data flow. It corresponds to the session layer of the OSI model. See also *System Network Architecture* and *OSI.*

Data Integrity A term that refers to how few errors are occurring in the transfer of transmission of data. The lower the error rate, the better the data integrity.

Data Link In an *SNA (IBM System Network Architecture)* network, basically the two parts are data links and nodes (Fig. D.5). There are many types of data links; some are simply a LAN connection over twisted pair and some incorporate additional hardware. The latter connect remote locations through the public telephone network, utilizing ISDN, frame relay, X.25, and other lower-level link protocols.

Data-Link Connection (DLC) A user connection to a frame-relay network that could be thought of as a virtual channel on a multiplexer. A DLC is a half duplex (one way) data channel. To get full duplex communications in frame relay, the DLCs from each user are combined to provide full-duplex communications ability. Hence, two DLCs constitute a *PVC (Permanent Virtual Circuit).* See also *Data-Link Connection Identifier.*

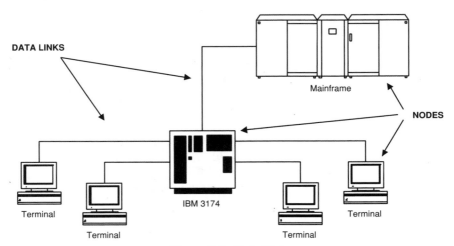

DATA LINKS

Mainframe

NODES

Terminal

IBM 3174

Terminal

Terminal

Terminal

Figure D.5 Data Link

Data-Link Connection Identifier (DLCI) A number given to a *DLC (Data-Link Connection)* on a frame-relay service (Fig. D.6). DLCIs are provided to customers by the frame-relay service provider. It is the part of the frame-relay packet header that provides space for 10 bits

SAMPLE DLCI ALLOCATION

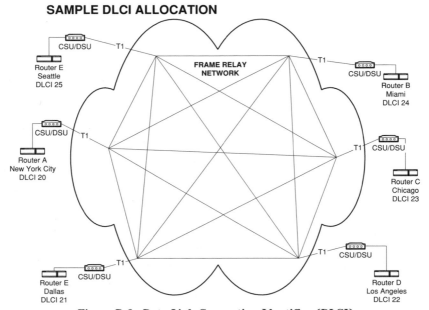

CSU/DSU T1
Router E
Seattle
DLCI 25

**FRAME RELAY
NETWORK**

T1 CSU/DSU
Router B
Miami
DLCI 24

CSU/DSU T1
Router A
New York City
DLCI 20

T1 CSU/DSU
Router C
Chicago
DLCI 23

T1 CSU/DSU
Router E
Dallas
DLCI 21

T1 CSU/DSU
Router D
Los Angeles
DLCI 22

Figure D.6 Data Link Connection Identifier (DLCI)

of data that is used to name the logical channel that the frame will be routed to. Furthermore, the DLCI is embedded into the data stream from the very beginning of the transmission to the very end. Routers use DLCIs to specify a *PVC (Permanent Virtual Circuit)* or *SVC (Switched Virtual Circuit)* that links to other routers or devices in a frame-relay network. In the basic frame-relay specification, DLCI values that are given to links or routes are different for every individual router in a network. Different routers can use different DLCI numbers to identify the same route (connection) in the same network. In the *LMI (Local Management Interface)* extended specification for frame relay, DLCIs are "globally significant" and the same DLCI numbers are given to all routes by all routers. The DLCI is very similar to (and performs the same task as) the *LCN (Logical Channel Number)* in X.25. For a table of DLCI allocations, see *DLCI.* See also *LMI.*

Data Link Control (DLC)   The part of a communications protocol that resides in the overhead and provides a user (or the protocol itself) a means to control connect, disconnect, error-correcting, transmission-speed, and other operation-crucial functions.

Data Link Layer   A layer in a communications protocol model. In general, the data link layer receives and transmits data over the physical layer media (twisted pair, fiber optic, etc.). The latest model (guideline) for communications protocols is the *OSI (Open Systems Interconnect).* It is the best model so far because all of the layers or functions work independently of each other. For a diagram of the OSI, SNA, and DNA function layers, see *Open Systems Interconnection.*

Data-Link Switching (DLSw)   An interoperability standard, described in RFC 1434, that provides a method for forwarding SNA and NetBIOS traffic over TCP/IP networks using data-link layer switching and encapsulation. DLSw uses *SSP (Switch-to-Switch Protocol)* instead of IBM *SRB (Source Route Bridging),* eliminating the major limitations of SRB, such as hop-count limits, broadcast traffic, timeouts, lack of flow control, and lack-of-prioritization schemes. See also *SRB (Source Route Bridging)* and *SSP (Switch-to-Switch Protocol).*

Data Network Identification Code (DNIC)   Part of an X.121 address. DNICs are divided into two parts. The first part specifies the country in which the addressed PSN is located and the second specifies the PSN itself. See also *X.121.*

Data Packet Switch   Also called a *packet switch* or *data switch* (Fig. D.7). It is a device that routes segmented transmissions between end users, using a connectionless protocol, such as X.25, Ethernet, frame relay, token ring, ATM, or TCP/IP. Data packet switching is often performed in levels, with one protocol carrying another. For example, an Ethernet-based transmission from a LAN could be routed over a frame relay or ATM connection via a data packet switch. Packet switches consist of *PDN (Public Data Networks)* or *PSN (Packet-Switching Networks)*, which are the basis of the frame relay, ATM, and X.25 services that public telecommunications companies offer.

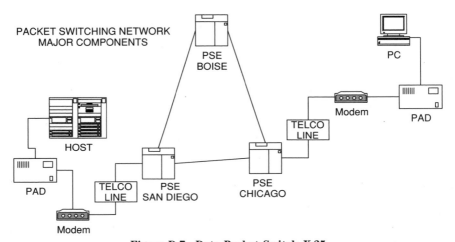

Figure D.7   Data Packet Switch–X.25

Data Service Unit (DSU)   Also called a *CSU/DSU (Channel Service Unit/Data Service Unit)*. A DSU is a hardware device that can come in many shapes and sizes. Rack-mount, shelf-mount, and stand-alone DSUs are available. A CSU/DSU has three main functions. The first function is to act as a demarcation point for a T1 (DS1) service from a local communications company. The second function is to provide line format and line-code conversion (B8ZS to AMI, SF or D4 to ESF, 135 V to 0 V) between the public network and the customer-premises equipment, if necessary. The third function is to provide maintenance or alarm services and loop-back for isolating problems with the T1 line or customer's equipment. For a photo, see *CSU/DSU*.

Data Set Ready (DSR)   Pin 6 of a DB25 connector wired for the RS232C protocol. This wire is used by a modem or SDI device to send a signal that acknowledges that is ready to receive data.

Data Sink   A loose term for *DTE (Data Terminal Equipment)*. It is a reference to network equipment that accepts data transmissions.

Data Span   Another name for a service purchased from a communications company. It can refer to any digital service, including T1, 56K, ISDN, or any other data-carrying service.

Data Stream   All data and overhead transmitted through a communications link in a single read, write, or file-transfer operation.

Data-Switching Exchange (DSE)   Also called *PSE (Packet Switching Exchange)*. A part of an X.25 packet switching network that receives packets of data from a PAD (Packet Assembler/Dissembler) via a modem. The PSE makes and holds copies of each packet, then transmits the packets one at a time to the PSE that they are addressed to. The local PSE then discards the copies as the far-end PSE acknowledges the safe receipt of the original.

Datagram   A logical grouping of information sent as a network layer unit over a transmission medium without prior establishment of a virtual circuit. The word *datagram* is used in place of *packet* or *frame* to clarify the connectionless protocol environment in which they exist. IP datagrams are the primary information units in the Internet. The terms *cell, frame, message, packet,* and *segment* are also used to describe units of electronic data at different layers of the OSI reference model and in various technology circles. In TCP/IP, the datagram contains application data, a TCP header, and an IP header. For a diagram of a datagram, see *IP Datagram.*

Datagram Delivery Protocol   The AppleTalk network layer protocol that is responsible for the socket-to-socket delivery of datagrams over an AppleTalk internetwork.

DATU (Direct-Access Test Unit)   Also called *MLT (Mechanized Loop Test)*. Equipment that is either added on or built into a central office switch. DATU allows a technician to dial the phone number of the DATU or MLT equipment and execute a test for shorts, opens, and grounds remotely. In response to a digital voice, the technician enters a password and a choice of options. The results of the test can be read back to the technician by a digital recording or sent to them via an alpha-numeric pager. DATU units can also send a locating tone on the technicians choice of TIP, RING, or both TIP and RING. The test unit can also short lines and remove the battery voltage for testing purposes.

DB (Decibel)   A decibel is 1/10 of a Bel. This is a measurement of increase or decrease of a signal that comes from the ratio of transmitted power to received power. To have a general idea of what a decibel is, remember that negative decibels represent a loss of power. Positive decibels represent an increase in power. If you compare decibels with the way your ear reacts to sound, every –3 decibels would cut the original loudness of the sound in half.

DB Connector (Data-Bus Connector)   A type of connector for serial or parallel communication ports incorporated into personal computers or telecommunications devices. The specific connector type is determined by the number of pins that the connector has (i.e., DB9, DB15, and DB25). The number of pins that are used and what purpose they serve is determined by the communications standard that is used across the DB physical interface. See also *RS-232*.

DB9   A connector used for data-connectivity applications. It has nine pins, and it can be configured for several protocols, including the popular RS-232 (Fig. D.8).

Figure D.8   DB9 Connector

DB15   A connector used for data-connectivity applications. It has 15 pins and can be configured for several protocols, including the popular RS-232.

DB25   A connector used for data-connectivity applications. It has 25 pins and can be configured for several protocols, including the popular RS-232 (Fig. D.9).

DBA (Dynamic Bandwidth Allocation)   A feature of nonconnection-oriented protocols at the physical layer, such as X.25, frame relay, and

Figure D.9 DB25 Connector

ATM. These telecommunications protocols permit a variable number of users to utilize the full bandwidth of a connection. One method is statistical time-division multiplexing used in X.25, which assigns packets to available channels, regardless of the user or the number of users. ATM and frame relay are also capable of allocating bandwidth to users until the local loop is fully utilized. This is a natural feature of these protocols because no user actually has to make a connection, or reserve a channel over the physical layer to make a transmission. See also *STDM, Connection-Oriented Protocol,* and *Connectionless Network Protocol.*

dBi (Decibels Isotropic)   Particularly in wireless Ethernet (IEEE 802.11b) and other radio design circles, antenna gain is an indicator of how well an antenna focuses RF energy in a preferred direction. Antenna gain is expressed in dBi, which is the ratio of the power radiated by an antenna in a specific direction compared to the power radiated in that direction by a nondirectional antenna fed by the same transmitter. Antenna manufacturers normally specify the antenna gain for each antenna they manufacture. A typical low-power Yagi antenna can have a 12dBi of directional advantage over an isotropic (nondirectional) antenna. This can mean a factor of up to five times the radio range.

DBm   Decibels below 1 milliwatt. This is a measure of power loss with 1 milliwatt as the transmission reference. As a common example if you receive a signal at 1 milliwatt, then you have a loss of 0 dBM. If you receive a signal that is 0.001 milliwatts, then you have a loss of 30 dBM. Some methods of testing analog phone lines include dialing a number that answers and provides a 1-milliwatt reference signal at 1000 Hz. The meter on the line measures the 1000-Hz signal on its end and displays a reading. Most POTS telephone lines are between –20 and –32 dBm.

DBrn   Decibels above reference noise. This is the same method of comparing transmitted and received signals by the log 10 of a ratio, only a reference is premeasured and used as the input (the denominator in the ratio).

DBU   Decibels below 1 microwatt. Just dBm at a smaller increment. This is a measure of power loss with 1 microwatt as the transmission reference. As a common example is: If you receive a signal at 1 microwatt, then you have a loss of 0 dBM. If you receive a signal that is 0.001 microwatts, then you have a loss of 30 dBM.

DC   The ASCII control code abbreviation for direct control. The binary code is 0001001 and the hex is 11.

DC (Direct Current)   DC is current that is induced by a voltage source that does not change direction from positive to negative. DC can fluctuate, and carry an analog signal by varying the DC current and voltage. DC can pulse, it can spike, and it can do many things. The one thing that DC cannot do is change direction. Common sources of DC are batteries, AC power adapters, and power rectifiers.

DCC (Data Communications Channel)   An overhead channel in an AT&T SONET ring. It allows the individual nodes to communicate control information to each other.

DCC (Data Country Code)   One of two ATM address formats developed by the ATM Forum for use by private networks. It is adapted from the subnetwork model of addressing in which the ATM layer is responsible for mapping network-layer addresses to ATM addresses. Compare with *ICD*.

DCE (Data Communications Equipment)   See *Data Communications Equipment*.

DCS (Digital Cross-Connect System)   A DCS is also called a *DACS (Digital Access Cross Connect System)*, depending on the manufacturer.

A digital cross-connect system is a fundamental part of a local and long-distance company's network because of the rapid deployment of broadband transmission equipment (SONET, DS3). Cross connecting broadband services can be cumbersome because larger circuits (OC-1) require coax. The DACS or DCS is a rack-mountable system that enables any circuit that interfaces with it to be electronically cross connected from one path to another within the network it is connected to. Circuits that can interface with a digital cross-connect system include DS0, DS1, DS3 (or T3), STS-1, and SONET OC-1. An incoming circuit can be rerouted by simply making path changes in DACS administrative software. For a diagram and photo of a DCS/DACS system, see *Digital Cross Connect System.*

DCV (Digital Compressed Video)   There are several types of DCV. The object of compressed video, in general, is to transmit an initial picture, then transmit only the parts of the picture that move. A good example is a video phone application, where only a person's mouth and facial features move. Everything else in the video phone picture stays the same.

DDM (Distributed-Data Management)   A software entity in IBM *SNA (System Network Architecture)* environments that provides peer-to-peer communication and file sharing. DDM is one of three SNA transaction services. See also *Document Interchange Architecture* and *SNA Distribution Services.*

DDP (Datagram Delivery Protocol)   The AppleTalk network layer protocol that is responsible for the socket-to-socket delivery of datagrams over an AppleTalk Internetwork.

DE (Discard Eligibility)   DE is one of the bits in the standard 16-bit header of a frame-relay frame. It is set to 1 to notify the network that its frame should be discarded at the first onset of network congestion. See also *Frame Relay.*

De Facto Standard   A standard that has come about because of consumer popularity, not because of formal approval of a standards committee.

Dead Spot   A dead spot is an area within a transmitter's range where the radio signal being transmitted cannot be received. Dead spots occur for many different reasons. Sometimes the signal is blocked or reflected, sometimes it is because you are located in a small valley that dips below the radio transmission.

Decibel (dB)   A decibel is $\frac{1}{10}$ of a Bel. This is a measurement of increase or decrease of a signal that comes from the ratio of transmitted power to received power. To have a general idea of what a decibel is, remember that negative decibels represent a loss of power. Positive decibels represent an increase in power. If you compare decibels with the way your ear reacts to sound, every –3 decibels would cut the original loudness of the sound in half. See also *DBm*.

Decimal-to-Binary Conversion   For a conversion table of binary to decimal and hexadecimal, see *Appendix E*.

Decimal-to-Hexadecimal Conversion   For a table on decimal-to-hexadecimal conversion, see *Appendix E*.

Decoder   A device or software program that converts a signal or transmission from one protocol to another.

Dedicated Access   Reference to a telephone line that is usually provided by an IXC (Inter Exchange Carrier, long-distance company) for exclusive dialing of long distance on their network. Sometimes the customer has the line installed themselves by a *LEC (Local-Exchange Carrier)* and gets billed separately for the dedicated-access line and the long-distance service. Some dedicated-access lines are capable of dialing local calls, but their long-distance service is dedicated to a specific IXC.

Dedicated Channel   A channel within a T1 or T3 that is dedicated to a specific customer. Other than that, it is a private line/dedicated circuit.

Dedicated Circuit   Also called a *private line*. A private line is a pair of wire or (two pairs of wire for a T1) that runs from your location to a location that you want to be connected to with a dedicated high-speed data connection. Once a private line is installed, it is there all day, every day. There is no dialing on a private line because it does not go through switching circuitry, although it does get regenerated (the data signal on the channel is received and retransmitted). Dedicated lines could be on copper, which they have been very much in the past, but since the deployment of SONET, it is possible to put hundreds of private lines and switched lines on a pair of optical fibers.

Dedicated Line   A telephone line from the phone company that is dedicated to one user or device. Most fax machines and modems are on dedicated lines. A dedicated line is not a trunk because a trunk is a line that multiple users share. A dedicated line can also be a dedicated circuit (a private data line), but it is less common.

Default Gateway   1. An address that identifies the route that leads outside of a subnetwork, or segment of a local-area network. Usually the default gateway IP address belongs to a router.   2. In IP routing (Layer 3 switching), default IP gateway routes IP packets that have unresolved destination IP addresses or addresses that are not in its local network/routing table. Setting the default gateway IP address tells the switch how to send packets to a device that is not on the local network. Defining multiple default IP gateways provides redundancy. If the primary default IP gateway fails, a layer 3 Ethernet switch or router sends packets to the secondary default IP gateways in the order in which they were configured.

Default Route   1. In data networking, a static (manually input) routing table entry that instructs the router where to send frames that have an address that is not identified within routing memory. See also *Routing Table*.   2. In voice networking, a reference to a trunk group that a call would be connected to if the peripheral device that the telephone system is attempting to transfer the call to is in a failed state. For example, if a PBX tries to transfer a call lto an *IVR (Interactive Voice Response)* system that has been powered down, the PBX would sense the outage because it would not get the proper signaling over the connecting trunks. The PBX would then send the calls to an alternative destination. this alternative destination would be a default route. This routing could terminate to a group of outbound trunks, a busy signal, or to a queue where an agent can answer the call.   3. Another name for default gateway. See *Default Gateway*.

Definity   A **PBX (Private Branch Exchange System)** manufactured by Lucent Technologies. For a picture, see *Private Branch Exchange.*

Degaussing Coil   Degaussing is to demagnetize. A degaussing coil is simply a long coil of wire that is bent into the shape of a circle. If a CRT (picture tube or monitor tube) becomes magnetized, you will notice an area of discoloration. By waving a degaussing coil around this area, you will demagnetize the screen of the CRT. Many monitors have degaussing coils built into them.

Degradation   Another term for attenuation. As a signal traverses down a wire, fiber-optic cable, or through the air, it loses power and becomes distorted. This phenomenon is referred to as *attenuation, loss,* or *degradation.*

De Jure Standard   A standard that exists because of its approval by an official standards body, as opposed to a de facto standard, which exists simply because of industry and market popularity. See also *De Facto Standard* and *Standard.*

DEL   The ASCII control code abbreviation for delete idle. The binary code is 1111111 and the hex is F7.

Delay   The time difference between when a signal is sent and when it is received.

Delay Variation   Another way to describe the signal distortion known in electronics circles as jitter. Over IP telephony networks, delay variation is the worst kind of distortion. Delay variation causes users' voices to sound robotic, if understandable at all.

Delayed Ring Transfer   A PBX and key-system feature that is just like call forwarding, except that before the call is forwarded, it rings a pre-selected number of times. If you are not in your office and someone calls you, you can have the call forwarded to another associate after a certain number of rings. This is delayed ring transfer.

Delta Channel   Another name for an *ISDN (Integrated Services Digital Network)* "D" channel, which is the data or control channel of an ISDN line.

Delta Modulation   A form of encoding analog signals to digital binary. Instead of sampling a signal and creating an 8-bit binary number, like the standard ADC, delta modulation samples the change (*delta* is the term for change in the physical sciences) of the signal (Fig. D.10). Delta modulation only looks for two changes in an analog signal, change higher and change lower. This higher or lower signal is sent to a far-end device and the signal is re-created or decoded with these simple higher or lower instructions. Included is a diagram of an analog signal and its delta-code equivalent. Notice that as the signal goes up in voltage, the delta-modulation technique registers a 1, when it goes down in voltage it registers a zero.

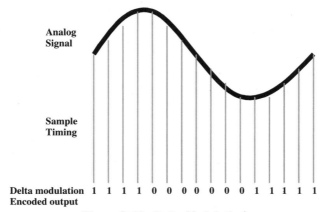

Figure D.10   Delta Modulation

Demarc (Dmarc, Demarcation Point) A Dmarc is where the local phone company hands off a telephone circuit. A Dmarc can be in the form of a standard network interface for a residential line, a DSX panel for a T1 or T3, an RJ212X for a business line, or an RJ45 for an ISDN line. The Dmarc separates customer-owned equipment from telephone company-owned equipment. It is also the place where responsibility for the circuit's performance is separated.

Demarcation Point See *Demarc.*

Demodulation The inverse of modulation. When a radio signal is received, the information that was sent over it (audio, video, or data) is still mixed in with the carrier. The process of demodulation removes the carrier signal from the information you want. Some different types of demodulators are AM, FM, and PM.

Demultiplex The inverse of *multiplex.* Multiplexing is the process of encoding two or more digital signals or channels onto one. The reason that channels are multiplexed together in communications is because it saves money. When we use all of the wires in a cable and need more, it costs less to add electronics on the ends of a cable than to install a new one (imagine the expense from LA to NY). A T1 encodes 24 channels into one by using frequency-division multiplexing. In a simpler explanation, a T1 makes it possible to place 24 lines that once needed 24 pairs on only 2 pairs. When a group of signals are multiplexed together, they are all sampled at a high rate of speed, faster than the combined speed of all the channels being multiplexed. For a diagram of the multiplexing process, see *Time Division Multiplexing.*

Demux Abbreviation for *de-multiplex.* To separate multiplexed channels from one transmission into their original individual channel.

Deregulation The transition of government authority and control away from specific business activities of telephone or cable TV companies. The complete deregulation of the RBOCs will be a gradual process and, depending on the way in which the telecommunications industry evolves, complete deregulation might never happen. The purpose of deregulation is to promote new technology, lower prices, improve service, and create a more abundant supply of telecommunications services.

Des (Designation Strip) (slang "Dez") A designation strip, or desi strip is the piece of paper or label that goes under a button on a phone. The type written on the desi strip identifies the feature with the programming for that button. Newer equipment incorporates a small LCD display for the button designation.

DES (Data Encryption Standard) A standard cryptographic algorithm developed by the United States National Bureau of Standards. It is often referred to by its newest version, which is the triple DES, or 3DES, which is a 168-bit secure encryption. DES is not too difficult for a hacker to break with the correct tools; however, 3DES is far more difficult to break or hack, but only with the support of supercomputer resources.

Designated Port See *Spantree.*

Designated Router (DR) In OSPF routing, a router that is elected by being in OSPF routing area zero, and having the highest IP address. A designated router must possess one link to a network's backbone (by design). The designated router is the router that all Area Border Routers (ABRs) in an autonomous system communicate link state status to and receive updated link state databases from. If a link in the primary backbone area (routing area zero) should fail, the ABR or DR receives a Link State Update (LSU) packet from the router that the changed link belongs to. In response to this packet, the ABR or DR sends a Link State ACK message and then stops forwarding packets. It recalculates its routing database with the new information, sends the routing database change to the other routers within its area, then begins forwarding packets again.

Designation Strip See *Des.*

Desk-Top Engineer A person that maintains personal computers/workstations and LAN network connectivity. They might also maintain station services on a PBX switch, which is the connecting, programming, and moving of telephone extensions throughout a network. Desk-top engineers do not usually do switching equipment upgrades or additions and do not do LAN administration. Desk-top engineers are also called *desk-top technicians.*

Desk-Top Technician See *Desk-Top Engineer.*

Destination Address The identifier of a network device that is receiving data. Also called by *MAC (Media-Access Control)* address. As data is sent from a device, that data works its way from layer to layer. During its transport, additional headers that contain addressing or routing information are added on and stripped off. In networking via the *Open Systems Interconnect (OSI)* model each independent layer can add its own addressing and routing information within the header for that corresponding layer. From start to finish, the *MAC (Media-Access Control)* address remains encapsulated within the original header, which is permanently burned into the memory of every network interface device or *NIC (Network Interface Card)*. This address is the identifier of the device that is to receive the data.

Destination MAC   The *MAC (Media-Access Control)* address specified in the address field of a packet. It is also called a *destination address*. A destination address is the identifier of a network device that is receiving data. As data is sent from a device, that data works its way from layer to layer. During its transport, additional headers that contain addressing or routing information are added on and stripped off. In networking via the *Open Systems Interconnect (OSI)* model, each independent layer can add its own addressing and routing information within the header for that layer. From start to finish, the MAC address remains encapsulated within the original header, which is permanently burned into the memory of every network-interface device or *NIC (Network Interface Card)*. See also *MAC Address*.

Detector   The circuit inside a radio receiver that detects fluctuations in the modulated carrier (radio signal). The detector simply filters the carrier (radio portion) out of the wanted end signal (audio and/or video). The simplest form of detector is the germanium diode/filter capacitor used in AM receivers.

Deutsche Industrie Norm (DIN)   A German national standards organization that develops de jure standards for Germany. Many standards developed or suggested by this organization are adopted by other organizations that are national and/or international. The original DIN connector is one of them. See also *DIN Connector.*

DHCP (Dynamic Host-Configuration Protocol)   A protocol that provides the specific service within a network of automatically configuring hosts/workstations within that network. DHCP is capable of automatically configuring the IP address, subnet mask, default gateway, and DNS addresses.

DIA (Document Interchange Architecture)   A method or format for transferring data files. DIA defines the protocols and data formats needed for the transparent interchange of data files in an IBM *SNA (System Network Architecture)* network. DIA is one of three SNA transaction services. See also *Distributed Data Management* and *SNA Distribution Services.*

Dial-By-Name Directory   A feature of voice-mail systems that enables a caller to be transferred from the automated attendant to a person (or department) by knowing that person's name, then dialing (spelling) the corresponding letters on the dial pad. Automated attendants are regarded as a poor impression to customers that are first-time callers. This is especially true for companies that deal in service or retail.

Dial String   A set of instructions that are sent to a device (such as a modem) that is capable of dialing a number on an analog phone line. The instructions in dial strings are the same that you take for granted. They include go off-hook, wait for dial tone, dial digits, wait for answer, etc.

Dial Tone   An analog method of signaling. When you pick up the handset of a telephone that is connected to the phone company or a PBX system, you hear a buzzing hum sound. That sound is a dial tone, a signal from the PBX or telephone company central office switch to go ahead and dial your number.

Dial-Tone Delay   The time from when you go off-hook and when you receive a dial tone from the host switch.

Dial-Up Line   A line that can be dialed into. Some dial-up lines include the *POTS (Plain-Old Telephone Service)* to your house, ISDN, and switched 56 data circuits.

Dial-Up Modem   A modem that is intended to be used on the public-switched telephone network. It is connected to a phone line and that phone line has a phone number that people can dial with their modems. These modems are the most common in personal computers. The other type of modem is a short-haul modem, which doesn't dial numbers—it just extends a digital signal (e.g., to the other side of a building for a printer).

Dial-Up Networking (DUN)   A computer-communication method where a POTS telephone line is used in conjunction with a modem for the physical and data-link layer connection and a specific protocol is used to transfer information. In Internet applications, *PPP (Point-to-Point Protocol)* or *SLIP (Serial Line Internet Protocol)* is used.

Diaper   A piece of plastic wrap that is temporarily placed over cable faults.

DID (Direct Inward Dial)   A phone line that comes from the local phone company and connects to your PBX switch. A DID line has a phone number (and DNIS or virtual directory number attached to it) that is targeted to ring directly to a phone on the PBX network without going to a console operator, or anywhere else first. The PBX system usually needs specific DID trunk (incoming line) hardware to make DID lines work.

Dielectric   A material that does not conduct electricity. Dielectric materials are used as insulating materials, such as the vinyl coating on copper wires. Good dielectric materials (more frequently called *insulators*) are glass, ceramic, rubber, and plastic.

Differential Encoding   A physical media-transmission format used in IEEE 802.5 and token-ring *LANs (Local-Area Networks)*. Differential encoding combines a clocking signal with the data stream. A binary 1 is denoted by a voltage increase and a binary 0 is denoted by a voltage decrease. The voltage reset between voltage increases and decreases represents the clocking or timing source. By design, differential encoding is less prone to attenuation, but is more sensitive to *RFI (Radio-Frequency Interference)* than *PCM (Pulse-Code Modulation)* formats. This format is also referred to as *Differential Manchester Encoding* or *Manchester Encoding*.

Differential Phase-Shift Keying (DPSK)   A pre-V-series modem standard. This phase-modulation method was used in Bell 201 standard modems. DPSK incorporated dibits (two bits represented by a phase shift). The code representation is as follows: A phase shift of the current carrier phase by 45 degrees represents a 00 binary value, a 135-degree phase shift from the present carrier phase represents a 01 binary value, a 315-degree phase shift from the present carrier phase represents a 10 binary value, and a 225-degree phase shift from the present carrier phase represents a 11 binary value.

Digital   A signal that has only two possible levels per cycle, in contrast to analog, which can have an infinite number of possible levels per cycle (Fig. D.11). The great thing about a digital signal is that it can be

ANALOG   DIGITAL

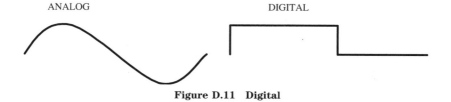

Figure D.11   Digital

regenerated easily. Even though it might pick up noise and RFI as it is transmitted along a wire, when it is regenerated, all the noise is cut out because the regenerating device looks for only two levels of signal to reproduce, 1 and 0. Therefore, all the other stuff, such as white noise and maybe even an unwanted radio station, are not regenerated.

Digital Announcer   A device that stores RANs (Recorded Announcements), and plays them to a specific line/trunk when instructed to do so by an ACD system (Fig. D.12).

Figure D.12   Digital Announcer

Digital Audio   Digital audio is analog audio that is stored in a digital code. It is good to store audio and other information digitally because when the signal is read, decoded, and converted to analog, unwanted noise and interference is filtered out. Other methods of audio storage are subject to unwanted electrical noise that is the electronic nature of the storage media itself. The nature of one of these older alternative technologies, such as the audio cassette, stores the analog signal directly onto a magnetic tape. The magnetic tape itself has inconsistencies in the metals and other materials used in their manufacture. As the magnetic tape glides against a playback head in a cassette player, it creates its own audio signal which resembles a light hiss sound. In the case of digital audio, the means of storage and encoding being a *Compact Disc (CD)* or a *Digital Audio Tape (DAT)* the imperfections in the storage media are ignored by the electronics that reads the digital code from the DAT or CD. This is because they only read ones and zeros, paying no attention to hisses or popping sounds. Even if the signal is muffled, it is a muffled signal of ones and zeros, which gets decoded into a clear reproduction.

Digital Cable TV   A service offered by telecommunications companies. Digital cable TV differs from analog cable TV in several aspects. Digital cable TV is a compressed digital signal that is transmitted on an analog carrier. The digital compression allows for more than five times the number of stations to be delivered via the same bandwidth. The additional channel capacity allows entertainment operators to deliver "near on-demand" programming by staggering the start times of programs on different channels. Because the carrier of digital TV is analog, it can also be delivered through the air, as in satellite TV systems. The digital format naturally eliminates unwanted noise and interference from programing, regardless of the media in which it is delivered.

Digital Compressed Video (DCV)   There are several types of DCV. The object of compressed video, in general, is to transmit an initial picture, then transmit only the parts of the picture that move. A good example is a video phone application, where only a person's mouth and facial features move. Everything else in the video phone picture stay the same during most of the transmission.

Digital Cross-Connect System (DCS)   A DCS is also called a *DACS (Digital-Access Cross-Connect System)*, depending on the manufacturer (Fig. D.13). A digital cross-connect system is a fundamental part of a local and long-distance company's network because of the rapid deployment of broadband transmission equipment (SONET, DS3). Cross-connecting broadband services can be cumbersome because larger circuits (OC-1/STS-1) require coax. The DACS or DCS is a rack-mountable system that enables any circuit that interfaces with it to be electronically cross-connected from one path to another within the network it is connected to. Circuits that can interface with a digital cross-connect system include DS0, DS1, DS3 (or T3), STS-1, and SONET OC-1. An incoming circuit can be rerouted by simply making path changes in DACS administrative software. In the DACS shown in the diagram, a T1 circuit coming into the input side (left side) could be cross-connected to exit as one of the channels in a DS3 on the right side.

Figure D.13   Digital Cross-Connect System (DCS)

Digital Frequency Modulation   Another term for *Frequency-Shift Keying.*

Digital Line Protection   *Digital line protection* refers to the protection of modems from the higher line voltages of digital lines. Lines connected to a digital line interface card on a PBX system are wired and look the same as normal phone lines from the phone company, but they operate at a higher voltage. Digital service lines from the phone company (T1) lines are also a higher voltage, about −135 VDC analog modems. Modems are not for digital lines, they are for 52-V analog lines. If an analog modem is connected to a digital line by mistake, it could be destroyed. Digital line protection is a feature designed into modems that protects them from a mistaken connection to a digital line.

Digital Loop Back   A feature of transmission equipment that allows a user to reroute a signal back to the source instead of into the termination or end equipment. By doing this, the user can see if the signal going into the equipment is good or bad. If the signal loops back and is good, but the signal is bad coming out of the equipment on the far end when the loop back is removed, then the trouble is most likely in the end equipment. In many modems and digital service units, the loop back can be controlled remotely.

Digital Microwave   Digital microwave has become a very economical way to bypass construction costs of broadband private line services. Many *CAPs (Competitive-Access Providers)* have access to microwave radio resources, such as licensing, equipment, and installation. Digital microwave is also called an *eyeball shot, 38 gig,* or is just referred to as *radio.* Most of the microwave being installed for private-line service today is in the 33-GHz to 39-GHz frequency range. These microwave units use an FM-FSK over two sidebands for transmitting at full duplex. They are available in T1, DS3, and STS-1 (which is a DS3 formatted for SONET). The 38-GHz microwave has a range that depends on the size of the antenna (dish) placed on the outdoor radio unit. The choices in antenna size are one or two feet in diameter. For example, a one-foot dish antenna has a maximum range of one to three miles, depending on the regional weather conditions (rainfall, snow, and especially fog drastically attenuate microwave transmissions). The two foot dish has a range of two to seven miles, also dependent on the weather in the region. For a diagram of a microwave application, see *Terrestrial Microwave.*

Digital Service Cross-Connect (DSX)   A reference to a digital service termination/patch panel that allows DS1 and DS3 circuits to be monitored by test equipment, such as a TTC Tberd or T-ACE. DSX panels are usually terminated via wire-wrap and accessed at the front via bantam-type test cords. For a photo of wire-wrap terminals, see *Wire-Wrap* (Fig. D.14).

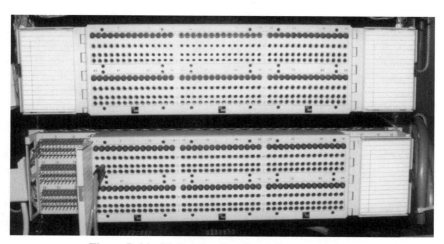

Figure D.14   Digital Service Cross-Connect (DSX)

Digital Signal Processor (DSP)   A hardware enhancement to Ethernet switch ports (most commonly a gateway port for IP Telephony applications) that gives them the ability to act as media termination points during the execution of user features on IP telephony/server-based PBX systems. Also referred to as an *MTP (Media Termination Point)*. The DSP within an Ethernet switch or router utilizes its built-in codecs and transcoders to temporarily hold a connection, whereas an IP telephony server locates an available bandwidth channel or waits for a user to execute feature keystrokes on a telephone.

Digital Simultaneous Voice and Data (DSVD)   A type of modem that allows users on either end to exchange data files and talk over the same phone line. Both users must be using a DSVD-compatible modem in order to use the DSVD features.

Digital Subscriber Line-Access Multiplexer (DSLAM)   The distribution device for *xDSL (x Digital Subscriber Loop)* service from a central office. The DSLAM combines and separates the different formats of communications contained in the xDSL carrier and routes them to their respective hosts. Because xDSL is capable of carrying voice, video, and data, the voice needs to be routed to a voice central office switch, the video needs to be routed to a CATV head-end and the data needs to be sent to a packet switch. The DSLAM performs all of these functions.

Digital Subscriber Loop (DSL)   See *xDSL, VDSL, HDSL, IDSL, RADSL, ADSL, QAM, DMT,* and *DSLAM*.

Digital Subscriber Line   Another name for an *ISDN BRI (Basic Rate Interface)*. See also *DSL*.

Digital-to-Analog Converter (DAC)   A device that performs the function of encoding analog voice or video signals into a stream of binary digits. The analog-to-digital converter within a T1 channel bank samples a caller's voice at a rate of 8000 times per second. Each sample's voltage level is measured and converted to one of 256 possible sample levels. The DAC converts all of the digital numbers back into an audio signal. See also *Analog-to-Digital Converter*.

Digital Versatile Disc (DVD)   A newer version of the common 650-MB CD (Compact Disc). The DVD is capable of storing 17 GB per side. DVD players can play newer DVD discs containing audio and video information, as well as your old audio-format CDs.

Dijkstra's Algorithm   Also referred to as *SPF (Shortest Path First)* or *Link State Algorithm*. A class of router operating software that enables routers to build their own complex address-routing tables that detail every router and node within their network. The routing-table building process is accomplished through information multicasts. The routing-table multicasts are referred to as *LSPs (Link State Packets)* and they consume payload bandwidth to transmit this information. The process of sending and receiving LSPs is called the *discovery process.* Multicasts are only sent when there is a change in the network, such as a circuit connection going down, or a new router or connection being added. Link-state algorithms use tremendous amounts of router-system memory (20 MB to 30 MB in a 30-node network), and consume significant processor resources within a router's circuitry. During the startup of a link-start network, the discovery process can take hours. The great advantage to this complex operating method is that routing loops are not created. See also *Distance Vector Routing Algorithm* and *Hybrid Routing Algorithm.*

DIMM (Dual Inline Memory Module) A *DRAM (Direct-Random Access Memory)* package that consists of a small circuit board with DRAM memory chips on it. The DIMM is an expansion of the original SIMM, providing an interface to a 64-bit bus, rather than a 32-bit bus. The DIMM also differs from a SIMM in that it has pin connectors on both sides of the small board. For an illustration of a DIMM, see *SDRAM.*

DIN Connector   A screw-on type connector that is installed on coaxial cable in RF/microwave applications. DIN connectors have better intermodulation suppression and power-handling capabilities than N-type and other coax connectors (Fig. D.15).

Figure D.15   DIN Connector

Diode   An electronic semiconductor device that simply put, only conducts electricity in one direction (Fig. D.16). Whether or not the device conducts is dependent on which direction the device is "biased." Diodes (or rectifiers) are used to change alternating current (AC) to direct current (DC). If a more positive voltage is applied to the anode lead of the diode, then the diode simply acts like a wire. If the more positive voltage is applied to the cathode lead, then it acts as if there is no connection.

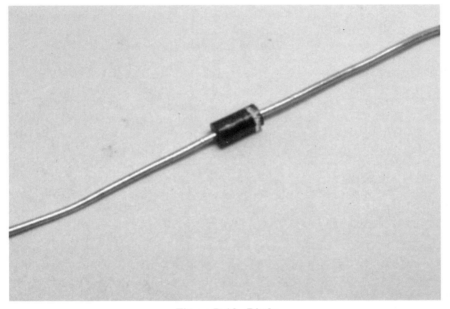

Figure D.16   Diode

DIP (Dual Inline Pin)   A way a component is physically made. A DIP component has two rows of pins (pins are a means to solder the component into a circuit). Many components are "packaged" this way. Some of them are DIP 8-segment display, DIP integrated circuit, and of course, DIP switches.

DIP Switch   A very small manual switch that comes in a DIP package (Fig. D.17). DIP is an abbreviation for *Dual-Inline Pin,* it is the way a component is physically made. See *DIP* for more information.

Figure D.17   DIP Switch

Dipole   A type of balanced antenna with two elements that are fed by transmission line. Dipoles are mildly directional. A common dipole antenna is the "rabbit-ear" style, is used with TVs.

Direct-Access Storage Device (DASD, "Dazzdee")   A technology that incorporates a memory-retrieval method where a disk drive or RAM can retrieve or save data directly to a specific address without having to scan through addresses in the medium to find it. CD-ROM is an example of a storage device that incorporates DASD technology.

Direct Current (DC)   DC is electrical current that is induced by a voltage source that does not change direction from positive to negative. DC can fluctuate, carry an analog signal by varying the DC current and voltage. DC can pulse, it can spike, it can do many things. The one thing that DC cannot do is change direction. Common sources of DC are batteries, AC power adapters, and power rectifiers.

Direct Inward Dial (DID)   A phone line that comes from the local phone company and connects to your PBX switch. A DID line has a phone number that has a DNIS or virtual directory number attached to it and is targeted to ring directly to a phone on the PBX network without going to a console operator or anywhere else first. The PBX system usually needs specific DID trunk (incoming line) hardware to make DID lines work.

Direct Inward System Access (DISA)   A feature of PBX phone systems where a user can dial a number that terminates into the PBX and either get another dial tone (with which to make long-distance calls) or to access their voice-mail system.

Direct Memory Access (DMA)   The transfer of data from a peripheral device, such as a hard disk drive, into *RAM (Random-Access Memory)* without that data passing through the microprocessor. DMA transfers data into memory at high speeds with no processor overhead.

Direct Outward Dial (DOD)   A feature of a PBX system that allows telephone stations to access outside dial tone or not access outside dial tone. If you pick up a phone on a PBX system and dial "9" for an outside dial tone, you might hear a siren sound instead. If you do, that means that the particular phone you are dialing on does not have DOD enabled. In most PBX systems, even though DOD is not enabled, an emergency "911" call will still go through. It is common to place special instructions on the phone to explain how to make an emergency call.

Direct Sequence Spread Spectrum (DSSS)   A type of radio modulation that carries binary data, consisting of voice, data, or video at rates up to 11 Mbps (Fig. D.18). Under the IEEE standard 802.11DS, which is a part of IEEE 802.11b, the DS modulation technique uses a "chipping sequence" of phase shifted transmissions in parallel for each bit. The sequence of bits is sent in parallel across a frequency spectrum consisting

Sample DSSS Modulation Chipping Sequence				
Sample Data Stream: **1 0 1 1 0**				
Chipping Code for 1 = 11001100				
Chipping Code for 0 = 01101101				
The DSSS Chipping Code Would be:				
11001100	01101101	11001100	11001100	01101101
1	**0**	**1**	**1**	**0**

Channel				
1		01101101	11001100		01101101
2	11001100	01101101		11001100	01101101
3	11001100		11001100	11001100	
Time				

Figure D.18   DSS Chipping

of 11 22-MHz wide channels that range from 2.400 Ghz to 2.483 GHz. Further, each chipping sequence is broadcast in duplicate over three alternating channels. So, when each bit is transmitted in its "chip" code format, one of the three channels is idle, while the other two transmit duplicate information. This is why 802.11DS allows for only three radios to operate in the same airspace—each radio needs 3 channels and there are only 11. The another common wireless LAN transmission method is FHSS (Frequency Hopping Spread Spectrum), which is limited to 2 Mbps, operates in the same frequency band, and does not have as long an operating range as DSSS.

Direct Station Selection (DSS)   A device that can be added to a PBX telephone set that has additional buttons on its face so that a user can see what extensions are in use (off hook) and which are free. When a call comes in to an answering agent, he/she can look at the direct station selection module attached to their phone and see whether the desired person is on their phone or not. Calls can be made and transferred to the extensions appearing on the DSS by pressing their associated button.

Directional Antenna   An antenna that is sensitive to the direction of the received or transmitted signal. Rabbit ears (dipole antennas) and "dish" type antennas are directional. Dish-type antennas are not only directional, but they focus a received transmission to the LNB/element (some call it the *stinger* because it is raised to the front of the dish). They also focus transmitted signals that are bounced into the dish and out to their destination.

Directional Coupler   A device that is engineered into a microwave antenna system that allows transmit and receive signals to be used on the same antenna. The device accomplishes this by differentiating the powerful transmit signal from the weak receive signal.

Dirty Power   Power that comes directly from the power company. The power in our homes is "dirty" power because it is not a pure 120V AC. As it travels cross country on power lines, it collects all kinds of EMI of all frequencies. Dirty power is also subject to unpredictable outages. Some devices that are used to clean up dirty power are UPS systems and surge suppressors.

DISA (Direct Inward System Access)   A feature of PBX phone systems where a user can dial a number that terminates into the PBX and either get another dial tone (to make long-distance calls on) or access their voice-mail system.

DISC (Disconnect) In X.25, a frame-layer (level two) command that is defined by the last three bits in a control byte of an unnumbered or "control" type of frame in the X.25 protocol being 010. It is used to notify another device that it is going offline. For more details, see *X.25 U Frame*.

Discard Eligibility (DE) DE is one of the bits in the standard 16-bit header of a frame-relay frame. It is set to 1 to notify the network that its frame should be discarded at the first onset of network congestion. See also *Frame Relay*.

Discard Eligible In reference to a committed information rate that a customer has paid for in conjunction with a frame-relay circuit, any data sent at a rate that exceeds the committed information rate is discard eligible, which means it will not be transmitted.

Disco (Disconnect) Many telephone and cable companies call their orders to disconnect a service "disco orders."

Disconnect Mode (DM) See *DM*.

Disconnect Supervision The ability of a PBX switch to recognize the disconnecting of the far end of a call. Keep in mind that in early switch days, people accessed trunks by picking up a phone and released them when the call was over. Machines do not know when the call is over because they are not the ones having the conversation. When you call someone, have a normal conversation, then hang up, you assume that the PBX system disconnected the path from your phone to the trunk and disconnected the link between that same trunk and the central office. Without disconnect supervision, the PBX does not know when to release (hang up) a central-office trunk. Without disconnect supervision, your trunks will soon all be busy, but no one will be on the phone!

Discovery Mode In Cisco Systems' routing methods, discovery mode is a reference to a mode/feature of a router operating system that builds and adjusts routing tables automatically. Router entries made within these operating systems can be static (manually entered by a user) or dynamic (automatic). It is also referred to as *dynamic configuration*. See also *Link State Algorithm*.

Discrete Multi Tone (DMT) A transmission format used by xDSL equipment that enables digital signals to be encoded and transmitted over variable-quality twisted copper pairs. DMT places 256 *QAM (Quadrature*

Amplitude Modulation) subchannels on a single copper pair. It is important to note that different xDSL equipment manufacturers have used (and still use) different line-coding techniques, including CAP and 2B1Q. ANSI standards (at the time of this writing) favor the DMT format. In DMT, each subchannel has a bandwidth of 4.3125 kHz. This is a total 1.104 MHz utilized within the pair. Depending on the type of QAM used within each channel, 2.048 Mbps (E1) is achieved easily. The characteristic of DMT that makes it such a great transmission method is that when it is initially attached to a copper pair, it goes through a loss and gain test cycle of all 256 channels. Any channel that is affected by noise, cross talk, poor attenuation from bridge taps, or other typical "bad" things about twisted copper plant, can be "turned off" and not used. The channels that have high throughput will be utilized. If there is a major change regarding the pair (for instance, if a new AM radio transmitter is installed in the area), the DMT transmission electronics actively monitor the channels affected and either modify the version of QAM on that channel or turn the channel off. See also *QAM*.

Dish   In telecommunications, a *dish* refers to a parabolic dish antenna. It has this name because its shape is a parabolic curve, so all radians from a single point are reflected into one direction. For a diagram, see *Parabolic Dish Antenna*.

Disk Controller   Either an expansion card in a personal computer or the electronics that are incorporated into a disk drive. A disk controller manages the hardware operations of disk drives and diskettes.

Disk Drive   A hardware device that is manipulated by a software program called a *disk operating system*. Some disk drives have a disk built in to them and some have interchangeable disks. The function of a disk drive is to store, read and write memory from the disks that are made for them. Different disk drives are capable of storing different amounts of data (measured in bytes). A 3.5" floppy disk is capable of storing up to 1.44 MB (1.44 million bytes). A CD-ROM disk is capable of storing up to 650 MB. Another kind of disk is a hard disk, which is built in to a hard-disk drive. These disks are not capable of being interchanged, but they can hold many times more memory. A common hard-disk drive in a personal computer can store more than 17 GB (17,000,000,000 bytes) of information. For photos, see *Hard Disk Drive* and *CD Rom Drive*.

Disk Mirroring   A data-storage technique where data files are written to and from two disk drives simultaneously using two separate disk controllers. The technique is used to provide redundancy in data storage so that the data would not be lost if a disk or driver were to fail.

Disk Operating System (DOS)   A software program that manipulates a disk drive. The DOS contains the information that tells the disk drive how to format, and read and write to disks. MS-DOS (Microsoft disk operating system) is probably the most common known disk operating system to PC users. Without a DOS program loaded onto your computer, it is virtually useless.

Disk Server   In a network environment, a disk drive within a PC or server that is accessible by users of the network. Data on the disk server can be shared by all users that have security privileges to the disk server.

Display   One of the several communications output interfaces to a user. Displays enable a computer or controlling device to communicate with a user visually. Displays are available in the form of monitors, LCD screens, and light emitting diodes. Other output devices that computers and controlling devices use to communicate with their users are printers, speakers, and lights (such as alarm indicators).

Distance-Sensitive Pricing   The pricing of communications services based on the distance between the two points connected by the phone company or service provider. Long-distance companies use a method of figuring the distance between two cities "airline mileage" (see *Airline Mileage* for more information). The price of the service is then based on the mileage that is calculated. The more mileage, the higher the price. Local exchange companies that provide a service across town figure the price by the number of central offices the line passes through. Each office requires a channel termination (also called *chanterm*) of the line as it enters and leaves each central office. Physically, a channel termination is a cross-connect from a channel of one transmission device to another.

Distance Vector Multicast Routing Protocol (DVMRP)   A software program that assists the operation of a router. DVMRP helps routers keep track of each other and the connections between them in a network. DVMRP uses *IGMP (Internet Group-Management Protocol)* to exchange routing datagrams with its neighboring routers. See also *Internet Group-Management Protocol* and *Distance Vector-Routing Algorithm.*

Distance Vector-Routing Algorithm   A class of routing algorithms used by network routers that repeat on the number of hops in a route to find a shortest path through a network. As instructed by the distance vector protocol, all routers in a network send their routing table information to their neighbors in what are called "updates." Distance vector-routing algorithms call for each router to send its entire routing table in each update, but only to its neighbors. The routers refer to the address tables that are developed by the updates to pass along data. Each of a router's ports has an associated table that contains a list of addresses that that port sends to and from. If a network that incorporates routers with distance vector-control software is configured incorrectly, routing loops can be formed, which can cause a network to crash. However, they are computationally simpler than link-state or hybrid routing algorithms. Distance vector-routing algorithms are also referred to as *Bellman-Ford routing algorithms.* See also *Link-State Routing Algorithm* and *Hybrid Algorithm.*

Distinctive Ringing   A feature of a PBX system that allows telephone sets to ring differently. This is a very nice feature if you would like outside calls to ring differently than internal calls from co-workers. This helps a user to know if they should say "hello" or "Emergency service, may I help you?" Distinctive ringing is also used in offices where many phones are in close proximity to each other. When one phone rings, all the people in the office know whose phone it is by tone, pattern, and pitch of the ring. Fax machines can also identify distinctive ring patterns and answer the ring type for which they are set.

Distortion   Any change to a signal's original waveform, except size (amplitude). Changing the size of a waveform is amplification (for larger) or attenuation (for smaller). The most common form of signal distortion is "clipping," which you can hear when the volume on a low-end stereo is turned up to high.

Distributed Data Management (DDM)   A software entity in IBM *SNA (System Network Architecture)* environments that provides peer-to-peer communication and file sharing. DDM is one of three SNA transaction services. See also *DIA* and *SNA Distribution Services.*

Distributed Queue Dual Bus (DQDB)   IEEE 802.6 Standard. A broadband protocol that is full duplex and implemented on fiber optic. The Distributed Queue Dual Bus is an architecture that is made of two serial busses, called *a* and *b,* which carry data transmissions in opposite directions simultaneously (hence full duplex). The busses can be implemented in a straight line or in a ring. If a fiber is cut for some reason, the node equipment reconfigures itself to accommodate for the disconnection. When this happens, the DQDB is divided into two networks and the individual nodes adjacent to the cut automatically restructure themselves as head ends (Fig. D.19).

DQDB open ended bus
with head ends
automatically reconfigured

Fiber Cut

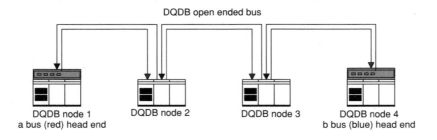

DQDB node 1
a bus (red) head end

DQDB node 2
Self reconfigured
as a "b" bus head end

DQDB node 3
Self reconfigured
as an "a" bus head end

DQDB node 4
b bus (blue) head end

DQDB open ended bus

DQDB node 1
a bus (red) head end

DQDB node 2

DQDB node 3

DQDB node 4
b bus (blue) head end

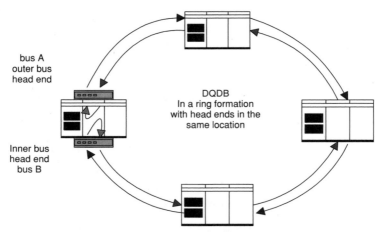

bus A
outer bus
head end

DQDB
In a ring formation
with head ends in the
same location

Inner bus
head end
bus B

Figure D.19   Distributed Queue Dual Bus

Distributed Routing Protocol   Also known as *Link State Routing Protocol, Interior Gateway Routing Protocol,* and *Shortest Path First.* A distributed routing protocol is a methodology used in router protocol design. This methodology enables routers within an autonomous network (i.e., corporate LAN) to identify each other and the status of their port connections. Distributed routing protocols create three databases within a router's memory: a neighboring router database, a link database, and a routing table. The routing table is created by applying Dykstra's algorithm to the first two databases. The most widely used interior gateway routing protocol is Open Shortest Path First (OSPF). See also *OSPF.*

Distribution Cable   Cable that connects a PBX switch or telephone company central office to its customers (Fig. D.20). It is the cable system

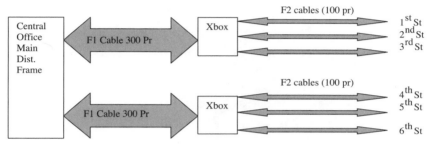

Figure D.20A   Distribution Cable: Shown as 300-Pair Feeds

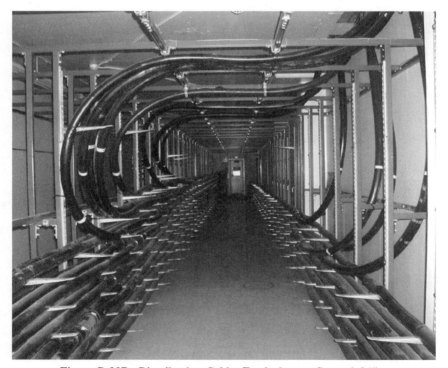

Figure D.20B   Distribution Cable: Feeds from a Central Office

of outside plant. Distribution cable usually has two parts, an F1 (facility 1) and an F2 (facility 2). The F1 goes from the central office to an access point (AP) or cross-connect point where it is then cross-connected to F2 pairs. The telephone cable that runs along neighborhood streets is usually F2 cable and the cables that run along main roads are usually F1 cables. For a picture of a cross box, see *Access Point*.

Distribution Frame   This is also called a *Main Distribution Frame (MDF)*. It is the place where all the wire, fiber optic, or coax for a network is terminated (Fig. D.21). The distribution frame is usually placed as close to the central office switch or PBX as possible.

Figure D.21A   Distribution Frame for a PBX

Figure D.21B   Distribution Frame for a PSTN Central Office

Distribution Layer   One of the three LAN network design layers. In LAN network design, the three switch layers are core layer, distribution layer, and access layer (Fig. D.22). The core layer provides redundancy for distributing traffic across multiple access layers. The access layer is generally a pair of high-performance switches that provide switching and routing among a group of access switches networked by trunks. The access layer provides switches where users connect, so high port quantity is desired in this layer.

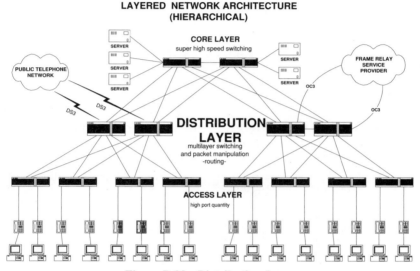

**LAYERED NETWORK ARCHITECTURE
(HIERARCHICAL)**

Figure D.22   Distribution Layer

Dithering   A variable error in a *GPS (Global Positioning System)* latitude and longitude signal. The error is purposely integrated into the positioning system to prevent anyone other than the government from having absolutely precise positioning information. A typical dithering error in a civilian-purchased GPS unit is about ±0 to 50 feet.

Divestiture   The break up of AT&T by the United States federal government due to a business practice was considered to be of monopolistic nature, effective December 30, 1983. AT&T and its 22 Bell operating companies were separated. The 22 Bell companies were combined into seven regional Bell operating companies (RBOCS). AT&T was legally limited to the long-distance business (although they were allowed to be in the computer business), and the seven RBOCs were limited to the local telephone business. Both AT&T and the seven RBOCS were restricted from manufacturing telecommunications equipment and from sharing any customer or market information. This judgment was made by Judge

Harold Greene, and it paved the way for competition in the U.S. telecommunications industry. The stipulations on the companies involved are now changing. AT&T and the RBOCS are allowed to enter each other's businesses and compete with each other, and other new communications companies (called *CLECs, Competitive Local-Exchange Carriers*, and *CAPs, Competitive-Access Providers*, and new long-distance companies) are building networks and offering communications services.

DLC (Digital Loop Carrier)   Equipment that is used to provide two dial tones over a single twisted copper pair (Fig. D.23). If a customer wants to have an additional telephone line installed and no more twisted pairs are left to feed their area, the telephone company can install a DLC, which will make two phone lines work on one pair. The DLC has two parts. The first part is a central office DLC unit, which is rack mountable. The two phone lines (the original one and the new one) are cross-connected into this unit. The second part is a modified network interface that contains electronics. It receives the two phone services transmitted from the central office DLC unit over one twisted pair and provides a hand-off to the customer's wire as two separate twisted pairs. The DLC accomplishes this by taking the two analog lines and digitally multiplexing them onto one pair. DLC works great for voice applications, but it can have abnormal effects on fax machines and modems.

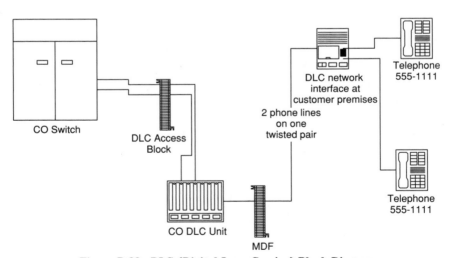

Figure D.23   DLC (Digital Loop Carrier) Block Diagram

DLC (Data-Link Connection)   A user connection to a frame-relay network that could be thought of as a virtual channel on a multiplexer. A DLC is a half duplex (one way) data channel. To get full duplex communications in frame relay, the DLCs from each user are combined to provide full-duplex communications ability. Hence, two DLCs constitute a *PVC (Permanent Virtual Circuit)*. See also *Data-Link Connection Identifier.*

DLCI (Data-Link Connection Identifier)   A number given to a *DLC (Data-Link Connection)* on a frame-relay service (Fig. D.24). DLCIs are provided to customers by the frame-relay service provider. It is the part of the frame-relay packet header that provides space for 10 bits of data that is used to name the logical channel that the frame will be routed to. Furthermore, the DLCI is embedded into the data stream from the very beginning of the transmission to the very end. DLCIs range from 0 to 1023. Routers use DLCIs to specify a *PVC (Permanent Virtual Circuit)* or *SVC (Switched Virtual Circuit)* that links to other routers or devices in a frame-relay network. In the basic frame-relay specification, DLCI values that are given to links or routes are different for every individual router in a network. Different routers can use different DLCI numbers to identify the same route (connection) in the same network. In the *LMI (Local Management Interface)* extended specification for frame relay, DLCIs are "globally significant" and the same DLCI numbers are given to all routes by all routers. The DLCI is very similar to (and

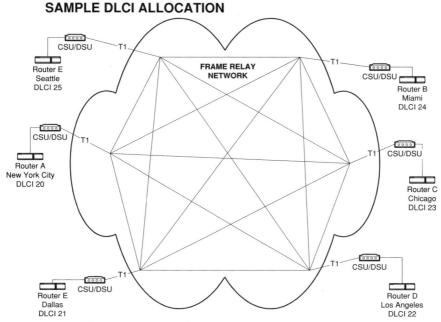

Figure D.24   DLCI (Data-Link Connection Identifier)

performs the same task as) the *LCN (Logical Channel Number)* in X.25. For a table of DLCI allocations, see *DLCI*. See also *LMI*.

DLE   ASCII control-code abbreviation for data link escape. The binary code is 0000001, the hex is 01.

DLL (Dynamic Link Library)   An executable subroutine stored as a file that is separate from the programs that could use it. DLLs allow for efficient use of memory and applications to share program resources. DLL file extensions are .DLL, .DRV, and .FON.

DLSw (Data-Link Switching)   An interoperability standard, described in RFC 1434, that provides a method for forwarding SNA and NetBIOS traffic over TCP/IP networks using data-link layer switching and encapsulation. DLSw uses *SSP (Switch-to-Switch Protocol)* instead of IBM *SRB (Source Route Bridging)*, eliminating the major limitations of SRB, such as hop-count limits, broadcast traffic, timeouts, lack of flow control, and lack-of-prioritization schemes. See also *SRB (Source Route Bridging)* and *SSP (Switch-to-Switch Protocol)*.

DM (Disconnect Mode)   A command defined by the last three bits in a control byte of an unnumbered or "control" type of frame in the X.25 protocol being 000. It is a code that a device sends when it has come into service and is ready to receive a reset SABM to synchronize into the link. For more details, see *X.25 U Frame*.

DMA (Direct Memory Access)   A technology incorporated into PC architecture that provides a means of transferring data from a peripheral device, such as a hard disk drive, into *RAM (Random Access Memory)* without that data passing through the microprocessor. DMA transfers data into memory at high speeds with no processor overhead.

Dmarc (Demarcation Point)   A Dmarc is where the local phone company hands-off a telephone circuit. A Dmarc can be in the form of a standard network interface for a residential line, a DSX panel for a T1 or T3, an RJ212X for a business line, or an RJ45 for an ISDN line. The Dmarc separates customer-owned equipment from telephone company-owned equipment. It is also the place where responsibility for the circuit's performance is separated.

DMT (Discrete Multi Tone)   A transmission format used by xDSL equipment that enables digital signals to be encoded and transmitted over variable-quality twisted copper pairs (Fig. D.25). DMT places 256

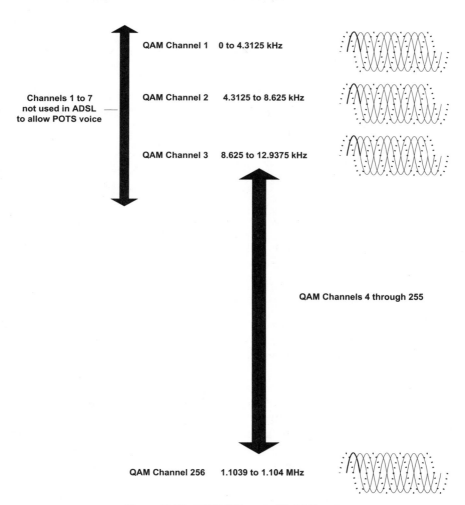

DMT
Discrete Multi Tone

QAM Channel 1   0 to 4.3125 kHz

Channels 1 to 7
not used in ADSL
to allow POTS voice

QAM Channel 2   4.3125 to 8.625 kHz

QAM Channel 3   8.625 to 12.9375 kHz

QAM Channels 4 through 255

QAM Channel 256   1.1039 to 1.104 MHz

Figure D.25   DMT (Discrete Multi Tone)

QAM (Quadrature Amplitude Modulation) subchannels on a single copper pair. It is important to note that different xDSL equipment manufacturers have used (and still use) different line-coding techniques, including CAP and 2B1Q. ANSI standards (at the time of this writing) favor the DMT format. In DMT, each subchannel has a bandwidth of 4.3125 kHz. This is a total 1.104 MHz utilized within the pair. Depending on the type of QAM used within each channel, 2.048 Mbps (E1) is achieved easily. The characteristic of DMT that makes it such a great transmission method is that when it is initially attached to a copper pair, it goes through a loss and gain test cycle of all 256 channels. Any channel that is affected by noise, cross talk, poor attenuation from bridge taps, or other typical "bad" things about twisted copper plant can be "turned off" and not used. The channels that have high through-put will be utilized. If there is a major change regarding the pair (for instance, if a new AM radio transmitter is installed in the area), the DMT transmission electronics actively monitor the channels affected and either modify the version of QAM on that channel or turn the chan-nel off. See also *QAM*.

DN (Directory Number)   In Nortel PBX administration, the reference to an extension number. The DN is a software entity. The DN extension number can be a button on a phone, an ACD agent, an ACD queue, or trunk route. See also *TN*.

DND (Do Not Disturb)   A feature of PBX telephone sets to disallow any calls or pages while the feature is activated. The feature is usually activated and deactivated by pushing the "do not disturb" (DND) but-ton on the phone.

DNIC (Data Network Identification Code)   Part of an X.121 address. DNICs are divided into two parts. The first part specifies the country in which the addressed PSN is located and the second specifies the PSN itself. See also *X.121*.

DNIS (Dialed Number Identification Service)   A service from your phone company that is similar to Automatic Number Identification (caller ID), except, instead of providing the caller's number, the number (or a four-five digit DID type routing DN is provided) the caller dialed is pro-vided. This is useful when a company receiving the call has several in-coming numbers. If certain numbers dialed by customers through a spe-cific DID trunk determine how the call should be handled, then an *ACD (Automatic Call Distributor)* system can use those digits to route the call to a certain extension. If 10 trunks have four different 800 numbers ring to them, the 800 number that is advertised in Spain can be routed to Spanish-speaking personnel and an 800 number advertised in France can be routed to French-speaking personnel.

DNS (Domain Name System)   Also called *domain name server.* An alternative way of identifying network addresses on the Internet or in LANs. The DNS and some network software tools give a network administrator the ability to identify a server as Astro1 or Beta2, or identify an Internet address of 225.225.225.10 as *www.domainnamesystem.com*. DNS makes addresses easier for people to remember and identify with. It stores and tracks these names and addresses, just so we can use names instead of numbers when sending messages. In a DNS environment, a ping command could be used with the DNS name as well as the hardware address.

Do Not Disturb   See *DND*.

Document Interchange Architecture (DIA)   A method or format for transferring data files. DIA defines the protocols and data formats needed for the transparent interchange of data files in an IBM *SNA (System Network Architecture)* network. DIA is one of three SNA transaction services. See also *Distributed Data Management* and *SNA Distribution Services*.

Documentation   Information regarding a network that is updated to allow others that are involved in managing a network do so in an efficient and timely manner. Managing a network without the proper documentation on the components and the way in which the components are connected can be very cumbersome, unreliable, and costly.

DOD (Direct Outward Dial)   A feature of a PBX system that allows telephone stations to access outside dial tone or not access outside dial tone. If you pick up a phone on a PBX system and dial "9" for an outside dial tone, you might hear a siren sound instead. If you do, the particular phone you are dialing on is not DOD enabled. In most PBX systems (even though DOD is not enabled), an emergency "911" call will still go through. It is common to place special instructions on the phone to explain how to make an emergency call.

Doghouse   A closure that contains cellular/PCS transmission equipment. Doghouses can come with heater/air-conditioner units (environmental control) and are about the size of a small doghouse. Near some cellular/PCS antennae is a small building, called a *hut*. Inside the hut is where the doghouse is located.

Domain   1. For Windows NT Environments, a group of resources (servers, printers, workstations, etc.) that are controlled by or authenticated (granted access) to the same Windows NT directory data base. Domains exist in Windows NT for security and access control.   2. For the Internet, a domain is an Internet address, such as 225.225.225.10. This address can be identified through *DNS (Domain Name System)* as a name, such as *www.domainnamesystem.com*. That name/address could attach to another network or Internetwork. 3. In general or other telecom environments (such as IBM SNA), a domain is a reference to a group of resources.

Domain Controller   For Windows NT, the server or computer on which the Windows NT software is running.

Domain Name   A name that identifies one or more IP addresses. A good example of a domain name is *mcgraw-hill.com*. When this name is entered into the address field of an Internet browser, *DNS (Domain Name System)* translates this to the IP address 208.243.114.191. The *HTML (HyperText Markup Language)* Web page that resides at this address will then be to transferred to the browser that requested it.

Domain Name System (DNS)   Also called *domain name server*. An alternative way of identifying network addresses on the Internet or in LANs. The DNS and some network software tools give a network administrator the ability to identify a server as Astro1 or Beta2, or identify an Internet address of 225.225.225.10 as *www.domainnamesystem.com*. DNS makes addresses easier for people to remember and identify with. It stores and tracks these names and addresses, just so we can use names instead of numbers when sending messages. In a DNS environment, a ping command could be used with the DNS name as well as the hardware address.

Domain-Specific Part (DSP)   A reference to the part of an *NSAP (Network Service Access Point)* format ATM address that contains an area identifier, a station identifier, and a selector byte. See also *Network Service Access Point*.

Dongle (Dongle Key)   A device for protecting copyrights on computer software that looks very much like a DB25 gender-changer/adapter. Inside the dongle is usually an encoded ROM circuit with a user-rights serial number burned into it. If the dongle is not plugged into the printer port of the PC that the software is loaded on, the software does not work (Fig. D.26).

DOS (Disk Operating System)   See *Disk Operating System*.

DOS Attack   A hacker method of bringing a network server down through overutilization of the server's CPU. The hacker opens multiple DOS sessions with the server until it cannot process all of the sessions simultaneously. The server ultimately freezes up or shuts down.

Dot Address   A reference to the common notation for IP addresses in the form *N.N.N.N.*, where each number "N" represents one byte

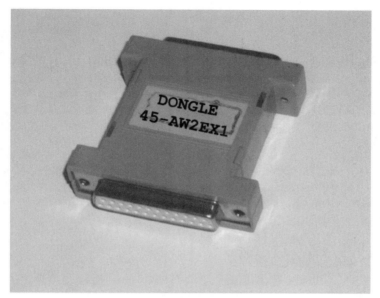

Figure D.26   Dongle Key

(converted to decimal) of the four-byte IP address. It is also called *dotted notation* or *four-part dotted notation*. After *DNS (Domain Name System)* was developed, it became necessary to specify which address was being referred to (the dot address or the name address). See also *Domain Name System*.

Dotted Notation   See *Dot Address*.

Dot Pitch   A measure of computer monitor resolution capability. In computer CRT tubes, phosphors are arranged in tiny triangular patterns. Each triangle is composed of a green, red, and blue dot. The dot pitch is the distance in millimeters between dots of the same color from one triangle to its neighbor. The smaller the dot pitch, the higher the detail capability of the monitor. See also *Stripe Pitch*.

Double Word   In computer memory, a word is 16 bits, which is one data unit processed by the bus. Newer computers and other processing systems are built with 32- and 64-bit busses, which gives them the ability to process double words (32 bits) and quad words (64 bits).

Downstream   In asymmetrical broadband transmissions, a reference to the bandwidth or information flow away from the service provider and toward the customer/subscriber. ADSL and cable-modem/Internet services are asymmetrical. They consist of a larger downstream and smaller upstream component. Asymmetrical transmissions are best suited for end-user Internet services.

DPSK (Differential Phase Shift Keying)   A pre-V-series modem standard. This phase-modulation method was used in Bell 201 standard modems. DPSK incorporated dibits (two bits represented by a phase shift). The code representation is as follows: A phase shift of the current carrier phase by 45 degrees represents a 00 binary value, a 135-degree phase shift from the present carrier phase represents a 01 binary value, a 315-degree phase shift from the present carrier phase represents a 10 binary value, and a 225-degree phase shift from the present carrier phase represents a 11 binary value.

Drag Line   A string or rope pulled into a conduit for making future wire or cable installation easier.

Drift   When a carrier frequency changes unintentionally because of a transmitter problem. Drift can occur because of a temperature change. Drift can also be caused by bad connections, or defective components. Crystal oscillators are the most drift-reliable circuits. Frequency drift can also be caused by temperature changes in the atmosphere, because of the diffraction of the radio signal as it travels through different densities of air. Looking down a road toward the horizon on a hot day you might notice that the road and other objects look like they are wet or wavy. This is a visual example of atmospheric diffraction, or drift.

Drive Bay   A preprovisioned slot located on the front of a personal computer that provides a space to install an internal disk drive, such as a CD-ROM or floppy disk drive.

Drive Ring   A ring with a nail attached to it, used to fasten or hold drop wire to telephone poles or sides of buildings (Fig. D.27).

Driver Software   A small software program that contains operating instructions for a computer to control a connected or integrated device, such as a printer, modem, or sound card.

Drop   Another term for *service wire.* The service wire is the aerial or underground wire that runs from your home or office to the terminal in your back or front yard. Abbreviations are *ASW (Aerial Service Wire)* and *BSW (Buried Service Wire).*   2. A twisted pair or fiber-optic line that connects a host to a node in a LAN environment.

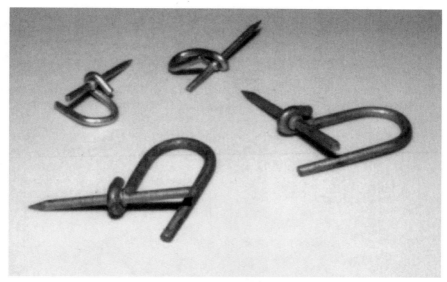

Figure D.27   Drive Rings

Drop Cable   Another term for *service wire*. See *Drop*.

Drop Clamp   A device that is used to attach an aerial service wire to a "J hook" or "Ram's horn," which is attached to a building or pole.

Drop Reel   A reel that is used to transport and distribute drop wire during installation (Fig. D.28).

Drop Wire   Another term for *service wire*. See *Drop*.

Dry T1   A T1 that is without the –135V DC battery voltage. A CSU/DSU has the ability to convert a wet T1 to a dry one. The T1 circuit is transmitted with a –135V DC voltage on the public telephone network to power repeaters and other conditioning equipment. A dry T1 is also called a *DS0*.

DS (Digital Service)   The prefix for digital service circuits. A comparison of the DS-level circuits and other carriers is in Fig. D.27.

DS (Direct Sequence)   In wireless LAN, a shortened version of the term *DSSS (Direct Sequence Spread Spectrum)*. A type of radio modulation that carries binary data, consisting of voice, data, or video at rates up to 11 Mbps. Under the IEEE standard 802.11DS, which is a part of IEEE 802.11b, the DS modulation technique uses a "chipping sequence" of

Figure D.28   Drop Wire Reel

phase-shifted transmissions in parallel for each bit. The sequence of bits is sent in parallel across a frequency spectrum consisting of 11 22-MHz wide channels that range from 2.400 Ghz to 2.483 GHz. Further, each chipping sequence is broadcast in duplicate over three alternating channels. So, when each bit is transmitted in its "chip" code format, one of the three channels is idle while the other two transmit duplicate information. This is why 802.11DS allows for only three radios to operate in the same airspace—each radio needs 3 channels and there are only 11. The another common wireless LAN transmission method is FHSS (Frequency Hopping Spread Spectrum), which is limited to 2 Mbps, operates in the same frequency band, and does not have as much operating range as DSSS.

DS0 (Digital Service Level Zero)   64 Kb/s. Equivalent of one voice (or one analog POTS) line. A DS0 is the basic building block of which all the other DS services are comprised. A DS0 can come in two flavors, in-band signaled and out-of-band signaled. The in-band signaled DS0 is best suited for carrying voice applications. It has a 56 Kb/s user bandwidth and an 8 Kb/s channel is bit-robbed from the total 64 Kb/s DS0 bandwidth. The signaling channel is for carrying dialed digits, of hook and dial-tone signals for a central office. With the in-band signaling format, 24 DS0s can be carried on a T1. The counterpart is an out-of-band signaled DS0, which has a bandwidth of 64 Kb/s. This format is best suited for data transmissions. You can get 23 out-of-band signaled DS0s on a T1. The 24th channel is used for the signaling of the other 23.

DS1 (Digital Service Level 1)   1.544 Mb/s. Another name for a T1 (Fig. D.29). The specific difference between a DS1 and a T1 is that the T1 is on copper and comes with a –135-V battery voltage, and the DS1 is a dry circuit, on copper or fiber-optic lines, with no battery voltage. Other than that, they are the same. A DS1 has a total bandwidth or transmission speed of 1.544 Mb/s. The 1.544 Mb/s is divided into 24 64 Kb/s channels. A DS1 (T1) is available in several different packages that offer different line formats and framing formats. The package that a customer requests from a phone company depends on what they want to use the DS1 for and what kind of equipment they have. Telecommunications customers use DS1 circuits as private lines to connect data devices from one geographical place to another or to transport large amounts of dial tone to the premises. DS1 circuits are also used to connect directly to a long-distance company for broadband WAN service. Telecommunications companies also use DS1 (they are T1 circuits within their own network) circuits to provide more telephone service where a shortage of twisted pairs is available (see *SLC96*).

DS1circuit/line types and applications

Line format/coding	framing format	signaling	Application
AMI	SF/D4	in-band	24 voice/modem channels
AMI	ESF	in-band	24 voice/modem channels
AMI	ESF	out-of-band	23 voice/modem or digital/data channels
B8ZS	SF/D4	in-band	24 voice/modem channels
B8ZS	ESF	in-band	24 voice/modem channels
B8ZS	ESF	out-of-band	23 voice/modem or digital/data channels

Figure D.29   DS1/T1 Line Coding and Framing Formats

DS1C (Digital Service Level 1)   A digital signal that combines two DS1 channels. The aggregate frequency is 3.152 Mb/s, which contains 3.088 Mb/s of payload, (two DS1s) and 64 Kb/s (one DS0) of overhead for transmission control.

DS2 (Digital Service Level 2)   6.312 Mb/s. A DS2 is four DS1 channels multiplexed together within a DS3 multiplexer. A DS2 is not available to customers. It is just a step in the creation of a DS3.

DS3 (Digital Service Level 3)   44.736 Mb/s. A DS3 is a circuit that is provided to customers by telephone companies (Fig. D.30). It is a transport for 28 T1 circuits, which adds up to 672 DS0 circuits (voice channels). Telecommunications customers use DS3 circuits as private lines to connect data devices from one geographical place to another or to transport large amounts of dial tone to the premises. DS3 circuits are also used to connect directly to a long-distance company for broadband WAN service. Telecommunications companies also use DS3 circuits to provide more telephone service where a shortage of twisted pairs is in their cable plant. Sometimes it is less expensive for a telephone company to install the DS3 electronics in areas, rather than long feeds of large twisted copper-pair cables. DS4 (Digital Service Level 4) 274 Mb/s. A DS4 is a transport for six DS3 circuits. Its capacity in DS1 circuits is 168. The capacity in DS0 circuits is 4032.

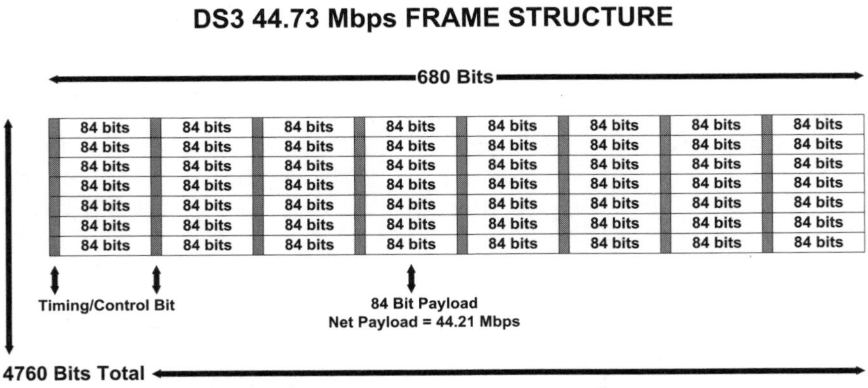

DS3 44.73 Mbps FRAME STRUCTURE

Figure D.30   DS3 44.73 Mbps Frame Structure

DSL (Digital Subscriber Loop)   See *xDSL, VDSL, HDSL, IDSL, RADSL, ADSL, QAM, DMT,* and *DSLAM.*

DSL Inline Filter   A device that isolates the background hiss on an xDSL line from telephones (Fig. D.31). For a diagram of how they connect, see *ADSL.*

DSLAM (Digital Subscriber Line-Access Multiplexer)   Pronounced "dee-slam." The distribution device for *xDSL (x Digital Subscriber*

Figure D.31   ADSL In-Line Filter

Loop) service from a central office (Fig. D.32). The DSLAM combines and separates the different formats of communications contained in the xDSL carrier and routes them to their respective hosts. Because xDSL is capable of carrying voice, video, and data, the voice needs to be routed to a voice central office switch, the video needs to be routed to a CATV head-end and the data needs to be sent to a packet switch. The DSLAM performs all of these functions.

DSP (Digital Signal Processor)   An enhancement to Ethernet switch ports (most commonly gateway ports in IP Telephony) that gives them the ability to act as media termination points during the execution of user features on IP telephony/server-based PBX systems.   Also referred to as an *MTP (Media Termination Point)*.   The DSP within an Ethernet switch or router utilizes its built-in codecs and transcoders to temporarily hold a connection while an IP telephony server locates an available bandwidth channel or waits for a user to execute feature keystrokes on a telephone.

DSP (Domain-Specific Part)   A reference to the part of an *NSAP (Network Service Access Point)* format ATM address that contains an area identifier, a station identifier, and a selector byte. See also *Network Service Access Point.*

DSR (Data Set Ready)   Pin 6 of a DB25 connector wired for the RS232C protocol. This is the wire that a modem or SDI device uses to send a signal that acknowledges that it is ready to receive data.

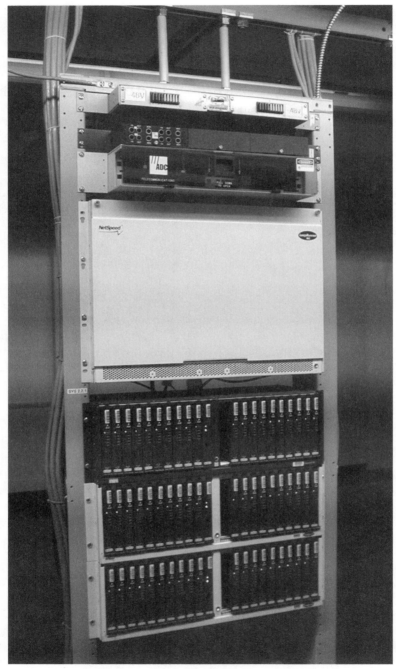

Figure D.32   DSLAM (Digital Subscriber Line-Access Multiplexer)

DSS (Digital Switched Service)   1. A service offered by telephone companies where a telephone line is switched while still in its digital form. Many telephone lines that leave a customer's premises via T1 are converted to analog when they reach the telephone company central office. DSS lines are run directly into the central office telephone switch in digital form (64 Kbp/s per line).   2. A device that can be added to a PBX telephone set that has additional buttons on its face so that a user can see what extensions are in use (off hook) and which are free. When a call comes in to an answering agent, he or she can look at the direct station selection module attached to their phone and see whether the desired person is on their phone or not. Calls can be made and transferred to the extensions appearing on the DSS by pressing the associated button.

DSSS (Direct Sequence Spread Spectrum)   A type of radio modulation that carries binary data, consisting of coded voice, data, or video at rates up to 11 Mbps. Under the IEEE standard 802.11DS, which is a part of IEEE 802.11b, the DS modulation technique uses a "chipping sequence" of phase-shifted transmissions in parallel for each bit. The sequence of bits is sent in parallel across a frequency spectrum consisting of 11 22-MHz wide channels that range from 2.400 Ghz to 2.483 GHz.   Further, each chipping sequence is broadcast in duplicate over three alternating channels. So, when each bit is transmitted in its "chip" code format, one of the three channels is idle while the other two transmit duplicate information. This is why 802.11DS allows for only three radios to operate in the same airspace—each radio needs 3 channels and there are only 11. The another common wireless LAN transmission method is FHSS (Frequency Hopping Spread Spectrum), which is limited to 2 Mbps, operates in the same frequency band, and does not have as much operating range as DSSS.

DSU (Data Service Unit, CSU/DSU, Channel Service Unit/Data Service Unit)   A DSU is a hardware device that is available in many shapes and sizes. Rack-mount, shelf-mount, and stand-alone DSUs are available. A CSU/DSU has three main functions. The first function is to act as a demarcation point for a T1 (DS1) service from a local communications company. The second function is to provide line format and line-code conversion (B8ZS to AMI, SF or D4 to ESF, 135 V to 0 V) between the public-network and the customer-premises equipment, if necessary. The third function is to provide maintenance or alarm services and loop-back for isolating problems with the T1 line or customer's equipment. For a photo, see *CSU/DSU.*

DSVD (Digital Simultaneous Voice and Data)   A type of modem that allows users on either end to exchange data files and talk over the same phone line. Both users must be using a DSVD-compatible modem in order to use the DSVD features.

DSX (Digital Service Cross-Connect)   A reference to a digital-service termination/patch panel that allows DS1 and DS3 circuits to be monitored by test equipment, such as a TTC Tberd or T-ACE. DSX panels are usually terminated via wire-wrap and accessed at the front panel by bantam-type connector jacks. For a photo of wire-wrap terminals, see *Wire-Wrap*. For a photo of a DSX panel, see *Digital Service Cross-Connect Panel*.

DTE (Data Terminating Equipment)   DTE is equipment that receives a communications signal. For a data connection to work between I/O (input/output) devices, one needs to be designated the "communications-sending" equipment and one the "communications-terminating" or DTE (data-terminating equipment). A computer's printer port is a DCE port, a printer is a DTE device. A practical way to classify the two is: DCE is the sender of data and the DTE is the receiver of data.

DTMF (Dual-Tone Multiple Frequency)   The tones that you hear when you dial a single-line push-button phone. The tones are a mixture of two frequencies. For a diagram, see *Dual-Tone Multiple Frequency*.

DTMF Cut Through   A feature of voice-response systems, voice-mail systems, and auto attendants to hear the digits that you dial and play a RAN (recorded announcement) at the same time. This feature reduces the frustration level for people who hate to listen to voice-message systems because when the listener makes their choice (pushes a key) the selection is executed and the RAN stops immediately.

DTR (Data Terminal Ready)   A light on the front of a modem or data-communications device that indicates that it is ready to receive a handshake signal from another communications device.

DTU (Digital Test Unit)   Some DTUs are stand-alone devices, and some are add-ons for integration into telecommunications equipment.

Dual Homing   Another term for *alternate routing, redundancy,* or *self-healing* that is used in telecommunications networking.

Dual Ring of Trees   A network topology that uses a dual ring topology as a backbone for other ring or star networks (Fig. D.33).

Dual-Tone Multiple Frequency (DTMF)   The tones that you hear when you dial a single-line push-button phone (Fig. D.34). The tones are a mixture of two frequencies. The frequencies are connected according to the diagram provided.

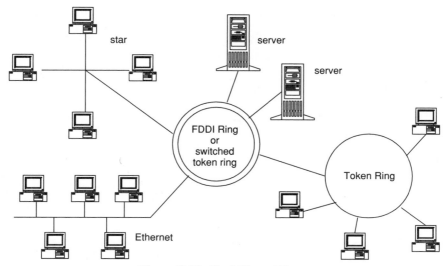

Figure D.33   Dual Ring of Trees

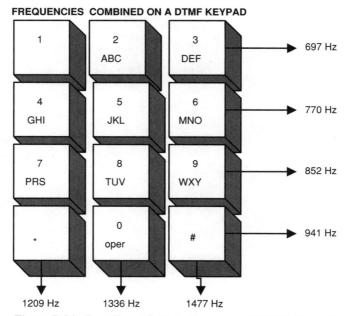

Figure D.34   Dual-Tone Multiple Frequency (DTMF) Keypad

Duct   A conduit or pipe that runs from one location to another, within a building, connecting buildings, or connecting cable vaults. It is a good practice to install a duct, then pull cable into it, rather than directly mounting or burying the cable.

Dumb Terminal   An I/O (input/output) communications terminal. They are called "dumb" because they do not do any processing of information, they just display it. Dumb terminals are used to display input/output information from ACD systems, PBX switches, SONET transport equipment, or other interface applications. Dumb terminals are not used as much anymore because "terminal-emulation" programs are available for personal computers. Examples of popular dumb terminals are the Wyse 50 and VT100 (Fig. D.35).

Figure D.35   WYSE 50 Dumb Terminal

Dummy Load   A device that is connected to electronic output equipment, such as radio transmitters, power supplies, and even stereo systems to test them under full strain. One dummy load for a home stereo system, for example, is two large 8-ohm resistors, one for each channel (left and right) in place of the speakers. The resistors allow a full volume-range (power output) test without having to endure the sound.

DUN (Dial-Up Networking)   A computer-communication method where a POTS telephone line is used in conjunction with a modem for the physical and data-link layer connection and a specific protocol is used to transfer information. In Internet applications, *PPP (Point-to-Point Protocol)* or *SLIP (Serial Line Internet Protocol)* is used.

Duplex Jack   A jack or connecting block with two jacks on its face.

Duplex Transmission (Full Duplex)   A communications protocol that has the ability to send and receive at the same time. A DS1 is a full-duplex protocol that carries other protocols. The alternative to full duplex is half duplex, where two communications devices take turns sharing a line. Humans speak half duplex, because it is too hard to have a meaningful conversation while both people are talking at the same time. CB radios and "Walkie Talkies" are also half duplex.

DVD (Digital Versatile Disk)   A memory-storage system that comes in the same footprint and hardware format as the CD-ROM. DVDs have a storage capacity of 4.7 GB to 17 GB of data on each side of the disk. DVD drives are able to access data at a rate of 1.3 Mbps.

DVD2 (Digital Versatile Disk, 2nd Generation)   A version of the DVD disk drive that can read CD-R and CD-RW disks, as well as audio CDs. See also *DVD*.

DVD-RAM   A standard for DVD disks that can be written to and erased over 10,000 times and have a storage capacity of 2.6 GB per side. DVD-RAM formatted disks are not compatible with DVD-RW disks and vice versa at the time of this writing.

DVD-ROM (Digital Video Disk Read-Only Memory)   A DVD standard that enables multimedia entertainment to be provided in improved formats over VHS video recordings, and can also play audio CDs, CD-I discs, and CD-R discs. The DVD-ROM standard disc can hold up to 17 GB of data. The standard compression format used to place video and data on them is MPEG-2.

DVD-RW (Digital Video Disc ReWriteable)   A standard for DVD disks that can be written to and erased more than 10,000 times, and have a capacity of 3 GB per side. DVD-RAM formatted disks are not compatible with DVD-RW disks and vice versa at the time of this writing.

DVMRP (Distance Vector Multicast Routing Protocol)   A software program that assists the operation of a router. DVMRP helps routers keep track of each other and the connections between them in a network. DVMRP uses *IGMP (Internet Group Management Protocol)* to exchange routing datagrams with its neighboring routers. See also *IGMP.*

DWDM (Dense Wave Division Multiplexing)   Another term for *WDM (Wave Division Multiplexing)*, which is to separate different transmissions by frequency, as AM and FM radio broadcasts are. WDM is a specific reference to separating different transmissions by using *different colors (or frequencies) of light* within the same fiber optic strand. See also *WDM.*

Dynamic Bandwidth Allocation (DBA)   A feature of nonconnection-oriented protocols at the physical layer, such as X.25, frame relay, and ATM. These telecommunications protocols permit a variable number of users to utilize the full bandwidth of a connection. One method is statistical time-division multiplexing used in X.25, which assigns packets to available channels, regardless of the user or the number of users. ATM and frame relay are also capable of allocating bandwidth to users until the local loop is fully utilized. This is a natural feature of these protocols because no user actually has to make a connection, or reserve a channel over the physical layer to make a transmission. See also *STDM, Connection-Oriented Protocol,* and *Connectionless Network Protocol.*

Dynamic Configuration   A reference to a mode/feature of a router's operating system that builds and adjusts routing tables automatically. Router entries made within these operating systems can be static (manually entered by a user) or dynamic (automatic). It is also referred to as *discovery mode* in Cisco Systems networks. See also *Link-State Algorithm* and *Dynamic Host Configuration Protocol.*

Dynamic Host Configuration Protocol (DHCP)   A protocol that provides the specific service within a network of automatically configuring hosts/workstations within that network. DHCP is capable of automatically configuring the IP address, subnet mask, default gateway, and DNS addresses.

Dynamic Link Library (DLL)   An executable subroutine stored as a file that is separate from the programs that could use it. DLLs allow for efficient use of memory and applications to share program resources. DLL file extensions are .DLL, .DRV, and .FON.

Dynamic Load Balancing   A feature of *ACD (Automatic Call Distributor)* systems to evenly distribute incoming calls to agents.

Dynamic Memory (DRAM)   RAM (random access memory) that holds its data as long as the power is on. The other popular RAM memory is static memory, which is slower, but holds its data when the power is interrupted. See also *NVRAM* and *RAM*.

Dynamic Routing   A reference to a type of operating system within a router that adjusts its routing/address tables automatically when the internetwork that it resides in changes. An example of a network change would be to add a new connection between two existing routers or to bridge two networks together. The function within the operating system that makes dynamic routing possible is called by several names, including *discovery, multicasting,* and *flash updating.* An example of a *dynamic routing protocol* is Cisco System's *Enhanced IGRP (Enhanced Interior Gateway-Routing Protocol).* See also *Link-State Algorithm.*

Dynamic VLAN   There are two kinds of VLANs: static and dynamic. Static VLANs are associated with switch ports, and dynamic VLANs are associated with the MAC addresses of devices attached to the switch. Dynamic VLANs allow users to move to another office, which could have a data connection installed. The switch would recognize the MAC address of the device and automatically include its traffic in the same VLAN as the previously connected switch port. Dynamic VLANs are not recommended in large hierarchical campus networks or multilayered switched enterprise networks due to troubleshooting complexity. Dynamic VLANs have an inherent nature to eventually overutilize a network backbone in campus environments. Dynamic VLANs are intended for use in single-office or single-building environments. See also *VLAN* and *Frame Tagging.*

E.164 Number   Another name for an ISDN 15-digit telephone number, referred to as *ITU-T E.164 addressing.* See also *SPID.*

E Link (Extended Link)   An *SS7 (Signaling System 7)* signaling connection between a signaling-end point translator and a signal-transfer point. SS7 is the protocol that controls call transfers between central offices in North America.

E-Mail (Electronic Mail)   A software program that you can load onto a computer network that allows the users on the network to write each other notes and send copies of documents. Lotus Notes and CCMail are two examples of this software. See also *X.400* and *POP.*

E&M (Ear and Mouth)   A type of loop signaling for analog telephone circuits. E&M technology dates back to the time telegraphs were used and is an outdated service no longer offered as a service by most telephone companies, however, E&M trunking still has special uses because of its simplistic nature. E&M interfaces come in handy in the PBX environment when there is a need to connect to an analog audio device such as an overhead paging system, or tape recorder. There are 5 types of E&M interfaces that can have either two or six wires in the loop. The most common type of E&M signaling is the Four Wire Wink Start E&M, and the next most used is the Four Wire Immediate Start E&M. The Wink Start E&M operates as follows: The call originating switch goes off-hook and then waits for a "wink" from the terminating or destination switch. When the destination switch provides the 200 ms off-hook "wink," then the originating switch sends dialed digits. After the dialed digits are received and a connection is made to the terminating loop by a handset being taken off hook, the same "off hook" condition is given over to the E&M trunk connecting the terminating switch to the destination switch. When one switch goes "on-hook" or hangs up, the other does as well. The most simple E&M signaling method is the Immediate Start E&M, where the originating end goes "off-hook," or provides a 1000 ohm short on the line and sends digits without regard to the other end. The originating switch stays off-hook until the receiving switch goes off-hook and then back "on-hook," or the call originator goes back "on-hook" or hangs up. The E&M immediate start is the better choice for interfacing external audio devices to PBX systems, and is the less appropriate choice for PBX trunking because if the terminating switch does not answer the call, and the originating switch does not manually hang-up or go back "on-hook", then the loop is left connected. This problem with Immediate Start E&M is the reason that Wink Start E&M was brought about. Wink Start E&M is also called E&M with Answer Supervision.

E1   A European standard that is the counterpart to an American T1. The E1 and T1 are not completely the same. They both use 64 Kbps channels, but the T1 has 24 and the E1 has 32. The following table compares E1 and T1. The European standards are used in all countries, except the United States, Japan, and Singapore (Fig. E.1).

	Total Bandwidth	Total number of 64-Kbps Channels	Number of Channels used for Out of Band Signaling
T1	1.544 Mbps	24	1
E1	2.048Mbps	32	2

Figure E.1   E1

E2   A European standard that carries four E1 circuits. It is used in similar applications in Europe that a T2 does in North America. A comparison of E2 and T2 circuits follows. The European standards are used in all countries, except the United States, Japan, and Singapore (Fig. E.2).

E3   A European version of a T3. An E3 has a smaller bandwidth and carries fewer sub-channels in comparison to a T3. The European standards are used in all countries, except the United States, Japan, and Singapore (Fig. E.3).

	Total Bandwidth	Total number of 64-Kbps Channels	Equivalent E1/T1 Carried
T2	6.312 Mbps	96	4
E2	8.448 Mbps	128	4

Figure E.2   E2

	Total Bandwidth	Total number of 64-Kbps Channels	Equivelent E1/T1 Carried
T3	44.736 Mbps	672	28
E3	34.368 Mbps	512	16

Figure E.3   E3

E911 (Enhanced 911)   Enhanced 911 service uses an *ANI (Automatic Number Identification)* signal that comes in with the call and cross references it to a database of addresses and displays the result on a *PSAP (Public Safety Answering Point)* agent's monitor screen. Standard 911 uses the ANI only. E911 is useful for the times that the caller is not capable or knowledgeable enough of the area to provide a correct address in a timely manner.

EA (Extended Addressing)   Extra bits reserved in a frame header for applications that require additional addressing information.

Ear and Mouth   See *E&M Trunk*.

Earth Bulge   When implementing terrestrial radio paths longer than 7 miles, the curvature of the Earth may become a factor in path planning and require that the antenna be located higher off the ground. The additional antenna height needed can be calculated using the following formula:

$$H = \frac{D^2}{8}$$

where H = height of "earth bulge" (in feet)   and D = distance between antennas (in miles).

Earth Ground   The electrical potential of the earth (0 V). To maintain a good earth ground, a metallic rod is driven into the ground (the length of a standard grounding rod can vary, depending on the geographical location). Any wire connected to that rod is "grounded." The power company installs a rod like this when they connect power to a home. The telephone and cable-TV companies wire their network interfaces (lightning protection) to the power company's earthground rod. The alternative to earth ground is a "floating ground." A floating ground is simply a reference point that is not "earth grounded." The negative terminal of your car battery is a floating ground and any home appliance that has a two-prong electrical plug is also a floating ground.

EAS (Emergency Alert System)   An emergency broadcast system that is incorporated into cable-TV network-distribution systems. Simply put, it is an override or "cut-in" to all cable-TV stations. In an emergency, the government can override the normally transmitted programming for the purpose of informing the public. For a photo, see *Emergency Alert System;* for a diagram, see *Cable-TV Head End.*

EBCDIC (Extended Binary-Coded Decimal Interchange Code)   The character code used widely in IBM mainframe environment computing systems (Fig. E.4). Personal computers predominantly use the ASCII code.

EBCDIC Character Table

		0000	0001	0010	0011	0100	0101	0110	0111	1000	1001	1010	1011	1100	1101	1110	1111
BITS 5,6,7,8		0	1	2	3	4	5	6	7	8	9	A	B	C	D	E	F
0000	0	NUL	DLE			SP	&	-									0
0001	1	SOH	SBA					/		a	j			A	J		1
0010	2	STX	EUA		SYN					b	k	s		B	K	S	2
0011	3	ETX	IC							c	l	t		C	L	T	3
0100	4									d	m	u		D	M	U	4
0101	5	PT	NL							e	n	v		E	N	V	5
0110	6				ETB					f	o	w		F	O	W	6
0111	7				ESC	EOT				g	p	x		G	P	X	7
1000	8									h	q	y		H	Q	Y	8
1001	9		EM							i	r	z		I	R	Z	9
1010	A					¢	!		:								
1011	B					.	$	,	#								
110	C		DUP		RA	<	*	%	@								
1101	D		SF	ENQ	NAK	(	)										
1110	E		FM			+	;	>	=								
1111	F		ITB		SUB				?	"							

Example: Character A is Hex C1, Binary 11000001

Figure E.4   EBCDIC-to-Binary Conversion Table

EBS (Emergency Broadcast System)   An alert system that is incorporated into cable-TV network-distribution systems. Simply put, it is an override or "cut-in" to all cable-TV stations. In an emergency, the government can override the normally transmitted programming for the purpose of informing the public. For a photo of an EBS character generator, see *Emergency Alert System.*

ECH (Enhanced Call Handling)   A reference to devices that provide service for telephone calls, such as voice-mail systems, integrated voice-response systems and *ACD (Automatic Call Distribution)* systems.

Echo Trail   A reference to echo delay detection as measured by the Codecs of voice-enabled routers. Echo is a byproduct of voice networks where the user hears their own voice. In circuit-switched telephony networks, echo is caused by a mismatch of impedance from the two-channel (four-wire) switch/transport environment to the single-channel (two-wire) local loop. A small echo is intentionally left in voice network design because it provides a confirmation to users that their voice is being transmitted. When echo exceeds 25 ms, it becomes a nuisance to the speaker. This echo is managed by echo cancellers. The Codecs within routers that do translations from VoIP networks to the public phone network or switch-based PBXs have built-in echo cancellers. The echo cancellers within Codecs are limited to the amount of delay they can "wait" for. This audio memory is called the *echo trail* and is limited to 40 ms. See also *G.711, G.729.*

ECP (Enhanced Call Processing)   A reference to a lucent technologies' voice-mail feature that allows callers to route themselves to their destination via a prompt/response system of messages.

(ECP) Extended Capabilities Port   A standard parallel port for PCs that supports bidirectional communications for printers at 10 times the speed of the original Centronics standard.

Eddy Current   Eddy current is the electrical current produced in the core (the central piece of metal that coils of wire are wrapped around) of a transformer. As the transformer's core magnetizes and demagnetizes in conjunction with the AC electricity flowing through the coils wrapped around it, magnetic fields are created. These magnetic fields cause electric current to flow in the core. The core heats up because of the current flow, and this heat is considered an inefficiency. This inefficiency is called *eddy current loss.*

Edge   A reference to the part of a network beyond a WAN link, where an end user is located. See also *Content Networking.*

EDO DRAM (Extended Data-Output Dynamic Random-Access Memory)   A type of DRAM that is faster than conventional DRAM (Fig. E.5). EDO DRAM is capable of fetching bytes of data while simultaneously sending bytes to the CPU. EDO DRAM has been outdated by SDRAM, which, at speeds greater than 100 MHz, is twice as fast. Illustrated is a 16-MB SIMM EDO DRAM. See also *SDRAM.*

Figure E.5   A 16-MB SIMM EDO DRAM Board

EEPROM (Electrically Erasable Programmable Read-Only Memory)
A type of *EPROM (Erasable Programmable Read-Only Memory)*. A microchip that contains circuitry that is capable of storing binary instructions, then being erased or reset. Read-only memory is where instructions that tell a CPU how to work are stored. The different types of EPROM devices include *EEPROM (Electrically Erasable Programmable Read-Only Memory)* and *UVEPROM Ultra-Violet Erasable Programmable Read-Only Memory)*, which can be erased by ultraviolet light exposure.

EIA Standards (Electronic Industries Association Standards)   EIA standards are available from the EIA's headquarters in Washington, DC.

EIA-530   Electronic Industries Association Standard 530 is the standard for two electrical implementations of RS-422 (for balanced transmission) and RS-423 (for unbalanced transmission). See also *RS-422, RS-423,* and *EIA/TIA-449.*

EIA/TIA-232   Electronic Industries Association/Telecommunications Industry Association Standard 232 is the common physical-layer interface standard that supports unbalanced circuits at signal speeds of up to 64 Kbps. It is interchangeable with the V.24 standard. It is also called *RS-232.*

EIA/TIA-449   Electronic Industries Association/Telecommunications Industry Association Standard 449 is a physical-layer interface developed by EIA and TIA. It is a version of EIA/TIA-232 capable of 2 Mbps and longer cable runs. It is also called *RS-449.*

EIA/TIA-586   Electronic Industries Association/Telecommunications Industry Association Standard that describes the characteristics and applications for various grades of *UTP (Unshielded Twisted Pair)* cabling. For more specific information, see *CAT 1, 2, 3, 5,* and *7.*

EIDE (Enhanced Integrated Drive Electronics)   Also called *Fast AT Attachment (Fast ATA).* An improvement over the standard IDE storage interface, which transfers data to and from a disk medium at speeds up to 16.6 Mbps, which is four times faster than IDE. EIDE is also capable of supporting drives that store up to 8 GB more than IDE disk drives.

Eight Click Rule/Eight Second Rule   A guideline that Internet website creators use. The objective is to get the user the information they need within 8 clicks and 8 seconds or less. It is believed that if an end user must wait longer than 8 seconds or make more than 8 clicks, they will move onto an alternative website.

Eighty/Twenty   1. A rule of thumb used by telephone companies whereby switch and transport facilities would be increased in a certain area when the utilization reached 80%. This rule worked in the former monopolistic business model well. By the time the remaining 20% was utilized,

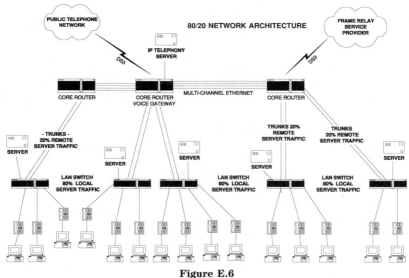

Figure E.6

there would be additional network facilities installed to accommodate growth. In the newer competitive business environment, additional network facilities are built based only on marketing forecasts and the revenue potential of the market that the network serves. This change in business attitude has had good effects on the vast majority of telecommunications service customers.   2. A design consideration in enterprise networks where servers are distributed throughout a network. The purpose of this design method is to keep 80% of the traffic on the same autonomous system (possibly a building or floor). This design was used when switching equipment for the backhaul of traffic to or through a core was very expensive. Since the drop in price for LAN switching equipment, the cost of managing an 80/20 network has proved to exceed the cost of the switching equipment that would provide one central core. The core design is called a hierarchical network design. See also *Hierarchical Network Architecture* and note the placement of servers (See Fig. E.6 on previous page).

EIR (Excess Information Rate)   This is the rate of data transfer in bits per second that exceeds the *CIR (Committed Information Rate)* on a frame-relay *DLC (Data Link Connection)*. These frames have their *DE (Discard Eligibility)* bit set to 1, which means that if congestion occurs in the network, these frames will be discarded. Most frame-relay service providers try to transport customer data at a rate equal to the EIR, but they only guarantee to transport it at the *CIR (Committed Information Rate)*, a high percentage of the time (95% to 99%). See also *CIR* and *AIR.*

EISA (Extended Industry-Standard Architecture)   A 32-bit bus interface for personal computers and some UNIX-based workstations and servers. EISA was an improvement to the 16-bit *ISA (Industry-Standard Architecture)*.

ELAN (Emulated LAN: Emulated Local Area Network)   (Pronounced E-LAN; also called LANE: LAN Emulation) A reference to an ATM (Asynchronous Transfer Mode) with added features to make it interface directly with desktop host/computers thus, "emulating" Ethernet. ATM is a switching method, which makes it difficult to scale and difficult to extend to the desktop. Because the core of ATM is a virtual private line switch technology and not a variable packet technology, it does not have the natural scalability that Ethernet does. The ELAN set of features are add-ons to ATM that make it a little more scalable and much easier to get to the desktop. The challenge of ELAN still remains in price and in continuing to be compatible with future network applications such as IP telephony.

Electrolysis   The use of electricity to change the properties of chemicals or electroplating. Many wires and conductors are electroplated with different metals to increase conductivity. This is the opposite of creating electricity with the use of chemicals (batteries).

Electromagnetic Interference (EMI)   Interference caused by a radio signal or other magnetic field, inducing itself onto a medium (twisted/

non-twisted pair wire) or device (telephone or other electronics). The world we live in is full of radio waves that are emitted from electric appliances, such as blenders, automobile engines, transmitters, and even fluorescent lights. Even though we take preventative measures to avoid picking up these unwanted signals, they sometimes find their way into places where they are not wanted. Electromagnetic interference is usually caused by one of two things. The first is when a wire connected to a device acts as an antenna and picks up the EMI, which is then passed on to the electronics inside the device and is amplified. The second is when an electronic component inside a device acts as an antenna because of poor design, poor shielding, or because the component is defective.

Electromotive Force (EMF)   Another name for *voltage,* which is the origination of the designator used in Ohm's Law formulas. In the formula for calculating voltage, E represents voltage (in volts), R represents resistance (in ohms), and I represents current (in amps). $E = I \times R$

Electronic Switching System (ESS)   A family of telecommunications switches manufactured by Lucent Technologies. The 5ESS is a common central office switch used by RBOCs.

Electronic Warfare   A professional field in the armed forces that specializes in the science of disabling communication and control equipment with the use of EMI, EMP, and by creating "ghost" or deceptive radar images.

Electrostatic Discharge (ESD)   Static electricity. ESD became a big deal when computer and electronics manufacturing companies started using *CMOS (Complementary Metal-Oxide Semiconductor)* electronic components in the devices that they make. Many microchips contain CMOS transistors called *MOSFETs (Metal-Oxide Semiconductor Field-Effect Transistors)*. CMOS components are in every PC made today. They are also used to make the circuitry for LCD watches, telecommunications equipment, home electronics, and many others. The advantage of CMOS components is that they use less power than other components (such as *TTL, Transistor/Transistor Logic*). This is why you can have a tiny battery power your wrist watch or calculator for months. However, the disadvantage of CMOS is that it is extremely sensitive to static electricity. Whenever handling CMOS components, be sure that your body is grounded (to drain off any static that might be on your body). CMOS components can be damaged by static fields, such as one that is created when you brush your hair. Even if the static doesn't arc out into a component, its field can still damage or weaken it. If CMOS components are weakened by static, they usually fail unpredictably in the future.

E-LMI (Enhanced Local Management Interface)   In frame relay applications, E-LMI simplifies the process of configuring traffic shaping on routers. Without the E-LMI feature, users must configure traffic shaping rate enforcement values, possibly for every virtual circuit. Enabling E-LMI reduces chances of specifying inconsistent or incorrect values when configuring network routers. It also enables a router to adjust the PVC values dynamically to changes made in the Frame Relay Network he is connected to. An example of such a change would be a change in CIR values of a PVC as a result of a change in the contract between a customer and its service provider. Using E-LMI, the frame relay switch automatically conveys the new CIR, Be, and Bc values of the PVC to the end routers.

EM   ASCII control code abbreviation for *End of Medium*. The binary code is 1001001, the Hex is 91.

Emergency Alert System (EAS)   An emergency broadcast system that is incorporated into cable-TV network-distribution systems (Fig. E.7). Simply put, it is an override or "cut-in" to all cable-TV stations. In an emergency, the government can override the normally transmitted programming for the purpose of informing the public. For a photo, see *Emergency Alert System*; for a diagram, see *Cable-TV Head End*.

EMF (Electromotive Force)   See *Electromotive Force*.

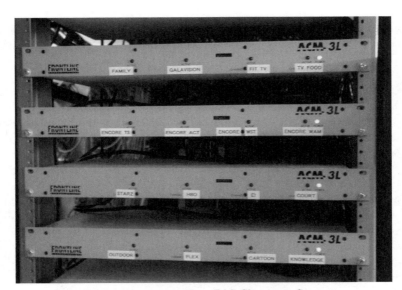

Figure E.7   Emergency Alert EAS Character Generator

EMI (Electromagnetic Interference)   See *Electromagnetic Interference.*

Emission   A reference to electromagnetic waves (this includes heat, radio, and light) radiating from a source. For example, the sun emits ultraviolet radiation and radio stations emit electromagnetic signals.

EMT (Electrical Metal Tubing)   The metal tubing that electricians use to encase electrical wire. EMT is also used to provide a path into and throughout buildings in telecommunications applications. EMT is 2 to 4 inches in diameter and protects the communications cable (copper or fiber) from being cut easily.

Emulated Local-Area Network (ELAN)   (Pronounced E-LAN; also called LANE: LAN Emulation) A reference to an ATM (Asynchronous Transfer Mode) with added features to make it interface directly with desktop host/computers thus, "emulating" Ethernet. ATM is a switching method, which makes it difficult to scale and difficult to extend to the desktop. Because the core of ATM is a virtual private line switch technology and not a variable packet technology, it does not have the natural scalability that Ethernet does. The ELAN set of features are add-ons to ATM that make it a little more scalable and much easier to get to the desktop. The challenge of ELAN still remains in price and in continuing to be compatible with future network applications such as IP telephony.

Emulation   The use of a PC to act or communicate as a dumb I/O terminal. To use a PC in this application (such as to plug into a microwave link and boost its power), it must be equipped with terminal-emulation software. The microwave-link device has its own microprocessor and only needs a device to communicate with, that device is usually a VT100 terminal. Terminal-emulation software allows your PC to "look like" a VT100 terminal to the microwave radio equipment.

Encapsulation   The process of receiving binary data of a particular format or protocol, and adding additional routing/signal bits to the beginning and/or end. The new bits are called a *header* (or *trailer,* if added to the end). As data moves toward the physical layer of a network (or "down levels"), it receives additional headers. As data moves toward the host application (or "up levels"), the headers are stripped off. Data units are given a different name, depending on which level of the *OSI (Open Systems Interconnect)* model where they currently reside. See also *Tunneling.*

Encapsulation Bridging   A method of transporting a frame of data from one physical-layer protocol (such as Ethernet) to another (such as FDDI) by adding an additional header and possibly a trailer. Another bridging method is called *translational bridging,* where the *MAC (Media-Access Control)* protocol address format within the frame is changed. Both bridging methods are accomplished by Cisco and Nortel router/routing equipment.

Encryption   The process of translating of data into a secret code. Encryption is the most effective way to achieve data security. To read an encrypted file, you must have access to a key or password that enables you to decrypt it. Nonencrypted data is called *plain text,* encrypted data is referred to as *cipher text.* Most data encryption methods are based on algorithms.

End Device/Instrument   A telephone, fax machine, modem, terminal adapter, PBX system, computer, or anything else that terminates a communications link.

End Office (EO)   The telephone company central office that serves or connects to the end user/customer. The telephone line in your home connects to an EO. In other telephony applications, a line might connect directly to a long-distance company's node, bypassing the EO. This is called a *Dedicated Access Line (DAL).*

End to End   A reference to the ability of a circuit to communicate/signal from one end user to another without altering service. Regular *POTS (Plain Old Telephone Service)* telephone lines are end-to-end signaled communication lines. After the circuit is established, you can still dial digits into a voice-mail system to reach an extension.

End User   The customer, the one that uses or consumes a product or device.

Enet   Abbreviation for Ethernet.

Engineer Furnish and Install (EF&I)   A way to purchase something. If you would like to install a SONET ring in your campus environment and you ask Nortel Networks for pricing, they will offer you the option of just buying the equipment or buying the equipment, and having them engineer and install it.

Enhanced 911   Enhanced 911 service uses an ANI signal that comes in with the call and cross references it to a database of addresses. It displays the result on a *PSAP (Public Safety Answering Point)* agent's monitor screen. Standard 911 uses the ANI only. E911 is useful for the times the caller is not emotionally capable or knowledgeable enough of the area to provide a correct address in a timely manner.

Enhanced DNIS   A step above standard *DNIS (Dialed Number Identification Service)*. Enhanced DNIS comes with *ANI (Automatic Number Identification)*, better known as *caller ID*.

Enhanced IGRP (Enhanced Interior Gateway Routing Protocol)   A distance/vector-based traffic-control program used by routers. Cisco Systems developed the first version of IGRP in the early 1980s. The original IGRP was an improvement over the widely used *RIP (Routing Information Protocol)* because it used less bandwidth to maintain routing tables. It accomplished this by only transferring new information about its routing tables, rather than the entire routing table. Enhanced IGRP is also able to encapsulate and convert AppleTalk, Novell IPX, and IP routing information. See also *IGRP.*

Enhanced Integrated Drive Electronics (EIDE)   Also called *Fast AT Attachment (Fast ATA)*. An improvement over the standard IDE storage interface, which transfers data to and from a disk medium at speeds up to 16.6 Mbps, which is four times faster than IDE. EIDE is also capable of supporting drives that store up to 8 GB more than IDE disk drives.

Enhanced Local Management Interface (E-LMI)   In frame relay applications, E-LMI simplifies the process of configuring traffic shaping on routers. Without the E-LMI feature, users must configure traffic shaping rate enforcement values, possibly for every virtual circuit. Enabling E-LMI reduces chances of specifying inconsistent or incorrect values when configuring network routers. It also enables a router to adjust the PVC values dynamically to changes made in the frame relay network he is connected to. An example of such a change would be a change in CIR values of a PVC as a result of a change in the contract between a customer and its service provider. Using E-LMI, the frame relay switch automatically conveys the new CIR, Be, and Bc values of the PVC to the end routers.

Enhanced Parallel Port (EPP)   A type of parallel port that has the same physical DB25 characteristics as any other parallel port, yet is twice as fast. EPP is the parallel port used on many laptop/portable computers. See also *ECP.*

ENQ   The ASCII control-code abbreviation for *enquiry*. The binary code is 0101000 and hex is 50.

Entity   A device attached to a network, such as a workstation, router, server, terminal, or printer. Devices are referred to as *entities* because, through the network architecture, devices are given different addresses or names, called *aliases*. All aliases translate through routing and address tables.

Envelope   1. Reference to the modulated carrier signal in a radio transmission.   2. A data block in a packet transmission network that contains addressing or other data in binary form.

Environment   Electronic equipment usually has specified requirements for the environment in which it is located. The requirements are usually listed in the literature that comes with the equipment. Common environmental requirements are: –35 degrees F to 85 degrees F, 20% to 60% humidity, and dedicated 120V AC power.

EO (End Office)   See *End Office*.

EOT   The ASCII control-code abbreviation for *end of transmission*. The binary code is 0100000 and hex is 40.

EPP (Enhanced Parallel Port)   A type of parallel port that has the same physical DB25 characteristics as any other parallel port, yet is twice as fast. EPP is the parallel port used on many laptop/portable computers. See also *ECP*.

EPROM (Erasable Programmable Read-Only Memory   A microchip that contains circuitry capable of storing binary instructions, then being erased or reset. Read-only memory is where instructions that tell a CPU how to work are stored. The different types of EPROM devices include *EEPROM (Electrically Erasable Programmable Read-Only Memory)*, and *UVEPROM (Ultra-Violet Erasable Programmable Read-Only Memory)*.

EQ (Equalization, Equalizer)   To adjust the tone or sound of a circuit by diminishing or augmenting specific frequency bands. The tone control on a radio is a type of equalizer. A radio transmitter might have a tendency to amplify low-end signals, such as the sound of a bass guitar or drums better than high-end signals, such as the sound of a voice or cymbals. An equalizer can be used to reduce or increase the amplification of either end of the broadcast for an even and accurate reproduction of the input.

Equalization   See *EQ.*

Equipment Cabinet   There are many types of equipment cabinets, but the most common is 7 feet high by 24 to 26 inches wide with a 22- or 19-inch wide mounting rack built into it. Quality equipment cabinets are equipped with blower fans to circulate air through them and have a locking door. Some have clear plastic doors so you can view the alarm/status lights on the front of your equipment. Any time equipment uses a fan to keep it cool, you should place it in the cleanest, dust-free environment possible. If you have a dusty environment, the fans will blow dust into your equipment, which will build up and act as an insulating blanket. This will cause overheating and failure of electronic components.

Erasable Programmable Read-Only Memory (EPROM)   See *EPROM.*

Erlang   A one-hour unit of telephone traffic. This can be one phone call that lasts for one hour, or two phone calls that last for 30 minutes each, etc. Erlangs consist of *CCS (Centum Call Seconds).*

ERP (Effective Radiated Power)   The actual power in watts radiated from a transmitter's antenna. A typical FM radio station has an ERP of 15,000 watts (15 kw).

Error Checking   The methods used by modems and other transmission equipment to detect errors in the data received in a transmission. Common error-checking methods are *VRC (Vertical Redundancy Checking)*, *LRC (Longitudinal Redundancy Checking)*, and *CRC (Cyclic Redundancy Checking)*. The most accurate error-checking method used in modems today is called CRC (cyclic redundancy checking). CRC is designed to check a frame-type protocol.

Here is a simplified version of CRC logic. Imagine that a binary number is read off (transmitted) to you. You hear the ones and zeros, and as you hear them you write them down (remember them). Then, you are given another number to divide the first number by *(FRC, Fame Check Sequence)*, then another number (algorithm) that should match up as the answer. If the answers match, then there is a 99.99995% chance that you heard the first number correctly and there are no errors. Drawn out in a very simplified form, CRC error-checking logic looks something like this:

```
Bit block to be transmitted: 11001000 (this is equal to 200 in
                                       decimal).
```

```
FRC frame: (added to the end)   1010 (this is equal to 10 in decimal).
```

```
Algorithm: (the answer)         10100 (this is equal to 20 in decimal).
```

So, the receiving equipment divides the block transmitted by the FRC frame and should get the algorithm. If the numbers don't match, then it requests a retransmission.

Error Rate   Some transmission test equipment (TTC Tberd), some computer software (Novell), and some network adapters (SMC Ethernet) have the separate ability to transmit a known group of packets over a network, then see how accurately they are returned. When they are returned, the equipment divides the total number of packets by the number of packets that have errors. The end number is a percentage, which is the error rate.

ESC   The ASCII control code abbreviation for escape. The binary code is 1011001 and the hex is B2.

ESD (Electrostatic Discharge)   See *Electrostatic Discharge.*

ESF (Extended Superframe Format)   A type of T1 line. An additional innovation of the D4/SF (Superframe) format. In a T1 circuit, each channel is sampled (which is an 8-bit DAC sample) and the bits are sent on down the line. If you take 8 bits and multiply it by 24, you get 192. If you add one "timing" bit to the end of the 192-bit chain, you get 193. A Superframe is 12 of these 193-bit frames, chained together. This allows each of the 12 framing bits (193rd bit) in each of the 193-bit sequences to mean something other than a timing signal. Different meanings include signaling, such as dial tone, digits dialed, and off-hook or busy. Then, in-band signaling comes into play, in which ESF can be configured as either in-band or out-of-band. Within the 24 channels, the least-significant bit (the one that will have the least effect on the accuracy of the DAC conversion) of the 6th and 12th samples of each frame are used for additional signaling, control, and maintenance. This is called *bit robbing,* and it is the reason why in-band signaled T1 lines have only 56 Kb/s of bandwidth, as opposed to "non-bit-robbed" (clear channel) T1 lines, which have 64 Kb/s of bandwidth.

ESS (Electronic Switching System)   A family of telecommunications switches manufactured by Lucent Technologies. The 5ESS is a common central-office switch used by RBOCs.

ETB   The ASCII control-code abbreviation for end-of-transmission block. The binary code is 0111001 and hex is 71.

EtherChannel   A Cisco Systems trademark for combining multiples of two or four Ethernet trunks between switches into one. Four gigabit

Ethernet links can be combined to form one 4 Gbps link. This type of trunking is commonly used in backbone and distribution environments.

Ethernet A family of *LAN (Local Area Network)* protocols. Ethernet is one of the oldest communication protocols for personal computers. When a LAN is mentioned, two things should immediately come to mind. Physical topology and the protocol the LAN uses to manage communications between devices. Ethernet can be implemented in a bus or star physical topology (Fig. E.8). The alternative family of LAN protocols is the token-passing type, which are configured as a ring topology. See *Token Ring*.

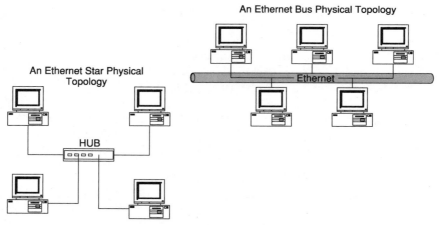

Figure E.8   Ethernet

In an Ethernet LAN, computers are given a means to communicate with each other called a protocol. A protocol is a set of rules and instructions for communicating. Within the protocol is a "logical topology." Even though a network can be connected as a star, it can still look like a bus to the communications equipment because all of the computers/ devices are connected to the same wire (in the star diagram, the hub is a device that connects all the wires together). The way Ethernet works is similar to the way people talk in a group. Instead of using wire to carry the binary coded information as Ethernet does, people use air to carry sound information. When there is a silence, then one of the persons in the group is able to speak. When the person speaks, they might say "Johnny, do you know the answer for 5 + 5?" Even though all the people in the group hear this message, they know it is for Johnny because the message was "addressed" to him. So, only Johnny will respond "10."

Then, imagine as Dawn and Vicki acknowledge a silence and try to speak at the same time. This is confusing and no one understands the information. Ethernet has the same problem and it is called a *collision*. Collision is the disadvantage of Ethernet. Because of the possibility of collisions (which happen very frequently), Ethernet is called a "contention based" protocol because all of the connected devices are contending for use of the network. Manufacturers have come out with new ways to avoid collisions, called *CSMA/CD* and *CSMA/CA*. Ethernet has many different types of wiring to connect devices, and many different *NICs (Network Interface Cards)* to select from that need to be installed in each computer or device on the network. The following is a list of Ethernet protocols and the type of wiring used for each.

Ethernet Autonegotiation   The IEEE 802.3u 100BaseT feature specification that provides for flow control (pause frames) and full-duplex operation. Full duplex allows for 100 Mbps send and 100 Mbps receive, for a total 200 Mbps Ethernet connection over a Cat5 twisted pair. The autonegotiation is an enhancement of the link integrity signaling method used in 10BaseT networks and is backward compatible with link integrity. Autonegotiation allows the NIC or the network device to adjust its speed to the highest speed that both ends are capable of supporting. To be able to use this feature, both the network device (switch port) and the NIC must contain the autonegotiation logic.

Ethernet Prioritization   In converged Ethernet environments, a reference to the IEEE 802.1p standard for prioritization of LAN traffic among Ethernet switches based on the switch port, MAC address, or IP address associated with the communicating end appliance (whether it is an IP phone, video monitor, host PC, printer, or server). Packets are tagged as belonging to a queue, which determines the priority of the packet. By the 802.1p standard, queue 0–3 is normal and 4–7 are high priority. 802.1p functions hand-in-hand with 802.1Q or VLANs.

Ethernet Switch   A hardware device that provides dedicated bandwidth to segments of the network that are attached to it. The primary purpose of a LAN switch is to provide additional bandwidth to users by reducing the number of users per segment. In doing this, collisions are eliminated and the 50% to 60% efficiency of Ethernet is increased to near 100% (collisions are still possible between the switch and the attached device). A LAN switch can be compared to a bridge with many ports (at a much lower cost per port) and extensive traffic manageability. Unlike a non-POTS voice switch or PBX switch, multiple hosts (computers, IP telephone extensions) can be connected to each port. This is because a LAN switch forwards frames based on the MAC address of the connected devices, rather than the switch port as in voice switch operation. Because basic LAN switches are OSI Layer 2 devices (meaning the forward packets based on a hardware address), they are capable of bridging unlike networks, such as Ethernet to FDDI and token ring. Larger LAN switches are capable of performing layer 3 *routing,* where the packet forwarding function happens based on a leading packet's IP address. LAN switches often replace Ethernet hubs (no wire replacement is needed, but fiber optic may be installed for backbone connectivity) when they are implemented into a network. LAN switches are the core enabler of IP telephony. See also *Bridge, Layer 3 Switch,* and *IP Telephony* (Fig. E.9).

Figure E.9 Ethernet Switch

ETX The ASCII control code abbreviation for end of text. The binary code is 0011000 and the hex is 30.

Even Parity A method of bit-stream checking. Parity is used in error correction. The number of logic "ones" is counted in a bit stream. There is "odd parity" and "even parity." Which is used depends on if you like odd or even numbers, or if the modem you are trying to connect with likes odd or even numbers. Parity is a part of error-checking protocol. It is simply the part of the protocol where the two devices are told if they are counting odd number bits or even number bits. In odd parity, if the number of ones is an odd number, then a parity bit is set to "one" at the end of the bit stream. This is *odd parity* because the parity bit is set to one when the number of "ones" is odd. In *even parity,* the parity bit is set to "one" when the number of "one" bits is even.

Exa See *ExaByte.*

ExaByte Exa is a unit of scientific/engineering notation. 1×10^{18} power. Or you could think of Exa as 18 zeros on the end of a number. Five ExaBytes (5EB) is equal to 5,000,000,000,000,000,000 bytes.

Excess Information Rate (EIR)   This is the rate of data transfer in bits per second that exceeds the *CIR (Committed Information Rate)* on a frame-relay *DLC (Data Link Connection)*. These frames have their *DE (Discard Eligibility)* bit set to 1, which means that if congestion occurs in the network, these frames will be discarded. Most frame-relay service providers try to transport customer data at a rate equal to the EIR, but they only guarantee to transport it at the *CIR (Committed Information Rate)*, a high percentage of the time (95% to 99%). See also *CIR* and *AIR*.

Excess Rate   1. In *ATM (Asynchronous Transfer Mode)*, the data-transfer rate in excess of the insured rate for a given connection. The excess rate equals the maximum rate minus the insured rate. For example, if an ATM connection has an insured rate of 1.5 Mbps, and it is actively transferring data at 2 Mbps, the excess rate is 0.5 Mbps. Excess traffic is delivered only if network resources are available and the excess data transferred is marked *Discard Eligible*.   2. The term excess rate is occasionally used as a reference to *Excess Information Rate (EIR)* in a frame-relay connection. In both cases, the term is a reference to a bonus that is given to a bandwidth subscriber that is above what they pay for because it costs the service provider nothing to allow this bandwidth to be used when it otherwise would not be. Compare with *Insured Rate* and *Maximum Rate*.

Exchange   The area that a single central office services. Soon, when number portability is fully implemented, an exchange will not be associated with a central office. It will be associated with an area and the legal regulations imposed on communications companies in that area. Currently, each central office is assigned a group of numbers that it can use. The numbers are the first three digits (not including the area code). The numbers (801)-355-xxxx, (801)-237-xxxx, and (801)-575-xxxx are assigned to the Quest Salt Lake City, Utah main central office.

Exchange Area   An area served by multiple communications companies and multiple central offices. Each exchange area has its own legal regulations regarding how companies can compete (or price and package their services) and what services they are required to provide.

Expansion   The reverse of compression in companding (compression-expansion). A pulse-code modulation technique that takes small samples of an analog signal the same way that is done in *(PAM) Pulse Amplitude Modulation*, except that the resultant PAM signal is converted to binary. Companding compresses the binary signal using a mathematical algorithm. This allows more analog channels within the same network bandwidth. Companding is used over *PCS (Personal Communications Service)* cellular radio networks. See also *Mu Law* and *A Law*.

Expansion Slots   Space allocated in a KSU for the addition of circuit cards in the future. Additional circuit cards needed might be trunk interface cards or station/extension interface cards.

Explicit Frame Tagging   In LAN switching, frame tagging is a method used to identify which Virtual LAN (VLAN) a packet belongs to. Within an Ethernet switch, a VLAN behaves as a single Ethernet segment, where all computers/hosts that are a part of a VLAN see each other's traffic. Traffic cannot traverse from one VLAN group to another unless it leaves the switch and is routed back in on another port. The router provides traffic security/management. Frames are marked by the switch in two methods: implicit and explicit. Implicit frame tagging is a method where the VLAN information is added within the packet. Explicit frame tagging is a method where an external VLAN header is added to the frame.

Explorer Packet   In IBM source-route bridging networks, a message sent through a network by an end device that gathers path and route information. When the message returns to the device that sent it, it contains a list of all of the entities in the network and the order that they are connected. From a digital or mathematical standpoint, this is a complete map of the network.

Extended Addressing (EA)   Extra bits reserved in a frame header (two in frame relay) for applications that require additional addressing information.

Extended Binary Coded Decimal Interchange Code (EBCDIC)   The character code used widely in IBM mainframe-environment computing systems. Personal computers predominantly use the ASCII code. For a table of EBCDIC characters and their binary values, see *EBCDIC*.

Extended Capabilities Port (ECP)   A standard parallel port for PCs that supports bidirectional communications for printers at 10 times the speed of the original Centronics standard.

Extended Industry-Standard Architecture (EISA)   A 32-bit bus interface for personal computers and some UNIX-based workstations and servers. EISA was an improvement to the 16-bit *ISA (Industry-Standard Architecture)*.

Extended Superframe Format (ESF)   See *ESF.*

Extensible Markup Language (XML)   A database format that is intended to enhance the capabilities of content services and delivery over the Internet. XML is more flexible than HTML because it allows for database formats to be self-defining. What makes XML able to work this way is that it does not have a fixed tag set or fixed semantics, which means that the semantics and tag sets are defined by the application programs that utilize the XML. XML could be thought of as a translation book where users can place a word, whatever word it may be in any language with any pronunciation or sound, and then place a picture or meaning next to the word; further, the pages of the book can be ordered any way the user likes. However, some connection to another book and page must be made for reference (this would be one of the other databases, such as a banking record). When the book is completed, it is, of course, a database in itself. XML also features a simpler method of set-up than HTML and is widely available. It is commonplace for XML to be used to access a portion of data from an Internet web page and transfer it to another display method, such as a stock ticker display, a pager or cellular phone display, or telephone display.

Extension   A telephone or equipment connection on a *PBX (Private Branch Exchange)* or key system. Extensions can be from two to seven digits long. An extension is sometimes referred to as a *DN (Directory Number).* An extension can be an electronic keyphone if it is connected to a digital PBX interface or it can be a modem, fax machine, or analog phone (like the ones made for home use) if it is connected to an analog interface.

External Modem   Also called a *stand-alone modem.* A modem that comes in its own package (case) and comes with a cable that plugs into a COM port/serial port on a computer or data device. External modems are popular for dial-up remote-access administration for PBX switches. The alternative to external modems are internal modems, which are popular in PCs and come in the form of circuit cards that plug into the PC's motherboard. Both do the same job equally as well, but most computers only have two COM ports. If you use an internal modem, then you can use your two COM ports for other applications, you don't have to have another power outlet, and internal modems are usually less money.

Extranet   A part of a company intranet that permits outside Internet users to view and interact with certain functions within the company network, such as corporate updates, employment pages, and product ordering.

Eyeball Shot   Another name for a terrestrial microwave link. The link is made by two radio transceivers equipped with parabolic dish antennas pointed directly at each other. Radio can carry point-to-point transmissions of many bandwidths, including DS1, DS2, DS3, STS1, and OC1. Their range can vary, depending on the size of the antenna (dish), weather in the region, and the amount of power emitted. Including all of the previous factors, a link can range from 0 to 50 miles. For a diagram, see *Terrestrial Microwave.*

F

F Connector   A connector for coax cable. The standard for cable-TV is
75-ohm coax with an F-type connector on the end (Fig. F.1). For other
types of coax connectors, see *Coax*.

Figure F.1   F Connector

Face Plate   1. Some telephone jacks come with separate face plates that snap on to the front of a telephone jack. The snap-on/snap-off design makes it easier to access the wire connections, but this style of jack is usually a poor choice for a long-term application.   2. Some telephone sets have face plates that fit over the buttons and display of the phone. Instead of buying the phone that is the color of your choice, you simply buy a universal phone, then buy the face plates with your choice of color.

Facilities   1. A reference to equipment required to complete a communications objective. For example, twisted copper pairs, coax, and fiber optic are considered facilities to local telecommunications providers. 2. The part of an X.25 packet header in the X.25 protocol that negotiates and defines window size, packet size, protocol, and transfer speed, to name a few.

Facilities-Based Carrier   A telephone company that has its own switches and communications facilities, unlike a reseller.

FACS (Facilities)   A portion of a phone company that tracks the use of facilities (cable pairs and central-office switch ports). If you are a telephone network technician and you find that a pair in a cable (cable pair) is bad, you call FACS and notify them that the pair is unusable. This way, they don't think that they have an extra pair for future service.

Facsimile   A fax machine. A machine that can dial a telephone number connected to another fax machine and generate a copy of a document fed into it on the far end.

Fan Out   A multiplexer. The device that breaks a DS1 or DS3 service down into the size that a customer wants. A fan out on a DS3 line breaks the 28 DS1 channels out for a customer and a fan out on a DS1 line breaks the DS1 into 24 DS0 channels.

FAP (Fuse Alarm Panel)   A power-distribution panel that is installed at the top of a relay rack. All equipment in the rack is wired to the panel for power. Each device has its own fuse within the panel to protect the rectifier from an "over-current" condition if a device fails or a wire shorts. If any of the fuses blow, an alarm indicator is displayed. For a photo, see *Fuse Alarm Panel.*

Far-End Cross-Talk (FEXT)   The uncommon phenomena of signals sent over twisted copper pairs bleeding onto each other via magnetic fields produced at cross connections, or within defective electronic equipment (Fig. F.2). The *far end* refers to the problem occurring between a remote node and customer interface (or customer equipment).

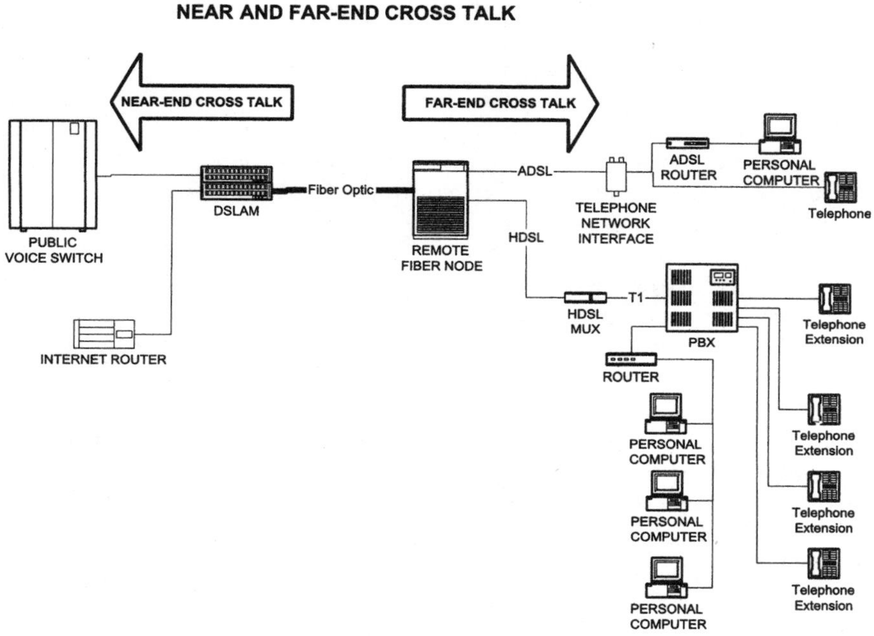

NEAR AND FAR-END CROSS TALK

Figure F.2   Far-End Cross-Talk

Farad   The standard unit of capacitance. A capacitor is an electronic device with two special properties. It only allows alternating current to pass through it, and it can store an electric charge. One of the many applications of capacitors is to filter AC out of DC power supplies and rectifiers. This is done by placing a capacitor from the DC output to ground. The capacitor appears as an easier path to voltage fluctuations and RFI, and an impossible path to direct current (DC). Physically, a capacitor is two plates of metal separated by an insulator (mylar is common). The physical size of a 1-F capacitor would be two sheets of tin foil the size of a football field, insulated (or separated) by a thin sheet of mylar. The farad is a huge unit of capacitance, so most capacitors are measured in microfarads (μF). For a schematic symbol of a capacitor and a photo, see *Capacitor*.

Fast ATA (Fast AT Attachment)   Also called *EIDE (Enhanced Integrated Drive Electronics),* it is an improvement over the standard IDE storage interface, which transfers data to and from a disk medium at speeds up to 16.6 Mbps, which is four times faster than IDE. EIDE is also capable of supporting drives that store up to 8 GB more than IDE disk drives.

FAT (File Allocation Table)   A computer file that tracks as a directory for the contents of a hard drive. See also *FAT16* and *FAT32*.

Fat Client   A reference to a network attached workstation that has adequate processing power for applications utilized within its realm of business. A fat client performs data processing via its own internal hardware, and relies on networked servers for data storage, retrieval, and sharing. A fat client environment emulates the evolving client/server computing strategy opposed to the thin-client type. As with all technologies, which is best economically and administratively depends on the user's application. See also *Thin Client*.

FAT16 (File Allocation Table Version 16)   A disk-drive file-tracking method that used 16 bits to address clusters, which limited the size of hard-drive partitions to 512 MB. FAT 16 uses 32 KB clusters to store data.

FAT32 (File Allocation Table Version 32)   A disk-drive file-tracking method that uses 32 bits to address memory clusters. FAT 32 supports hard drives as large as 2 TB (terabytes). FAT32 uses 4-KB memory clusters to store data.

FAX   See *Facsimile*.

Fax Gateway   A routing device that uses H.323 Codec for digital to analog conversion, or standard T.38 phase 2, which is a Cisco Systems developed fax gateway protocol that is now an ITU-T standard, ITU-T T.38. A fax gateway is a device/port that allows a single telephone line to receive both faxes and voice calls. The fax gateway answers all incoming calls. It listens for a fax handshake tone, and if it does not hear one, it rings the telephone connected to it. T.38 is a feature that can be enabled on individual router FXS ports. A fax gateway performs the same function as a fax switch, only in a LAN application.

Fax Jack   A device that connects to a phone line and has two jacks on the other end, one for a fax machine and one for a telephone. These devices are for users that want to use only one phone line for faxes and voice calls. When a call arrives, the fax jack answers the line immediately and waits for a mechanical tone from another fax machine. If it does not hear the tone, it assumes that the call is a person and not a fax machine. It then rings the phone on its other end. If the fax jack would have "heard" a tone, it would have connected the line to the fax machine plugged into its other end. The only bad thing about fax jacks or other line-sharing devices is that when they seize the line immediately, they block caller-ID (ANI) signals.

Fax on Demand   A feature of voice-response systems that enable a caller to listen to a recorded message that gives them a selection of information that they can receive via a fax. After making a selection, the caller is then prompted by another recording to enter the number that they would like the information faxed to.

Fax Switch   See *Fax Jack.*

FCC (Federal Communications Commission)   An organization of the federal government that was set up by the Federal Communications Act of 1934. The FCC works in conjunction with the 50 state Public Service Commission bodies and Congress. It has the legal authority to regulate the following three areas of communications. Communications being defined as radio, video, telephone, and satellite communications within the United States.   1. Regulate who is permitted to manufacture and sell telecommunications equipment and service.   2. Regulate the price of interstate long distance.   3. Determine the electrical standards for telecommunications, such as operating frequency of transmitting devices.

FCC Complaints   1-888-225-5322 is a public service to initiate complaints regarding slamming, radio interference, illegal monitoring activities, and other FCC-regulated issues.

FCC Tariff   A ruling on a type of communications service. A tariff defines a service and the price that certain companies are allowed to charge. Tariffs usually restrict RBOCs and AT&T from being competitive by forcing them to sell service at higher prices than the companies wish. If the FCC did not impose these tariffs on the communications giants, it would be impossible for new smaller companies to become established and compete. Incumbent providers would simply drop their rates so low that the other companies would be driven out of business or be driven to being bought out by one of the larger companies.

FCS (Frame Check Sequence)   A 16-bit cyclic redundancy check used in X.25. The FCS has its own section of overhead in the X.25 frame. FCS is proven to be over 99.998% effective in detecting errors. For more information, see *X.25.*

FDDI (Fiber-Distributed Data Interface)   LAN backbone protocol that requires its own fiber-optic cabling, NIC cards, and software to configure them. FDDI is a 100 Mb/s protocol that came about when Ethernet and token ring were in their 10 Mb/s and 16 Mb/s infanthood. Since then, Ethernet has developed 1000 Base-T, which is a 1,000 Mb/s protocol. Token Ring has now developed into Switched Token Ring, which has a maximum throughput of 80 Mb/s. The maximum throughput of FDDI is a true 100Mb/s, and its self-healing ring capability is only matched by Switched Token Ring and exceeded by SONET. FDDI is a great LAN backbone architecture, but it is twice as expensive as its Ethernet and Token Ring competition, and it requires fiber-optic cabling. If your backbone architecture will carry important data that cannot have any downtime, FDDI and Switched Token Ring are your options. If you have crucial data flow and long distances (a ring of more than a 2-mile circumference) or an environment where EMI is abundant (such as a factory with large electrical equipment), then FDDI on fiber optic will be your prime backbone architecture. A FDDI ring has a maximum circumference of 62 miles (100 km), but a repeater or node must be spaced every 1.25 miles (2 km). Up to 1000 nodes can be placed on a FDDI ring. Switched FDDI has become available recently. For a diagram of a FDDI application, see *Fiber Distributed Data Interface*.

FDM (Frequency-Division Multiplexing)   Multiplexing is the process of encoding two or more digital signals or channels onto one. Channels are multiplexed together in communications to save money. When we use all of the wires in a cable and need more, it costs less to add electronics on the ends of a cable than to install a new one (imagine the expense from LA to NY). A T1 encodes 24 channels into one by using time-division multiplexing. Many radio stations are put in the same airspace by using frequency-division multiplexing.

FDMA (Frequency-Division Multiple Access)   Another name for *frequency-division multiplexing* that the cellular telephone industry uses.

FDX (Full Duplex)   1. A full-duplex line or communications path is able to communicate both directions, transmit and receive, at the same time. A T1 is a full-duplex line, with one pair used for transmit and the other used for receive. Full duplex can be accomplished on one pair of wires by using two multiplexed channels, one for receive, and one for transmit. There are two other types of transmissions. One is half duplex, where transmit and receive are sent one at a time. A CB radio works in half-duplex mode: one person talks while another listens, and vice versa. The other type of transmission is simplex, where communication is one way only. An FM radio station or TV broadcast is simplex.   2. In LAN switching, a feature that enables speed to be doubled by having simultaneous two-way conversation. For example, a full-duplex 100BaseTX connection between a switch port and a host/computer would have simultaneous 100 Mbps capability for transmit and 100 Mbps for receive, for a total of 200 Mbps throughput. Full-duplex Ethernet operation is a byproduct of LAN switch architecture. LAN switches must be implemented to obtain full-duplex Ethernet. Full-duplex functionality does not work across hubs.

Feature Buttons   Buttons on PBX telephone sets that activate features such as hands free, call forward, do not disturb, transfer, and speed dial. Feature buttons can be changed or customized to a user's liking. For example, if John likes to have the top feature button as "hands free" and the bottom button as "transfer," then the buttons can be programmed for those functions. If Sally likes the bottom button to be her sister's "speed call" button, then the feature button on the bottom of the phone can be programmed that way.

Feature Cartridge   A cartridge containing ROM or RAM that permits a phone system (PBX) to use different features. I always thought that if you buy the system, why not put all the software features on it, and make it simple. If you would like to have the features, you need additional software, and software licenses are expensive. Feature cartridges give customers the option of buying what they want.

Feature Code   If you want to use a feature on a PBX system, but don't have a button on your phone that performs that feature, you can use a feature button in combination with a code. If you would like to transfer a call on a phone system, but you don't have a transfer button on your phone, you can transfer the call by pushing the feature button and then entering a code (e.g., 52) on the dial pad. Feature codes are also used in the public telephone network. If you are a USWest customer in Utah, you can activate a feature that calls the party that just called you (if you did not get to the ringing phone in time to answer it). The feature code for this service is *69. Just pick up the phone and dial *69. There is a charge for this service just like there is a charge for directory assistance.

Feature Group   In PBX systems, programming individual telephone extensions can be very tedious if you have 300 of them on your system. Some PBX manufacturers build a feature into their programming called a *feature group*. You can assign features to a feature group, such as: transfer on button 1; speed dial on buttons 5, 6, 7, 8, 9; do not disturb on button 2; voice mail on button 3; etc. Then you can assign extensions to the feature group. Each extension assigned to a feature group will have the features of that feature group. This is much easier than programming every button on all 300 phones.

Feature Phone   A reference to a PBX telephone set that is capable of select features, such as do not disturb, call forwarding, speed dial, transfer, etc.

FEC (Forward Error Correction)   See *Forward Error Correction.*

FECN (Forward Explicit Congestion Notification)   A bit in the overhead of a data packet travelling in a frame-relay network. This bit is set to 1 if the packet travels through an area of network congestion that is the same direction (or forward) of its flow. This bit is a signal from the frame-relay network to higher-level protocols within DTE and DCE to take appropriate flow-control action. An example of flow-control action would be to not attempt to exceed the *Committed Information Rate (CIR)* for the connection. This would prevent the unnecessary transmission of low priority frames that would surely be discarded. The counterpart of FECN is *BECN (Backward Explicit Congestion Notification)*.

Federal Communications Commission (FCC)   See *FCC*.

Federal Telecom Standards Commission   An organization established in 1973 to assist in the development in telecommunications interface standards. Federal Standards begin with FEDSTD.

Federal Universal Service Fee   A Federal tax placed on telecommunications services provided by telephone companies.

Feedback   The reintroduction of an amplifier's output signal back to its input. If you have been at a public gathering or speech and heard the microphone make a loud squeal sound through the loudspeakers, you have heard a type of feedback. This type is caused by the sound from the speakers finding its way back into the microphone and the amplifier.

Feedhorn   A part of a parabolic dish apparatus that is suspended in front of the dish. The feedhorn is comprised of a cover and an LNB converter, which contains the actual antenna (Fig. F.3). See also *LNB Converter*.

Figure F.3   A Feedhorn with LNB Converter Attached Above the Satellite Dish

FEP (Front-End Processor)   A communications "front-end" device that can be loaded with a "firewall" to prevent unwanted users from accessing the communications network. An FEP can also perform routing, and differentiate between different communications protocols, depending on the software that runs on it. For a diagram of an FEP, see *Front-End Processor*.

FER (Frame Error Rate)   Some transmission test equipment (TTC Tberd), some computer software (Novell), and some network adapters (SMC Ethernet) have the separate ability to transmit a known group of packets over a network and then see how accurately they are returned. When they are returned, the equipment divides the total number of packets (or frames) by the number of packets that have errors. The end number is a percentage, which is the error rate. See also *Telnet* and *Ping*.

Ferric Oxide   A compound that is sometimes used as a coating on magnetic tapes. It (and some other compounds) can be magnetized.

Ferrule   1. A part of a fiber-optic connector that holds the connector ends in alignment when they are connected together.   2. A metal ring that is sometimes found on power cords. It helps to reduce RFI passing from the power company into the device using the cord.

FET (Field-Effect Transistor)   The two different varieties of transistors, bipolar and field-effect, are designed to manipulate electricity flowing through them in different ways. Bipolar transistors are current-controlled devices and field-effect transistors are voltage-controlled devices. The advantage of field-effect transistors is that because they are voltage controlled, they can switch from one to zero and draw hardly any current. Current is what drains batteries and field-effect transistors help make batteries last a long time. Bipolar transistors are composed of different types of silicon stacked on top of each other. Field-effect transistors are composed of one piece of silicon, with a different type of silicon added to the sides (Fig. F.4). Field-effect transistors are available

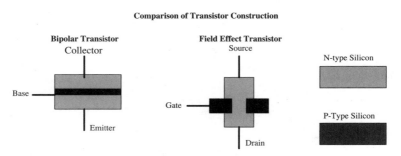

Figure F.4   FET (Field-Effect Transistor)

in different types: Junction, MOS (Metal-Oxide Semiconductor), IS (Ion-Sensitive), DE (Depletion-Enhancement) MOSFET, and E (Enhancement-Only) MOSFET.

FEXT (Far-End Cross-Talk)   The uncommon phenomena of signals sent over twisted copper pairs bleeding onto each other via magnetic fields produced at cross connections, or within defective electronic equipment. The *far end* refers to the problem occurring between a remote node and customer interface (or customer equipment).

FF   ASCII control code abbreviation for form feed. Binary code is 1100000, Hex is C0.

FFDI (Fast Fiber Data Interface)   PlusNet Phoenix AZ.

FHSS (Frequency Hopping Spread Spectrum)   A wireless LAN radio modulation technique where (in the case of 802.11FH) 79 1-MHz radio channels occupy a spectrum of 2.400 to 2.483 GHz. As data is transmitted over the radio, the radio alternates synchronously with the receiver over the 79 channels, via an algorithm or other method used by the radio manufacturer, a minimum of two times per second. The maximum FHSS throughput is 2 Mbps. See also *DSSS* (which is newer and provides up to 11 Mbps when deployed under the 802.11DS standard).

Fiber-Distributed Data Interface (FDDI)   See *FDDI,* and Fig. F.5 for a diagram.

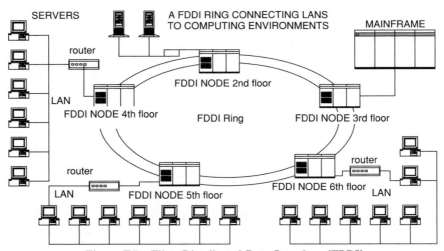

Figure F.5   Fiber Distributed Data Interface (FDDI)

Fiber Optic   A thin strand of tiny layers of glass that have different re-fractive properties. The layers of material that have different refractive properties enable the thin strand to channel light through it by bending the rays of light. The light travels through the core of a fiber, and is bent back toward the core when it enters the cladding. Fiber-optic cable can be: multi-mode or single mode. Multi-mode fiber optic has a larger core than single-mode (Fig. F.6). Single mode is better for transmitting long distances, and multi-mode is better for transmitting multiple colors of light (or sending more than one signal on a single fiber). Single-mode fiber is much more widely used in telecommunications than multi-mode cable. SONET is a fiber-optic based protocol standard that uses single-mode, graded-index fiber optic (Fig. F.7).

CROSS SECTION OF FIBER OPTIC TYPES

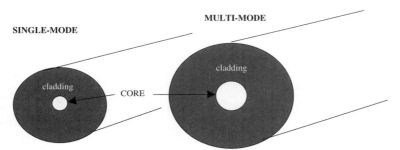

Figure F.6   **Multi-mode Fiber Optic is Larger in Diameter Than Single-Mode Fiber Optic**

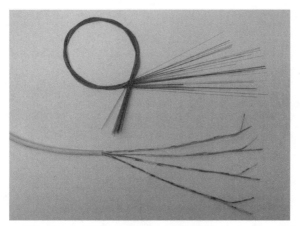

Figure F.7   **Forty-Eight Strands of Fiber Optic in Comparison to a Four-Pair UTP Cable**

Fiber-Optic Attenuator A small device with two connectors, one on each side. A fiber-optic attenuator works like your sunglasses, it reduces the level of light passing through it, just as sunglasses reduces the level of light entering your eyes so that you can see more effectively.

Fiber-Optic Buffer The plastic coating on individual fibers. The color of the buffer colors distinguishes fibers from each other. The 12 different colors for buffers are shown in *Appendix F*.

Fiber-Optic Cleaver A device used to splice fiber-optic strands. The cleaver is a precision tool that cuts the end of a fiber strand to a perfectly smooth 90-degree flat end (Fig. F.8). Without having the ends of a fiber cleaved during splicing, the fibers do not line up exactly. If this occurs, light will not transfer from the core of one fiber to the next properly. This is called *fresnel loss*. Typical fresnel loss on a single-mode fiber that has been cleaved and fusion spliced correctly is 0.01 dB.

Figure F.8   Fiber-Optic Cleaver

Fiber-Optic Color Code   See *Appendix F.*

Fiber-Optic Connector   The three main different types of fiber-optic connectors are: SC, ST, and FT (Fig. F.9).

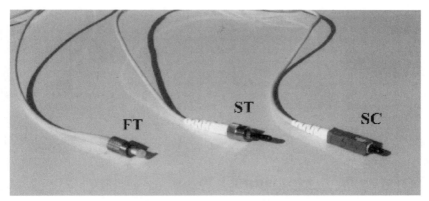

Figure F.9   Fiber-Optic Connectors

Fiber-Optic Distribution Panel   Also called a *fiber-optic distribution bay,* this unit is a fiber-optic termination device and organizer (Fig. F.10). It also houses fiber-optic splice trays, where the connector plugs (called *pigtails*) are spliced to the ends of fiber-optic cables.

Figure F.10   Fiber-Optic Distribution Panel

Fiber-Optic Splice Closure   A housing for a fiber-optic splice that can be buried or placed in a handhole/vault environment (Fig. F.11). These closures are designed specifically for fiber optic and have a splice tray incorporated within them. Fiber-optic splice closures are airtight when assembled correctly.

Figure F.11   Fiber-Optic Splice Closure

Fiber Remote   A solution made by Northern Telecom that extends IPE (intelligent peripheral equipment, Northern Telecom's name for PBX telephone station equipment) a distance of up to five miles (Fig. F.12).

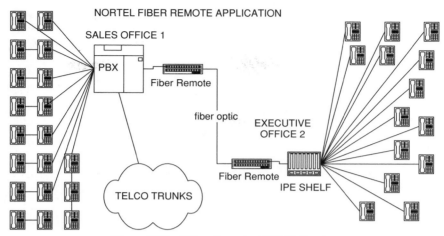

NORTEL FIBER REMOTE APPLICATION

Figure F.12   Fiber Remote: PBX Extension

Fiber To The Cabinet (FTTCab)   A network architecture where an optical fiber connects the telephone switch to a street-side cabinet, where the signal is converted to feed the subscriber over a twisted copper pair. This is a cost-effective way for telephone companies that have an existing twisted copper pair plant to implement digital services, such as those in the xDSL family.

Fiber Transmitter   A device used to extend cable-TV services over fiber optic. Fiber transmitters could be thought of as a radio-wave to light-wave converter. The signal carried by a fiber optic connected to a fiber transmitter is modulated the same way as the signal carried by coax. Fiber-transmitter technology makes it possible for cable-TV companies to deliver a broader spectrum of services to areas that are distant from the head end/node.

Fiber Tube Splitter   Also called a *Loose Tube Splitter*. A tool used to open the shell that encases fiber-optic strands. For a photo, see *Loose Tube Splitter*.

FIFO (First In First Out)   This means that the first bit into a memory (for temporary storage) is the first out when it is retrieved. In some memory-handling schemes, the last bit in is the first out (LIFO). You can portray LIFO by imagining that you have a box. As you stack books into the box, and then remove them one by one, the first book you put in is the last one out. If you fill the box in the same manner and then turn it over and open the bottom to remove the books, you have a FIFO input and retrieval scheme.

Fifth Generation   A reference to the Lucent 5E family of electronic telephone service switching systems, which is far more advanced and flexible than the 4E and previous switches.

Filament   1. A part of an electronic vacuum tube (Fig. F.13). The filament is often called a *heater*. The filament in a tube heats up and causes electrons to be emitted from the cathode.   2. The part of an incandescent light bulb that heats and lights. Filaments in light bulbs are often made of tungsten. The light bulb no longer lights when the tungsten filament breaks.

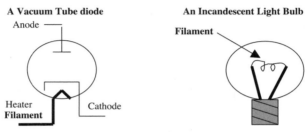

Figure F.13   Filament

File-Allocation Table (FAT)   A computer file that tracks as a directory for the contents of a hard drive. See also *FAT16* and *FAT32*.

File Server   A computer in a client-server environment that stores files or data. All the information regarding customer account transactions could go to a *file server*. When a PC on a network wants to run a report about customer accounts, the program will retrieve the data from the file server. The other type of server is an *application server*. This means that the program (application is another word for computer program) that runs the report is located on (or in) the server. The PC (client) requests the application server to calculate reports, then the server downloads the reports to the PC or a printer.

File-Transfer Protocol (FTP)   See *FTP*.

Filtering   A function of bridges where if a frame is received from a network segment/port that has a destination address that is on that same segment/port, then it is not processed through the bridge. Bridges transport frames that must move from one segment/port to another.

FireWire   A specification for an external serial bus that can transfer data at 400 Mbps and handle 63 simultaneous devices. This bus is intended for use with DVD players, graphics and audio peripherals, and other multimedia devices. FireWire is an innovation of Apple Computers, as is defined under IEEE 1394.

Firmware   Software instructions set permanently in ROM that are only able to be changed by replacing and reburning the ROM chip that contains the information. See also *EEPROM (Electrically Erasable Programmable Read-Only Memory)*.

Fish Tape   A tool for pulling wire through walls, conduit, or anywhere else that wire needs to go. It is a coil of thin and stiff metal tape encased in plastic. It has an opening where the stiff wire tape can be pulled out and extended through a conduit (or other place) where the wire needs to be installed. The wire is then attached to the end of the fish tape so that it can be pulled through. As the person holding the fish tape is pulling the wire through the conduit, they recoil it back into the plastic case. This tool is necessary for electricians and telecommunication wire installers.

FITC (Fiber To The Curb)   A reference to a hybrid fiber/copper public telephone network design, where broadband services are carried by fiber optic to a terminal near a customer's premises. The drop line that runs from the terminal is twisted pair or coax.

FITH (Fiber To The House)   A reference to an all fiber-optic public telephone network design, where broadband services are delivered to the customer premises/network interface by fiber optic.

Fixed Wireless   A microwave band of radio services that use modulation schemes similar to PCS/cellular only permitted to transmit at higher power. Fixed wireless is an alternative to the twisted-pair and coax method of providing local telephone service.

Flag   1. A single bit used in a communication header to signal a maintenance or other condition.   2. In X.25, a byte sequence of 01111110 put at the start and end of an X.25 frame. When X.25 lines are idle or if no data is being sent over them, the two ends continuously send each other flags to keep themselves synchronized.   3. A reference to an "if-then" logical condition that is used in lower-level computer programming languages.

Flapping   A network problem that is caused by a faulty circuit connection (or a similar condition) that fools routers into thinking that the circuit is continuously being added or dropped from the network. This causes the routers to continuously readvertise the route as being available or unavailable in the network. Depending on the routing protocol used (distance vector, link state, or hybrid), the problem can affect only the link that is made by the circuit or the entire network.

Flash   A form of telecommunications signaling. To send a flash signal, press the switch-hook of a telephone briefly. If you have call waiting on your telephone line and another call comes in (you hear the beep), you briefly push in the switch hook on your telephone to switch to your other call. When you want to revert back to the original call, briefly press the switch hook on your phone again. Some telephones have a flash button on them, which is a more convenient and less cumbersome way to send a flash signal than flipping the telephone's switch hook.

Flash Crowd   In Internet environments, an overrun on a network server by legitimate users. Publicly advertised events often cause a huge demand for a particular website. A good example is when the NASA Mars probe was transmitting pictures of Mars that could be viewed on the Internet at NASA's website.

Flash Memory   Digital storage that can be electrically erased and reprogrammed so that software can be stored, booted, and rewritten, as necessary. Flash memory retains data when power is removed. It was developed by Intel and is licensed to other semiconductor companies.

Flash Update   A hand-over of address/routing information from one router to another in response to a change being made within the network. Flash updates occur automatically in link-state and hybrid router operating systems.

Flat Network   A LAN deployment technique that incorporates all network hosts and servers into one broadcast domain, or effectively the same Ethernet segment. Applications that are chatty or broadcast/multicast intensive should not be implemented on flat networks. When flat networks are segmented, active hubs are replaced with LAN switches, which provide multiple VLAN (Virtual LAN segment) services and inherent security features.

Flexible Ringing   Also called *distinctive ringing*. A feature of a PBX system that allows telephone sets to ring differently. This is a very nice feature if you would like outside calls to ring differently than internal calls from co-workers. This helps a user to know if they should say "hello" or "Emergency service, may I help you?" Distinctive ringing is also used in offices where many phones are close to each other. When one phone rings, all the people in the office know whose phone it is by tone, pattern, and pitch of the ring.

Flicker   A characteristic of computer monitors, yet considered a form of video distortion by some. Flicker is caused by the screen refresh rate. If the refresh rate is slower than 70 refresh cycles per second (or 70 Hz), it will most likely be noticeable and straining to the human eye.

Flip Flop   The individual devices that dynamic computer or data-device memories are made of. One flip-flop can store one bit of information. The RS-flip flop is the main component of a 555 timer (an electronics industry-standard circuit). A flip flop is a member of the logic component family (Fig. F.14). It is comprised of two transistors that turn on and off inversely to each other and four bias resistors. On the schematic symbol, S represents for set and R represents reset. Output Q is positive when a positive pulse finds its way to the S lead of the RS flip-flop; the Q output stays positive and "remembers" the pulse. The Q output is reset to low when the R lead receives a positive pulse.

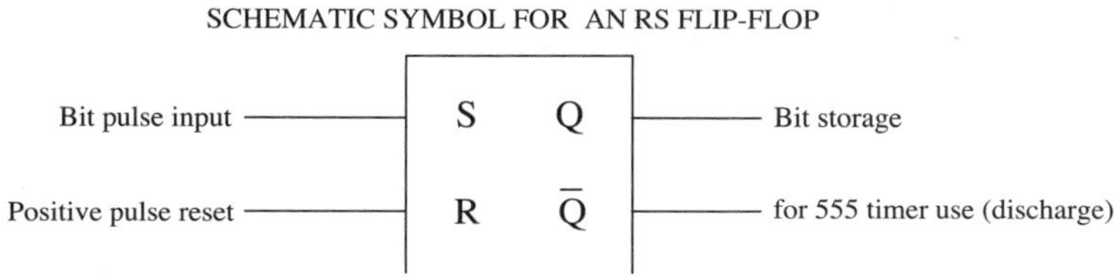

SCHEMATIC SYMBOL FOR AN RS FLIP-FLOP

Bit pulse input ———————| S   Q |———— Bit storage

Positive pulse reset ——————| R $\overline{Q}$ |———— for 555 timer use (discharge)

Figure F.14   Flip Flop

Flooding   What routing/bridging/LAN switches do when they do not recognize a frame's destination address. When a frame containing a recipient address that has not been seen before is received from a port/segment, the LAN switch transmits the frame out on all of its ports (within the same Virtual LAN). When and if a response is generated from a device, the sender address is recognized in the frame header and stored in the routing table with its associated port. This method of locating/discovering which port an unknown user is attached to is called *flooding*. RIP, EIGRP, RIPV2, and BGP use this method.

Floppy Disk   A magnetic disk used for storing digital data. A floppy disk is literally a floppy plastic disk, coated with iron oxide (or another magnetic compound) inside of a plastic case. A 3.5" floppy disk is capable of storing up to 1.44 megabytes (1,440,000 bytes) of data.

Flow   A reference to the rate of data travelling through a LAN, WAN, or multi-user environment. More specifically, the data flowing from an origination device, such as a LAN host (or networked PC) across an active virtual channel within a network and to a destination device. Multiple flows can exist on a single circuit.

Flow Control   1. A reference to the technique of ensuring that a transmitting device, such as a modem, does not overwhelm a receiving device with data. When the incoming memory buffer on the receiving device is full, a message is sent to the sending device to suspend the transmission until the data in the buffer has been processed. IBM calls their flow-control method *pacing,* but most people just refer to it as *flow control,* regardless of the network being discussed.   2. In full-duplex LAN switching, a method used by switches to slow the transmission of traffic. This is done using a pause frame to stop the transmission of a host. The switch contains the pause frame to one specific port, therefore not stopping all traffic through the entire network.

Flush Jack   A telephone or data-connection jack that is mounted in a wall in the same manner as an electrical outlet. Flush jacks can have as many as six connections/plugs on its face. AMP, Siecor, Leviton, Lucent, and Amphenol are all companies that manufacture a wide variety of flush jacks and other connectivity solutions.

Flux   A material that is used in the center of solder rolls. Some solders have acid cores, which cleans the connection to be soldered. If the connection is dirty, then the solder will not adhere to the metal.

FM (Frequency Modulation)   See *Frequency Modulation.*

FNC (Federal Networking Council)   The council that coordinates the communications networking among Federal agencies, such as NASA and the Department of Defense.

FOD (Fax on Demand)   A feature of voice-response systems that enables a caller to listen to a recorded message that gives them a selection of information that they can receive via a fax. After making a selection, the caller is then prompted by another recording to enter the number that they would like the information faxed to.

Fog Attenuation   In wireless networks, the diminishment of a radio signal due to dispersion caused by water droplets in the air. For specific details, see *Rain Attenuation.*

FOIRL (Fiber Optic Inter Repeater Link)   A fiber-optic version of Ethernet developed on the IEEE 802.3 fiber-optic specification. FOIRL is the early version of the 10BaseFL specification, which is for replacing FOIRL. See also *10BaseFL.*

Football   Another name for an aerial service wire splice. For a photo, see *Aerial Service Wire Splice.*

Footprint   1. The path that a satellite makes across the Earth's surface. If you drew a line of the satellite's path on the face of the Earth, that line would be its footprint.   2. A reference to a cabinet or the space a device takes up. If a sales person tells you that you can have a bunch of new features in the same footprint, then you don't have to change the cabinet, you might only need to change some software and a card or two.

Forced Account Code Billing   A service offered by long-distance companies and a feature of PBX systems. The way the feature works is that every time an individual wants to make a long-distance call they must input (or dial) an account code (and sometimes a password) after they dial a 1 (1 is the first number dialed in long-distance calls). After they dial the account code and/or password, they hear a confirmation tone, then they continue by dialing the long-distance phone number. When the telephone bill comes, each call on each account code is itemized. This is a convenient way to keep track of who is making which long-distance phone calls, to where, and when. Then, you will have the ability to go to the person and ask why.

Foreign Exchange Service   A telephone number that is served by another exchange. If you are on the south side of town, that area of town is probably serviced by a different central office exchange than the north side of town. Let's say that your business moves to the north side of town. The telephone company notifies you that you will need to change your phone number because you are moving to a different exchange area. (Each central office exchange has its own number plan for the first three of seven digits.) You cannot change your phone number because you have paid a fortune to advertise it. You don't want your clients to be hassled by getting an intercept message saying that "The number you have dialed xxx-xxxx has been changed. The new number is xxx-xxxx." So, the telephone company offers you a service where your telephone lines are forwarded from your old office (south) to your new office. This service is called *foreign exchange service.* Your lines feed from a different (or foreign) central office than the office you are actually served by. Your customers then call you on the same number and never know that you moved, and they don't have to listen to a recording. You will probably only need to make your advertised number's foreign exchange (abbreviated *FX*) lines or trunks. The rest can be regular phone lines that are serviced from your exchange area. Soon, FX lines will be a thing of the past. Because of the Telecommunications Act of 1996, all phone numbers within an area code must be portable from one exchange or service provider to another. This is called *number* *portability.* This means that in the previous scenario, you could have had your numbers moved to any exchange within the area code and even changed phone companies without having to change your phone number.

FORTRAN (Formula Translation)   A high-level computer programming language.

Forward   A feature of PBX systems and a service offered by local, long-distance, and cellular telephone companies. The way that the Forward feature (also called *Call Forwarding*) works is if you know you are not going to be at your phone, you can make all your calls ring at a different number.

Forward Error Correction (FEC)   A method of recouping lost data over a transmission where the receiver of the damaged data uses a code to attempt to repair or correct the damage. This method saves bandwidth by avoiding retransmissions. The other method of error correction is called *BEC (Backward Error Correction),* which operates by having the receiver of a damaged data stream simply requests a retransmission of the data.

Forward Explicit Congestion Notification (FECN)   A bit in the overhead of a data packet travelling in a frame-relay network. This bit is set to 1 if the packet travels through an area of network congestion that is the same direction (or forward) of its flow. This bit is a signal from the frame-relay network to higher-level protocols within DTE and DCE to take appropriate flow-control action. An example of flow-control action would be to not attempt to exceed the *Committed Information Rate (CIR)* for the connection. This would prevent the unnecessary transmission of low priority frames that would surely be discarded. The counterpart of FECN is *BECN (Backward Explicit Congestion Notification).*

Forwarding Rate   In bridges or routers, the amount of data in packets per second that a device such as a switch/router/bridge can transfer traffic in on one port and out on another.

Four-Part Dotted Notation   The term given to the way that TCP/IP addresses and subnet masks are written in the form *N.N.N.N.,* where each number "N" represents one byte (converted to decimal) of the four-byte IP address. The decimals are only there to make the number easier for people to understand, the same way dashes are inserted into telephone numbers. Routing and other network equipment/software do not recognize the dots. An example of a four-part dotted notation address is: 255.255.255.248. It is also called a *Dot Address*.

FDPL (Foreign Data Processor Link)   Rockwell's version of a serial data interface from a PBX switch to a PC or other I/O equipment. Northern Telecom calls FDPLs *MSDLs (Meridian Serial Data Link)*. It is usually an additional card that plugs into a PBX system's common equipment-expansion area, which is equipped with an RS-232 communications port.

Fractal Compression   A data-compression method used in graphics and other files by converting them into mathematical equations, rather than straight binary values.

Fractional T1   A way to configure a T1 (DS1) service so that two or more of the DS0 circuits are joined together to make a larger data communications channel. If a full T1 is too big and expensive, but a 56K line is too small, then a fractional T1 might offer a good mid-range solution.

Fractional T3   A private line service where 4, 8, 12, or 16 T1 circuits (multiples of DS2) are combined together for a private data communications line.

FRAD (Frame Relay Access Device)   A device located on the customer's premises that acts as the Network Interface for frame relay services.

Fragment   An incomplete data frame.

Fragmentation   The process performed by a router or other network device whereby packets or datagrams are treated as files themselves and broken into smaller-sized data units for transmission over a different access protocol. For example, Ethernet packet size permits up to 1500 bytes of user data. X.25 running a common configuration only allows 128 bytes. A router would fragment the Ethernet packets to transmit them over the X.25 network. In TCP/IP, the IP layer is responsible for both fragmenting and reassembling fragmented packets.

Frame   A unit of data at the data-link and physical layers of the *OSI (Open Systems Interconnect)* model. A frame is the means of transport for packets, segments, and *PDUs (Packet Data Units)*, which are means of transporting the payload data through their according layers of the OSI (Fig. F.15). ATM, frame relay, X.25, and T1 all have different framing formats at their respective physical layers. A frame consists of two sections. They are called the *header* and the *data*, or the *overhead* and the

Each box represents an 8-bit sample for one of 24 channels. The last box represents a timing bit.

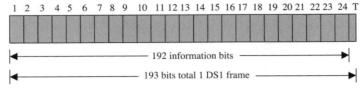

Figure F.15   A DS1 Frame

payload, respectively. The purpose of the header is to carry address and routing control information; the purpose of the data section is to carry data. See also *Data Encapsulation.*

Frame   One complete sampling and conversion cycle of a multiplexed data transmission. A frame is sometimes confused with a packet. A packet is an envelope that contains data and an address that the data is sent to. A packet contains data to be transmitted, error-correcting information for the data in the packet, an address, timing information, and other bits of data, depending on the protocol that the packet was formed under. A frame is a momentary picture of a multiplexed data transmission, containing bits of data, or samples from each channel.

Frame-Check Sequence (FCS)   A part of Cyclic Redundancy Checking error correction in data transmissions. See *Error Checking* for more information.

Frame Forwarding   A reference to the process by which frame-based traffic, such as *HDLC (High-Level Data-Link Control), SDLC (Synchronous Data-Link Control),* and more evolved protocols (such as X.25 and frame relay) traverse an ATM network.

Frame Rate (FPS)   Frames per second. The number of frames transferred in a second.

Frame Relay   A packet-forwarding standard protocol. Frame relay is a private networking service offered by many telecommunications companies to accommodate the need for virtual WAN connections (Fig. F.16). The gross data-transfer speeds for frame relay start at 56 Kbps, and the protocol allows speeds beyond 2 Gbps, although the highest speeds offered to customers are limited to 44 Mbps (DS3/STS1). To connect to the service, a *FRAD (Frame-Relay Access Device)* is placed at the customer's premises. The FRAD is a router in most cases. Each FRAD within a customer's network is assigned a *DLCI (Data-Link Connection Identifier),* and each connection has a *CIR*

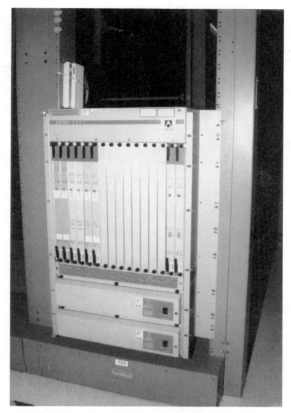

Figure F.16   Frame Relay Switch/Mfg. by Ascend

(Committed Information Rate), among other service parameters. Frame-relay networks are commonly referred to as *Public Data Networks (PDNs).*

Frame-Relay Compatible Protocols   TCP/IP, X.25, SNA, Novell, Banyan Vines, Decnet, NFS, Novell IPX, Appletalk, and more.

Frame-Relay Frame Format   The frame format for frame relay has five fields. In order, they are the flag, the header, the user information, the FCS, and the flag (Fig. F.17). The flags in frame relay are exactly as they are in the HDLC standard X.25, consisting of the following octet: 01111110. They exist for the sole purpose of separating frames. The header is comprised of two octets, and is the core of the frame. It defines the DLCI high order, C/R, EA, DLCI low order, FECN, BECN, DE, and another EA.

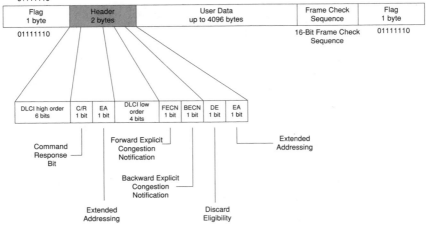

All Frames Start and End
with a Flag Sequence of
01111110

Flag 1 byte	Header 2 bytes	User Data up to 4096 bytes	Frame Check Sequence	Flag 1 byte
01111110			16-Bit Frame Check Sequence	01111110

DLCI high order 6 bits	C/R 1 bit	EA 1 bit	DLCI low order 4 bits	FECN 1 bit	BECN 1 bit	DE 1 bit	EA 1 bit

Command
Response
Bit

Forward Explicit
Congestion
Notification

Backward Explicit
Congestion
Notification

Extended
Addressing

Extended
Addressing

Discard
Eligibility

Figure F.17   Frame-Relay Frame Format

Frame Tagging   In LAN switching, frame tagging is a method used to identify which Virtual LAN (VLAN) a packet belongs to. Within an Ethernet switch, a VLAN behaves as a single Ethernet segment, where all computers/hosts that are a part of a VLAN see each other's traffic. Traffic cannot traverse from one VLAN group to another unless it leaves the switch and is routed back in on another port. The router provides traffic security/management. Frames are marked by the switch in two methods: implicit and explicit. Implicit frame tagging is a method where the VLAN information is added within the packet. Explicit frame tagging is a method where an external VLAN header is added to the frame. See also *802.1Q*.

Framing Bit   A bit added to a multiplexed data stream that separates each complete cycle of sampling, DAC conversion through all the channels being multiplexed. A framing bit can also be called a *timing bit*. See *Frame* for more details.

Franchise   A right (license) to provide telecommunications service to a community, county, or city. Telephone companies and cable-TV companies must have a franchise agreement before they sell their services. Copies of franchise agreements are public documents and you can get one from your local PSC office.

Free Phone   The European name for toll-free service that is similar to "800" service in the United States. Calls made to these numbers are billed to the called party or the owner of the number.

Free-Space Communications   A reference to radio communications, such as terrestrial microwave, satellite, and cellular.

Free Space Path Loss   In radio transmissions, signals degrade as they propagate through space. Better said, they weaken as distance increases in a predictable manner. Free space path loss is a factor in calculating the link viability. Free space path loss is easily calculated for miles or kilometers using one of the following formulas:

$$Lp = 96.6 + 20 \log_{10} F + 20 \log_{10} D$$

where Lp = free space path loss between antennas (in dB), F = frequency in GHz, D = path length in miles; or

$$Lp = 92.4 + 20 \log_{10} F + 20 \log_{10} D$$

where Lp = free space path loss between antennas (in dB), F = frequency in GHz, D = path length in kilometers.

Frequency   *Frequency* is a measure of the number of cycles per second, and its unit is the hertz (Hz). One hertz is equal to one cycle in one second. Two Hz is two cycles in one second, 1000 Hz is 1000 cycles in one second *(CPS)*. Cycles are used as a reference to measure the frequency of a waveform or signal. Cycles are usually referred to as a number of cycles per unit of time. CPS and Hz are measurements of the number of cycles you get per second in an analog transmission. BPS is measurement of how many "square wave" clock sample sequences are being read from a digital transmission. For a diagram of one cycle in a waveform, see *Cycle*.

Frequency Band   A range of frequencies that a certain class of radio communications operates within. For example, FM radio operates within 87.9 MHz and 108 MHz, which falls within the VHF frequency band. For a general list of frequency ranges, see *Band, Frequency*.

Frequency Band Division   In broadband fixed wireless, each system or "hop" is a full-duplex system. Two frequency bands are used to achieve this two-way operation, with the higher frequency band considered the "high" band in the link, and the lower frequency considered the "low" band. The transmitter at one end of the link uses the high band; the transmitter at the other end uses the low band.

Frequency-Division Multiple Access   Another name for frequency division multiplexing that the cellular telephone industry uses.

Frequency-Division Multiplexing (FDM)   See *FDM*.

Frequency Hopping   A method of radio transmission where the carrier frequency changes at fixed intervals.

Frequency Hopping Spread Spectrum (FHSS)   A wireless LAN radio modulation technique where (in the case of 802.11FH) 79 1-MHz radio channels occupy a spectrum of 2.400 to 2.483 GHz. As data is transmitted over the radio, the radio alternates synchronously with the receiver over the 79 channels via an algorithm or other method used by the radio manufacturer a minimum of two times per second. The maximum FHSS throughput is 2 Mbps. See also *DSSS* (which is newer and provides up to 11 Mbps when deployed under the 802.11DS standard).

Frequency Modulation (FM)   See *FM* (Fig. F.18).

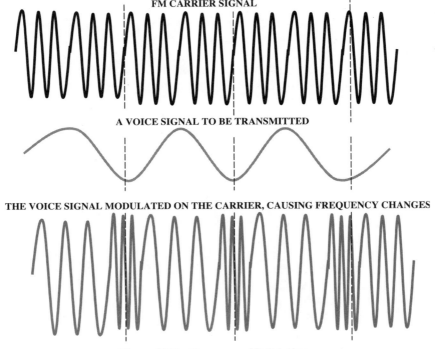

Figure F.18   Frequency Modulation

Frequency Response   A range of frequencies. A high-end quality home stereo system might have an output frequency response of 20 Hz to 20 kHz (the range of human hearing). A telephone line has a frequency response of 500 Hz to 3500 Hz (the range of human voice).

Frequency-Shift Keying (FSK)   A method of binary signal modulation. FSK is the way that modems send bits over the telephone lines. Each bit is converted to a frequency. A 0 is represented by a slower frequency, and a 1 is represented by a higher frequency. Morse Code is a method of binary keying, where all signals have two states (binary), long and short. FSK is the same technology, only two different frequencies are used, instead of two lengths of beeps or lines (Fig. F.19).

FSK conversion of bits to frequencies

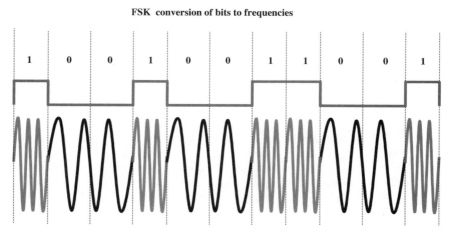

Figure F.19   Frequency Shift Keying

Fresnel Loss   A loss of signal in a fiber optic because of a splice or a crack in the glass of the fiber. The splice or crack causes a type of reflection, called a fresnel reflection. This causes the light to scatter in different directions, rather than "focus" its way down the fiber.

Fresnel Reflection   1. A reflection of light because of a splice or a crack in a fiber optic (Fig. F.20). 2. A reflection of a terrestrial radio signal off of an object (such as a building, hill, or body of water) from its fresnel

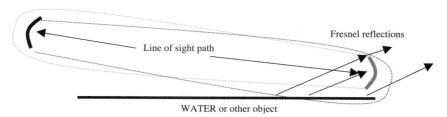

Figure F.20   Fresnel Reflection

zone. Fresnel reflections can cause a reduction in signal strength or a total loss of the signal altogether.

Fresnel Zone   The area within a terrestrial microwave dispersion pattern, but outside of the line-of-sight path. The fresnel zone should be clear of any obstructions. Obstructions in the fresnel zone can cause fresnel reflections.

FRF (Frame Relay Forum)   An association of corporate members consisting of vendors, carriers, users, and consultants committed to the implementation of frame relay in accordance with national and international standards. For more information, check their website, *www.frforum.com*.

FRF.11   The frame relay forum implementation agreement for voice over frame relay (May 1997). This specification defines multiplexed data, voice, fax, DTMF digit-relay, and CAS/Robbed-bit signaling frame formats, but does not include call setup, routing, or administration facilities.

FRF.12   The convergence of voice and data over the same packet structure within frame relay. The FRF.12 Implementation Agreement (also known as FRF.11 Annex C) was developed to allow long data frames to be fragmented into smaller pieces and interleaved with real-time frames. In this way, real-time voice and non-real-time data frames can be carried together on lower speed links without causing excessive delay to the real-time traffic. After this standard was released, voice over frame still had the same primary drawback of difficult end-to-end management of traffic. The frame relay network isolates end points from a management perspective when the packets being carried are broken up and loaded directly into the frame. IP voice is often carried over frame relay because it gives the user the best of all three worlds, convergence of voice and data, end-to-end network visibility, and cost-effective transport.

Friction Electricity   Another name for *static electricity*. The opposite of static electricity is *dynamic electricity*, like that from a power outlet. Static electricity consists of free electrons (electrons that are not a part of the valence shell of an element). The free electrons have no means to flow to a positively charged media, so they build up and form a charge or a voltage. The voltage is not measurable because when a measuring device is introduced to the static field of electrons, it provides a path for them to flow. When the electrons start flowing, they are no longer static. Typical static or frictional charges on a dry day can reach beyond 30,000 volts.

FRMR (Frame Reject)   A command defined by the last three bits in the control byte of an unnumbered or "control" type frame in the X.25 protocol being 001. A frame reject is a response to an invalid or meaningless control frame. After the FRMR has been sent, a *SABM (Set*

Asynchronous Balanced Mode) frame should be returned, which will reset the link. For more information on X.25 frame-level signaling, see *X.25 U Frame* and *X.25 Control Frame*.

Front-End Processor (FEP)   A communications "front end" device that can be loaded with a "firewall" to prevent unwanted users from accessing the communications network (Fig. F.21). An FEP can also perform routing and differentiate between different communications protocols, depending on the software that runs on it.

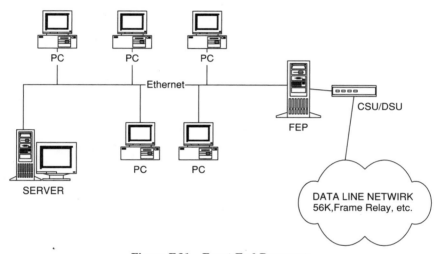

Figure F.21   Front-End Processor

FS   The ASCII control code abbreviation for form separator. The binary code is 1100001 and the hex is C1.

FSK (Frequency-Shift Keying)   See *Frequency-Shift Keying.*

FT Connector   A metallic, screw-on fiber-optic connector. For a photo, see *Fiber-Optic Connector.*

FTP (File Transfer Protocol)   A set of instructions for transferring files on a network that uses TCP/IP. TELNET is an example of an FTP. There are two parts to an FTP: the server part and the client part. The server part runs on a larger computer (or server) and the client part is run by hosts that copy or transfer files from the servers. FTP is very useful because it allows communicating machines of different types and operating systems to transfer information. FTP is also capable of doing some data-format conversions (i.e., ASCII to binary).

FTTC (Fiber to the Curb, FITC)   See *FITC*.

FTTCab (Fiber to the Cabinet)   A network architecture where an optical fiber connects the telephone switch to a street-side cabinet, where the signal is converted to feed the subscriber over a twisted copper pair. This is a cost-effective way for telephone companies that have existing twisted copper pair plant to implement digital services, such as those in the xDSL family.

FTTH (Fiber to the House)   Another variation of FITH. It just depends on which country you are in or which service provider/manufacturer you are working with.

Full Duplex   1. See *FDX*.   2. In LAN switching, a feature that enables speed to be doubled by having simultaneous two-way conversation. For example, a full-duplex 100BaseTX connection between a switch port and a host/computer would have simultaneous 100 Mbps capability for transmit and 100 Mbps for receive, for a total of 200 Mbps throughput. Full-duplex Ethernet operation is a byproduct of LAN switch architecture. LAN switches must be implemented to obtain full-duplex Ethernet. Full-duplex functionality does not work across hubs.

Functional Entity   An entity, if it is physical or logical, that performs a task. An example of a functional entity is the data link layer of the OSI protocol. It is not necessarily a thing, it is just a phase in a process.

Functional Signaling   Signaling of an ISDN line or other communication where the signaling of the circuit is performed in a manner that the user understands as well as the machines that make the communications work. An ISDN signal that is functional is a display reading that says "incoming call from John Doe," the ISDN could just give a ring or some other notification, but the call-signaling information bits in the ISDN circuit are actually decoded and passed on to the user.

Fuse   A current-sensitive protection device. A fuse is designed into equipment so that if a component within that equipment should fail, it will blow the fuse. When a component fails, it draws excessive current. Excessive current causes excessive heat, which causes fire and ruins other components (not to mention danger to people). Always replace a fuse with the correct size. Replacing a blown fuse with one that has a higher amperage rating could damage the equipment.

Fuse Alarm Panel (FAP)   A power-distribution panel that is installed at the top of a relay rack (Fig. F.22). All equipment in the rack is wired to the panel for power. Each device has its own fuse within the panel to protect the rectifier from an over-current condition if the device fails or a wire is shorted. If any of the fuses blow, an alarm indicator is displayed.

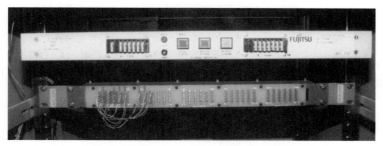

Figure F.22   Fuse Alarm Panel (FAP)

Fusion Splicing   A method of splicing fiber-optic cable. Fusion splices have less fresnel refraction when they are complete and, therefore, less loss than mechanical splices. The way a fusion splicer works is after the two fiber ends to be spliced are inserted into the splice housing, the splicer cleaves the ends, butts them together and thermally fuses the ends together. Fujikura is a popular manufacturer of fusion-splicing equipment (Fig. F.23).

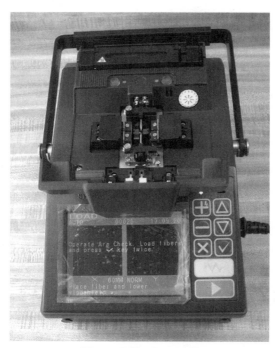

Figure F.23   Fusion Splicer

FX Line   See *Foreign Exchange Service.*

FX Trunk   See *Foreign Exchange Service.*

FXO (Foreign Exchange Office)   1. A telephone company central office that is of another exchange, which means that it has a different numbering plan. See *Foreign Exchange Service.*   2. An analog telephony interface that is intended to connect to a central office or PBX. FXO interfaces come in several types, including ground start, loop start, and E&M. For trunks that are larger in capacity, T1 and ISDN are used rather than single FXO interfaces (Fig. F.24).

Figure F.24   FXO Interface Module

FXS (Foreign Exchange Station End)   A telephony connection that is intended to be connected to an end device, such as an analog 2500 telephone. FXS interfaces come in several data link layer types, including ground start, loop start, and E&M. They are also available in RJ11- and Amp/Centronics-style connectors for the physical interface (Fig. F.25).

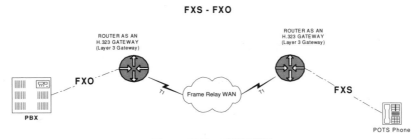

Figure F.25   FXS/FXO

G

G.703/G.704 A physical connection standard. The ITU-T G.703/G.704 standards are electrical and mechanical specifications for connections between telephone company equipment and customer DTE using BNC connectors and operating at E1 data rates.

G.704 The ITU-T electrical and mechanical specifications for hierarchical interfaces between telephone company equipment and customer-owned *DTE (Data Terminating Equipment)* using BNC connectors for E1.

G.706 The ITU-T standard for frame alignment and CRC for hierarchical digital frame structures listed in G.704 (E1/T1,3).

G.707 The ITU-T standard for European SDH bit rates (STM rates).

G.708 The ITU-T standard for network-mode interface for European SDH.

G.709 The ITU-T standard for synchronous transfer mode multiplexing.

G.711 A Codec standard. The G.711 ITU-T standard for voice compression describes the 64 Kbps PCM voice coding technique. In G.711, encoded voice is already in the correct format for digital voice delivery in the PSTN or through PBXs. Using this type of compression in a VoIP network reduces delay and improves voice quality. It also reduces the load placed on routers performing the digital signal processing. The disadvantage is that it is a "bandwidth guzzler" in terms of connections over a WAN link. The bandwidth per conversation required, including Ethernet and other framing overhead, can approach 128 Kbps. The most important aspect of G.711 and other Codec methods is that they provide a true convergence of traffic for VoIP networks, which makes for better end-user management of bandwidth. See also *G.729*.

G.721   The ITU standard for 32 Kbps ADPCM.

G.722   The ITU standard for 7-kHz audio coding for DS0 channels.

G.723   A Codec standard. The ITU-T G.723.1 standard describes a compression technique that can be used for compressing speech or audio signal components at a very low bit rate as part of the H.324 family of standards. This Coder/Decoder has two bit rates associated with it: 5.3 and 6.3 Kbps. The higher bit rate is based on ML-MLQ technology and provides a somewhat higher quality of sound. The lower bit rate is based on code exited linear prediction (CELP) and provides system designers with additional flexibility. G.723 is not a VoIP Codec method because of the true bandwidth advantages of G.729, and the fact that IP telephony products run the H.323 session protocol.

G.725   The ITU-T standard for the use of a 7-kHz audio codec over DS0.

G.726   A Codec standard. The ITU-T G.723.1 standard describes ADPCM coding at 40, 32, 24, and 16 Kbps. ADPCM-encoded voice can be interchanged between packet voice, PSTN, and PBX networks if the PBX networks have ADPCM interface capability. Some long-haul interface circuit switch designs incorporate this, but is not a preferred design as contrasted to G.711 and G.729 in cost.

G.728   A Codec standard. The G.728 ITU-T standard describes a 16-Kbps low-delay version of CELP (Code Excited Linear Prediction) voice compression. G.728 CELP voice coding must be translated into a public telephony format by a router or voice gateway for delivery to the PSTN. This is not a popular method used in IP telephony or VoIP networks due to a lack of advantage in voice quality, router load, or bandwidth.

G.729   A Codec standard. The G.729 ITU-T standard describes CELP (Code Excited Linear Prediction) compression where voice channels are coded into 8 Kbps streams (Fig. G.1). There are two variations of this standard (G.729 and G.729 Annex A) that differ primarily in computational complexity; both provide near-excellent speech quality similar to 32 Kbps ADPCM. The G.729 standard is preferred in a VoIP network because it conserves bandwidth on an 8 : 1 ratio. One voice connection with overhead (Ethernet and other framing) consumes about 11 Kbps over a WAN link (i.e., T1 private line). All G.729 CELP voice coding must be translated into a public telephony format by a router or voice gateway for delivery to the PSTN. G.729 has some drawbacks, including the inability to carry modem traffic to any significant extent, and it also degrades DTMF tones and music to a degree noticeable to a keen listener. See also *MOS (Mean Opinion Score)*.

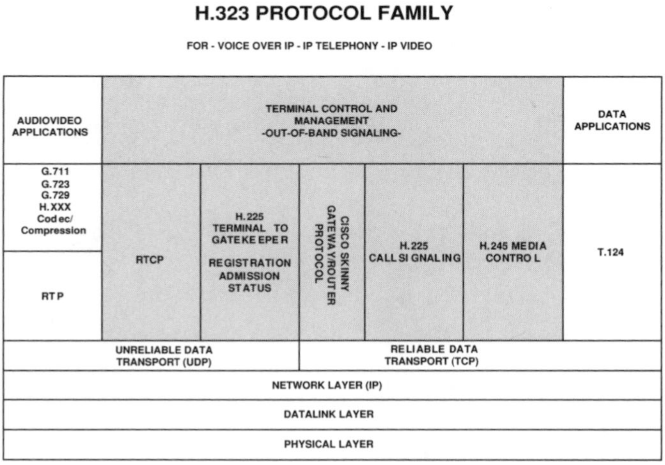

H.323 PROTOCOL FAMILY

FOR - VOICE OVER IP - IP TELEPHONY - IP VIDEO

AUDIOVIDEO APPLICATIONS	TERMINAL CONTROL AND MANAGEMENT -OUT-OF-BAND SIGNALING-					DATA APPLICATIONS
G.711 G.723 G.729 H.XXX Codec/ Compression	RTCP	H.225 TERMINAL TO GATEKEEPER REGISTRATION ADMISSION STATUS	CISCO SKINNY GATEWAY/ROUTER PROTOCOL	H.225 CALL SIGNALING	H.245 MEDIA CONTROL	T.124
RTP						
	UNRELIABLE DATA TRANSPORT (UDP)		RELIABLE DATA TRANSPORT (TCP)			
NETWORK LAYER (IP)						
DATALINK LAYER						
PHYSICAL LAYER						

Figure G.1   G.711 and G.729

G.747   The ITU-T standard for second-order multiplexing at 6.3 Mbps to carry three 2.048-Mbps E1s.

G.755   The ITU-T standard for multiplexing three 44.736-Mbps DS3 channels to one 139.264-Mbps channel.

G.802   The ITU-T standard for connecting networks that are based on different digital hierarchies and different voice-encoding methods.

G.804   The ITU-T framing standard that defines the mapping of *ATM (Asynchronous Transfer Mode)* 53-byte cells from the data-link layer to the physical layer. See also *Asynchronous Transfer Mode*.

G.811   The ITU-T standard for timing requirements at the outputs for primary reference clocks suitable for plesiochronous operation of international communications links.

G.812   The ITU-T standard for timing requirements at the output of slave clocks suitable for plesiochronous operation of international communications links.

G.824   The ITU-T standard for the control of signal distortion in T1 and T3 connections.

G.826   An ITU-T listing/publication that contains definitions for "availability" and related terms used to describe link quality in point-to-point radio applications that include licensed and unlicensed paths. See also *P.530*.

G.901   The ITU-T standard for the general considerations on digital-line systems.

G.921   The ITU-T standard for digital lines, based on the E1 hierarchy.

G.960   The ITU-T standard for digital sections for ISDN BRI.

G.961   The ITU-T standard for digital transmission systems on twisted copper for ISDN BRI.

Gaff   What telecommunications and power company personnel wear to climb wooden telephone and power poles. The official name for these devices are *lineman's climbers.* They are also called *climbers, hooks,* and *spurs.* They consist of a steel shank that has straps on it so that it can be strapped to a person's leg. On the inside of the shank is a spike that is used to stab into the pole. For a photo, see *Climbers.*

Gain, Antenna   Antenna gain is an indicator of how well an antenna focuses RF energy in a preferred direction. Antenna gain is expressed in dBi (the ratio of the power radiated by the antenna in a specific direction to the power radiated in that direction by a non-directional antenna fed by the same transmitter). Antenna manufacturers normally specify the antenna gain for each antenna they manufacture.

Gallium Arsenide   A new innovation in semiconductor technology. Gallium arsenide is more tolerant of heat than silicon, and it also has the special ability to convert light into electricity and vice versa.

Game Port   The de-facto standard interface for personal computers, which incorporates a DB15 connector for attaching joysticks and other interactive devices.

GAN (Global-Area Network)   A network classification. Other network classifications are: LAN *(local-area network)*, MAN *(metropolitan-area network)*, and *WAN (wide-area network)*. American Express utilizes a GAN to provide financial services to its customers worldwide.

Gas Carbon   A device the telephone companies use for lightning protection. When telephones were first used, there was a big problem with lightning striking telephone wires and subsequently electrocuting people or burning homes down. Lightning protectors give lightning an easier path to ground compared to IW or a person. The early lighting protectors were made of carbon. When they were hit by lightning, they would short to ground, then the phone line would be out of service until a telephone technician came and replaced them. The new lightning protectors are made with a gas. When the lightning hits them, they temporarily short to ground, then re-enable the phone line automatically. This innovation greatly reduced the number of bad phone lines that a telephone company would have after a thunderstorm. Gas carbons have no carbon in them, they are just called that because the old lightning protectors were made of carbon. The new gas lightning protectors are the same shape and size as the old ones, so they can easily fit into older network interfaces. For photos of lightning protectors, see *Lightning Protector* and *2 Line Network Interface.*

Gate   An electronic logic device. The three different functional types of logic gates are: the AND gate, the OR gate, and the inverter (NOT circuit). Gates are reactive devices. For a certain input, they react and produce an output. These small, simple circuits and other circuit devices (such as latches and flip-flops) make up all microprocessors and control devices. Thousands of logic gates and other components are used to make a single microprocessor or control chip.

Gate Array   A circuit, usually on one microchip, that contains many gates. The gates are connected together to perform a function, such as decoding.

Gatekeeper   In IP telephony, an entity on a network (usually a router) that provides address translation and access control. It also provides bandwidth control information to IP telephony servers involved in WAN call connections. For instance, if a first IP telephony server would like to connect a call to a second IP telephony server that is on the other side of a WAN network, the first server inquires with the Gatekeeper about bandwidth availability to carry the call packets. If the bandwidth is available, the gateway signals the telephony server the bandwidth availability associated with all routers and gateways assigned that route. The telephony server then selects the prioritized route and signals the call. If there is not adequate bandwidth available, the gatekeeper notifies the calling telephony server of this, and the server then commands the IP phone to play a fast-busy wave file to the user.

Gateway   In IP telephony, a gateway is the point at which a circuit-switched call is encoded and repackaged into IP packets. Gateways are hardware devices that are similar to routers except that they only perform layer 2 (bridge) functions (Fig. G.2). Gateways connect PSTN services such as ISDN, POTS lines, and T1 channelized voice circuits to IP telephony networks. Gateways incorporate protocols such as MGCP, H.323 to communicate with central control devices on a respective IP telephony network such as an IP telephony server.

Figure G.2   IP Telephony Gateway

Gateway   The former name for a router. When gateways began to perform services beyond what the Internet required, such as LAN protocol conversion and optimization, they were given the name *routers* by the companies that manufactured them. In some applications, a router is referred to as a *gateway router* because it provides a link to a foreign network and a specific (often limited) level of security.

Gauge (AWG, American Wire Gauge)   A measurement standard for copper wire. The gauge rating is the thickness of a solid copper wire. The larger the gauge, the smaller the wire (go figure). Most telephone wire ranges from 19 AWG to 26 AWG. Cat3 is commonly 24 AWG. The electrical wire in your home is probably 12 AWG.

Gaussian Distortion   Also called *white noise.* Gaussian distortion occurs at the physical layer of telecommunications circuitry and is distinguishable in audio applications as a hissing sound. It has several causes, including the random movement of electrons in an electronic circuit, and certain kinds of radio frequencies being magnetically induced into the affected circuitry.

Gender Bender   See *Gender Changer.*

Gender Changer   A small device used to mate two plug ends of the same type (Fig. G.3)

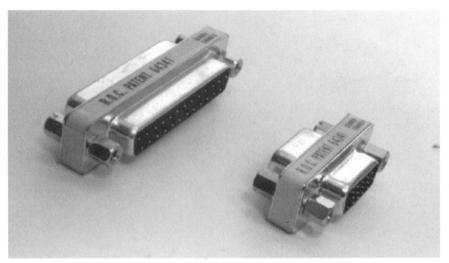

Figure G.3   Gender Changers: DB25 (Left) and High-Density HD15 (Right)

Gender Mender   See *Gender Changer.*

Genderless Connector   See *Data Connector.*

General Format Identifier (GFI)   In X.25, a type of signaling frame called a *control packet* is intended for the packet layer to use. The first four bits of the first byte of this packet are called the *general format identifier* and they are always 0001 or 0010. See also *X.25.*

General Packet Radio Service (GPRS)   The third generation of wireless technology offered as a service in Europe and other countries with a posture toward European standards. GPRS provided by telecommunications companies provides users with a standards-based IP application interface, so multimedia and data can be transmitted over the radio link as well as voice. Transmission rates for GPRS are expected to mature to an excess of 1 Mbps, be able to simulate a constant connection, and be able to share services with CDMA2000 networks because the IP layer provides a standard application regardless of the radio links used on either end of the call.

General-Protection Fault   A warning in Microsoft Windows operating systems that notifies a user that an application has tried to access a portion of memory that it is not allowed to. This usually means the program is attempting to perform a function that it is not able to, sometimes because of a corrupted data address within RAM. General-protection faults often require a user to exit the application and reboot the PC.

GFCI (Ground Fault Circuit Interrupt)   A type of circuit breaker. In newer homes, GFCI electrical outlets are required within a certain distance of a sink or bathtub. Unlike a normal circuit breaker that simply disconnects power when a current to AC common is exceeded (e.g., 15 A is common), a GFCI breaks the circuit when current to ground (instead of AC common) is detected. The regular common circuit breakers are located in the circuit breaker box in your home. The GFCI breakers are located inside of the electrical outlet itself.

GHz (Gigahertz)   *Giga* means billion. 38 GHz means 38 billion Hz or 38,000,000,000 Hz.

Giant   In LAN networking, an invalid Ethernet frame that is more than 1518 bytes long. These frames are discarded when detected by LAN switches/router devices.

Giga (G)   Engineering notation for billion. One gigabyte is equal to 1 billion bytes. Hard disk-drive memories with gigabytes of capacity are now becoming very affordable.

Gigabit   One billion bits.

Gigabyte   One billion bytes.

Gigabit Ethernet   The 802.3z, 1000BaseX specification. Gigabit Ethernet is defined for fiber optic multimode and single mode, at 500 meters and 2 km, respectively. There is also a copper version that runs distances of 25 meters. Gigabit Ethernet is interfaced with SC fiber optic connectors, and there is a wide variety of laser-diode adapters available from manufacturers (i.e., the Cisco Systems GBIC connector) in the LX, LS, and LH range. Gigabit Ethernet standards use 8B/10B encoding and decoding schemes (Fig. G.4).

			GIGABIT ETHERNET			
			802.3z and 802.3ab distances			
STANDARD	SPECIFICATION	Wavelength L nanometers	Fiber Type	Modal Bandwidth MHz-km	Recommended Maximum Distance	
802.3z	1000BaseLH*	1300nm	9/10 Single Mode	n/a	10km	32,810ft
802.3z	1000baseLX	1300nm	5um Single Mode	n/a	3km	9,843ft
802.3z	1000BaseLX	1300nm	62.5/125um Multimode	500	550m	1804ft
802.3z	1000BaseLX	1300nm	9um Single Mode	500	5km	16,405ft
802.3z	1000BaseSX	850nm	62.5/125um Multimode	160	220m	722ft
802.3z	1000BaseSX	850nm	62.5/125um Multimode	200	275m	902ft
802.3z	1000BaseSX	850nm	50/125um Multimode	400	500m	1640ft
802.3z	1000BaseSX	850nm	50/125um Multimode	500	550m	1804ft
802.3ab	1000BaseT	n/a	Cat5 UTP	n/a	100m	328ft
802.3ab	1000BaseCX	n/a	balanced copper	n/a	25m	82ft
802.3u	100BaseFX	850nm	62.5/125um Multimode	400	400m	1,312ft
	*Proposed 802.3z Cisco Systems Proprietary as of 10/2000.					

Gigabit Ethernet 802.3 Distances

Figure G.4   Cisco Systems Gigabit Ethernet Interface 802.3z (GBIC-Gigabit Interface Card)

GII (Global Information Infrastructure)   Standards are still being set for GII by the ANSI at this writing.

Glare   In Telephone Central Office environments, a trunking condition where both end devices of an E&M loop try to seize, or go "off hook" at the same time.

Global-Positioning System (GPS)   A system developed by the U.S. Department of Defense. The GPS system uses geostationary satellites to triangulate the position of a GPS receiver located on the face of the earth. Location is given in latitude, longitude, and altitude. The GPS receiver can even calculate bearing (direction) and speed. GPS receivers are available to the general public in many electronics magazines. You can even buy a GPS at your local Radio Shack. The receivers that are sold on the civilian market do not decode the dithering of the satellite signal and don't work indoors. *Dithering* is a random error that is introduced to the carrier frequency so that the GPS system would be slightly inaccurate. The deviance in location and altitude is anywhere from ±50 feet. This is so hostile entities cannot use the Department of Defense's location tool against the United States with known accuracy.

GNE (Gateway Network Element)   An interface on SONET-node equipment that enables it to interface and exchange network data with other SONET nodes located within a network of SONET rings.

Goat   Another name for a *craft* test set. For a photo and explanation, see *Craft Test Set.*

Golden Cookie   In Internet applications, a cookie is a file given to Internet users by servers that the application interacts with. A "Golden Cookie" is a slang term that refers to a user who has made a purchase at the site they are currently surfing, at which time they made the purchase, they were given the golden cookie. Golden cookies can give users premium processor/bandwidth allocation for surfing a particular site if the site is equipped with layer 7 networking ability. This helps Internet websites retain customers that actually make purchases by giving them faster network service or performance.

GPRS (General Packet Radio Service)   The third generation of wireless technology offered as a service in Europe and other countries with a posture toward European standards. GPRS provided by telecommunications companies provides users with a standards-based IP application interface, so multimedia and data can be transmitted over the radio link as well as voice. Transmission rates for GPRS are expected to mature to an excess of 1 Mbps, be able to simulate a constant connection, and be able to share services with CDMA2000 networks because the IP layer provides a standard application regardless of the radio links used on either end of the call.

GPS (Global-Positioning System)   See *Global-Positioning System.*

Grade-1 Cable   A type of twisted-pair cable that is designed for PBX, telephone, RS-232, and other low-speed (1Mb/s or less) data applications.

Grade-2 Cable   A type of twisted-pair cable that is designed for transmissions up to 4 Mb/s. It is good for older IBM 3270 applications, IBM PC networks, and ISDN.

Grade-3 Cable   A type of twisted-pair cable that is designed for older LAN networks like 10BASE-T and 802.5 Token Ring.

Grade-4 Cable   A type of twisted-pair cable designed for Ethernet speeds up to 10 Mb/s. See *CAT4* and *CAT5*.

Grade-5 Cable   A type of twisted-pair cable designed for speeds up to 100 Mb/s. Used on IBM Token-Ring networks. This is not CAT5 cable. Grade 5 cable is two twisted pairs of stranded copper.

Graded-Index Fiber   The core of graded-index fiber optic is made of many layers of glass, consisting of many refractive indexes that cause the light to gradually bend as it approaches the outside of the fiber (Fig. G.5). Graded-index fiber (like stepped-index fiber) is available in multi-mode or single-mode, and it is more expensive. The alternative to graded-index fiber is stepped-index fiber, which has a core made of glass, consisting of one refractive index.

LIGHT AS IT TRAVERSES THROUGH THE CORE OF A FIBER OPTIC

Stepped index fiber   Graded index fiber

Figure G.5   Graded-Index Fiber Vs. Stepped-Index Fiber

Graphic Equalizer   To adjust the tone or sound of a circuit by diminishing or augmenting specific frequency bands. The tone control on a radio is a type of equalizer. A radio transmitter might have a tendency to amplify low-end signals, such as the sound of a bass guitar or drums better than high-end signals, such as the sound of a voice or cymbals. An equalizer can be used to reduce or increase the amplification of either end of the broadcast for an even and accurate reproduction of the input.

Graphics-Controller Card   An add-on circuit card that controls and monitors advanced functions. The graphics card determines the resolution, refresh rate, and number of colors possible. See also *SVGA* and *AGP.*

Gray, Elisha   The possible true inventor of the telephone. Alexander Graham Bell beat him to the patent by a few hours.

Greenwich Mean Time (GMT)   Also known as *zulu time*. This now-obsolete term was named after the very accurate clock standards in Greenwich, England. The clocks are incredibly accurate because of the cesium timing reference standard, which is the time-keeping element in "bits clocks" (timing devices used in central office nodes to synchronize SONET equipment. Greenwich Mean Time is now known as *Universal Time Coordinated (UTC)*.

Grid   The part of an electronic vacuum tube that the input signal is fed to. The grid manipulates current flow between the filament and the plate (anode). See *Fig. G.6*.

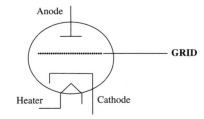

Figure G.6   The Grid of a Vacuum Tube

Ground   Earth Ground. The electrical potential of the earth is 0 V. To maintain a good earth ground, a metallic rod four to six feet long is driven into the ground. Any wire connected to that rod is "grounded." The power company installs a rod like this when they connect power to your home. The telephone and cable-TV companies wire their network interfaces (lightning protection) to the power company's ground rod. The alternative to earth ground is a *floating ground*. A floating ground is simply a reference point that is not "earth grounded." The negative terminal of your car battery is a *floating ground* and any home appliance that has a two-prong electrical plug is also a floating ground (in newer homes that are wired correctly).

Ground Clamp   A clamp or strap that is used to make a secure connection to a water pipe or grounding rod. The ground clamp then provides a way to connect a wire to earth ground (Fig. G.7).

Ground Fault   A trouble on a telephone line that is caused by one of the twisted-pair phone wires coming into contact with electrical ground.

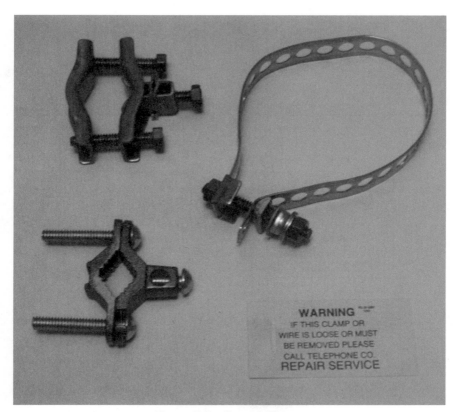

Figure G.7   Ground Clamps

A ground fault causes a hum on a telephone line. A ground fault can ocur in many places on a telephone line. Ground faults are commonly caused by a portion of the phone line being wet or by a bad lightning protector.

Grounding Field   An array of grounding rods placed in the ground and connected together around an antenna site or central-office site. The idea of a grounding field is to provide the best possible earth ground for electronic equipment.

Ground Start Trunk   A phone line that uses a ground instead of a short (loop-start trunks use a short between tip and ring) to signal the central office for a dial tone. Some PBX telephone systems require the use of ground-start lines (trunks). Most newer PBX systems use loop-start (individual or on a T1) trunks or ISDN PRI trunks.

Group-3 Protocol   An international standard for facsimile protocols that defines how two communicating entities will send and receive graphic facsimile information over telephone lines. The Group-3 protocol enables an 8.5" by 11" page document to be transferred in about 20 seconds.

Group-4 Protocol   An international standard for facsimile protocols for transmission by ISDN devices over ISDN connections. See also *Group-3 Protocol.*

Group Address   Also known as a *Multicast Address.* A single address that transmits to multiple network devices. IP multicast addresses range from 224.0.0.0 to 239.255.255.255. See also *IP Multicast* and *Broadcast.*

GS   The ASCII control code abbreviation for group separator. The binary code is 1101001 and the hex is D1.

GS Trunk (Ground-Start Trunk)   See *Ground-Start Trunks.*

GSTN (Global Services Telephone Network)   The term used by the ITU-T and their engineering recommendation circles when referring to the PSTN (Public Services Telephone Network).

GTE (General Telephone and Electronics)   An independent telephone company (independent of the RBOCs). GTE owns and operates smaller local telephone companies across the United States.

GTP (General Telemetry Processor)   A device manufactured and implemented to receive and process telecommunications equipment alarming protocols, such as *TBOS (Telemetry Bit-Oriented Serial).*

Guard Band   A frequency band separator between radio channels. A guard band prevents multiple or nearby radio stations on the dial from being received simultaneously.

Guest Mailbox   A temporary voice mailbox set up in a voice-mail system. It can be attached to a phone or it can be a "virtual mailbox," where the user simply has a phone number that other people can call to leave messages.

Guy Hook   A hook that is bolted to power/communications poles and used to attach guy wires. See *Fig. G.8.*

Figure G.8   Guy Hook

Guy Thimble   A device used to attach a guy wire strand to a bolt, which, in turn, attached to an anchor in the ground. See *Fig. G.9.*

Figure G.9   Guy Thimble

Guy Wire   A steel cable that provides lateral support for a vertical structure, such as a telephone pole or radio tower/antenna.

H Channel (High-Speed Channel) A full-duplex ISDN primary-rate channel operating at 384 Kbps. Compare with B channel, D channel, and E channel.

H PAD (Host Packet Assembler/Dissembler) A PAD that is specifically located on the host end of a communications link. Even though PADs are the same on each end, sometimes technicians refer to them in a specific manner. The H PAD is a device that is located at the host end of a virtual communications link in a frame protocol environment that provides for statistical time division multiplexing. The HPAD also adds and removes address, envelope, and HDLC information.

H.225 In the ITU-T H.323 family (or species) of protocols, H.225 provides support for incorporating security methods and features that can be built on as needed. One example is the "cryptoH323Token," which is incorporated into the H.235 method of "password with hashing" security scheme.

H.235 Security support for H.323. H.235 incorporates features such as password requests that require communicating devices such as routers to identify themselves and provide a password. This is a method to help prevent devices from networks other than your own to connect. It also incorporates end user ID and PIN (Personal Identification Number) features. These features are enabled by Crypto-Tokens defined in H.225.

H.245   An ITU-T feature addition to H.323 that provides the capability for exchanging DTMF signals out-of-band or in CCS T1 or ISDN circuits. Another important function of H.245 is the compression/decompression (Codec) type negotiation, such as G.711 and G.729, between the calling and called end devices. H.245 also performs the UDP port negotiation between the two ends of the network.

H.261   The compression standard specified under the H.320 videoconferencing standard. H.261 is similar to but not compatible with MPEG (H.261 is more CPU efficient). H.261 is the video coding and decoding method for the moving picture component of audiovisual services at the rates of 1 to 30 frames per second. It describes the video source coder, the video multiplex coder, and the transmission coder. H.261 defines two picture formats. The first is the CIF (Common Intermediate Format), which specifies 288 lines of luminance information (with 360 pixels per line) and 144 lines of chrominance information (with 180 pixels per line). The second is the QCIF (Quarter Common Intermediate Format), which specifies 144 lines of luminance (with 180 pixels per line) and 72 lines of chrominance information (with 90 pixels per line).QCIF provides a better video quality when the number of frames per second is less than 3. H.261 also provides for two different coding methods that are used depending on the available bandwidth. The H.261 algorithm includes a mechanism for optimizing bandwidth usage by trading picture quality against motion so that a quickly changing picture has a lower quality than a relatively static picture. When used in this way, H.261 is a constant-bit-rate encoding rather than a constant-quality, variable-bit-rate encoding.

H.263   A backward-compatible update to the H.261 standard that enhances picture quality using a required half-pixel motion estimation technique, predicted frames, and a Huffman coding table optimized for low bit rate transmissions.

H.320   An ITU-T standard that defines connection-based videoconferencing using ISDN PRI, 56K, and fractional T1 circuits to connect the end users. H.320 equipment was made obsolete by H.323 standards, which enable video to be sent over IP and provide standards-based methods of integrating the older H.320 and newer H.323 end user equipment.

H.323   An ITU-T standard that describes packet-based video, audio, and data conferencing. H.323 is an umbrella standard that describes the architecture of the conferencing system and refers to a set of other standards (H.245, H.225.0, and Q.931) to describe its actual protocol. The

H standards are used for initial call signaling set-up and monitoring across a QoS or voice enabled network. The actual IP telephony voice conversation is carried by RTP (Fig. H.1). See also *H.323v1* and *H.323v2*.

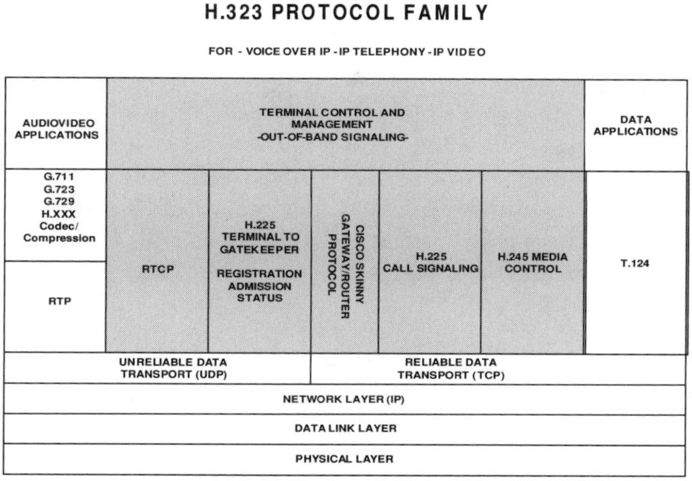

H.323 PROTOCOL FAMILY

FOR - VOICE OVER IP - IP TELEPHONY - IP VIDEO

AUDIOVIDEO APPLICATIONS	TERMINAL CONTROL AND MANAGEMENT -OUT-OF-BAND SIGNALING-					DATA APPLICATIONS
G.711 G.723 G.729 H.XXX Codec/ Compression	RTCP	H.225 TERMINAL TO GATEKEEPER REGISTRATION ADMISSION STATUS	CISCO SKINNY GATEWAY/ROUTER PROTOCOL	H.225 CALL SIGNALING	H.245 MEDIA CONTROL	T.124
RTP						
	UNRELIABLE DATA TRANSPORT (UDP)		RELIABLE DATA TRANSPORT (TCP)			
	NETWORK LAYER (IP)					
	DATA LINK LAYER					
	PHYSICAL LAYER					

Figure H.1   Family of Protocols

H.323 Audio   In IP telephony, audio signals contain digitized, compressed sound. H.323 supports proved ITU standard audio Codec algorithms, including G.711 for speech, which transmits voice at 56 or 64 Kbps. Support for other ITU voice standards (G.722, G.723, G.728, G.729) is optional, because each one reflects tradeoffs between speech quality, bit rate, computing power, and signal delay.

H.323 Video   In H.323, video capabilities are optional. Video-enabled H.323 terminals must support the H.261 Codec, with optional support for the H.263 standard. Transmission is no greater than that selected during a capability exchange process during call setup. Because both H.261 and H.263 support QCIF, communication between different terminals is possible.

H.323v1   For IP telephony purposes, the ITU-T H.323 provides POTS line signal translations between an IP WAN link and a POTS telephone. H.323v1 has the ability only to connect and disconnect a call. Signaling such as hook-flash required to initiate supplementary features such as hold, transfer, and conference cannot be carried. H.323v1 could be viewed from the legacy telephony perspective as being an in-band signaling method.

H.323v2   A modified version of the ITU-T H.323 family of protocols that enables the RTP/UDP/IP transport stream to be rerouted via hook-flash type signaling on the POTS end or the IP telephony end. Inherently, with the elimination of the need for the software media termination point, H.323v2 enables compressed voice over the IP portion of the network such as G.723 and G.729. Typically, layer 2 gateways incorporate H.323v2.

H.324   An ITU-T enhancement to the original H.320 standard that defines videoconferencing using plain old telephone system (POTS) lines.

H.450   A feature supportive of ITU-T H.323 that enables supplementary services such as call transfer. H.450 allows an H.323 end point to redirect an answered call to another H.323 endpoint. Within the standard, it is referred to as *call deflection.* The end user with call forward enabled signals the gateway of the new end point.

Half Duplex   Two-way communications, one direction at a time. CB radio is an example of half-duplex operations; two people take turns transmitting and receiving. The two other types of transmissions are *full duplex* and *simplex.* A full-duplex line or communications path is able to communicate both directions, transmit and receive at the same time. A T1 is a full-duplex line, with one pair used for transmit and the other used for receive. Full duplex can be accomplished on one pair of wires by using two multiplexed channels, one for receive, and one for transmit. Simplex is one-way communication only. An FM radio station or TV broadcast is simplex.

Half-Wave Antenna   An antenna that is one-half the wavelength of the frequency that is designed to receive or transmit. For example, the wavelength of an antenna for a 38-GHz microwave signal is equal to:

$$Wavelength = \frac{Speed \ of \ light \ (\text{in meters/second})}{Frequency \ (\text{in CPS})}$$

$$= \frac{300,000,000}{38,000,000,000} = 0.0079 \ \text{meters}$$

$$.0079 \ \text{meters} \ = \ 7.9 \ \text{mm}.$$

A full-wave antenna for 38 GHz would be 7.9 mm. A half wave would be half of that, 3.9 mm. It's a small antenna because it is for a small wavelength. That's why they call it *microwave.*

Hand Hole   A small cable vault that is basically big enough to get your hands into. It is used for outside cable plant splices. Hand holes are most common in the fiber-optic realm of outside plant. A typical hand hole is about two feet wide, three or four feet long, and two feet deep (Fig. H.2).

Figure H.2 Hand Hole

Hand Off   To connect a phone call or other service from one telephone company to another. Hand-offs usually happen in a place called a *co-location*.

Hand Shake   The initial connection set-up part of a protocol for modems. When you dial out on a modem, often you can hear a beep after the modem on the far end picks-up. This is a *handshaking signal*. After the handshaking signal the modems go through the handshaking process, which is to exchange information specific to what speed they will transmit and what error-detection protocol will be used. After the handshake is complete, data transmission begins.

Hands Free   Also called *speaker phone*. Hands free is a feature of PBX telephones and 2500 (single-line standard) telephones. It allows the user to talk on the telephone as if it were an intercom, not using a handset. Speaker phones have a microphone and a speaker built into the telephone set itself.

Handset   The device attached to a telephone that you hold to your head during a telephone call. It has a speaker and a microphone.

Hard Call Forward   A type of call transfer or routing where an incoming telephone call does not ring at the redirecting party's number or station (the line that is forwarded). It immediately rings at the line or station to which it is destined. Hard call forward is usually set by the line or station user, as opposed to a soft call forward, which is usually set by a voice mail, PBX, or CO switch administrator.

Hard Disk Drive (HDD)   Also called *hard drive*. A means of random-access magnetic memory. Within larger hard disk drives are actually several magnetic disks (Fig. H.3). Magnetic heads read and write data to the disks. Hard disk drives were first used in computers by IBM in the early 1970s. Several standards provide a means for hard disk drives to interface with other computer components, such as the CPU and RAM. The most common are *IDE (Integrated Drive Electronics* or *Intelligent Drive Electronics)* and *SCSI (Small-Computer System*

Figure H.3   A 17.3-GB Hard Disk Drive

Interface). Hard disk drives are available in varied memory sizes that range from several megabytes (MB) to many gigabytes (GB). Illustrated is a 17.3-GB IDE hard disk drive that is commonly used in personal computers.

Hardware   Physical electronic equipment. Most electronic equipment needs software, which are the programs and instructions that tell it how to run.

Hardware Address   See *MAC Address.*

Hardwire   To be physically connected by wire, either with cross connects or customized cables. The alternative to hardwiring equipment is to use modular equipment, such as a modular jack. A modular jack is equipped with a plug so that devices can be easily attached and detached. Old jacks, which can still be found in old homes, are *hard wired,* which means that the telephone cord had to be permanently affixed to the terminals inside the jack with screws. The same went for nonmodular or hardwired telephones. If you wanted to have a longer cord, you couldn't buy one at the store and just plug it in. You had to call the phone company and they would send a telephone technician out to install a longer cord for you.

Harmonic   A frequency that is a multiple of a lower frequency. For example, 6000 Hz is a harmonic of 3000 Hz. 12,000 Hz and 15,000 Hz are also harmonics of 3000 Hz. Multiply 3000 or any other number by any positive integer and that is a harmonic frequency.

Harmonic Distortion   The tendency of a circuit to amplify and pass harmonics of an input signal. Feedback is an example of harmonic distortion. Feedback is the squealing sound you often hear when a person approaches a microphone at a public speech. See also *Feedback.*

Harmonica Adapter   An adapter that converts a 25-pair cable plug into 12 four-conductor RJ11 plugs or 24 two-conductor RJ11 plugs. Harmonica adapters are frequently used as an alternative connectivity (as opposed to hardwired 66M150 blocks) on temporary (and sometimes an inexpensive permanent) installations of key or PBX telephone systems. For a photo, see *Modular Adapter* and *258A Adapter.*

HBD3 (High-Density Binary 3 or High-Density Bipolar 3)   A line-coding type used on E1 circuits. This line coding prevents timing loss on E1 transmissions when excessive zeros are transmitted.

HDD (Hard Disk Drive)   See *Hard Disk Drive.*

HDLC (High-Level Data-Link Control)   A communications protocol that was developed by the ISO, based on the *SDLC (Synchronous Data-Link Control)* method created by IBM in 1970. HDLC was reviewed and modified by the ITU-T (then the CCITT), and then called *LAP (Link-Access Procedure).* LAP was included as a part of the 1976 version of X.25. LAP was refined further by the ISO and the ITU-T (then the CCITT) and renamed *LAPB (Link-Access Procedure Balanced mode).* LAPB is the frame-layer protocol in X.25, and the data-link layer protocol by OSI standards.

HDSL (High Bit-Rate Digital Subscriber Line)   A physical layer carrier that delivers data networking up to 1.544 Mbps over two copper pairs and up to 2.048 Mbps (E1) over two or three pairs at a base transmission range of 20,000 feet from a central office. It is similar to SDSL and has symmetrical transmission capabilities. The transmission range for HDSL can be extended to more than 30,000 feet using repeaters, and can be extended even further using fiber optic. Line format and coding is not an issue in HDSL implementation, as it is with legacy T1 and E1. All HDSL is clear channel. HDSL is a service provided by telephone companies. See also *xDSL.*

HDSL2   An enhancement to HDSL that incorporates *OPTIS (Overlapped PAM Transmission with Interlocking Spectra)* line coding. It is intended for long reach (to 18,000 feet) single-pair T1 and long-reach two-pair T1, without the use of repeaters or doublers. The HDSL2 standard provides a significantly more-reliable connection than HDSL in noisy/crosstalk environments. HDSL2 is not intended to replace HDSL, but to provide an additional means of providing reliable service where HDSL cannot.

HDT (Host Digital Terminal)   In cable-TV networks, HDT is a node that modulates many digital DS0 voice signals into two-way upstream and downstream RF signals (Fig. H.4). The HDT is the heart of cable telephony. It has the ability to interface directly with SONET nodes or voice-switching equipment, including those in the Nortel DMS family. The HDT-modulated voice signals are combined onto the traditional cable-TV transmission and demodulated at the customer premises by a voice port. A voice port is a far-end receiver/modem that provides customer end modulation/demodulation of cable voice service. The voice port is then connected to customer's telephone wiring. Arris/Nortel is a manufacturer of HDT cable telephony equipment.

Figure H.4   HDT (Host Digital Terminal)

HDX (Half Duplex) Half duplex is two-way communications, one direction at a time. CB radio is an example of half-duplex operations; two people take turns transmitting and receiving. The two other types of transmissions are full duplex and simplex. A full-duplex line or communications path is able to communicate both directions, transmit and receive at the same time. A T1 is a full-duplex line, with one pair used for transmit and the other used for receive. Full duplex can be accomplished on one pair of wires by using two multiplexed channels, one for receive and one for transmit. Simplex is one-way communication only. An FM radio station or TV broadcast is simplex.

Head End Where cable-TV signal processing takes place. Located at the head end is the array of satellite dishes that the cable-TV company uses to pick-up their programming transmissions. They re-channelize the stations (e.g., put WGN on channel 8 and HBO on channel 14), add in local commercials and local programming. When all the signal processing is done, the cable-TV company then sends the broadband (a multi-channel signal) TV signal down its coax cables so that it can be distributed to subscribers. See also *Cable TV.*

Head End   A device in a network that has multiplexing, demultiplexing, and/or switching capability in a broadband network. All stations transmit toward a head end.

Header   Binary address and signaling information that is added to the front of a data packet that instructs network components how and where to send the packet.

Headless (Slang)   A device that cannot be managed directly by connecting a terminal to it. It must be connected to a device that has that capability. For example, a trunk card or router services card in a PBX system cannot be managed directly because they do not have a console or terminal port built into them. They must be plugged into a system to be configured. The industry more typically refers to these types of devices as being *unmanaged devices.*

Health Insurance Portability and Accountability Act (HIPAA)   A comprehensive law passed by the United States Congress in 1999 that drives the development of electronic data interchange (EDI) for specified administrative and financial health care transactions in the United States. This act is intended to save health care providers $9 billion annually in administrative costs and an estimated $4 billion annually in fraud prevention. For telecom network managers of hospital and medical insurance corporations, as of 2001, there are standards for security and EDI that must be followed.

Held Order   A telephone or cable-TV service that cannot be installed because of a shortage of equipment. The shortage could be because the equipment is not installed yet (such as telephone cables in a new neighborhood) or the available facilities have run-out (i.e., all the pairs in a particular area are used, or the central-office switch has utilized all of its line interfaces).

Henry   The unit of inductance. Inductance is also referred to as reactance. An inductor, or coil of wire, is a reactive device. Reactance is the resistance that a component gives to an AC or fluctuating DC current. The two components that cause reactance: inductors (coils) and capacitors. The difference between resistance and reactance is that resistance is always the same, regardless of the voltage amplitude or frequency applied to the resistive device. The reactance of a component changes along with frequency changes, or the speed at which an AC current changes direction. The higher the frequency, the higher the reactance or resistance to that frequency. The reason that coils of wire cause reactance is that as electricity flows through them, they force the electricity to create a magnetic field every time it changes direction. A perfect inductor has zero reactance to a DC current, and has a specific reactance or resistance to every AC current. Each coil or inductor has a value in henrys. The higher the number of henrys, the more it will resist AC or fluctuating DC. Coils are used to filter out ("choke" out) DC fluctuations in power supplies. They are also used to help tune in radio or other frequencies.

Reactance is also caused by other electronic conditions where it is not useful. All wire and electronic components possess a small amount of reactive properties (e.g., the reason twisted-pair wire causes attenuation of signal strength is because of the inductance of the copper wire and the capacitance of the two wires being next to each other.)

Hermaphroditic Connector   A genderless connector developed by IBM that is usually called a "Data Connector." The Data Connector does not need complementary plugs (male and female) to make a connection, like all other known communications modular connecting systems. The Data Connector is specifically designed and used for switched token ring backbone applications.

Hertz (Hz)   A measure of the number of cycles per second in a waveform. One Hz, or Hertz, is equal to one cycle in one second. Two Hz is two cycles in one second, 1000 Hz is 1000 cycles in one second, or Cycles Per Second (CPS). Cycles are used as a reference to measure the frequency of a waveform or signal. Below is a diagram of two waveforms and one cycle of each singled out. Cycles are usually referred to as a number of cycles per unit of time. Cycles per second and Hertz are measurements of the number of cycles you get per second in an analog transmission. Bits per second is measurement of how many "square wave" clock sample sequences are being read from a digital transmission. For a diagram of one cycle (hertz), see *Cycle.*

Heterodyne   To mix a radio-frequency signal with an audio or other signal to be carried in a transmission.

Heterogeneous Network   A network that connects more than one type or manufacturer's operating platform. A network that is comprised of equipment/software that is made by different manufacturers. IBM's SNA *(System Network Architecture)* is a complex networking standard because it was made to connect every large heterogeneous network known at its time of inception.

Hexadecimal   A number system based on 16 numbers, instead of 10, like the one we count with in our everyday lives. Hexadecimal is a convenient shortcut to inputting 16-/bit binary numbers during machine-language programming. Hexadecimal counts as follows: 0, 1, 2, 3, 4, 5, 6, 7, 8, 9, A, B, C, D, E, and F. For a conversion table of hexadecimal, binary, and decimal numbers, see *Binary-to-Decimal Conversion.*

Hexadecimal-to-Binary Conversion   For a conversion table for binary (base 2), hexadecimal, and decimal (base 10) numbers, see *Binary-to-Decimal Conversion.*

Hexadecimal-to-Decimal Conversion   For a conversion table for binary (base 2), hexadecimal, and decimal (base 10) numbers, see *Binary-to-Decimal Conversion.*

HF (Hands Free)   Also called speaker phone. Hands free is a feature of PBX telephones and 2500 (single-line standard) telephones. It allows the user to talk on the telephone as if it were an intercom (not using a handset). Speaker phones have a microphone and a speaker built into the telephone set itself.

HFC (Hybrid Fiber/Coax)   A reference to the modern cable-TV/telecommunications distribution plant. Fiber optic is used to extend mass amounts of bandwidth to communities. The fiber optic terminates into a fiber node. Then services are delivered to nearby subscribers via coax.

Hierarchical Network Architecture   An enterprise LAN/WAN design methodology where all traffic leads to a central core, where servers and other computing services are located. In hierarchical design, there are three parts to a network that include the core layer, the distribution layer, and the access layer (Fig. H.5). This design method is used most recently in comparison to the 80/20 method used in the past. The availability of super high-speed LAN switches with backplanes that are capable of switching at a total bandwidth in excess of 30 Gbps have made hierarchical designs more practical than the distributed 80/20 computing method. The largest advantage of hierarchical designs is that they are highly scalable and expandable as well as flexible for implementing new services. The older 80/20 designed networks eventually grow to a point where no new services can be added without creating bottlenecks that are very expensive to remedy. See also *Core Layer, Distribution Layer,* and *Collapsed Core.*

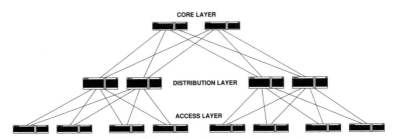

Figure H.5 Hierarchical Network Architecture

Hierarchical Routing   A method used to reduce the complexity of routing tables within a large network. This is accomplished by separating a large network into several smaller-sized networks, which are connected by routers, hub routers, or switches.

Hi-Fi (High Fidelity)   An attempt a reproduction of an audio signal that is as close as possible to the original using the most advanced technology possible within a price range.

High and Dry   A test result of direct-access test units attached to central office switching equipment that means a twisted pair is clear of shorts, grounds, and equipment, including telephones and sometimes load coils (depending on how the DATU is configured).

High Bit-Rate Digital Subscriber Line (HDSL)   A physical layer carrier that delivers data networking up to 1.544 Mbps over two copper pairs and up to 2.048 Mbps (E1) over two or three pairs at a base transmission range of 20,000 feet from a central office. It is similar to SDSL and has symmetrical transmission capabilities. The transmission range for HDSL can be extended to more than 30,000 feet using repeaters, and can be extended even further using fiber optic. Line format and coding is not an issue in HDSL implementation, as it is with legacy T1 and E1. All HDSL is clear channel. HDSL is a service provided by telephone companies. See also *xDSL*.

High Fidelity (Hi-Fi)   See *Hi-Fi*.

High-Level Data-Link Control (HDLC)   A communications protocol that was developed by the ISO, based on the *SDLC (Synchronous Data-Link Control)* method created by IBM in 1970. HDLC was reviewed and modified by the ITU-T (then the CCITT), and then called *LAP (Link-Access Procedure)*. LAP was included as a part of the 1976 version of X.25. LAP was refined further by the ISO and the ITU-T (then the CCITT) and renamed *LAPB (Link-Access Procedure Balanced mode)*. LAPB is the frame-layer protocol in X.25, and the data-link layer protocol by OSI standards.

High-Level Language   A computer programming language that interfaces meaningful instructions (to humans) to lower-level programming languages, such as machine language. FORTRAN, COBOL, C, BASIC, and SAS are high-level programming languages.

High-Pass Filter   An electronic device that eliminates frequencies below a specified frequency. The two categories of frequency filters are: active and passive. Active filters use active devices that require power, such as transistors and op amps to amplify the desired signal and attenuate the undesired signal. Passive filters are made with components that do not require external power, such as capacitors and inductors. Capacitors and inductors have reactive properties that cause them to resist or pass an AC signal.

High-Performance Parallel Interface (HIPPI)   Newer versions of HIPPI are used as a backbone to connect supercomputers. Multiple copper pairs or fiber-optic pairs are used to achieve transfer speeds of 6.4 Gbps (HIPPI 6400). HIPPI has been used as an intra-room computer connection, but with the new incorporation of fiber optic to the standard, HIPPI could be used to transmit data across campus environments.

High-Speed Serial Interface (HSSI)   A physical connection between a DTE and DCE device, developed by Cisco Systems and T3Plus networking, that enables connection speeds well above the 64-/256-Kbps range, to which other standards are limited. Other standards include V.35 and RS-232C serial interface. Among other uses, these interfaces are used to connect between a public telephone service provider's CSU/DSU and a customer's local network device, such as a router.

HIPAA (Health Insurance Portability and Accountability Act)   A comprehensive law passed by the United States Congress in 1999 that drives the development of electronic data interchange (EDI) for specified administrative and financial health care transactions in the United States. This act is intended to save health care providers $9 billion annually in administrative costs and an estimated $4 billion annually in fraud prevention. For telecom network managers of the hospital and medical insurance industry, as of 2001, there are standards for security and EDI that must be followed.

HIPPI (High-Performance Parallel Interface)   Newer versions of HIPPI are used as a backbone to connect supercomputers. Multiple copper pairs or fiber-optic pairs are used to achieve transfer speeds of 6.4 Gbps (HIPPI 6400). HIPPI has been used as an intra-room computer connection, but with the new incorporation of fiber optic to the standard, HIPPI could be used to transmit data across campus environments.

Hit   A user accessing an application or server on a network. A hit can be thought of as a completed and ended phone call, only in a packet switch environment rather than a circuit switched environment (as in a POTS voice call).

HIVR (Host Interactive Voice Response)   A telecommunications and data-processing technology that interfaces a person to information held in a computer by using a phone line. If you have ever called your bank and entered your account number, a password, and a prompt so that a computerized voice can read back your bank account balance, then you have used HIVR.

Hold Recall A feature of PBX telephone systems and key systems that makes calls put on hold ring back to the person who put them on hold. The hold-recall time interval can be set to a time specified by the system administrator (e.g., 15, 30, 45, or 60 seconds).

Hold Down An operational state/condition for routes in a data network that is similar to a busy state for voice trunks. Hold down is a state that a route is placed into so that routers will neither advertise the route nor accept advertisements about the route for a specific length of time (the hold-down period). Hold down is used to flush bad information about a route from all routers in the network. A route is typically placed in hold down when a link in that route fails.

Hollow Pipeline A term that is usually used to describe a private line out-of-band signaled (CCC, Clear Coded Channel) DS1. There is no timing, no framing, no error correction. You input your bit stream on one end and out they come out the other end in the same order. There is only a maximum speed you can transmit. For a DS1, it is 1.536 Mb/s. This is 1.544 Mb/s less the framing overhead of 8 Kb/s.

Home Run A telephone or data communications wiring scheme that means the wire that is installed runs from a jack to a point where it can be cross-connected or terminated to DCE equipment or a telephone NI. Each station (PC or telephone must have its own dedicated wire). CAT 5 wiring and PBX wiring must be installed in a home run manner. Token ring is installed station to station. See *Fig. H.6.*

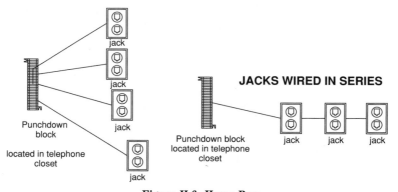

Figure H.6 Home Run

Homologation   The conformity of a product or specification to meet international standards, such as those set by ITU-T. Homologation enables portability and compatibility between corporate and international boundaries.

Hook Flash   Hook flash or "Flash" is a form of telecommunications signaling. To send a flash signal, press the switch-hook of a telephone briefly. If you have call waiting on your telephone line and another call comes in (you hear the beep), you briefly push in the switch hook on your telephone to switch to your other call. When you want to revert back to the original call, you briefly press the switch hook on your phone again. Some telephones have a flash button on them, which is a more convenient and less cumbersome way to send a flash signal than flipping the telephone's switch hook.

Hook Switch (Switch Hook)   The switch that is pressed when you hang up a telephone handset (Fig. H.7).

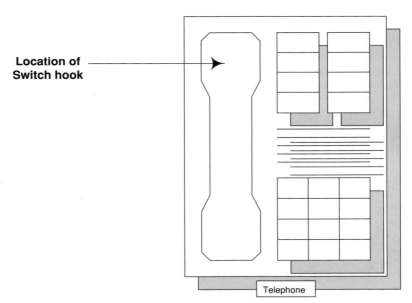

Figure H.7   Hook Switch

Hooks   Also known as *climbers, lineman's climbers, spurs,* and *gaffs.* Climbers for use on power poles have a shorter blade on the shank than climbers made for tree climbing (lumber/trimming applications). For a photo, see *Climbers.*

Hop   1. A terrestrial microwave radio link.   2. The passage of two data packet between two separate network nodes (for example, two routers).

Hop Count   A routing metric (measure) used in routing algorithms. It is the number of routers or switching devices between a router's port and the destination address on a network. Hubs are not counted in a hop count. The *RIP (Routing Information Protocol)* uses a hop count to determine the shortest path for a packet to be routed.

Horizontal Blanking   As the beam inside a picture tube or monitor scans across the screen, it must be turned off while it retraces back to the starting point of the next line that it is going to "paint." The turning off of the beam is called *blanking*.

Horizontal Output   The power amplifier that amplifies the horizontal output sync signal in a TV or monitor. The output is fed into a deflection yolk, which creates the magnetic fields that control tracing of the CRT beam in a sideways direction. There is also a vertical output amplifier, which does the same for the up and down tracing of the CRT beam. The horizontal output frequency for a standard TV is 15.73425 kHz. When you turn on some TVs, you can faintly hear the very high pitched "dog whistle" sound of the horizontal output circuitry.

Horizontal Polarization   The pointing of a microwave dish antenna so that the transmission dispersion is in a sideways, or horizontal, pattern. The headlights on cars are polarized in a horizontal manner, so the light dispersion is spread across the horizontal surface of the road. The other kind of polarization is vertical, where the transmission dispersion is in an up-and-down pattern. The two antennas or dishes employed in a point to point application need to be polarized the same way.

Horsepower   In electronics, it is sometimes useful to compare wattage power to horsepower to get an idea of what a watt of power really is capable of. One horsepower is equal to 746 W. Some electric motors are stamped with horsepower rating.

Host   A computer, server, printer, or workstation that has a hardware address and resides on a network. Additionally, a host is similar to a node in its network existence, except that the term *host* usually implies that the system processes received and transmitted data through all layers of the network architecture used, reaching the application.

Host Address   The IP address assigned to a workstation, server, router, printer or other networked device. Host addresses are four bytes long.

An example of a host address is 124.234.120.12. Each host address is translated to a *MAC (Media-Access Control)* address within the network software *ARP (Address-Resolution Protocol)* files. Host IP addresses never end in "0" because addresses that end in "0" are network addresses (subnet masks). See also *Network Address, Subnet Mask, MAC Address, ARP,* and *IP address.*

Host Computer   Usually a reference to a mainframe computer (although many hosts are actually servers) that controls the storage and retrieval of data in extremely large databases. Host computers can simply retrieve data for another computer (PC or Server) or do calculations and summary on data itself.

Host Digital Terminal (HDT)   In cable-TV networks, HDT is a node that modulates many digital DS0 voice signals into two-way upstream and downstream RF signals. The HDT is the heart of cable telephony. It has the ability to interface directly with SONET nodes or voice-switching equipment, including those in the Nortel DMS family. The HDT-modulated voice signals are combined onto the traditional cable-TV transmission and demodulated at the customer premises by a voice port. A voice port is a far-end receiver/modem that provides customer end modulation/demodulation of cable voice service. The voice port is then connected to customer's telephone wiring. Arris/Nortel is a manufacturer of HDT cable telephony equipment.

Host Interactive Voice Response (HIVR)   See *HIVR.*

Hot Line   A telephone that rings another phone with no dialing. Hot lines are created by using devices called *hot-shot dialers,* which automatically dial a phone number. Telephones connected to hot-shot dialers usually have no dial pad. Hot lines are also called *ring-down circuits.*

Hot Pluggable   A reference to a circuit card or I/O device which can be installed or removed without turning off the system power.

Hot-Shot Dialer   A device used to create a hot line or ring-down circuit. A hot line is a telephone that rings another telephone with no dialing required by the user. The hot-shot dialer automatically dials the phone number when the handset is lifted.

Hot Standby Router Protocol (HSRP)   A Cisco Systems proprietary feature that enables a redundant router to take over the tasks of a primary should it fail to send an "I'm OK" signal to the standby, or redundant router. HSRP can be configured to provide nearly instant switchover of packet forwarding functions from one router on a segment/switch to another.

HSRP is often configured creatively with STP to provide failover and redundancy at layer 2 (Ethernet switch) and layer 3 (network) OSI levels.

Hotel Console A PBX console specially designed for hotel front-desk use. Consoles are available in two types: business and hotel.

HSDL (High-Speed Digital Subscriber Line) An older term for ASDL. This is a service in the making to provide video to the home over twisted-pair telephone lines. Its current line format is T1 AMI, 16 Kb/s to the CO, (for control to change the channel) and 1.528 Mb/s to your TV. The twisted pairs are incorporated with adaptive digital filtering to help correct attenuation and noise.

HSRP (Hot Standby Router Protocol) A Cisco Systems proprietary feature that enables a redundant router to take over the tasks of a primary should it fail to send an "I'm OK" signal to the standby, or redundant router. HSRP can be configured to provide nearly instant switchover of packet forwarding functions from one router on a segment/switch to another. HSRP is often configured creatively with STP to provide failover and redundancy at layer 2 (Ethernet switch) and layer 3 (network) OSI levels.

HSSI (High-Speed Serial Interface) A physical connection between a DTE and DCE device, developed by Cisco Systems and T3Plus networking, that enables connection speeds well above the 64-/256-Kbps range, to which other standards are limited (Fig. H.8). Other standards include V.35 and RS-232C serial interface. Among other uses, these interfaces are used to connect between a public telephone service provider's CSU/DSU and a customer's local network device, such as a router.

Figure H.8   HSSI Port

HT The ASCII control code abbreviation for horizontal tab. The binary code is 1001000 and the hex is 90.

HTML (Hyper Text Markup Language)   A text- and graphics-formatting software that uses coding to indicate how a received part of a document should be presented by a viewing application, such as an Internet Web browser. Netscape and Internet Explorer are good examples of Web-browser software that utilize HTML. See also *Hypertext* and *Web Browser.*

HTTP (Hypertext-Transfer Protocol)   The control software used by Web browsers and Web servers on the Internet to provide support to lower-level OSI protocols when transferring *HTML (Hyper Text Markup Language)* based files, which contain text and graphics information.

Hub   In Ethernet, a device that connects many segments to one. Hubs can be active (where they regenerate signals sent through them) or passive (where they do not regenerate, but merely "pass" signals sent through them). Ethernet Hubs re-broadcast all information sent through them to all connections. See *Fig. H.9.*

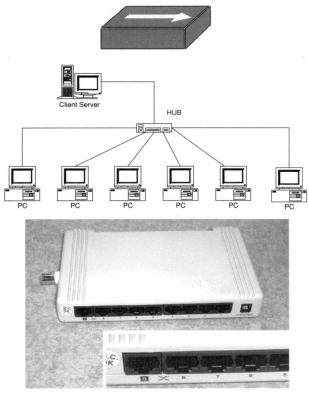

Figure H.9 Hub

Hunt   A telephone line feature. If you have several phone lines that are answered by a group of people, the telephone company can make those several phone lines work together. If a call comes in to a line that is associated with other phone lines in a hunt group, the call will rotate from line to line until it finds a line that is not busy. If all the lines are busy, then the caller will get a busy signal.

Hunt Group   A number of telephone lines that are associated together by the telephone company central office or a PBX system. When a call comes in to a hunt group, it cycles through the group of lines until it finds one that is not busy, then it rings that phone (or extension, if it's a PBX system).

Hybrid Cable   A communications cable that consists of two different types of media. A cable that contains twisted-pair copper and coax, or twisted-pair and fiber optic, etc., would be a hybrid cable.

Hybrid Key System   A telephone switching system that enables the user to choose which lines appear or don't appear under specific keys of electronic telephones. This is typical of a PBX system, except in the case of a hybrid key, you still have to select a line to dial out. When using a PBX system, the line is automatically selected from a pool when you dial 9.

Hybrid Network   From an administrative level, an internetwork made up of more than one type of network topology (star and ring), including LANs and WANs.

Hybrid Routing Algorithm/Protocol   Also called *balanced hybrid routing*. This router operating software combines distance vector routing and link-state routing methods. Hybrid routing techniques incorporate topology changes to trigger routing table updates and better metrics for calculating routes through a network. These methods help produce efficient routing, timely routing table updates, and a more-reliable network. Also, they do not require as much processor or bandwidth overhead as link-state protocols. Two examples of hybrid routing protocols are Cisco System's *Enhanced IGRP (Enhanced Interior Gateway Routing Protocol)* and the OSI standard *IS-IS (Intermediate System to Intermediate System)* routing protocol. See also *Distance Vector Routing Algorithm* and *Link-State Algorithm*.

Hypertext   The name given to a software feature that enables application users to move from one file or address to another by "clicking" on a word (called a *link*). If a user is reading a document about telecommunications and the word *hypertext* is colored blue, the user could click on that word and a new document about hypertext would be displayed. The de-facto standard color for hypertext word links is blue. Microsoft Word is an application that has hypertext capability. Internet browsers, such as Netscape Navigator, have hypertext link capability. Internet Web page creators have made use of hypertext links extensively.

Hypertext Markup Language (HTML) A text- and graphics-formatting software that uses coding to indicate how a received part of a document should be presented by a viewing application, such as an Internet Web browser. Netscape and Internet Explorer are good examples of Web-browser software that utilize HTML. See also *Hypertext* and *Web Browser.*

Hypertext Transfer Protocol (HTTP) The control software used by Web browsers and Web servers on the Internet to provide support to lower-level OSI protocols when transferring *HTML (Hyper Text Markup Language)* based files, which contain text and graphics information.

Hysteresis 1. A phenomenon in transformer cores or other inductive devices where an electrical current is formed in the metallic core of the device. It causes the core to heat up and increases the inductive value of the device.   2. Hysteresis is also a technique to reduce noise in digital circuitry. In this case, a comparator is designed to receive a feedback from its output. This gives it a better reference to switch from. A comparator with a hysteresis loop is better known as a *Schmitt trigger.*

Hz (Hertz) See *Hertz.*

I Frame (Information Frame)   In X.25, a type of control header. When the first bit of the control header is a 0, the X.25 equipment knows that the rest of the data in this frame identifies the window number of the frame received and the number of the next frame. The I frame is the majority of frames sent in X.25 because these are the frames that carry application data. The other types of frames are *S Frames (Supervisory Frames)* and *U Frames (Unnumbered Frames)*. See also *X.25, S Frame,* and *U Frame.*

I&M   Abbreviation for *installation and maintenance.*

I/O (Input/Output)   A class of devices that interface humans with computers. Some examples of user-interface devices are keyboards, monitors, terminals, and printers.

I&R   Abbreviation for *installation and repair.*

IAB (Internet Architecture Board)   A collection of internetwork researchers who discuss issues pertinent to Internet architecture. This group is responsible for appointing a variety of Internet-related groups, such as the Internet-Assigned Numbers Authority. The IAB members are appointed by the trustees of the Internet Society. See also *Internet Society.*

IANA (Internet-Assigned Numbers Authority)   An organization operated under the support of the *ISOC (Internet Society)* as a part of the Internet Architecture Board. IANA delegates authority for IP-address allocation and domain name assignment to the InterNIC. IANA also maintains a data base of protocol identifiers they assign for the TCP/IP stack, including autonomous system numbers. See also *Internet Architecture Board, Internet Society,* and *InterNIC*.

IBDN   Nortel Network's name for the CAT5 horizontal wiring standard.

IC (Integrated Circuit)   Another name for microchip. An *LSI (Large-scale Integration)* or *VLSI (Very Large-Scale Integration)* device that often comes in a DIP (dual in-line) package. IC circuits are designed and manufactured to do general functions or for specific applications. Some ICs are made for amplifying audio signals, some are made for storing binary memory (i.e., RAM), etc.

ICE Age (Information Communications and Entertainment Age)   What telecommunications companies call today's trend in telecommunications services. The striving for the ability to provide customers data connectivity, cable-TV service, and voice service on one line, with one company, with one bill at the end of the month.

ICMP (Internet-Control Message Protocol)   A network-layer Internet instruction set that routers and other networked equipment use to sense each other and share diagnostic information about the network. It is used by hosts and routers to send feedback to each other regarding traffic routing, retransmission, and other control or notification. For example, when damaged datagrams are discarded, ICMP is the part of the TCP/IP stack that sends this information to the sender. PING is a common "echo request" that is generated by ICMP. ICMP messages are encapsulated within IP datagrams.

ICS-7750   Perhaps the first IP telephony-based PBX in a cabinet to be introduced to the open telecommunications marketplace. The ICS-7750 truly integrates voice into Ethernet (802.3) networks. Manufactured by Cisco Systems, it was released in September 2000. Its major components are the cabinet, the SSP (System Switch Processor), the SAP (System Alarm Processor), the MRP (Multi-function Router Processor), the SPE (System Processing Engine), and the Power Supply. This system is difficult to compare part for part to a circuit-based PBX system because of its flexibility and purpose. The purpose of an IP telephony system is not to provide voice circuit switching services as in a traditional PBX, but to integrate voice into existing packet data flows by providing a *central source of management* and a *distributed method of* *interconnecting* to the public telephone network and voice-enabled network appliances.

IDE (Integrated Drive Electronics or Intelligent Drive Electronics)
A common and relatively less-expensive interface technology for magnetic- and optical-storage devices. This technology incorporates the drive-control electronics as a part of the hard drive, as other newer drive technologies. IDE is also known as *ATA (AT Attachment)* for the popularity of the AT motherboard at the time of its development by the *SFF (Small Form Factor) Committee*. Newer versions of the ATA/IDE interface are called *EIDE (Enhanced IDE)*, *Ultra-ATA*, and *ATA/66*. See also *SCSI*.

IDI (Initial Domain Identifier)   The part of a *NSAP (Network Service Access Point)*-format ATM address that specifies the address type, allocation, and administration control for upper OSI layers. See also *Network Service Access Point*.

IDN (Integrated Digital Network)   A term that refers to a network that has operability between many different devices through a standard digital protocol.

IDSL (ISDN Digital Subscriber Line)   A physical-layer transport method that provides symmetric download and upload speeds from 64 to 144 Kbps on a single pair of copper wires. The maximum range of IDSL from a central office is 18,000 feet, but this can be doubled with a repeater. IDSL uses 2B1Q line coding, the same kind of line-modulation technique employed in SDSL and ISDN. IDSL is used for transporting ISDN. This allows the xDSL family of technologies to be backwards compatible with ISDN. For a table of the DSL family of carriers, see *xDSL*.

IEC (Inter Exchange Carrier, IXC)   IEC is a long-distance company, like AT&T, Sprint, Worldcom, or MCI.

IEEE (Institute of Electrical and Electronics Engineers)   A professional organization whose activities include the development of telecommunications and networking standards. IEEE LAN standards, such as the Ethernet 802 family, are the predominantly implemented LAN standards today.

IEEE 802 Ethernet   The family of standards that define 802 Ethernet. See Fig. I.1 and the individual definitions.

IEEE 802 FAMILY OF ETHERNET STANDARDS

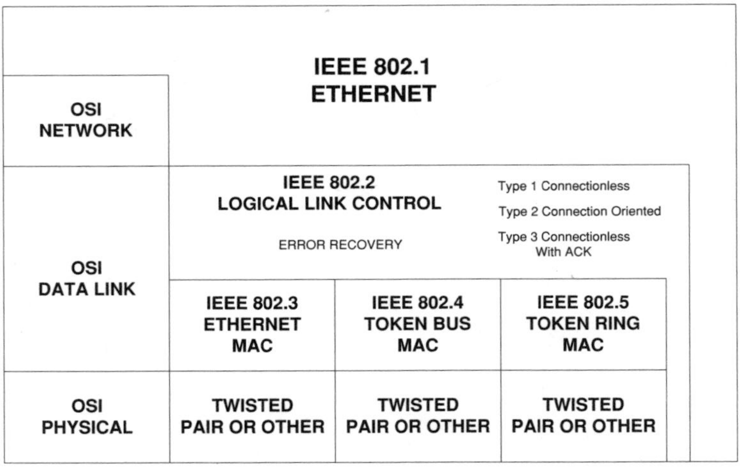

	IEEE 802.1 ETHERNET		
OSI NETWORK			
OSI DATA LINK	**IEEE 802.2 LOGICAL LINK CONTROL** ERROR RECOVERY	Type 1 Connectionless Type 2 Connection Oriented Type 3 Connectionless With ACK	
	IEEE 802.3 ETHERNET MAC	IEEE 802.4 TOKEN BUS MAC	IEEE 802.5 TOKEN RING MAC
OSI PHYSICAL	TWISTED PAIR OR OTHER	TWISTED PAIR OR OTHER	TWISTED PAIR OR OTHER

Figure I.1   IEEE 802 Ethernet

IEEE 802.1   EEE specification that defines a software algorithm that prevents network loops by creating a spanning tree. The original version of
the algorithm was invented by Digital Equipment Corporation (DEC).
The IEEE 802.1 algorithm is a modification of the DEC version and they
are not compatible. See also *Spanning Tree*.

IEEE 802.1d   The IEEE standard for spanning tree algorithm that prevents
loops in redundantly connected LAN switches. Spantree is automatically
enabled when redundant bridges are connected. If redundant bridges were
connected to a network without Spantree enabled, the dual connected
bridges would forward the same frames to each other in an endless loop.
This condition saturates bandwidth immediately, and renders all devices
associated with the loop useless. The way that Spantree works is that when
bridges are initialized (powered on), they send a signal to other networked
devices called a Bridge Protocol Data Unit (BPDU). When bridges/switches
receive these BPDUs from other devices, they become "aware" that other
bridges are connected to the network and whether any are connected in
redundancy to them. Using BPDU information, bridges on the network
elect a "root bridge" and a "designated bridge." Depending on the way the
bridges are physically connected, all ports are "blocked" or partially disabled except for "root ports," and "designated ports," which are bridge
ports closest (by number of hops) to a designated or root bridge. If a link
is lost, an alternate port then becomes the root port. New BPDU messages
are sent to notify other bridges of the status change. Most makers of

bridging hardware set the default to automatically send BPDUs and enable Spantree to ON. This is so that if a network is unknowingly connected with bridges in parallel, it will not bring the network down. The 802.1d standard evolved from Digital Equipment Corporation's (DEC) Spantree algorithm. 802.1d and the original *DEC Spantree* are not interoperable. Further, when incorporated with 802.1Q (VLANs), one instance of spanning tree must be set up for *each and every* VLAN.

IEEE 802.1p   The IEEE standard for prioritization of LAN traffic among Ethernet switches based on the switch port, MAC address, or IP address associated with the communicating end appliance (be it an IP phone, video monitor, host PC, printer, or server). Packets are tagged as belonging to a queue, which determines the priority of the packet. By the 802.1p standard, queues 0–3 are normal and 4–7 are high priority. 802.1p functions hand-in-hand with 802.1Q or VLANs.

IEEE 802.1Q   The IEEE standard that evolved from Cisco Systems' ISL (Inter-Switch Link) protocol. ISL and 802.1Q are not interoperable. The reference 802.1Q is better known as the VLAN or tag switching standard. It is a feature on post-1998 LAN switches that makes selected ports behave as if they were attached to the same segment or hub. Another good name for this feature would be *V-segment* or *virtual-segment*. Devices/users that exchange a large amount of information are usually placed within the same virtual LAN segment. This helps make the operation of the LAN switch more efficient, keeping traffic contained within specified ports. This allows other ports on separate VLANs to carry other nonrelated traffic simultaneously. VLANs are configured by a network engineer, network analyst, or network administrator. When IP telephony is implemented over an Ethernet-switched network, the telephone devices connected to the network are best placed into their own VLAN. Most switches that are 802.1Q compatible can recognize more than 1,000 VLANs. Further, there are two kinds of VLANs: static and dynamic. Static VLANs are associated with switch ports, and dynamic VLANs are associated with the MAC addresses of devices attached to the switch. Dynamic VLANs allow users to move to another office, which could have a switch port connection preinstalled. The switch would recognize the MAC address of the device and automatically include its traffic in the same VLAN as the previously connected switch port. See also *Frame Tagging*.

IEEE 802.11b   Wireless LAN standard update for increased speed to 11 Mbps at an operating frequency of 2.4 GHz. The modulation technique used in 802.11b is DSSS (Direct Sequence Spread Spectrum). WEP (Wired Equivalent Privacy) is also an addition in the 802.11b standard, which allows manufacturers to implement security up to and including 128-bit key encryption.

IEEE 802.11DS   The 802.11 wireless LAN standard for the direct sequence method of line coding. In the standard, there are 11 22-MHz wide stationary channels. This allows for an 11 Mbps throughput, with up to three nonoverlapping radio channels operating in the same area, which means three separate radio units (called *access points* in some wireless circles) can operate in the same area without interfering with each other.

IEEE 802.12   The IEEE LAN standard that defines the physical layer and the *MAC (Media-Access Control)* portion of the data link layer. IEEE 802.12 uses the demand-priority media-access scheme at 100 Mbps over a variety of physical media.

IEEE 802.14   The IEEE standard for the operation of cable telephony modems that enables cable-TV networks that are coax and hybrid fiber-coax in composition to carry Ethernet 802 traffic as well as ATM-based traffic. There are multiple MAC layer interfaces defined in 802.14 to make cable telephony services equally as flexible to the end user as traditional services enabled by DSL or ATM.

IEEE 802.2   A LAN protocol that defines an implementation of the *LLC (Logical Link Control)* portion of the data link layer. 802.2 processes errors, framing, flow control, and the network-layer (layer 3) software interface. 802.2 is used in IEEE 802.3 and IEEE 802.5 LANs. See also *IEEE 802.3* and *IEEE 802.5*.

IEEE 802.3   A LAN protocol that defines an implementation of the physical layer and the *MAC (Media-Access Control)* portion of the data-link layer. 802.3 uses *CSMA/CD (Carrier Sense Multiple Access/Collision Detection)* access at a variety of speeds over a variety of physical media. Extensions to the 802.3 standard define implementations for Fast Ethernet. Older physical variations of the 802.3 specification include 10Base2, 10Base5, 10BaseF, 10BaseT, and 10Broad36. Newer physical variations for fast Ethernet include 100BaseT, 100BaseT4, and 100BaseX. See also *CSMA/CD*.

IEEE 802.3ab   Ten gigabit Ethernet over copper UTP and fiber standard (10,000BaseT). Still in the process of standardization as of this writing, it is expected that the technology being implemented in the 802.3ab standard and/or its revisions will extend the Ethernet more than 40 km.

IEEE 802.3u   An Ethernet 100BaseT feature specification that provides for flow control (pause frames) and full-duplex operation. Full duplex allows for 100 Mbps send and 100 Mbps receive, for a total 200 Mbps Ethernet connection over Cat5 twisted pair. The autonegotiation is really an enhancement of the link integrity signaling method used in 10BaseT networks and is backward-compatible with link integrity. Autonegotiation allows the NIC or the network device to adjust its speed to the highest speed that both ends are capable of supporting. To be able to use this feature, both the network device (switch port) and the NIC must contain the autonegotiation logic.

IEEE 802.4   A LAN protocol that defines an implementation of the physical layer and the *MAC (Media-Access Control)* portion of the data-link layer. IEEE 802.4 uses logical token-passing access over a bus topology and is based on the token-bus LAN architecture.

IEEE 802.5   A LAN protocol that defines an implementation of the physical layer and *MAC (Media-Access Control)* portion of the data-link layer. 802.5 uses token-passing access at 4 or 16 Mbps over twisted-pair cabling and is similar to IBM token ring. See also *Token Ring*.

EEE 802.6   A metropolitan/campus network specification based on *DQDB (Distributed Queue Dual Bus)* technology. IEEE 802.6 supports data rates of 1.5 to 155 Mbps. *FDDI (Fiber-Distributed Data Interface)* is a high-speed LAN backbone protocol that is used mostly in favor of DQDB. See also *DQDB* and *FDDI*.

IEEE 1394   More commonly known as *FireWire*. A specification for an external serial bus that can transfer data at 400 Mbps and can handle 63 simultaneous devices. This bus is intended for use with DVD players, graphics and audio peripherals, and other multimedia devices. FireWire is an innovation of Apple Computers.

IEEE Radar Band Designation   The frequency ranges and names given by the IEEE for radio communications (Fig. I.2).

IEEE Radio Band Designations

L Band	1-2 GHz
S Band	2-4 GHz
C Band	4-8 GHz
X Band	8-12 GHz
Ku Band	12-18 GHz
K Band	18-27 GHz
Ka Band	27-40 GHz
V Band	40-75 GHz
W Band	75-110 GHz
mm Bandz	110-300 GHz
μmm Band	300-3000 GHz

Figure I.2   IEEE Radio Band Designations

IF (Intermediate Frequency)   An amplifier stage in radio/TV receivers that is designed to amplify and isolate a large range of frequencies. The IF amplifier receives the specific frequency to amplify from the detector stage of the tuner. The tuner is the device or circuit with which the user of the receiver selects the desired channel.

IGMP (Internet Group Management Protocol)   In packet networking, IGMP is one of the standards for IP multicasting in the Internet and in private enterprise networks. IGMP is used to establish host memberships, in particular, multicast groups on a single network. The mechanisms of the protocol allow a host to inform its local router, using host membership reports, that it wants to receive messages addressed to a specific multicast group. See also *Multicast* and *PIM*.

IGRP (Interior Gateway Routing Protocol)   A network traffic-control program used by routers. Cisco was the developer of IGRP in the early 1980s. IGRP determines the best path through an Internet by examining the bandwidth and delay of the networks between routers. IGRP converges faster than *RIP (Routing Information Protocol),* thereby avoiding the routing loops caused by disagreement over the next routing hop to be taken. IGRP is also not limited to a hop count of 16 like RIP is. IGRP has a new version that has been improved. It is called *Enhanced IGRP*.

IIH (IS-IS Hello, Intermediate System to Intermediate System "Hello")   A routing protocol message that is transmitted between routers and other network equipment that are using an IS-IS algorithm within their operating system. This message is used to update routing tables, network adjacencies, and other network-status information. See also *Intermediate System to Intermediate System*.

IISP (Information Infrastructure Standards Panel)   A charter of the ANSI to develop *GII (Global Information Infrastructure)* standards.

IMAP (Internet Message Access Protocol)   In integrated voice, email, and fax messaging environments, IMAP is used because when a message is retrieved from the message server, it is not deleted as it is in a POP (Post Office Protocol) environment. The additional copy left on the server enables users to access their messages from one point via multiple devices, such as their computer, cell phone, IP telephone, or PSTN voice access. The purpose of integrated messaging is to enable end users access to email or voice mail through any network appliance. If an email is accessed by telephone, the system will read the email to the user, and if a voice mail message is accessed by a PC, the PC will play the message through its speakers.

Immediate Start   A reference to E&M signaling for analog voice circuits. E&M technology dates back to the time telegraphs were used and is an outdated service no longer offered as a service by most telephone companies however, E&M trunking still has special uses because of its simplistic nature. E&M interfaces come in handy in the PBX environment when there is a need to connect to an analog audio device such as an overhead paging system, or tape recorder. There are 5 types of E&M interfaces that can have either two or six wires in the loop. The most common type of E&M signaling is the Four Wire Wink Start E&M, and the next most used is the Four Wire Immediate Start E&M. The Wink Start E&M operates as follows: The call originating switch goes off-hook and then waits for a "wink" from the terminating or destination switch. When the destination switch provides the 200 ms off-hook "wink," then the originating switch sends dialed digits. After the dialed digits are received and a connection is made to the terminating loop by a handset being taken off hook, the same "off hook" condition is given over the E&M trunk connecting the terminating switch to the destination switch. When one switch goes "on-hook" or hangs up, the other does as well. The most simple E&M signaling method is the Immediate Start E&M, where the originating end goes "off-hook," or provides a 1000 ohm short on the line and sends digits without regard to the other end. The originating switch stays off-hook until the receiving switch goes off-hook and then back "on-hook," or the call originator goes back "on-hook" or hangs up. The E&M immediate start is the better choice for interfacing external audio devices to PBX systems, and is the less appropriate choice for PBX trunking because if the terminating switch does not answer the call, and the originating switch does not manually hang-up or go back "on-hook," then the loop is left connected. This problem with Immediate Start E&M is the reason that Wink Start E&M was brought about. Wink Start E&M is also called E&M with Answer Supervision.

Impedance   A term used to replace resistance in AC or signal-serving circuits. For example, an 8-ohm speaker is a measure of impedance (resistance to a signal being fed into it). If you measure the DC resistance of the speaker with an ohmmeter, it is about 1 ohm. This is because impedance is a measure of resistance to an AC signal, not a DC signal (ohmmeters use a DC battery to measure resistance with). If you put an impedance meter on the speaker, it will measure the resistance of the speaker with an AC signal, which is resisted by reactive components, such as coils (a speaker is a coil of wire electrically) and capacitors. The impedance meter will read an impedance of 8 ohms.

Implicit Frame Tagging   In LAN switching, frame tagging is a method used to identify which Virtual LAN (VLAN) a packet belongs to. Within an Ethernet switch, a VLAN behaves as a single Ethernet segment, where all computers/hosts that are a part of a VLAN see each other's traffic. Traffic cannot traverse from one VLAN group to another unless it leaves the switch and is routed back in on another port. In some switch designs this is accomplished by an on-board router in an add-on module form. The router provides traffic security/management. Frames are marked by the switch in two methods: implicit and explicit. Implicit frame tagging is a method where the VLAN information is added within the packet. Explicit frame tagging is a method where an external VLAN header is added to the frame.

Impulse Distortion   A short and bursty distortion that occurs at the physical layer of signal transmissions. Impulse distortion is caused by slightly different voltage potentials being suddenly connected in switching equipment. In the audio realm of communication, impulse distortion sounds like a series of snap or pop sounds.

In-Band Signaling   In telephone circuits (DS1 to be specific) signals can be sent in two different ways: in-band and out-of-band. Signals are digits that you dial, dial tone, the phone being off hook, ringing, etc. An in-band telephone line is like the one in your home; the digits that you dial and the ringing are carried within the channel you talk on. Out-of-band signaling is a method that telephone companies and businesses use for larger PBX applications and data-transfer applications. An out-of-band, signaled DS1 has 24 multiplexed channels. The 24th channel carries the signaling for the other 23 channels or phone lines. The advantage of out-of-band signaling is that each channel has an increased capacity to carry data (8 Kb/s more) and the 23 channels are not used to find out if a line is busy (both directions, in and out). The off-hook sensing and busy signaling is done in the 24th channel. If you have a system that gets thousands of calls per day, this can reduce traffic.

Incoherent Light   Light that consists of many frequencies and wavelengths. A light bulb emits noncoherent light. Coherent light is light that consists of only one frequency or very close to one frequency. Coherent light looks to the human eye as a very pure color. Lasers and LEDs (like the one that lights when the hard drive in your computer is running) emit light that is very close to being coherent.

Index Of Refraction   A reference to how much light bends when it travels through a specific substance. When light travels through a swimming pool it refracts and makes everything appear wavy and distorted. When fiber optic is designed, it is planned to have different types of glass that have different refractive indexes. As the light travels from the core of the fiber toward the outer edge (cladding), it is bent back inward because of the increasing indexes of refraction of the glass it is passing through.

Inductance   A physical characteristic of conductors, semiconductors, and other electronic components. Inductance is the formation of an electromagnetic field around a device as electricity flows through it. Inductors are coils of wire that add this effect with each winding of the coil. Inductance is measured in henrys (H). An inductor, or coil of wire, is a reactive device. *Reactance* is the resistance that a component gives to an AC or fluctuating DC current. The two components that cause reactance are inductors (coils) and capacitors. The difference between resistance and reactance is that resistance is always the same, regardless of the voltage amplitude or frequency applied to the resistive device. The reactance of a component changes along with frequency changes, or the speed at which an AC current changes direction. The higher the frequency applied to an inductor, the higher the reactance or resistance to that frequency. Coils of wire cause reactance because as electricity flows through them, they force the electricity to create a magnetic field every time it changes direction. A perfect inductor has zero reactance to a DC current and has a specific reactance or resistance to every frequency of AC current. Each coil or inductor has a value in henrys. The higher the number of henries, the more it will resist AC or fluctuating DC. Coils are used to filter out ("choke out") DC fluctuations in power supplies. They are also used to help tune in radio or other frequencies.

Reactance is also caused by other electronic conditions where it is not useful. All wire and electronic components possess a small amount of reactive properties (that is, the reason that twisted-pair wire causes attenuation of signal strength is because of the inductance of the copper wire and the capacitance of the two adjacent wires).

Inductive   A reference to the electromagnetic (as opposed to the electrostatic of capacitors) reactive properties of a device. See also *Inductance.*

Inductive Coupling   The use of an inductor (coil of wire) to connect one amplifier or circuit stage to another. Inductive coupling is advantageous if low frequencies are the crucial part of the composite signal to be passed. High-end audio/home-stereo equipment is designed with inductive coupling.

Inductive Pick Up   A microphone or sensing device that uses the changes in a magnetic field or creation of a magnetic field within a component because of the vibrations in the air or of a nearby object.

Inductive Tap   A tap on a telephone line that does not connect to the pair of wires. Instead, it picks up the tiny magnetic field created around the pair and amplifies it, the same way that a radio transmitter receives electromagnetic waves from the air. An inductive tap must be within an inch or two of the pair to successfully receive an electromagnetic signal.

Industry-Standard Architecture (ISA)   A 16-bit bus standard used in personal computers in the late 1970s and early 1980s. *ISA* is a term mostly referred to by PC users when they are looking for a new expansion circuit board, such as an internal modem. ISA was outdated by the *EISA (Extended Industry Standard Architecture)* 32-bit bus. In 1993, Intel and Microsoft developed a new version of the ISA spec called *Plug and Play ISA*. This newer system enables the computer's operating system to configure the hardware aspect of the circuit board automatically without a user having to set jumpers or DIP switches. ISA slots/sockets on a motherboard are identified by their black color and larger pin size. See also *PCI*.

INE (Intelligent Network Element)   A network element, such as a router, node or hub, that has the ability to be electronically reconfigured (manually or remotely) or perform additional functions, such as protocol conversions.

Information Technology   The study of improving information and data processing with the use of newer and better devices/machines.

Infrared (IR)   Light waves that humans cannot see. Their wavelength (or frequency) is just below that of red light. Heat radiates infrared lightwave radiation.

INIC   ISDN Network Identification Code.

Initial Domain Identifier (IDI)   The part of a *NSAP (Network Service Access Point)* ATM address that specifies the address type, allocation, and administration control for upper OSI layers. See also *Network Service Access Point*.

Inline Power   In Ethernet networks, DC power provided to desktop devices via the same conductors that provide the transmit and receive signal (Fig. I.3). Inline power does not use the additional spare conductors in a CAT-5,7 installation. This enables an additional Ethernet link to be installed over the two spare pairs.

Figure I.3   Inline Powered Ethernet Switch

Inner-duct   A flexible plastic conduit that is placed within larger conduits. Inner-duct is used where multiple communications companies use or lease conduit space within the same conduit (Fig. I.4).

Figure I.4   Innerduct

Input-Output Address (I/O Address)   An address used to direct traffic to and from devices attached to a computer's serial (denoted *COM*) and parallel (denoted *LPT*) ports and/or expansion cards/ slots. COM1 has usually been set in new PCs with a default I/O address of 3F8.

Input/Output (I/O)   A class of devices that interface humans with computers. Some examples of user-interface devices are keyboards, monitors, terminals, and printers.

Inside Dial Tone   The dial tone provided by a PBX system. When you pick up the handset of an electronic telephone that is served by a PBX, you get an inside dial tone, which allows you to dial an extension that begins with a number other than 9 (internal extensions should never begin with a 9). When you dial 9, you get an outside dial tone or a dial tone that is served by a telephone company central office.

Inside Plant   Electronic equipment located inside buildings, including central-office switches, PBX switches, broadband transmission equipment, distribution frame, power supply/rectification equipment, and anything else you can find inside a central office. *Inside plant* does not include telephone poles, cable, terminals, cross boxes and cable vaults, or anything else you might find outdoors.

Inside Wiring (IW)   The telephone wire that is on the customer side of the Telephone Network Interface. IW includes jacks and any wiring in or attached to the outside of the house, as long as it is electrically on the customer side of the network interface.

Installer's Tone   Also called a test tone. A small box that runs on batteries and is used to put an *RF (radio frequency)* tone on a pair of wires. If a telephone technician can't find a pair of wires by color or binding post, they attach a tone to one end, then go to the other end and use an inductive amplifier (also called a *banana* or *probe*) to find the beeping tone (Fig. I.5).

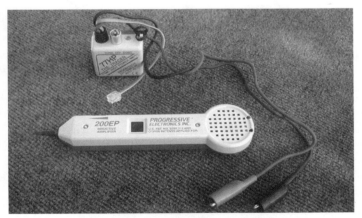

Figure I.5   Installer's Tone and Probe

Insured Burst   In an *ATM (Asynchronous Transfer Mode)* network, the largest burst of data above the insured rate that will be temporarily allowed on a *PVC (Permanent Virtual Circuit)* and not tagged by the traffic-policing function for dropping in the case of network congestion. Unlike the insured rate, the insured burst is not a bandwidth or a data rate. It is a quantity of data, specified in bytes or cells. See also *Insured Rate* and *Maximum Burst.*

Insured Rate   In an *ATM (Asynchronous Transfer Mode)* network, the guaranteed bandwidth or data-transfer speed that a bandwidth subscriber will receive when using a particular *PVC (Permanent Virtual Circuit)* connection. If the network that the subscriber is using has a low volume of traffic congestion, then the network will automatically allow additional speed. This additional speed or bandwidth is called the *excess rate.* Another bandwidth parameter addressed at the point of sale to a subscriber is *maximum rate.* The maximum rate is equal to the insured rate combined with the excess rate. See also *Cell Loss Priority, Cell Loss Ratio,* and *Maximum Rate.*

Integrated Circuit (IC)   See *IC.*

Integrated Drive Electronics/Intelligent Drive Electronics (IDE)   A common and relatively less-expensive interface technology for magnetic- and optical-storage devices. This technology incorporates the drive-control electronics as a part of the hard drive, as other newer drive technologies. IDE is also known as *ATA (AT Attachment)* for the popularity of the AT motherboard at the time of its development by the *SFF (Small Forum Factor) Committee.* Newer versions of the ATA/IDE interface are called *EIDE (Enhanced IDE), Ultra-ATA,* and *ATA/66.* See also *SCSI.*

Integrated IS-IS (Integrated Intermediate System to Intermediate System)   A link-state routing protocol based on the OSI routing protocol IS-IS, but with new support for IP and other protocols. Integrated IS-IS sends only one set of routing table updates, which is more efficient than two separate implementations. It was formerly called *Dual IS-IS.* For more information on link-state routing, see *Link-State Algorithm.*

Integrated Services Digital Network (ISDN)   ISDN is a service that first evolved in 1979. It brings the features of PBX systems and high-speed data-transfer capability to the telephone network. The only thing that makes ISDN complicated is the many available features. The two kinds of ISDN lines are *Primary Rate Interface (PRI)* and *Basic Rate Interface (BRI)* (Fig. I.6). Two types of channels are contained within an ISDN circuit. The B (bearer) channel carries the customer's communications, and a D (data) channel provides control and signaling for the B channels. The *BRI (Basic Rate Interface)* ISDN line has two B channels and one D channel. A PRI has 23 B channels and one D channel.

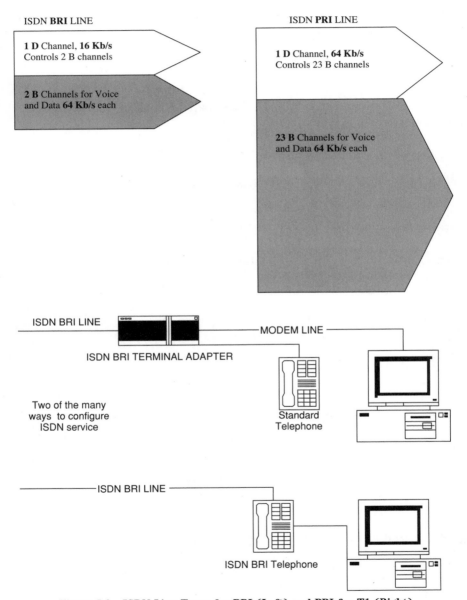

ISDN **BRI** LINE

1 **D** Channel, **16 Kb/s**
Controls 2 B channels

2 **B** Channels for Voice
and Data **64 Kb/s** each

ISDN **PRI** LINE

1 **D** Channel, **64 Kb/s**
Controls 23 B channels

23 **B** Channels for Voice
and Data **64 Kb/s** each

ISDN BRI LINE

ISDN BRI TERMINAL ADAPTER

MODEM LINE

Two of the many
ways to configure
ISDN service

Standard
Telephone

ISDN BRI LINE

ISDN BRI Telephone

Figure I.6   ISDN Line Types for BRI (Left) and PRI for T1 (Right)

The separate control of the ISDN line over the D channel is what enables the broad flexibility and features available with ISDN. When you are talking or sending a data transmission over an ISDN line, the voice and/or data is carried by the B channels. While you are talking on your ISDN line, you can still dial digits (signal the central office) to change or alter the state of your service because of the separate D channel. For example, imagine you want to arrange a meeting with a client. You dial the client's telephone number on your ISDN telephone to reach the client. While you are speaking with the client, you can dial up an Internet access on your computer and put two baseball tickets in at the ticket counter using the same phone line. Then you can fax your client directions by downloading a map provided by the baseball ticket office, disconnect and redial your client's fax number. All of this occurs while talking to your client the entire time. Through the advanced convenience and flexibility of ISDN, you can send different types of data and messages to different places at the stroke of a few buttons, and at a much faster speed than a regular telephone line. If you are interested in ISDN, call your local phone company. They can help you decide on what kind of terminal adapter (equipment that connects your computer and phone equipment to the ISDN line) to buy and what kind of features to subscribe to. ISDN is not yet available everywhere.

Intelligent Hub   A hub that has the ability to be electronically reconfigured (manually or remotely) and perform additional functions, such as protocol conversions and bridging functions. For a diagram of where a hub fits into a network, see *Hub*.

Inter LATA (Inter Local Access Transport Area)   Simply stated, a LATA is an area code. Inter LATA refers to services that go from one area code to another, like long-distance telephone calls. *Intra LATA* refers to services that originate and terminate in the same area code.

Interactive Terminal Interface (ITI)   Another name for the X.28 interface. A reference to the software functions that make *PADs (Packet Assembler/Disassemblers)* convert asynchronous (stop/start or dumb) terminal data into *STDM (Statistical Time-Division Multiplexing)* packets used in X.25.

Interactive Voice Response (IVR)   A telecommunications and data-processing technology that interfaces a person to information held in a computer by using a phone line (Figs. I.7 and I.8). If you have ever called your bank and entered your account number, a password, and a prompt so that a computerized voice can read back your bank account balance,

Figure I.7   Interactive Voice-Response System

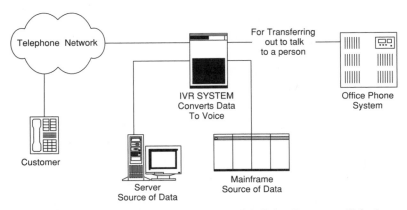

Figure I.8   A Network Diagram for Interactive Voice Response. Telephone Connections are T1 and Data Connections are Ethernet

then you have used IVR. IVR systems are capable of sending fax information as well.

Interconnect Agreement   This is also known as a *co-location agreement*. The two types of co-location agreements are physical and virtual. A *physical co-location* is an interconnection agreement and a physical place where telephone companies hand off calls and services to each other. This is usually done between a CLEC and an RBOC. The CLEC installs and maintains interconnection equipment usually consisting of optical carrier (SONET) equipment and a digital cross-connect system. There are other types of co-locations. Alarm companies like to have their alarm-signaling equipment located in the local central office for security and convenience of connecting alarm circuits. Long-distance companies co-locate with local telephone companies as well.

A *virtual co-location* is an interconnection agreement and a physical place where telephone companies hand off calls and services to each other. This is usually done between a CLEC and an RBOC. A virtual co-location is when telephone company A (the CLEC) requests that their phone company's network be connected to telephone company B's (the RBOC's) network. Telephone company B charges company A lots of money. Company B owns, installs, and maintains the equipment. To company A, the interconnection is virtual, because they never physically do anything to it when and after it is installed. Company B likes this, because company A does not get free access to their premises.

Interarea Routing   The term used to describe routing between two or more logical areas. Compare with *Intra-Area Routing*.

Interface   A device or software program that connects two separate entities. The two entities can be virtual (software), hardware/electronic devices, or distinguish a separation of responsibility between two parties (telephone network interface).

Interior Gateway Protocol   Also known as *Link State Routing Protocol, Distributed Routing Protocol,* and *Shortest Path First*. An interior gateway routing protocol is a methodology used in router protocol design. This methodology enables routers within an autonomous network (i.e., corporate LAN) to identify each other and the status of their port connections. Interior gateway routing protocols create three databases within a router's memory: a neighboring router database, a link database, and a routing table. The routing table is created by applying Dykstra's algorithm to the first two databases. The two most widely used interior gateway routing protocol is Open Shortest Path First (OSPF). See also *OSPF*.

Intermediate Session Routing (ISR)   The first routing algorithm used in *APPN (Advanced Peer-to-Peer Networking)*. ISR (where still used) provides node-to-node connection-oriented routing. Network outages cause sessions to fail because ISR cannot provide nondisruptive rerouting around a failure. ISR was replaced by *HPR (High-Performance Routing),* which has been made obsolete by newer routing algorithms that have been incorporated into link-state, distance-vector, and hybrid routing protocols.

Intermediate System to Intermediate System (IS-IS)   An *OSI (Open-System Interconnect)* link-state routing protocol that is based on DECnet Phase-V routing. One version of IS-IS, called *Integrated IS-IS,* supports IP-based networks. For more information on link-state protocols, see *Link-State Algorithm.*

International Dialing   To dial international long distance from the United States, dial: 011-county code city code number. For a listing of country codes, see Appendix B. To dial the United States from another country that is a part of the NANP (North American Numbering Plan), simply dial the area code the same way you would call long distance to another state. To call the United States from another country that is not a part of the NANP, consult your long-distance company. The United States has different country codes/access codes for almost every country that is not a part of the NANP.

International GateWays   International telecommunications are done through gateway central offices. Gateway central offices (class 5 central offices) connect communications to other countries. The gateway does the translation from T1 to E1, T3 to E3, and vice versa.

International Organization for Standardization (ISO)   A consortium of worldwide telecommunications experts that has created a wide range of standards, including those relevant to data transport and data networking. ISO developed the *OSI (Open Systems Interconnect) Reference Model,* which is a mainstay for the data networking industry.

International Telecommunication Union (ITU-T)   A worldwide standards organization through which public and private organizations develop telecommunications standards for hardware and software. The ITU was founded in 1865 and became a United Nations agency in 1947. It is responsible for adopting international treaties, regulations, and standards governing telecommunications. The standardization functions were formerly performed by a group within the ITU called CCITT. After a 1992 reorganization, the CCITT no longer exists as a separate body.

Internet   A network of computers that originated as ARPANET, an information communications project of the United States Department of Defense. Over time, many other organizations, private and public, have utilized the project by connecting their computers to it. Its primary protocol is TCP/IP. Today, many Internet service providers can offer access to the Internet for as little as $15 per month. The Internet is growing exponentially as more service providers and customers gain access to it. It currently links millions of computers, with which users find and exchange information, buy and sell services or products, and play games.

Internet Address   A 32-bit dotted notation address that identifies a host on a network and the network it is a part of. The five types of Internet addresses are defined by classes. The network and host identification within the 32-bit address is determined by the subnet mask that is used in conjunction with the host IP address.

Internet-Assigned Numbers Authority (IANA)   An organization operated under the support of the ISOC (Internet Society) as a part of the Internet Architecture Board. IANA delegates authority for IP address allocation and domain name assignment to the InterNIC. IANA also maintains a data base of protocol identifiers they assign for the TCP/IP stack, including autonomous system numbers. See also *Internet Architecture Board, Internet Society,* and *InterNIC.*

Internet Browser   Also called a *Web browser* or *browser.* A computer program that allows users to download World Wide Web pages for viewing on their computers. Two popular browser programs are Netscape Navigator and Microsoft Internet Explorer. The first browser program was called *Mosaic,* and it was a text browser, as opposed to the newer graphical browsers.

Internet Control Message Protocol (ICMP)   A network-layer Internet instruction set that routers and other networked equipment use to sense each other and share diagnostic information about the network. It is used by hosts and routers to send feedback to each other regarding traffic routing, retransmission, and other control or notification. For example, when damaged datagrams are discarded, ICMP is the part of the TCP/IP stack that sends this information to the sender. PING is a common "echo request" that is generated by ICMP. ICMP messages are encapsulated within IP datagrams.

Internet Group Management Protocol (IGMP)   In packet networking, IGMP is one of the standards for IP multicasting in the Internet and in private enterprise networks. IGMP is used to establish host memberships, in particular, multicast groups on a single network. The mechanisms of the protocol allow a host to inform its local router, using host membership reports, that it wants to receive messages addressed to a specific multicast group. See also *Multicast* and *PIM.*

Internet Message Access Protocol (IMAP)   A protocol that provides e-mail users with a way to retrieve messages. IMAP performs expanded functions of POP. IMAP allows users to view/delete mail before downloading it to their PC. IMAP uses *SMTP (Simple Mail-Transfer Protocol)* to transfer mail messages from the user to the mail server.

Internet Operating System (IOS)   The software program developed by a group of scientists at Stanford University. The team was led by a married couple, Leonard Bosack and Sandy Lerner. They incorporated their software into a Digital Equipment computer so that it could route traffic to and from other computers. This was the first bridge/router. The IOS software invented and developed at Stanford is the operating system on which the company Cisco Systems, Inc. was founded. For more details about IOS and what it does, see *OSPF, EIGRP,* and *BGP.*

Internet Protocol (IP)   A network layer protocol in the TCP/IP protocol stack that offers a connectionless or packetized internetwork service. IP packets are delivered on a best-effort basis. If a packet or datagram cannot be delivered successfully, it is discarded. When this happens, the *ICMP (Internet Message Control Protocol)* section of the protocol stack notifies the sender that a datagram has been discarded. IP provides features for addressing, type-of-service specification, fragmentation, reassembly, and security. See also *TCP/IP.*

Internet Server   A server that users access for Internet services. Popular Internet services include access to the Internet, e-mail, news updates of the subscriber's choice, Web pages, etc. An Internet server is owned by an Internet service provider.

Internet Service Provider (ISP)   An Internet service provider purchases direct access to the Internet through an Internet company, such as UUnet and resells the service to smaller subscribers via dial-up modem (or to large customers via frame relay or private line T1). The ISP adds other services of their own, such as e-mail, news updates of the subscriber's choice, Web pages, etc. (Fig. I.9).

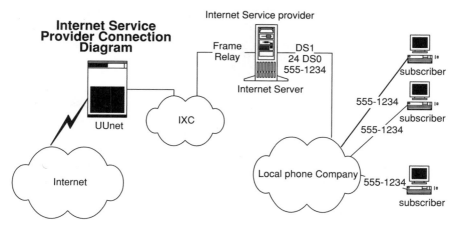

Figure I.9   Internet Service Provider

Internet Society (ISOC)   An international nonprofit organization that was founded in 1992. Its members coordinate the evolution and use of the Internet. In addition, the ISOC delegates authority to other groups related to the Internet, such as the IAB. The Internet Society world headquarters is located in Reston, Virginia, USA. See also *InterNIC*.

Internetwork Packet Exchange (IPX)   A Novell NetWare network layer protocol used to transfer data from servers to workstations. IPX has an extended address range that includes the *MAC (Media-Access Control)* address, which eliminates the need for an *ARP (Address-Resolution Protocol)*.

InterNIC (Internet Network Information Center)   An organization that serves the Internet community by providing user assistance, documentation, training, registration of Internet domain names, and other services. See also *Internet Society*.

Inter-Switch Link (ISL)   This proprietary Cisco Systems Ethernet switch protocol evolved into the IEEE 802.1Q standard. ISL and 802.1Q are not interoperable. The reference 802.1Q is better known as the VLAN or tag switching standard. It is a feature on post-1998 LAN switches that makes selected ports behave as if they were attached to the same segment, or hub. Another good name for this feature would be *V-segment*, or *virtual-segment*. Devices/users that exchange a large amount of information are usually placed within the same VLAN segment. This helps make the operation of the LAN switch more efficient, keeping traffic contained within specified ports. This allows other ports on separate VLANs to carry other nonrelated traffic simultaneously. VLANs are configured by a network engineer, network analyst, or network administrator. When IP telephony is implemented over an Ethernet-switched network, the telephone devices connected to the network are best placed into their own VLAN. Most switches that are 802.1Q compatible can recognize more than 1,000 VLANs. Further, there are two kinds of VLANs: static, and dynamic. Static VLANs are associated with switch ports, and dynamic VLANs are associated with the MAC addresses of devices attached to the switch. Dynamic VLANs allow users to move to another office, which could have a switch port connection preinstalled. The switch would recognize the MAC address of the device and automatically include its traffic in the same VLAN as the previously connected switch port. See also *Frame Tagging*. Eventually became the 802.1p and 802.1Q standard for Ethernet VLANs.

Interoffice   A reference to a telephone connection, a call, or service that originates in one central office and terminates in another central office within the same area code.

Interstate Long Distance   A long-distance service or call that originates in one state and terminates in another state.

Intranet   An intercompany or other organization-based network that is based on TCP/IP protocols. An intranet is accessible through the Internet, but only by those who have authorization (passwords, decryption keys, etc.). Intranets are less expensive alternatives to private networks, and are a form of *VPN (Virtual Private Network)*.

Intrastate Long Distance   A long-distance service or call that originates and terminates in the same state. A call from San Francisco, CA to San Diego, CA is an intrastate call.

Intra Office   A reference to a connection, a call, or service that originates and terminates in the same central office. If you call your neighbor, you are making an intra-office call.

Inverter   1. A device that converts DC to AC. Inverters are commonly used where data or computer equipment is used in a central office (Fig. I.10). The central office is equipped with –52-VDC power for telephone and transport equipment. The computer equipment, or any other equipment requiring 120 VAC is connected to an inverter, which converts the power. By using an inverter instead of just running a 120-V outlet, the equipment is protected from power outages on the same system as the telephone equipment (Fig. I.11). 2. A logic circuit that reverses a positive (logic 1) state to a 0 state.

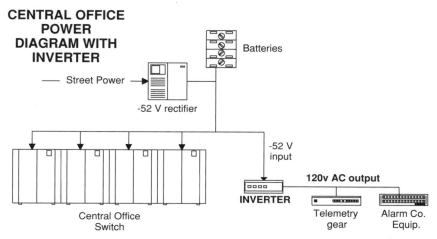

Figure I.10 Inverter Connection Diagram

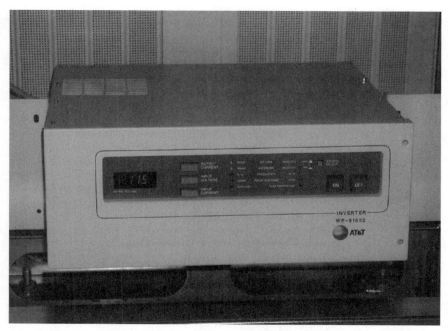

Figure I.11 An Inverter Manufactured by Lucent Technologies (Formerly AT&T)

Inward Trunk   A shared phone line that provides a path for incoming calls to a PBX system. A true inward trunk subscribed from a local phone company cannot be dialed out.

In-WATS (Inward Wide-Area Telephone Service)   An inward toll-free dialing service (800/888 lines) offered by telephone companies. In-WATS lines are priced and set up for incoming-only calls, and usually calls from a certain area. You can also subscribe to Out-WATS service as well. In-WATS can be for interstate and intrastate long distance. If you call an 800 number, you are most likely calling an in-WATS service line that a company has set up for customers. The time to start checking into WATS service is when your long distance to or from a specific area exceeds $200 per month.

IOD (Identification of Outward Dialing)   This is a call-accounting system feature of PBX and some key systems that captures every number dialed by a specific telephone extension and prints it out on a report for accounting and cost-tracking purposes.

Ion   An atom or molecule that has lost or gained one of its valence electrons and is no longer electrically neutral.

Ionosphere   A region of thin air that exists from 60 to 600 miles above the Earth's surface. Radio waves between 2 to 50 MHz are reflected (actually refracted in the same manner that light is refracted through a graded-index fiber optic) back to earth 500 to 3000 miles from where the transmission originated (Fig. I.12). The radiation from the sun causes air to be ionized. The ionosphere is different at night than it is during the day, causing some radio signals to be refracted only during the night, or vice versa. Because of ionospheric refraction, it is not unusual to pick up a Utah AM radio station in northern Mexico.

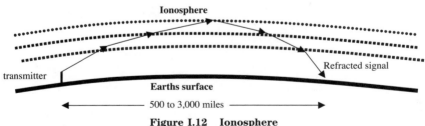

Figure I.12   Ionosphere

IOS (Internet Operating System)   The software program developed by a group of scientists at Stanford University. The team was led by a married couple, Leonard Bosack and Sandy Lerner. They incorporated their software into a Digital Equipment computer so that it could route traffic to and from other computers. This was the first bridge/router. The IOS software invented and developed at Stanford is the operating system on which the company Cisco Systems Inc. was founded.

IP (Internet Protocol)   A network-layer protocol in the TCP/IP protocol stack that offers a connectionless or packetized internetwork service. IP packets are delivered on a best-effort basis. If a packet or datagram cannot be delivered successfully, it is discarded. When this happens, the *ICMP (Internet Message Control Protocol)* section of the protocol stack notifies the sender that a datagram has been discarded. IP provides features for addressing, type-of-service specification, fragmentation, reassembly, and security (Fig. I.13). See also *TCP/IP.*

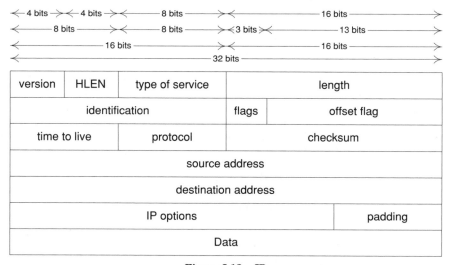

Figure I.13   IP

IP Address   In a TCP/IP message transmission, the datagram (or packet) is routed via the IP address. It consists of two parts: the network identifier and the host identifier.

IP-Based PBX   See *IP Telephony*.

IP Broadcast   A data frame whose subnet address mask is 255.255.255. 255. IP broadcast frames are not passed through routers because routing protocols use the zeros at the end of the subnet mask number to identify the subnet. Because 255.255.255.255 translates to 11111111.11111111. 11111111.11111111 in binary, there are no end zeros, and, therefore, no subnet (as far as the router's operation is concerned). See also *Multicast*.

IP Datagram   The TCP/IP name for a frame. It consists of the IP header and its associated TCP segment. The TCP segment contains the application data being sent and the TCP header. The IP header contains other overhead information, such as source and destination address, time to live, route recording, and more (Fig. I.14). For more details on the IP header, see *IP Header*. See also *IP Address*.

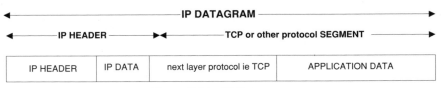

Figure I.14   IP Datagram

IP Header   The overhead in an IP packet or datagram. It carries routing and control information that gets the data it is carrying to where it needs to go. The different bit fields that make up an IP header are explained and illustrated in the following section (Fig. I.15).

- *Version*   The four-bit version field contains the version number of the sender's IP protocol software used to create the datagram.
- *HLEN/Header Length Field*   This four-bit field contains the length of the IP header in 32-bit words. This information is used by the receiving device to know where the header bits end and where the data bits start.
- *Type of Service*   This eight-bit field indicates the class of service desired for the data in its datagram. The three current service types are D/Delay, T/Throughput, and R/Reliability.
- *Total Length*   This 16-bit field indicates the length of the header and data in number of octets. The length of the data in the datagram can be figured by subtracting the value in the HLEN field from the value in the Total Length field.
- *Identification*   The datagram ID number. Datagrams can be broken down into smaller datagrams of the same format, if needed, during transmission. All fragments from a specific datagram take the same ID number with them. This is done to satisfy protocol changes.

- *Flags*   This three-bit field is also used in the fragmentation/defragmentation process.

- *Fragment Offset*   This 13-bit field is used to indicate the sequence of fragments.

- *Time to Live*   The number of router relays allowed before the datagram is discarded.

- *Protocol Number*   Defines the protocol of the data that IP is delivering.

- *Header Checksum*   This 16-bit field is the result of an algorithm calculation performed on the header by the sender. The receiver performs the same calculation and compares results as an error check.

- *Source and Destination Address*   The network ID number and the host number for the source of the data and the destination. Routers use this information to route and deliver the datagrams. There are four classes of IP addresses. See *IP Address Classes* for an explanation and diagram.

- *IP Option*   Used for many types of optional features in the IP protocol. Different types of optional data are security, Source Routing, Route Recording, and Timestamping.

- *Padding*   The padding field is where extra bits added to ensure that the header ends is an even multiple of 32 bits.

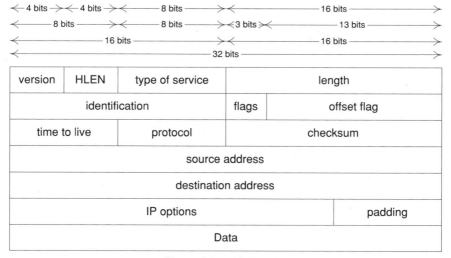

Figure I.15   IP Header

IP Multicast (Internet Protocol Multicast)   A data-transmission technique that allows IP packets to be sent to a multiple destinations. Rather than sending one packet to each destination, one packet is sent to a multicast group that is identified by a single IP destination group address. See also *Multicast Address*.

IP Security Option   A U.S. Government specification that defines an optional field in the IP packet header. It defines hierarchical packet security levels on a per-interface basis.

IP Telephone   An Ethernet (802.3) network appliance. Some IP telephones have switch ports built into them for the connection of a PC or other network device. The voice traffic that is initiated by an IP telephone is most widely sent via RTP and managed by MGCP, H.323, or proprietary protocols such as Cisco Systems' Skinny Gateway protocol. (Fig. I.16)

Figure I.16   IP Telephone

IP Telephony   Feature-rich PBX telephone service over a QoS (Quality of Service) enhanced LAN/WAN packet network (Fig. I.17). At its first inception, even many large telephone equipment manufacturers did not think that IP telephony could evolve as fast as it has—that is, to provide the same quality of voice service that is enjoyed on a circuit switched network, only on a congested private packet network. The drive behind IP telephony is the fact that open standards–based equipment costs less and is less expensive to maintain than the proprietary circuit–based PBXs that the telephone industry initially provided. In describing the PBX architecture in a packet-switched environment, the word *distributed* comes to mind. Rather than having a single PBX chassis in one room, an IP telephony network (or IP-based PBX) is geographically distributed within a building or a campus. This also makes a feature of PBX systems called *geographic redundancy* much easier to implement. Geographic redundancy means that PBX traffic can reroute itself to multiple routes and be controlled by multiple voice servers in multiple locations. Comparing circuit-based, traditional PBXs to their newer IP-based counterparts goes something like the following. What was once a station card or station interface in a PBX is equivalent to an Ethernet switch. These are the devices that provide network connectivity to end network appliances such as IP telephones or computer workstations. Trunk cards or trunk interfaces are replaced with routers or gateways. These devices provide connectivity to the Internet and public telco service. Circuit-based PBX core processors/CPUs (along with their associated software) are counterparted by open standards– based servers and open standards–based software in the IP telephony network. PBX features and integrated services (such as voice mail and IVR) are now easily added as a network application. In an IP telephony environment within a data network, there can be many servers performing support activities for voice services. The telephony server or voice server provides a central place to manage the voice traffic in the data network. Because IP telephony is enabling the less expensive creation of PBX systems, this reduces difficulty of entering the PBX manufacturing industry. Keep in mind that just like any other industry, it opens the doors for the creation of superior products as well as inferior ones. The most important part of an IP-based PBX system is the 802.1Q/802.1p (modern Ethernet) network infrastructure, which extends the formerly confined switching infrastructure of circuit-based PBX systems to the

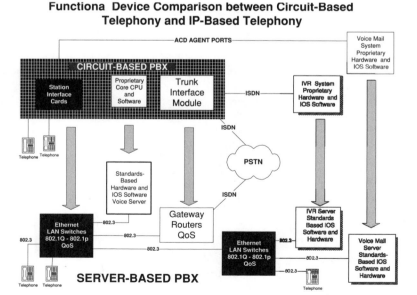

**Functiona Device Comparison between Circuit-Based
Telephony and IP-Based Telephony**

Figure I.17   Traditional PBX to IP Telephony Functional Comparison

desktop. IP telephony is a sophisticated concept. If anyone is having trouble moving into this technology, I recommend that they think of it as a "virtual PBX," where voice traffic is managed at a central location called a *voice server* or *call manager server*. For starters, I also recommend the review and further study of the following terms/topics: Ethernet; IP; STP; Layer 2,3,4 Switch; DHCP; DNS; Link State Routing; MTP; MGCP; H.323; G.711; and RTP, which carries the IP telephony voice packets.

IP Video   There are three categories of video, whether they are in a public broadcast, satellite, cable-TV, or IP or analog radio format. In IP, or packetized video, the same demands for video exist as in legacy technologies. The first category is broadcast video, where there is one sender and many receivers. The broadcast is scheduled, and then broadcast via IP multicast to network users; this can be live or previously recorded. The second category is video on demand, where there is one end user that desires to view a previously stored video file as an IP unicast transmission. In this case, there is a prerecorded video that is streamed from a storage server file. The third category of video is the interactive live video, which would also include videoconferencing, where all users are live. This is the most complex type of video and requires both an IP unicast transmission and, in some cases, an integrated multicast, depending on each individual scenario. See also *H.323, H.320, RTP, Multicast,* and *Unicast.*

IP Voice   (Also called *voice-over IP.*) The transmission of timing-sensitive packetized voice utilizing a stack of enhanced protocols originally designed for data internetworking. I refer to two types of IP voice: voice-over IP and the newer IP telephony. Voice-over IP encompasses the *point-to-point* integration of voice and data over IP networks that may ride on other networks that are frame relay or ATM based. On both sides of the data network, the voice is unpacketized and handed off to a circuit-switched environment. The other type of IP voice is the more sophisticated IP telephony, where the voice is packet switched throughout the entire local network and possibly the wide area network. The voice packets are translated to a circuit-based switching platform only if they go to the public network, or a traditional circuit-based PBX is integrated locally. See also *H.323, MGCP,* and *Skinny Protocol.*

IPE (Intelligent Peripheral Equipment)   One of Nortel Networks' terms that refers to a card that interfaces with PBX phones (Fig. I.18). Other references to equipment are *CM (Core Module,* which is the CPU) and *CE (Common Equipment,* which interfaces the different parts of the PBX switch together).

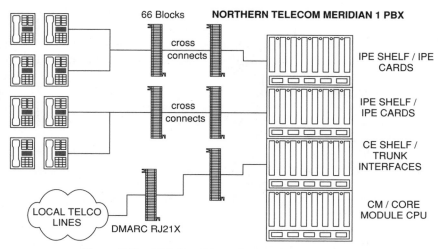

Figure I.18   IPE (Intelligent Peripheral Equipment)

IPSO (IP Security Option)   A U.S. Government specification that defines an optional field in the IP packet header. It defines hierarchical packet security levels on a per-interface basis.

IPT (IP Telephony)   Feature-rich PBX telephone service over a QoS (Quality of Service) enhanced LAN/WAN packet network. At its first inception, even many large telephone equipment manufacturers did not think that IP telephony could evolve as fast as it has—that is, to provide the same quality of voice service that is enjoyed on a circuit switched network, only on a congested private packet network. The drive behind IP telephony is the fact that open standards–based equipment costs less and is less expensive to maintain than the proprietary circuit–based PBXs that the telephone industry initially provided. In describing the PBX architecture in a packet switched environment, the word *distributed* comes to mind. Rather than having a single PBX chassis in one room, an IP telephony network (or IP-based PBX) is geographically distributed within a building or a campus. This also makes a feature of PBX systems called *geographic redundancy* much easier to implement. Geographic redundancy means that PBX traffic can reroute itself to multiple routes and be controlled by multiple voice servers in multiple locations. Comparing circuit-based, traditional PBXs to their newer IP-based counterparts goes something like the following. What was once a station card or station interface in a PBX is equivalent to an Ethernet switch. These are the devices that provide network connectivity to end network appliances such as IP telephones or computer workstations. Trunk cards or trunk interfaces are replaced with routers or gateways. These devices provide connectivity to the Internet and public telco service. Circuit-based PBX core processors/CPUs (along with their associated software) are counterparted by open standards–based servers and open standards–based software in the IP telephony network. PBX features and integrated services (such as voice mail and IVR)are now easily added as a network application. In an IP telephony environment within a data network, there can be many servers performing support activities for voice services. The telephony server or voice server provides a central place to manage the voice traffic in the data network. Because IP telephony is enabling the less expensive creation of PBX systems, this reduces difficulty of entering the PBX manufacturing industry. Keep in mind that just like any other industry, it opens the doors for the creation of superior products as well as inferior ones. The most important part of an IP-based PBX system is the 802.1Q/802.1p (modern Ethernet) network infrastructure, which extends the formerly confined switching infrastructure of circuit-based PBX systems to the desktop. IP telephony is a sophisticated concept. If anyone is having trouble moving into this technology, I recommend that they think of it as a "virtual PBX," where voice traffic is managed at a central location called a *voice server* or *call manager server* (Fig. I.19). For starters, I also recommend the review and further study of the following terms/topics: Ethernet; IP; STP; Layer 2,3,4 Switch; DHCP; DNS; Link State Routing; MTP; MGCP; H.323; G.711; and RTP, which carries the IP telephony voice packets. For additional illustrations, see *IP Telephony*.

Figure I.19   IP Telephony Server

IPTV   The Cisco Systems trademark for on-demand video over IP. See *IP Video*.

IPX (Internetwork Packet Exchange)   A Novell NetWare network layer protocol used to transfer data from servers to workstations (Fig. I.20). IPX has an extended address range that includes the *MAC (Media-Access Control)* address, which eliminates the need for an *ARP (Address-Resolution Protocol)*. IPX is a connectionless protocol. It does not have a mechanism incorporated to acknowledge packets that have

been received or to request the retransmission of lost/discarded packets. Furthermore, IPX has its own addressing scheme, which enables
data to travel through several networks. The address format has 80 bits:
32 bits for the network number and 48 bits for the node/host number.
The node/host number is the MAC address of the device. The network
address range is from 00.00.00.01 to FF.FF.FF.FF. A complete example
of an IPX address is 2a.3b.4c.5d + (48-bit MAC device address).

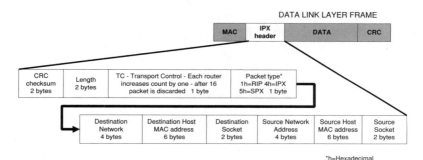

Figure I.20   Novell IPX Header Format

IPXWAN (IPX Wide-Area Network)   A part of the Novell network protocol stack that automatically negotiates end-to-end parameters for new
links. When a new connection to the network is made, the first IPX packets sent across are IPXWAN packets that negotiate the options for the
link. When the IPXWAN options are successfully determined, normal IPX
transmission begins. See also *IPX*.

IR (Infrared)   Light waves that humans cannot see. Their wavelength
(or frequency) is just below that of red light. Heat produces infrared
lightwave radiation.

ISA (Industry-Standard Architecture)   A 16-bit bus standard used in
personal computers in the late 1970s and early 1980s. *ISA* is a term
mostly referred to by PC users when they are looking for a new expansion circuit board, such as an internal modem (Fig. I.21). ISA was outdated by the *EISA (Extended Industry Standard Architecture)* 32-bit
bus. In 1993, Intel and Microsoft developed a new version of the ISA
specification called *Plug and Play ISA*. This newer system enables the
computer's operating system to configure the hardware aspect of the circuit board automatically without a user having to set jumpers or DIP
switches. ISA slots/sockets on a motherboard are identified by their black
color and larger pin size. See also *PCI*.

Figure I.21   ISA 56 K Modem Card

ISDN (Integrated Services Digital Network)   See *Integrated Services Digital Network*.

ISDN Digital Subscriber Line (IDSL)   A physical-layer transport method that provides symmetric download and upload speeds from 64 to 144 Kbps on a single pair of copper wires. The maximum range of IDSL from a central office is 18,000 feet, but this can be doubled with a repeater. IDSL uses 2B1Q line coding, the same kind of line-modulation technique employed in SDSL and ISDN. IDSL is used for transporting ISDN. This allows the xDSL family of technologies to be backwards compatible with ISDN. For a table of the DSL family of carriers, see *xDSL*.

ISDN Terminal Adapter   A device that interfaces an ISDN line to a customer's equipment. The *Terminal Adapter (TE)*, is purchased and usually connected by the customer. Many different terminal adapters enable the use of different features of an ISDN line. TEs come in a wide price range as well. For a diagram showing where an ISDN terminal adapter is connected, see *Integrated Services Digital Network*.

IS-IS (Intermediate System to Intermediate System) An *OSI (Open-System Interconnect)* link-state routing protocol that is based on DECnet Phase-V routing. One version of IS-IS, called *Integrated IS-IS*, supports IP-based networks. For more information on link-state protocols, see *Link-State Algorithm*.

IS-IS Hello (IIH) A routing protocol message that is transmitted between routers and other network equipment that are using an IS-IS algorithm within their operating system. This message is used to update routing tables, network adjacencies, and other network-status information. See also *Intermediate System to Intermediate System*.

ISL (Inter-Switch Link) This proprietary Cisco Systems Ethernet switch protocol evolved into the IEEE 802.1Q standard. ISL and 802.1Q are not interoperable. The reference 802.1Q is better known as the VLAN or tag switching standard. It is a feature on post-1998 LAN switches that makes selected ports behave as if they were attached to the same segment, or hub. Another good name for this feature would be *V-segment* or *virtual segment.* Devices/users that exchange a large amount of information are usually placed within the same VLAN segment. This helps make the operation of the LAN switch more efficient, keeping traffic contained within specified ports. This allows other ports on separate VLANs to carry other nonrelated traffic simultaneously. VLANs are configured by a network engineer, network analyst, or network administrator. When IP telephony is implemented over an Ethernet-switched network, the telephone devices connected to the network are best placed into their own VLAN. Most switches that are 802.1Q compatible can recognize more than 1,000 VLANs. Further, there are two kinds of VLANs: static and dynamic. Static VLANs are associated with switch ports, and dynamic VLANs are associated with the MAC addresses of devices attached to the switch. Dynamic VLANs allow users to move to another office which could have a switch port connection preinstalled. The switch would recognize the MAC address of the device and automatically include its traffic in the same VLAN as the previously connected switch port. See also *Frame Tagging.* Eventually became the 802.1p and 802.1Q standard for Ethernet VLANs.

ISO (International Organization for Standardization) A consortium of worldwide telecommunications experts that has created a wide range of standards, including those relevant to data transport and data networking. ISO developed the *OSI (Open Systems Interconnect) Reference Model,* which is a mainstay for the data networking industry.

ISO 3309 The HDLC procedures developed by ISO. There are two main specifications. ISO 3309:1979 specifies the HDLC frame structure for use in synchronous environments. ISO 3309:1984 specifies proposed modifications to allow the use of HDLC in asynchronous environments. See also *HDLC*.

ISO 9000   A set of international quality-management standards defined by ISO. The standards, which are not specific to any country, industry, or product, allow companies to demonstrate that they have specific processes in place to maintain an efficient quality-tracking system.

ISOC (Internet Society)   An international nonprofit organization that was founded in 1992. Its members coordinate the evolution and use of the Internet. In addition, the ISOC delegates authority to other groups related to the Internet, such as the IAB. The Internet Society world headquarters is located in Reston, Virginia, USA. See also *InterNIC*.

Isochronous   A transmission with no delay, such as a voice conversation between two people. Sending a letter through the mail is non-isochronous.

Isotropic   In antennas for wireless communications, *isotropic* refers to a nondirectional antenna, or better said, an antenna that transmits in all directions evenly. A monopole and dipole are common physical isotropic antenna configurations. See also *dBi*.

ISP (Internet Service Provider)   See *Internet Service Provider.*

ISR (Intermediate Session Routing)   The first routing algorithm used in *APPN (Advanced Peer-to-Peer Networking)*. ISR (where still used) provides node-to-node connection-oriented routing. Network outages cause sessions to fail because ISR cannot provide nondisruptive rerouting around a failure. ISR was replaced by *HPR (High-Performance Routing)*, which has been made obsolete by newer routing algorithms that have been incorporated into link-state, distance-vector, and hybrid routing protocols.

ISSI (Inter-Switching System Interface)   The standard physical interface between SMDS switches.

IT (Information Technology)   The study of improving information and data processing with the use of newer and better devices/machines.

ITI (Interactive Terminal Interface)   Another name for the X.28 interface. A reference to the software functions that make *PADs (Packet Assembler/Disassemblers)* convert asynchronous (stop/start or dumb) terminal data into *STDM (Statistical Time-Division Multiplexing)* packets used in X.25.

ITU-T (International Telecommunication Union)   A worldwide standards organization through which public and private organizations develop telecommunications standards for hardware and software. The ITU was founded in 1865 and became a United Nations agency in 1947. It is responsible for adopting international treaties, regulations, and standards governing telecommunications. The standardization functions were formerly performed by a group within the ITU called CCITT. After a 1992 reorganization, the CCITT no longer exists as a separate body.

ITU-T T.38   (Also called T.38 phase 2.) A Cisco Systems–developed fax gateway protocol that is now an ITU-T standard. A fax gateway is a device that allows a single telephone line to receive both faxes and voice calls. The fax gateway answers all incoming calls. It listens for a fax handshake tone and if it does not hear one, it rings the telephone connected to it. T.38 is a feature that can be enabled on individual router FXS ports.

IVR (Interactive Voice Response)   See *Interactive Voice Response.*

IW (Inside Wire)   The telephone wire that is on the customer side of the telephone network interface. IW includes jacks and any wiring in or attached to the outside of the house, as long as it is electrically on the customer's side of the network interface.

IXC (Inter Exchange Carrier, IEC)   A long-distance company, like AT&T, Sprint, Worldcom, and MCI.

J

J Box (Junction Box)   A metal or plastic box used as an access for cable or wire (coax, fiber, UTP, STP). When communications companies build their networks into buildings, the building management usually requires a J-box close to each entry of the building. The J-box allows other companies that want to gain access use of the same conduit.

J Hook   A spike with a hook on the end, specially designed to be pounded into wooden telephone/power poles. The installed J hook is a means to hang telephone or other aerial service wire. To hang a service wire on a J hook, a drop-clamp or wire-vise is placed on the service wire. The drop-clamp or wire-vise is designed with a loop that fits over the J hook (Fig. J.1).

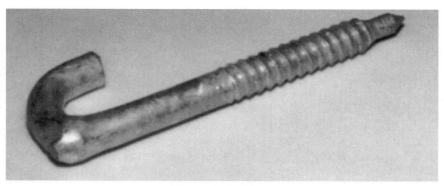

Figure J.1   J Hook

Jabber   A condition where a failed or failing device attached to or on a network continually transmits a random, meaningless data signal onto the network.

Jack   Also called a *connecting block*. A device that has a form of connectivity (a plug) used to terminate a wire/coax/fiber run. The jack provides access to the terminated wire/coax/fiber run via the connector(s)-plug(s) on its face. Some of the many types of jacks include flush jacks, which fit into a wall like an electrical outlet; baseboard (or biscuit) jacks, which are shaped like a box and mount to a baseboard near the floor; and duplex jacks and fourplex jacks, which have more than one plug on the face of them, with different types of connectors in each plug. For more information on the types of plugs available on jacks, see *R* (Fig. J.2) *J11, RJ45, RJ21X, BNC, RCA plug, ST,* and *SC*.

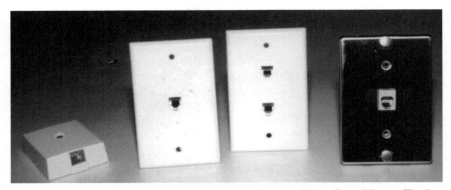

Figure J.2   Assorted Telephone Jacks. From Left to Right: Base Mount, Flush Mount, Duplex Flush Mount, and Wall Mount

Jacket   The outer covering of a multiple-wire or fiber-optic cable. The most common jacketing materials are PVC (Polyvinyl Chloride), Plenum (polyvinylidene diflouride), and ALPETH (aluminum/polyethylene). Older cable jacket types include lead and cloth. For a photo of ALPETH and lead cable, see *ALPETH*.

Jamming   A reference to the intentional interference of a receiver's ability to receive, detect, or demodulate a radio signal of any frequency. A signal of identical (or nearly identical) carrier frequency to the one being jammed is transmitted with a meaningless signal modulated on it. The receiver picks up both signals, which causes a hard-to-understand blur of noise. Directional transmissions and directional antennas are less susceptible to signal jamming.

Java   A programming language designed to write programs that users can safely download from the Internet to their computers without fear of viruses. Using small Java programs called *applets,* Internet Web pages can include animation, sound, simple games, and other interactive functions. Web browsers, such as Netscape Navigator and Microsoft Internet Explorer, have incorporated Java interpreters so that they are able to run JAVA applications downloaded from Internet Web pages. Java is an innovation of Sun Microsystems.

JDBC (Java Database Connectivity)   A Java *API (Application Programming Interface)* that enables Java programs to execute *SQL (Structured Query Language)* statements. This allows Java programs to interact with any SQL-compliant data base. Because nearly all *Relational Database Management Systems (DBMSs)* support SQL and because Java itself runs on most platforms, JDBC makes it possible to write a single database application that can run on different platforms and interact with different DBMSs. JDBC is similar to *ODBC (Open Data Base Connectivity),* but is designed specifically for Java programs, whereas ODBC is language independent. JDBC was developed by JavaSoft, a subsidiary of Sun Microsystems. See also *Open Database Connectivity.*

Jitter   An analog or digital communications line distortion caused by the carrier signal varying from its reference timing positions. Jitter can cause data loss, particularly at high speeds.

Johnny Ball   A device used to link two steel-strand loops together (Fig. J.3). Also called a *Knuckle-Buster.*

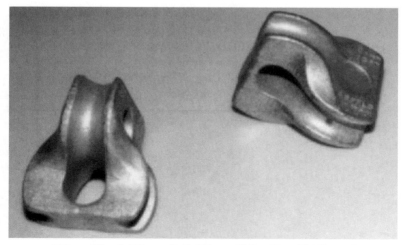

Figure J.3   A Johnny Ball (Also called a *Knuckle Buster*)

Joule (J)   A unit of energy. If you combine power and time, you get energy. One Joule is equal to one watt for one second. Running a 100-W light bulb for one hour expends 360,000 Joules of energy. Running a 5-HP (five horse-power) lawn mower at full strength through tall grass for one hour expends 13.428 MJ = 13,428,000 Joules (746 watts = 1 horsepower). What is useful about Joules? Joules are a way to compare different forms of energy. After calculating the cost per Joule of gasoline energy and electrical energy, it can be determined that 1 Joule of electrical energy costs nearly twice as much as 1 Joule of gasoline energy. (Based on gasoline at $1.50/gal, the lawnmower consuming $\frac{1}{5}$ of a gallon in one hour, and the cost of one kW/hour is 15 cents.)

Julian Dating   A date-expression format that does not use months of the year, only days. For instance, January 1st is simply 1, February 1st is simply 32, and December 31st is 365.

Jumper   1. Another name for a cross connect. A cross connect is the connection of one circuit path to another via a physical wire. Telephone cable pairs are terminated or "punched down" onto a termination block (usually a 66M150 or an AT&T 110 block) that has extra connections available for each pair so that jumper wires can be easily connected and rearranged between them.   2. A section of coax used to connect a transmitter or transmission line.

Jumper Wire   See *Jumper.*

Junction Box   See *J Box.*

K Band The band of frequencies designated by the IEEE between 18 GHz amd 27 GHz (1.67 cm to 1.11 cm). For a table, see *IEEE Radar Band Designation.*

K-Style Handset The type of handset typically found on PBX and key telephones. The ear piece and receiver are square shaped.

Ka Band The band of frequencies designated by the IEEE between 27 GHz and 40 GHz (1.11 cm to 7.5 mm). For a table, see *IEEE Radar Band Designation.*

Karn's Algorithm An algorithm implemented into routing protocols/ operating systems that improves round-trip time estimations by helping transport-layer protocols distinguish between good and bad round-trip time samples.

KB (Kilobyte) A measure of computer memory. See also *Kilobit.*

Kbit (Kilobit, Kb) 1000 bits. Not to be confused with KB, which is kilobyte. The speed of data transmission is usually measured in Kb/s and memory is measured in Kb.

Kbps (Kilobits per Second) A reference to how fast data is being transferred on a communications path.

Keep-Alive Interval   The period of time between each keep-alive message sent by a network device. Keep-alive intervals range from fractions of a second to several seconds, depending on the network device and the protocol used.

Keep-Alive Message   When no user data is being transmitted over a link, a packet sent between network devices (such as routers) at periodic time intervals that verifies the virtual and physical connection between them is still active.

Kerberos   In the TCP/IP protocol suite, a name given to a security system that authenticates users of a network and grants access authorization.

Kermit   A file-transfer protocol that works over phone lines and is known for its high accuracy and slower speed.

Kernel   The heart of an operating-system program that manages a computer's hardware and program resources. Kernels are comprised of short machine-level instruction sets.

Kevlar   A fine, stranded yellow fiber used to make bulletproof vests and built into fiber-optic cable to reinforce it.

Key   A button on a telephone that executes a feature or accesses a line on a telephone.

Key Pad   A dial pad.

Key Service Unit (KSU)   The main part of a key telephone system. The KSU contains the electronics that control which line is directed to which phone. The KSU is usually mounted in a closet or near the telephone companies demarcation point (where the phone lines come into the building).

Keyset   An electronic telephone that works only when connected to a proprietary KSU via the correct wiring. Keysets do not work if you connect them directly into a normal telephone line. If you like the idea of having Keyset features on a normal public telephone line, see *ISDN*.

Key System   See *Key Telephone System.*

Key Telephone System   The less-expensive and less-flexible alternative to a PBX system (Fig. K.1). On a key system, each telephone line appears under a key (button) on the phone. To access an outside line or answer an incoming call, you press the key associated with that line. Key systems are digital and have six major parts, the *KSU (Key Service Unit)*, line interface, station interface, power supply, connectivity, and the key sets (telephone sets). Most key systems are very user friendly, and can be installed by a user that knows little about telephony. The KSU is the cabinet that contains the electronics that controls the switching between key sets and phone lines. Some key systems have detachable line and station interfaces that plug into the KSU. The system usually comes with a specific number of incoming line interfaces (6) and a specific number of station interfaces (16). This size system is commonly referred to as a 4-16 (four/sixteen). The power supply is often a "power-adapter" type. Some manufacturers offer a UPS back-up option specially designed for the key system. The key-set telephones are sold individually, in a variety of choices that include 10-key (10 button), 20-key (20 button), display, and hands-free.

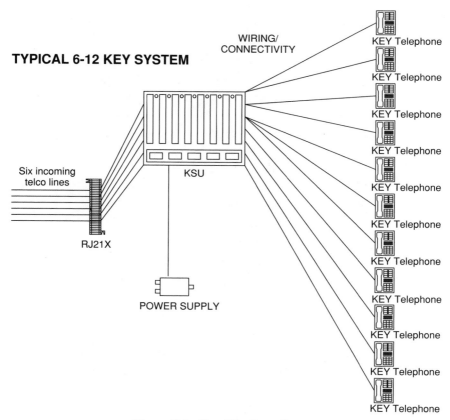

Figure K.1   Key Telephone System

Keyed RJ45　An RJ45 jack or plug that has a small protrusion on its side that acts as a key to prevent the wrong kind of equipment being plugged into the wrong jack. Some modems come with keyed RJ45 plugs to help prevent them from being plugged into an RJ45 that contains voltages intended for other equipment, which could cause damage.

KHz (Kilohertz)　One kHz is equal to 1000 Hertz, or 1000 Hz. Kilo is just an easier way of saying "one thousand," and k is a shortcut to having to write three zeros.

Knuckle-Buster　A device used to connect two steel strand loops together. Also called a *Johnny Ball*. For a photo, see *Johnny Ball*.

KSU (Key Service Unit)　See *Key Service Unit* (Fig. K.2).

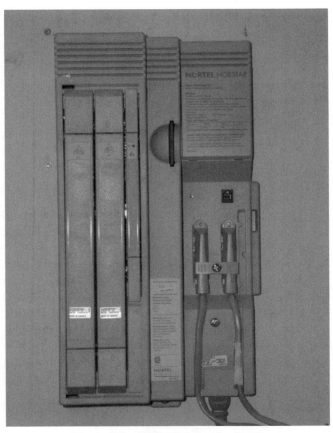

Figure K.2　KSU (Key Service Unit)

KTS (Key Telephone System)   See *Key Telephone System.*

Ku Band   The band of frequencies designated by the IEEE between 12 GHz and 18 GHz (2.5 cm to 1.67 cm). For a table, see *IEEE Radar Band Designation.*

KWH (Kilowatt Hour)   A unit of energy equal to 1000 watts of power for one hour. This is equivalent to operating ten 100-watt light bulbs for one hour. The average cost of a KWH in the USA is about 14 cents. For more information about energy, see *Joule.*

L

L Band   The band of frequencies designated by the IEEE between 1 GHz and 2 GHz (30 cm to 15 cm). For a table, see *IEEE Radar Band Designation*.

L Carrier   A long-haul TDM carrier that is transmitted over two coax cables. It is still in use in sub-oceanic applications. L carrier has a capacity of 13,300 voice channels, 3 kHz in bandwidth (each).

L1   A reference to the level-1 cache memory in personal computers. Level-1 cache is a memory bank that resides within a computer's CPU chip. It is used to store frequently executed machine-level instructions.

L2   A reference to the level-2 cache memory in personal computers. L2 cache memory can be incorporated into the design of the CPU or it can be mounted externally. Cache memory is used to store frequently executed machine-level instructions.

Label swapping   An older routing algorithm used by IBM *APPN (Advanced Peer-to-Peer Networking)* systems in which each router that a message passes through on its way to its destination independently determines the best path to the next router. Routing protocols have evolved to be much "smarter" and provide many more services, such as knowing the least-expensive path and using it first. See also *Link-State Algorithm*.

Ladder Diagram   A representation tool used to analyze communications signaling events between a DCE and DTE (Fig. L.1).

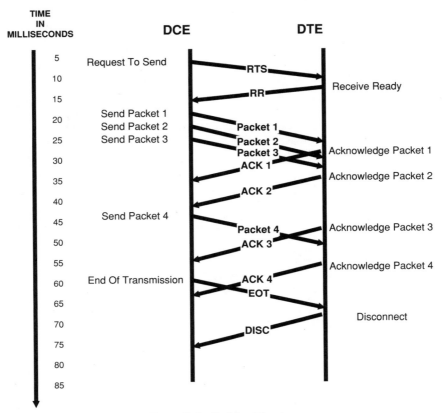

Figure L.1   Ladder Diagram

LADT (Local-Area Data Transport)   A reference to digital carrier (any carrier, T1, 56 K, etc.) in the local-carrier twisted-copper plant network.

Lag Current   A reference to the current flow in a capacitive electronic circuit carrying a signal (or AC power), the voltage builds up before the current actually flows. This is caused by *capacitive reactance.* In a purely capacitive electronic circuit, the current lags behind the voltage by 180 degrees.

Lambda (λ)   The symbol that is standard for representing the wavelength of a frequency. Wavelength is equal to: $\lambda = (300,000,000 \text{ m/s})/frequency$ (Hz). 300,000,000 m/s is the speed of light in a vacuum.

LAN (Local-Area Network)   A group of computers connected together within a building or campus. LANs are the most detailed of computer networks because they deal with the applications and operating systems of computers. The distinguishing thing among LANs is the way that the computers are connected and the protocol at which they communicate over the media that connects them. The two major LAN protocols (logical topologies) are Ethernet (star and bus) and Token Ring. Both Ethernet and Token Ring have evolved into Switched Ethernet and Switched Token Ring, which are different in operation and speed than their predecessors. Larger LANs use an operating system to control the LAN environment, such as Novell or Windows NT. Another component of LANs is the server, which can either run programs or store data for computers connected to the network. Servers are simply other computers (usually with more processing power and memory) that are configured by an operating system, such as Novell or Windows NT, to perform a specific function. For more information, see *Client Server, Token Ring, Ethernet, Switched Token Ring, Switched Ethernet, FDDI,* and *DQDB.*

LAN Adapter   Also known as a *NIC (Network Interface Card).* Typical LAN adapters are made by Bay Networks, SMC, US Robotics, and others (Fig. L.2). They are a circuit card that plugs into an expansion slot located on the mother board of a PC. They have an RJ45 plug and

Figure L.2   LAN Adapter Card (Also Called a *NIC*)

usually a BNC for connecting to the twisted-pair or coax network wiring in an office area (or wherever the network is).

LAN Switch A hardware device that provides dedicated bandwidth to segments of the network that are attached to it. The primary purpose of a LAN switch is to provide additional bandwidth to users by reducing the number of users per segment. In doing this, collisions are eliminated and the 50% to 60% efficiency of Ethernet is increased to near 100% (collisions are still possible between the switch and the attached device). A LAN switch can be compared to a bridge with many ports (at a much lower cost per port) and extensive traffic manageability. Unlike a non-POTS voice switch or PBX switch, multiple hosts (computers, IP telephone extensions) can be connected to each port. This is because a LAN switch forwards frames based on the MAC address of the connected devices rather than the switch port, as in voice switch operation. Because basic LAN switches are OSI layer 2 devices (meaning they forward packets based on a hardware address), they are capable of bridging unlike networks, such as Ethernet to FDDI and token ring. Larger LAN switches are capable of performing layer 3 *routing,* where the packet forwarding function happens based on a leading packet's IP address. LAN switches often replace Ethernet hubs (no wire replacement is needed, but fiber optic may be installed for backbone connectivity) when they are implemented into a network. LAN switches are the core enabler of IP telephony. See also *Bridge* and *IP Telephony.*

Land Line A regular POTS switched telephone line over twisted copper or other land-based facility. Land lines are noncellular or radio.

LANE (Local-Area Network Emulation) An added feature of ATM networks that allows *SVCs (Switched Virtual Circuits)* within them to be configured to function as a LAN backbone. In its raw form, ATM does not provide data-link layer and network-layer protocol support. LANE fills in the gap by providing multicast and broadcast support, address mapping (MAC layer to ATM), SVC management, and a usable packet format. The LANE ATM specification also defines Ethernet and token-ring-emulated LANs. See also *ELAN.*

LAP (Link-Access Procedure) The frame layer (*data link,* by OSI terms) protocol that was derived from *HDLC (High-Level Data-Link Control)* for X.25. LAP evolved into LAPB *(Link-Access Procedure Balanced mode)* and *LAPB Extended (LAPBE).* LAPB Extended allows for a 128-packet transmission window, rather than an eight-packet transmission window.

LAPB (Link-Access Procedure Balanced)   LAPB is the set of rules for the packet layer (equivalent to the network layer in the OSI) of the X.25 protocol to perform such functions as Receive Ready RR (the equivalent of Acknowledge ACK) and Reject (the equivalent of Retransmit). It defines the bit-by-bit structure of the frames that makes this layer serve the packet layer (network layer, if you like OSI terminology). LAPB allows for a eight-packet transmission window. LAPB extended allows for larger transmission window sizes, referred to as *modulo levels*. For example, LAPB modulo 128 would be an X.25 connection that operates at a 128-packet transmission window. For an example of a data packet, see *X.25 Data Packet.*

LAPB Extended (Link-Access Procedure Balanced Mode Extended)
A version of the X.25 incorporated LAPB protocol that allows larger window sizes and variable address lengths. The original LAPB allowed for a seven-packet transmission window that suited most terrestrial applications. When telephone company facilities became less error prone and faster, it made sense to modify the LAPB part of the X.25 data-link layer protocol to have less overhead. Another push for the change came from satellite communications, which have a delay that exceeds the eight-packet "in-route" window time. LAPB Extended provides the ability to have more packets in transit between the sender and receiver. This prevents the two ends from waiting for acknowledgements during transmission delays.

LAPD (Link-Access Procedure on the D channel)   ISDN data-link layer protocol for the D channel. LAPD was derived from the LAPB protocol and is designed primarily to satisfy the signaling requirements of ISDN basic access. It is defined by ITU-T recommendations Q.920 and Q.921.

LAPM (Link-Access Procedure for Modems)   The ARQ used by modems implementing the V.42 protocol for error correction. See also *ARQ* and *V.42.*

LASER (Light Amplification by Stimulated Emission of Radiation)   A laser is a device that emits only coherent light that is in phase. Coherent light is of one frequency (one pure color, not always visible to the human eye). Lasers create light similar to the way a fluorescent light does, by exciting a gas that emits photons (light particles) with electricity. When the particles are exited, they are trapped between two mirrors. One mirror has a small spot that allows light travelling in a very straight line to escape. This light is the laser beam.

LASS (Local-Area Signal Service)   A signaling feature that incorporates the # and * keys, which allow central-office switches to provide features to residential phone-line subscribers that could only be found on PBX systems before. The services (at an additional cost) include: Automatic Call Back (also called *Last Call Return*), which is a service that by dialing * and two digits the customer can hear a message that tells them what the last number was that tried to call their phone. It also gives them the option of dialing back the number or returning the call automatically. Automatic recall or last number redial allows the caller to redial the previously dialed number by entering a code. Nuisance-call trace or last-call trace, allows the user to trace the last call made to them and automatically file a report against the caller. Each trace usually costs about $2. This is done so that people only use it when needed, not when they want to play a joke on a friend. Caller ID or incoming call ID, allows a telephone-service subscriber to connect one caller-ID unit to their phone line to view the calling part's telephone number (and often the name) before answering the phone. Many other services offered by different names vary, depending on the telephone company offering them. Different names are required for each service by different companies— even though the services are identical. The FCC implemented this rule so that the RBOCs couldn't trademark a well-known name for each feature, thus allowing the smaller phone companies to be equally as competitive by having to use a different name for each service.

Last In First Out   See *LIFO*.

Last-Number Redial   A feature incorporated into PBX systems, Key systems, and also included on home use single line type telephones.

LATA (Local-Access Transport Area)   Simply stated, a LATA is an area code. *Inter LATA* refers to services that go from one area code to another, like long-distance telephone calls. *Intra LATA* refers to services that originate and terminate in the same area code.

Latency   1. In Ethernet, the delay between the time that a device requests access to a network and the time that it is granted permission and transmits. 2. In network routing or switching, the amount of time a device possesses a frame. More specifically, the delay between the time that a routing or switching device receives a frame and the time that frame is forwarded out the destination port. 3. A general reference to transmission delay or slow response time in a data network because of large amounts of traffic.

Layer 2 Switch   A bridge, router, or LAN switch that decides where to forward packets based on the MAC address, which is the OSI (Open Systems

Interconnect) model layer 2 address (Fig. L.3). This is also the address that is permanently burned into every NIC card and network device manufactured. Layer 2 switches were known as *bridges* and *LAN switches*. Now that bridges and LAN switches can forward packets based on information from layer 2 (MAC address), *Layer 3* (i.e., IP address), and *Layer 4* (i.e., TCP layer), it is necessary to distinguish which type of forwarding the network traffic device is configured for. This is said by calling the device a layer 2, layer 3, or multilayer switch. See also *Layer 3 Switch, Layer 4 Switch, Store and Forward Switching,* and *VLAN.*

Figure L.3   Layer 2 Ethernet Switch

Layer 3 Multicast   IP address class D addressing. See *PIM Dense Mode* and *PIM Sparse Mode.*

Layer 3 Switch   A layer 3 switch performs functions as well as those that were traditionally accomplished exclusively by a router (Fig. L.4). The major difference between a layer 3 switch and a router is that a router performs its core function in software and a switch performs its primary forwarding/routing function in silicon logic devices (hardware). Performing this function in electronics rather than software drastically increases speed. Layer 3 switches forward frames and packets based on MAC addresses (layer 2 address) *and* IP or IPX addresses (layer 3 addresses) when necessary. Layer 3 switches are capable of performing line format conversion, which is inherent from their layer 2 header stripping and reassembly. They have the ability to convert protocols, regulate traffic, and provide security services within the LAN. The true operation of a layer 3 LAN switch is as follows: a switch port receives the first frame of a transmission. If the frame is not addressed to the same VLAN within the switch, it is forwarded to an on-board router. The router forwards the packet to the appropriate LAN switch/VLAN. The LAN switch that sent the frame in the first place remembers to what port the router sent that first packet, then sends the rest of the frames to that port. See also *VLAN, Store and Forward Switching,* and *Multi-Layer Switching.*

Figure L.4   Layer 3 Ethernet Core/Distribution Switch (Left)
Multi-Layer Switch Symbol (Ethernet) (Right)

Layer 7 Switching   A reference to server load sharing and content networking. The idea behind layer 7 switching is that two objectives are accomplished. The first is that the network is given some type of intelligence that enables it to "predict" traffic. This is done by a content network server placed on a contiguous network. The server is configured by a user to "push" or transfer files from frequently used application servers to a cache server that is closer to the end user, or better said, the "edge" of the network. The second is to incorporate load sharing ability among multiple servers. This is so that if an application does receive a massive burst of traffic, the network can compensate by delivering packets evenly across many servers and provide the option to have the end user interact with the same server during the entire transaction. This eliminates the need for communicating applications to reauthenticate to different servers. Content that users "pull" from across the WAN link is also stored for others to access. See also *Content Networking.*

Layered Network Architecture   1. Communications is accomplished in steps. Each layer in a network architecture is a step toward the goal of moving data or voice from one place to another. The *OSI (Open Sysems Interconnect model)* is a layered architecture.   2. (Also called a *hierarchical network architecture.*) An enterprise LAN/WAN design methodology where all traffic leads to a central core, where servers and other computing services are located. In hierarchical design, there are three parts to a network: the core layer, the distribution layer, and the access layer. This design method is used most recently in comparison to the 80/20 method used in the past. The availability of super high-speed

LAN switches with backplanes that are capable of switching at a total bandwidth in excess of 30 Gbps have made hierarchical designs more practical than the distributed 80/20 computing method. The largest advantage of hierarchical designs is that they are highly scalable and expandable, as well as flexible for implementing new services (Fig. L.5). The older 80/20 designed networks eventually grow to a point where no new services can be added without creating bottlenecks that are very expensive to remedy. See also *Core Layer, Distribution Layer,* and *Collapsed Core.*

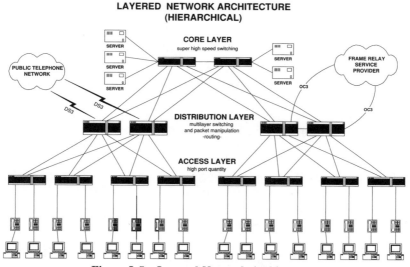

Figure L.5   Layered Network Architecture

LCD (Liquid Crystal Display)   LCDs are electronic display devices that operate by polarizing light so that a nonactivated segment appears invisible against a background. An activated segment does not reflect light (absorbs it) and therefore appears darker than the background. TTL is not generally used to drive LCD displays because it does not completely "deactivate the segment." A few tenths of a volt are present— even when a TTL logic device is completely in the "off" state. CMOS is the best type of device for driving LCD displays because of its ability to turn completely off and have no remaining bias voltage. LCD displays consume very little power, in contrast to other display methods (CRT, LED), but are sensitive to heat and need external light to be viewed in the dark. LCD displays consume little space compared to CRTs and are used in laptop computers. See also *Active Matrix Display.*

LCI (Logical Channel Identifier)   Another name for *logical channel number.* See *Logical Channel Number.*

LCN (Logical Channel Number)   In X.25, an identifier of a virtual circuit that will be used for all transmissions within the duration of a call. X.25 can support up to 4095 LCNs total in both directions for transmitting and receiving. LCNs exist for the same purpose in X.25 as a *DLCI (Digital-Link Connection Identifier)* exists in frame relay. It is a temporary virtual path used between network routing devices. The LCN is not a permanent virtual circuit, such as the type that a multiplexer would provide. The LCN represents one of the temporary users of a permanent virtual circuit. See also *X.25 Control-Packet Header Structure.* 2. The channel number address that rides along with a packet of data in statistical time-division multiplexing.

LCR (Least-Cost Routing)   A feature of PBX systems that enable them to be programmed to associate a dialed area code with a specific trunk (Fig. L.6). That trunk will be the least-cost route for that area code. Some PBX systems are capable of having a rate-table database with a call-accounting system. The rate-table database provides the PBX with cost per minute information for dialing certain area codes. If a user has an agreement with MCI (e.g., WATS service to Santa Clara, CA area code 408), then the PBX is programmed to connect anyone that dials "9-1-408-xxx-xxxx" to the MCI WATS trunk. The WATS TRUNK is cross connected to trunk interface/port number X. (X could be any number that the PBX uses in its trunk-numbering scheme.)

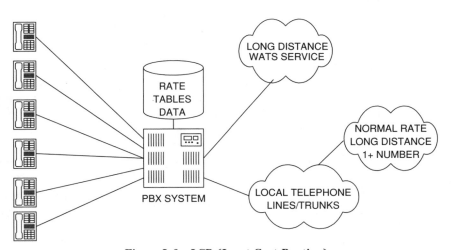

Figure L.6   LCR (Least-Cost Routing)

LD (Long Distance)   An abbreviation commonly used in PBX and key-system operator manuals.

LDAP (Lightweight Directory Access Protocol)   A centralized data-base directory that can be used by many devices, such as an IP tele-phone system, e-mail directory, IVR system, and ACD system. Since LDAP's development at the University of Michigan in 1997, it has be-come a de facto standard due to its wide use as a means for accessing X.500 (and more recently, non-X.500) directory systems.

LDM (Limited-Distance Modem)   Also called a *short-haul modem* or "line driver." Short-haul modems are commonly used to extend the dis-tance of a printer or other *DTE (Data-Termination Equipment)* de-vice from its host. One example is to extend the printer dedicated to printing call-accounting records from a PBX to an accountant's office. For a diagram, see *Limited-Distance Modem.*

Lead Cable   Before plastic (polyethylene) was invented, telephone cable was insulated with paper and jacketed in lead (Fig. L.7). The RBOCs still have some of this cable in use, and it does have one advantage over ALPETH (aluminum polyethylene) jackets. It is very heavy, so it is

Figure L.7   Lead Cable: 100 Pair with Pulp (Paper) Insulation

nonbuoyant in under-water applications, even when pressurized (see *Air-Pressure Cable*). Lead cable is currently being removed when at all possible because of the poisonous effects of lead on the environment (the lead cable shown has pulp-insulated pairs).

Lead Current   A reference to the current flow in an inductive circuit. Believe it or not, in an inductive circuit carrying a signal (or AC power), the current flows before the voltage is actually built up as a result of "inductive reactance." Current lead is measured in degrees or radians. In a purely inductive circuit, such as an AC motor, current leads voltage by 180 degrees.

Leaf Internetwork   The connection of separate LANs into a star topology with a core router connecting them.

Leased Circuit   Also called a *leased line* or *private line*. A leased line is a telephone service that is permanently connected from one point to another. Leased circuits include 56K analog and DS1. A leased circuit acts like a pipeline that carries data from one point to another. If you put a bit in one side, the same bit pops out on the other side. It can carry data across town, across the country, or around the world. Leased lines are relatively expensive. Because leased lines have been offered, new services, such as frame relay have evolved. Frame relay is a cost-effective solution for long-haul/long-distance data-transfer applications (Fig. L.8).

56K Analog Leased-Line/Private-Line Application

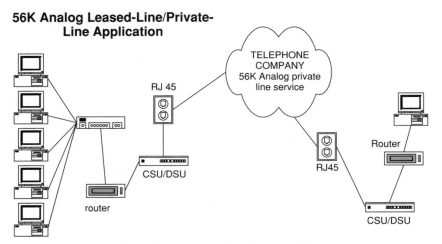

Figure L.8   Leased Circuit/Leased Line

Leased Line   See *Leased Circuit*.

Leased-Line Modem   A modem that uses a private line to interconnect. They have two or four wires, depending on the standard (ITU).

Leased-Space Agreement (LSA)   Also called a *colocation* or *co-lo*. A leased-spaced agreement is an arrangement that communications companies or communications services vendors (such as alarm companies or voice-mail service providers) make with telecommunications companies to use an area of a central office or node to place their network-interface equipment (Fig. L.9). See also *Colocation*.

Figure L.9   Leased Space for an Alarm Service Provider's Equipment

Least-Cost Routing (LCR)   See *LCR* (Fig. L.10).

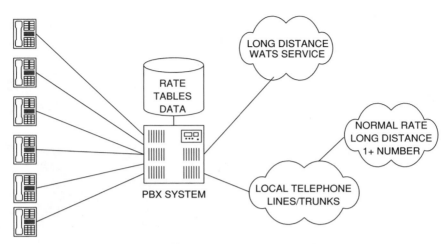

Figure L.10   Least-Cost Routing (LCR)

LEC   1. *LAN-Emulation Client.* On an ATM network, a LAN emulation client is an entity or software program that emulates/translates Ethernet and/or token ring and other protocols for the ATM network. The LEC performs data forwarding, address resolution, and other control functions for a single-end device (such as a personal computer) within a single *ELAN (Emulated LAN).* Each LEC is identified by a unique ATM address, and is associated with one or more MAC addresses reachable through that ATM address. See also *ELAN* and *LECS.*   2. *Local Exchange Carrier.* Most commonly a reference to one of the seven RBOCs. A telephone company that provides telecommunications services to end users and bills them for it. A local exchange carrier has switching networks and outside plant or cellular service to serve its customers. Examples of local-exchange carriers are: NYNEX, USWest, GTE, PAC BELL, Southern Bell Telephone, and a myriad of cellular-telephone service providers. The newest breed of local exchange carriers are CLECs (Competitive Local Exchange Carriers). They have all the same services as the RBOCs, except that they are generally only located in large metropolitan areas. Some larger CLECs include ELI (Electric Lightwave Inc.), Worldcom, AT&T local, and MFS (Metropolitan Fiber Systems). CLECs use (almost exclusively) SONET in conjunction with DCS as the foundation of their network architecture.

LECS (Local-Area Network Emulation Configuration Server)   In an *ATM (Asynchronous Transmission Mode)* network. A device or software program that assigns individual *LANE (LAN Emulation)* clients to particular networks by directing them to the corresponding server. There is, logically, one LECS per administrative domain. See also *LANE*.

LED (Light-Emitting Diode)   See *Light-Emitting Diode*.

Leg Iron   Also called *spurs* and *climbers*. A device that network technicians wear on their legs to climb wooden telephone poles. Leg irons are a hook-shaped stirrup/shank, with an iron bar that straps to the technician's inner shin. Each shank is equipped with a spur that points out from the inner ankle.

LEOS (Low Earth-Orbit Satellite)   A nonstationary satellite that orbits the earth at an altitude range of 300 to 500 miles. LEOS are used for many communications services, including paging, mobile radio communications, and data uplink.

Level-1 Cache   A memory bank that resides within a computer's CPU chip. It is used to store frequently executed machine-level instructions.

Level-2 Cache   A memory bank that exists in addition to Level-1 cache memory. It might be incorporated into the design of the CPU or it could be mounted externally. Cache memory is used to store frequently executed machine-level instructions.

LF   The ASCII control code abbreviation for line feed. The binary code is 1010000 and the hex is A0.

LFI (Link Fragmentation and Interleave)   In IP telephony environments, a QoS method of managing voice packets over links that are slower than 768 Kbps. Because these links are considered slow in relation to the 100 Mbps LAN Ethernet speeds, they require additional QoS attention to enhance their efficiency.

LIFO (Last In First Out)   A method of clocking memory bits into and out of a memory bank. It means that the last bit in is the first one out. You can imagine this concept by stacking books into a box for storage. The last book you put in the box is the first one out when you remove them.

Light-Emitting Diode (LED)   A diode that emits light when it is forward biased (Fig. L.11). If you have a PC, your hard-drive indicator and power indicator on the front are both LEDs. LEDs emit coherent light, which is light that is very close to one frequency (one color).

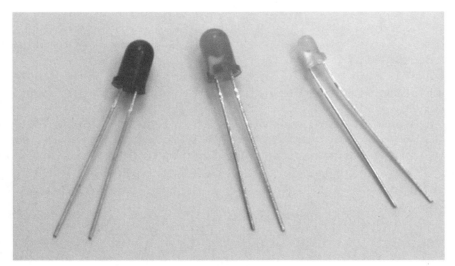

Figure L.11   Light-Emitting Diodes (LEDs)

Lightning Protector   A device used in telephone-company network interfaces that provides an easier path for lightning to travel to ground, compared to a telephone user, or inside wiring (Fig. L.12). Before lightning protectors, houses burned down because of lightning striking the telephone lines. The two types of lightning protectors are carbon and

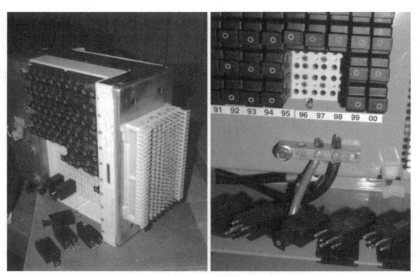

Figure L.12   Lightning Protector (100 Pair)

gas. The carbon protectors are simply a piece of carbon that connects tip and ring to ground. The gas protectors are the same, except that a gas is inserted instead of solid carbon. The good thing about gas-lightning protectors is that after they are hit by lighting, they do not need to be replaced and are reusable.

Lightweight Directory Access Protocol (LDAP)   A centralized data-base directory that can be used by many devices, such as an IP tele-phone system, e-mail directory, IVR system, and ACD system. Since LDAP's development at the University of Michigan in 1997, it has be-come a de facto standard due to its wide use as a means for accessing X.500 (and more recently, non-X.500) directory systems.

Limited-Distance Modem (LDM)   Also called a "short-haul modem" or "line driver." Short-haul modems are commonly used to extend the dis-tance of a printer or other *DTE (Data-Termination Equipment)* device from its host (Fig. L.13). One example is to extend the printer dedicated to printing-call accounting records from a PBX to an accountant's office.

SHORT-HAUL MODEM APPLICATION

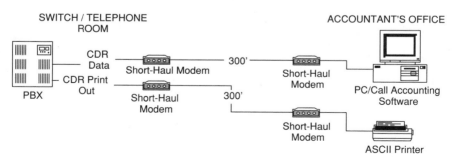

Figure L.13   Limited-Distance Modem (LDM), Also Called *Short-Haul Modem*

Limited-Resource Link   A connection, private line, or switched line that is only active when being used for transmission. *Limited-resource link* can be a reference to a parameter setting on a router's port or a reference to a connection that is shared by a single device or multiple devices.

Limited Router Explorer Packet   See *Spanning Explorer Packet.*

Line   *Line by itself* refers to a POTS (plain old telephone service) line. Any other telephone service line is preceded by the line type, such as, 56K private line, ISDN line, and 56K switched-service line.

Line Card   A circuit board that is inserted into a PBX or Hybrid Key system so that additional telephone lines can be interfaced with the network. Typical line cards give an expandability of 4, 8, or 16 additional line ports on the PBX system. Most PBX and Key KSU cabinets come with a standard number of card slots, and the users can buy however many cards they need. After inserting a new card into a system, each new trunk needs to be configured. Usually the CPU in the system does not recognize the additional line ports until it is programmed or "told" to. After it is programmed it will traffic outgoing/incoming calls accordingly.

Line Conditioning   A term that refers to modifying a twisted copper pair in an outside plant network so it can carry a digital data signal instead of an analog voice signal. Twisted pair circuits (a circuit is a local loop, which is the pair that connects the central office to the customer) are conditioned by adding noise filtering electronic components to them.

Line Current   The average current of a telephone line when the receiver is off hook is about 35 milliamps, which is 0.035 amps.

Line Driver   Also called a *short-haul modem* or *limited-distance modem.* Short-haul modems are commonly used to extend the distance of a printer or other *DTE (Data-Termination Equipment)* device from its host. One example is to extend the printer dedicated to printing call-accounting records from a PBX to an accountant's office. For a diagram, see *Limited-Distance Modem.*

Line Equipment   Also referred to as *OE (Office Equipment).* The line equipment is the actual interface port (from a circuit card) on the central office switch in a telephone-company central office. It is the equivalent to a station or *IPE (Intelligent Peripheral Equipment)* card in a PBX system (TN, for Nortel Specifics). Each telephone line has an associated line-equipment or office-equipment interface. That particular port is what defines the telephone service provided to the customer connected to it via the OSP network. The CPU (core) of the central-office switch associates a phone number with a line-equipment port or OE. When a customer of the phone company calls and requests that their phone number be changed, the service order eventually finds its way to a central-office technician or service translator that reprograms the line equipment with a new phone number.

Line Lock Out   A reaction of a central office when a phone line is left off hook. If you leave the phone off the hook, you get a loud *Ringer Off Hook* alert signal. The signal is transmitted for about a minute, then the telephone company central office locks the line out. This doesn't put the line out of service, it just stops sending the current down the line that it normally would. When a phone is off hook, the central office sends 20 to 30 milliamps down the line. If a phone is off hook, it wastes power, so the central office locks the line out. This uses only 1 to 2 mA of current. The central-office switch keeps the line locked out until it can sense the phone has been placed back on the hook.

Line Loop Back (LLB)   A troubleshooting function of CSU/DSU equipment and smart jacks, where the receive pair of a circuit is connected directly back into the transmitter (or a person manually disconnecting DTE and connecting receive to transmit). The object is to test the transmission line. If the transmission equipment transmits a signal that is "looped back" to it and it receives its own signal with no errors, then the line is ok. If there is a problem, it is inside or it is beyond the receiving equipment (Figs. L.14 and L.15).

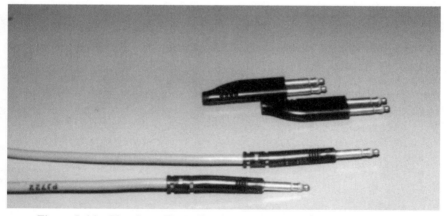

Figure L.14   Line Loop Back (LLB): Bantam Loop-Back Plugs for DSX Panel Loop Backs

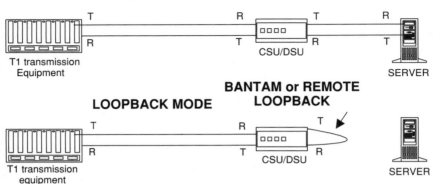

Figure L.15   Line Loop Back (LLB): Normal Operating Mode Vs. Loopback Mode

Line of Sight   Another name for a terrestrial microwave link, also called an *eyeball shot*. The link is made by two radio transceivers equipped with parabolic dish antennas pointed directly at each other. Radio can carry point-to-point transmissions of many bandwidths including DS1, DS2, DS3, STS1, and OC1. Their range can vary, depending on the size of the antenna (dish), weather in the region, and the amount of power emitted. Including all of the previous factors, a link can range from 0 to 50 miles.

Line Pool   A group of telephone lines or trunks configured in a PBX (in a PBX, a telephone is assigned to a pool, rather than a button) or key system for specific call purposes. Some specific purposes include grouping or "pooling" several lines under one button on a telephone. When that button is pushed, the user will access one of the trunks from the line pool that it is assigned. The line pool could be a WATS service or a group of lines dedicated for outgoing calls, so incoming calls are not blocked because of too many agents making outgoing calls at one time.

Line Powered   Telephone equipment that is powered by the telephone company central office-battery and ring voltage. Standard 2500 telephones (noncordless analog phones, such as the ones in your house that have no answering machine or are hands free) are line powered. They require no battery and no power adapter.

Line Protocol   The organized processes and rules that communications equipment use to transfer bits and bytes (data). There are many communications protocols, and layers of protocols that carry other protocols (called *protocol stacks*), including ISDN, Ethernet, token ring, POTS signaling, DS1, ATM, frame relay, and SONET.

Line Queuing   The opposite of call queuing (as in an ACD system). Some telephone systems have a feature called *line queuing*. If you try to dial out and you cannot get an outside line, you are put in queue, a waiting line for the next available trunk. Some systems can give music as if you were on hold for the line and some can ring your phone back. Back to call queuing, ACD systems place incoming calls in queue for the next available agent.

Line Ringing   A feature of telephone systems that enables a user to enable a phone to ring when specific lines are called. When the user programs or configures a specific telephone extension/station/set, a prompt usually says "line ringing." The user then enters the lines that they would like to ring on that phone. The user would then enter the corresponding line ports of the telephone system, (i.e., 1, 2, and 5 for the three associated line ports. Line 555-1234 terminates on port 1, line 555-4321 terminates on port 2, line 555-1111 terminates on port 5). When any of these lines are called, the telephone extension/station/set will ring.

Line Switching   See *Circuit Switching*.

Line Termination   A reference to a demarcation point that gives a telecommunications customer access to their service.

Line Turn Around (LTA)   The time required (in milliseconds) for a two-wire telecommunications circuit to reverse its transmission direction. A typical LTA is 1.5 ms. See also *Network Transit Time Delay*.

Line Turn-Around Time   The delay between transmit and receive in a transmission.

Line Voltage   A reference to the voltage on a pair. T1 line voltage is −135 volts, loopstart line voltage (POTS) is −52 volts, and the line voltage for most PBX and key systems is −24 volts.

Line Wrench   A wrench designed to fit the tool slot of a safety belt (Fig. L.16). Its primary use is to install/remove pole steps and pole attachments.

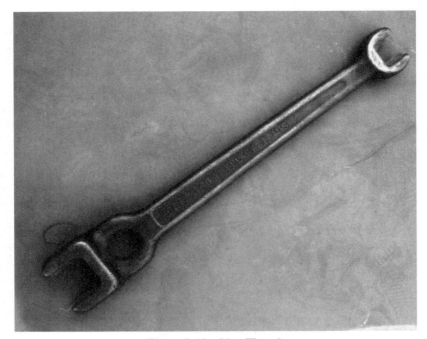

Figure L.16   Line Wrench

Linear Distortion   Usually referred to as Non-Linear Distortion. Non-linear distortion is caused in amplifier circuits when the collector current in the amplifying transistor is insufficient to make the transistor work within the optimal range (the linear range) of its transconductance curve. The resultant distortion is that the top half of the signal (or positive portion) is larger than the bottom half (negative portion). In some cases, the negative portion of the signal can actually be attenuated, rather than amplified.

Linearity   The consistent regeneration of a signal through an electronic circuit. If a circuit amplifies a 5-volt signal two times, 10 volts should be the resultant output. If the circuit is truly linear, raising the 5-volt input to 7 volts would give a resultant output of 14-volts.

Lineman   An obsolete term that used to refer to a person that maintained and installed telephone lines and services. The new term is *network technician*.

Lines of Force   Imaginary lines drawn around a magnetic or magnetized object that represent the direction and polarity (north and south) of the magnetic field around it.

Link   1. A connection, logical or physical, that connects two communications entities and allows them to exchange information. The entities can be either hardware devices (such as a workstation or microwave dish) or software applications (such as an Internet Web browser or software accounting program). See also *Data-Link Connection Identifier*.   2. In network routing, an interface on a router.

Link Budget   In radio transmissions, a link budget is a rough calculation of all known elements of the link to determine if the signal has the proper strength when it reaches the other end of the link. To make this calculation, the following information is generally used: frequency of the link, free space path loss, power of the transmitter, antenna gain, length of transmission cable and its loss, and path length. See also *Earth Bulge, Fresnel Zone,* and *Free Space Path Loss.*

Link Fragmentation and Interleave (LFI)   In IP telephony environments, a QoS method of managing voice packets over links that are slower than 768 Kbps. Because these links are considered slow in relation to the 100 Mbps LAN Ethernet speeds, they require additional QoS attention to enhance their efficiency.

Link-Layer Address   Another name for *MAC Address (Media-Access Control Address).* A standardized *OSI (Open-Systems Architecture model)* data-link layer identifier that is manufactured into every port or device that connects to a LAN. Network-control devices use MAC addresses to create and update routing tables and network data structures. MAC addresses are six bytes long. The IEEE delegates MAC address number ranges to manufacturing companies worldwide. MAC addresses are also called *hardware addresses, MAC-layer addresses,* or *physical addresses.* See also *Network Address.*

Link Level   The actual telecommunications line or the connection between two modems. *Link level* is generally a reference to an OSI physical-layer transport method, such as DS1, DSL, or V.35.

Link-State Advertisement (LSA)   Also called an *LSP (Link-State Packet).* It is a broadcast packet used by routers in a link-state network routing environment. These packets provide address information to neighboring routers about paths and costs. LSAs are the resource within a link-state routing protocol that gathers information used to build and maintain routing tables. See also *Link-State Routing Algorithm.*

Link-State Packet (LSP)   Usually called a *link-state advertisement.* It is a broadcasted data unit used by routers in a link-state network routing environment. These data units provide address information to neighboring routers about paths and costs. LSPs are the resource within a link-state routing protocol that gathers information used to build and maintain routing tables. See also *Link-State Routing Algorithm*.

Link-State Routing Algorithm   The basis of link-state routing protocols. Also referred to as *SPF (Shortest Path First)* or *Dijkstra's Algorithm*. A class of router operating software that enables routers to build their own complex address routing tables that detail every router and node within their network. The routing-table building process is accomplished through routers exchanging link-state advertisements. See also *Dijkstra's Algorithm*.

Link-State Routing Protocol   Also known as *Shortest Path First* (SPF), *Distributed Routing Protocol,* and the specific routing protocols IGRP and *Open Shortest Path First* OSPF. A link state protocol is a methodology used in router protocol design. This methodology enables routers within an autonomous network (i.e., corporate LAN) to identify each other and the status of their port connections. Link state routing protocols create three databases within a router's memory: a neighboring router database, a link database, and a routing table. The routing table is created by applying Dykstra's algorithm to the first two databases. See also *OSPF.*

Link State Update (LSU)   In OSPF routing, a packet sent to an ABR (Area Borer Router) or DR (Designated Router) when a link status changes. In response to this packet, the ABR or DR sends a link state ACK message and then stops forwarding packets. It recalculates its routing database with the new information and sends the link state database change to the other routers within its area. Then, during a short delay, it recalculates its routing database and begins forwarding packets again. This same process happens for the rest of the routers within its area.

Liquid Crystal Display (LCD)   See *LCD*.

Little Endian   A method of transmitting or storing data where the least-significant bit or byte is presented first. The opposite is big-endian, where the most-significant bit or byte is presented first.

LLB (Line Loop Back)   A troubleshooting function of CSU/DSU equipment and Smart Jacks where the receive pair of a circuit is connected directly back into the transmit (or a person manually disconnecting DTE and connecting receive to transmit). The object is to test the transmission line. If the transmission equipment transmits a signal that is "looped back" to it and it receives its own signal with no errors, then the line is ok. If there is a problem it is inside or it is beyond the receiving equipment. For a diagram see *Line Loop Back*.

LLC (Logical Link Control)   The upper half or "sublayer" of the *OSI's (Open Systems Interconnect model)* data-link layer. The LLC sub-layer handles error control, flow control, framing, and the addressing for the lower half of the data-link layer *MAC (Media-Access Control)* addressing. The most prevalent LLC protocol is IEEE 802.2, which includes both connectionless and connection-oriented variants. See also *Open Systems Interconnect* and *Media-Access Control*.

LMI (Local Management Interface)   LMI is a set of software enhancements to the basic frame-relay specification. The enhancements include the following: the ability to integrate with a keep-alive mechanism, which verifies that connections (DLCIs) are working; a multicast mechanism, which provides the network server with its local *DLCI (Data-Link Connection Identifier);* the multicast DLCI, global addressing, which gives all DLCIs in a network the same value for all routers; and a status mechanism, which provides an on-going status report on the DLCIs known to the switch. The customer's application sends status inquiries to the service provider's network, and the network returns the requested information. The LMI data rides within the customer-data part of the frame-relay frame, using the DLCI values 0 or 1023 (see *DLCI*). The more-advanced LMI application is called *CLLM (Consolidated Link-Layer Management)*. CLLM has the ability to report congestion control for each individual DLCI on the network.

LMOS (Line-Maintenance Operating System)   A computer program that RBOCs use to track outside and inside telephone facilities.

LMSS (Land-Mobile Satellite Service)   A satellite communications service that utilizes low-level satellites for communications over a widespread geographical area.

LNB Converter (Low-Noise Block Converter)   The device that houses the actual antenna on a satellite dish. The LNB also converts the Ku band (extremely high frequency) to C band, which can easily be carried from a satellite dish to a nearby head-end building over coax (Fig. L.17).

Figure L.17 LNB (Low-Noise Block) Converter: Ku Band to C Band

Load Balancing A reference to the designing/engineering a PBX or central-office switch so that each network group shares the traffic work load or the design/engineering of a data network to evenly share multiple communications paths. Some equipment (in both telephone and data) will automatically compensate for a lost communications path or network group.

Load Coil A load coil is a voice-amplifying device for twisted-pair wire. A load coil is usually placed on each twisted pair used for a voice line every 3000 feet past a central office. Coils are usually located in vaults, with twisted-pair splices. A typical load coil has an inductance of 30 mH. Other coils, used for other applications, are usually referred to as *choke coils*.

Load-Coil Detector A test device used to detect unseen load coils on a pair of wire (Fig. L.18).

Figure L.18 Load-Coil Detector

Loaded Line   A twisted copper pair that has load coils in place. Loaded lines are for voice telephone lines. To make a digital signal, such as a T1, work on a twisted pair, the telephone company must first remove the load coils.

Local-Access Transport Area (LATA)   Simply stated, a LATA is an area code. *Inter LATA* refers to services that go from one area code to another, like long-distance telephone calls. *Intra LATA* refers to services that originate and terminate in the same area code.

Local Air Time Detail   The list of phone calls itemized on a cellular or PCS telephone bill.

Local-Area Network (LAN)   See *LAN*.

Local-Area Network Emulation (LANE)   An added feature of ATM networks that allows *SVCs (Switched Virtual Circuits)* within them to be configured to function as a LAN backbone. In its raw form, ATM does not provide data-link layer and network-layer protocol support. LANE fills in the gap by providing multicast and broadcast support, address mapping (MAC layer to ATM), SVC management, and a usable packet format. The LANE ATM specification also defines Ethernet and token-ring emulated LANs. See also *ELAN*.

Local-Area Signal Service (LASS)   See *LASS*.

Local Call   A telephone call that originates and terminates within the same carrier's network or generally within the same area code. A call that does not incur additional charges for long distance. The determining factor between whether calls will be long distance or local is the tariff (laws that regulate cost of telephone service made by the FCC and local governing PUC) that the local telephone company operates.

Local Central Office (LSO)   Local Serving Office. A central office that performs telecommunications switching for a specific number-plan area. The number-plan area is currently defined by the first three digits of a seven-digit telephone number. When number portability takes effect, a local central office will no longer be defined as its number-plan area. It will be defined by the laws set forth by the PUC and the area its outside plant reaches. Typical switching systems installed in central offices in North America are Lucent Technologies' 5ESS and Northern Telecom's DMS family of switches. There are five classes of central offices and a local central office is a class five. There are five major parts to a central office. As a whole, these parts are referred to as *inside plant*. For a diagram, see *Central Office*.

Local Circuit Carrier Identification   In the United States, to identify the carrier of a local/in-state toll calls on a telephone circuit, dial 1 (area code) 700-4141 from the line in question. See also *Long-Distance Carrier Identification*.

Local Distribution Frame   Also called *Main Distribution Frame (MDF)*. The place where all the wire, fiber optic, or coax for a network is terminated. The distribution frame is usually placed as close to the central-office switch or PBX as possible.

Local Exchange   A reference to the serving area of a central office. Until number portability takes effect, the first three digits of your seven-digit telephone number defines the exchange (specific central office) you are located in. After number portability takes effect, an exchange will be defined as the tariffs and laws set forth by the local governing PUC. Your telephone number will be associated to you, rather than a central-office equipment port or address. You and your phone number will be tracked in a national database that allows you to take your phone number anywhere you move to, or any phone company you switch to, within an area code. Eventually, you will even be able to transfer your number to a cellular or PCS phone.

Local Exchange Carrier (LEC)   Most commonly, a reference to one of the seven RBOCs. A telephone company that provides telecommunications services to end users and bills them for it. A local exchange carrier has switching networks and outside plant or cellular service to serve its customers. Examples of local exchange carriers are: NYNEX, USWest, GTE, PAC BELL, Southern Bell Telephone, and a myriad of cellular telephone-service providers. The newest breed of Local Exchange Carriers are CLECs (Competitive Local Exchange Carriers). They have all the same services as the RBOCs except they are generally only located in large metropolitan areas. Some larger CLECs include AT&T local, ELI (Electric Lightwave, Inc.), MCI/Worldcom, and MFS (Metropolitan Fiber Systems). CLECs use (almost exclusively) SONET in conjunction with DCS as the foundation of their network architecture.

Local Explorer Packet   A signal generated by a workstation or other end-user device in an *SRB (Source-Route Bridging)* network to find another host/end-user device connected to the local ring. If the local explorer packet fails to find the sought host without passing through a router, then either a spanning explorer packet or an all-routes explorer packet is sent. The latter packets then search the larger LAN or WAN for the sought device.

Local Loop   The pair of wires that extends from the local telephone company's central-office main-distribution frame to the customer's premises.

Local Management Interface (LMI)   LMI is a set of software enhancements to the basic frame-relay specification. The enhancements include the following: the ability to integrate with a keep-alive mechanism, which verifies that connections are working; a multicast mechanism, which provides the network server with its local *DLCI (Data-Link Connection Identifier);* the multicast DLCI, global addressing, which gives all DLCIs in a network the same value for all routers; and a status mechanism, which provides an on-going status report on the DLCIs known to the switch.

Local Service Area   The geographical area that a customer can make calls without being billed additionally for long distance.

Local Tandem   A telephone company central-office switch that has the ability to connect or "switch" calls from two different central offices (Fig. L.19).

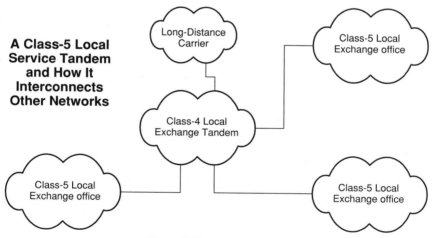

Figure L.19   Local Tandem

Local Traffic Filtering   A process by which a bridge or router recognizes and does not forward (drops) frames whose source and destination MAC addresses are located on the same interface port, thus preventing unnecessary traffic from being forwarded across the bridge. The standard for this is defined as a part of a spanning tree in IEEE 802.1.

Local Trunk   A trunk that is fed by a local central-office switch, as opposed to a long-distance carrier WATS service.

Lock Code   A code that a cellular telephone user can dial into their phone to prevent any calls from being made on it. After the code is entered again, the phone is able to make calls.

Logic   A mathematical process first developed by the Irish mathematician George Boole in the 1850s. The premises of logic is to know if a certain statement is true or false. An example is "the light is on." This statement can only be true or false. Logic couples this statement with others, such as "the switch is on, the power is on; therefore, the light must be on." If the switch is off and the power is on, then the light is off. If the switch is on and the power is off, then the light is off. These statements depict the truth table for an AND electronic logic gate, which is a primary building block of microprocessors. In the table depicted, the light switch would be "A," the power would be "B," and the light bulb would be "C." Truth tables are written in ones and zeros, rather than ons and offs. The science of this math is called *Boolean Algebra*. It is a book in itself and is usually explained quite well in textbooks that cover digital electronics (Fig. L.20).

TRUTH TABLE

A	B	C
0	0	0
0	1	0
1	0	0
1	1	1

AND GATE

Figure L.20   Logic

Logical Address   Another name for *network address*. See *Network Address*.

Logical Bus   A LAN logical topology. The logical topology defines the way that a LAN communicates. The physical topology defines the way that a LAN is physically wired. For example, even though an Ethernet network might be physically wired into the formation of a star, it really works as though it were a bus. The wire is just physically laid out and connected differently and the electronics are a little different.

Logical Channel   A multiplexed channel or frame-switched media (such as frame relay), where there is no physical wire, fiber, radio or coax path, but within the protocol stack is a communications path.

Logical Channel Identifier (LCI)   Another name for *logical channel number.* See *Logical Channel Number*.

Logical Channel Number (LCN)   In X.25, an identifier of a virtual circuit that will be used for all transmissions within the duration of a call. X.25 can support up to 4095 LCNs total in both directions for transmitting and receiving. LCNs exist for the same purpose in X.25 as a *DLCI (Digital-Link Connection Identifier)* exists in frame relay. It is a temporary virtual path used between network routing devices. The LCN is not a permanent virtual circuit, such as the type that a multiplexer would provide. The LCN represents one of the temporary users of a permanent virtual circuit. See also *X.25 Control-Packet Header Structure.* 2. The channel number address that rides along with a packet of data in statistical time-division multiplexing.

Logical Link   See *Logical Channel*.

Logical Link Control (LLC)   The upper half or "sublayer" of the *OSI's (Open Systems Interconnect model)* data-link layer. The LLC sublayer handles error control, flow control, framing, and the addressing for the lower half of the data-link layer *MAC (Media-Access Control)* addressing. The most prevalent LLC protocol is IEEE 802.2, which includes both connectionless and connection-oriented variants. See also *Open Systems Interconnect* and *Media-Access Control*.

Logical Ring   A LAN that operates as a ring-type protocol (logical topology)—even though it is physically wired as a star (physical topology).

Long Distance   A telephone call or telecommunications service that originates within one local service area (usually an area code) and terminates in another.

Long-Distance Carrier Identification   In the United States, to identify the long distance out-of-state call carrier for a telephone circuit, dial 1-700-555-4141 from the line in question. See also *Local Circuit Carrier Identification*.

Long Haul   See *Long Distance*.

Long Haul Modem   A modem that is capable of transmitting beynd distances of one mile. See also *Short-Haul Modem*.

Long Reach   A reference to SONET fiber-optic spans or links longer than 25 kilometers.

Longitudinal Redundancy Check (LRC) A method of checking for errors in communications transmissions by combining vertical error checking and longitudinal error checking (Fig. L.21). A transmission device sends data in bytes that are logically stacked on top of each other. The stack forms a block. The last bit of each line is used to form a check sequence. LRC is about 85% accurate in detecting and retransmitting blocks that contain errors. The newer method of error checking is *CRC (Cyclic Redundancy Checking)*.

DATA BLOCK

byte 1	1	0	1	1	0	1	1	1	0	Even parity
byte 2	1	0	0	1	1	0	1	0	0	
byte 3	0	1	1	0	0	0	0	1	1	
byte 4	1	1	1	1	1	1	1	1	0	
byte 5	0	0	0	0	0	0	0	0	0	
byte 6	1	1	1	1	0	0	0	0	0	
byte 7	0	0	1	1	0	0	1	1	0	
byte 8	1	0	1	0	1	0	1	1	1	
byte 10	1	1	0	0	1	0	1	1		

Longitudinal Parity Sequence (byte 9) SHADED
Vertical Parity Sequence (byte 10) Bottom
Row

LRC BIT
STREAM

byte 1	byte 2	byte 3	byte 4	byte 5	byte 6	byte 7	byte 8	byte 9	byte 10
10110111	10011010	01100001	11111111	00000000	11110000	00110011	10101011	100001	11001011

Figure L.21 Longitudinal Redundancy Chart (LRC)

Look-Up Table A translation table in a PBX system. Translation tables convert "dialed number" protocols into numbers that the public telephone network can recognize.

Loop Also called a *Local Loop*. The pair of wires that extends from the local telephone company's central-office main-distribution frame to the customer's premises.

Loop Antenna A directional antenna used mostly for UHF receptions. Some older TVs use small UHF loop antennas to receive broadcast channels from 13 to 83.

Loop Back Also called *Line Loop Back*. A troubleshooting function of CSU/DSU equipment and smart jacks, where the receive pair of a circuit is connected directly back into the transmit (or a person manually disconnecting DTE and connecting receive to transmit). The object is to test the transmission line. If the transmission equipment transmits a signal that is "looped back" to it and it receives its own signal with no errors, then the line is ok. If there is a problem, it is inside or it is beyond the receiving equipment. For a diagram and photo, see *Line Loop Back*.

Loop Down   A command function that terminates a loop-back state of a public-provided circuit. The loop-back mode enables a user to test the circuit by receiving its own transmission signal. *Loop up* is a command or term used to place a circuit in loop-back mode. See also *Loop Up*.

Loop Extender   An add-on device for a PBX switch or central-office switch that allows operation over an abnormally long loop or twisted pair, usually over 12,000 feet for a central-office switch and 1500 feet for a PBX. Loop extenders are also called *OPX (Off Premises Extension) adapters* for extending a station/extension to a remote location (over 1500 feet).

Loop-Start Line   A line that comes from a central office. The type of line determines which type of signaling the line requires to work. If a line is dedicated to one phone or group of phones (like in your house), it is a "line." If the line is going to be shared among many devices connected together by a PBX or key system, then the line is called a *trunk*. A loop-start line is a two-wire central-office trunk or dial-tone line that recognizes an "off hook" situation when a telephone switch-hook puts a 1000-ohm short across the tip and ring when the handset is lifted. This is the most common type of line. It is also called a *POTS line* and *plain-service line*. Other types of lines or trunks are: ground start and E&M trunks, ISDN PRI, and ISDN BRI.

Loop-Start Trunk   A trunk is a line that comes from a central office. The type of trunk determines which type of signaling the line requires to work. A loop-start trunk is a two-wire central-office trunk or dial-tone line that recognizes an "off hook" situation when a telephone switch hook puts a 1000-ohm short across the tip and ring when the handset is lifted. This is the most common type of line. It is also called a *POTS line* and *plain-service line*. Other types of trunks are ground start, and E&M trunks, ISDN PRI, and ISDN BRI.

Loop Up   A command function performed by a technician and/or test equipment, which places a public-service-provided circuit in loop-back mode. Loop-back mode enables a user to test the circuit by receiving its own transmission signal. *Loop down* is a command function that removes the loop back. See also *Loop Down*.

Loop Up/Loop Down   To loop up is to put a CSU/DSU in loopback mode. To loop down is to remove the loop back and resume normal operation.

Loose Tube Buffer   A PVC tube that is about as big around as a drinking straw that has up to twelve optical fibers within it. The idea behind a loose-tube buffer is that when a cable is bent, the fibers inside will have slack and freedom to move and naturally adjust to the bend. If a filler is inside the loose tube, it would crush or crack (fracture) the optical fibers when the cable is bent. This would render them useless.

Loose Tube Splitter   A tool used to open the shell that encases fiberoptic strands (Fig. L.22).

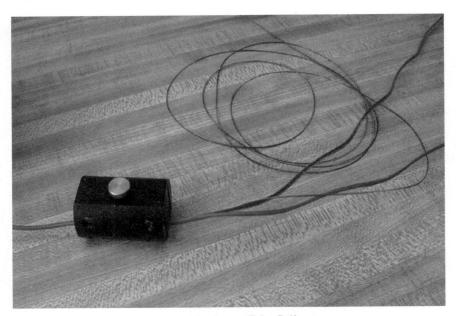

Figure L.22   Loose Tube Splitter

Loss   The reduction of a signal's voltage level as it travels down a line, measured in decibels. Attenuation is also called *Loss* because some signal is always lost through resistance and reactance. Optical lightwave signals are also attenuated when they traverse through a fiber optic because of impurities in the fiber optic and the fact that light intensity decreases with distance.

Lossy   A characteristic of a network that is prone to lose packets when it experiences high levels of traffic. There are many reasons for networks to mysteriously discard packets that shouldn't be lost in times of congestion. A simple cause would be different network nodes or switches having varying traffic-shaping parameter settings or having network nodes that define traffic parameters differently.

Lost Call   A call that did not complete or was blocked because of a lack of switching facilities.

Loudspeaker Paging   A feature of PBX and key systems that allows a user to connect the telephone system to an external paging amplifier and speakers. The interface is usually broken out onto a 66 block, then cross connected to the input of a paging amplifier. When a person wants to page someone, the telephone system will prompt the user to choose *internal* or *external* (or zone 1, 2, or 3). If the user chooses *external*, then the page is heard over the loudspeakers driven by the external amplifier instead of the telephone set speakers.

Low Frequency   The range of frequencies between 30 and 300 kHz.

Low-Pass Filter   An electronic device that eliminates frequencies above a specified frequency. The two categories of frequency filters are active and passive. Active filters use active devices that require power, such as transistors and 541 op amps to amplify the desired signal and attenuate the undesired signal. Passive filters are made with components that do not require external power, such as capacitors and inductors. Capacitors and inductors have reactive properties that cause them to resist or pass an AC signal.

LPT Port   A logical designation for a group of I/O addresses that "tells" a computer which "plug" to send the printer communications to. LPT ports are usually designated LPT 0, LPT 1, and LPT 2.

LRC (Longitudinal Redundancy Check)   See *Longitudinal Redundancy Checking*.

LSA   1. *Leased Space Agreement*. Also called a *colocation* or *co-lo*. A leased-spaced agreement is an arrangement that communications companies or communications services vendors (such as alarm companies or voice-mail service providers) make with telecommunications companies to use an area of a central office or node to place their network-interface equipment. See also *Colocation*.   2. *Link-State Advertisement*. Also called an *LSP (Link-State Packet)*. It is a broadcast packet used by routers in a link-state network routing environment. These packets provide address information to neighboring routers about paths and costs. LSAs are the resource within a link-state routing protocol that gathers information used to build and maintain routing tables. See also *Link-State Routing Algorithm*. For a photo, see *Leased-Space Agreement*.

LSB (Least-Significant Bit)   The bit in an octet that carries the least value. You can better understand this by comparing it to our base-10 numbering system. Imagine a "least-significant number." If you are 43 years old, the 4 is the most significant number and the 3 is the least-significant number. If the 4 were lost, then you would only be three (a very significant difference), it is the most significant compared to the 3. If the 3 were lost, you would still be 40. LSB is another way of saying "least-significant digit," which is used to round numbers off in elementary mathematics. In T1 in-band signaled circuits, least-significant bits are robbed from the bit stream of the 6th and 12th sample in each channel. The voice-sample bits are replaced with signaling information and maintenance information bits.

LSI (Large-Scale Integration)   Microchip ICs are classified as *SSI (Small-Scale Integration), MSI (Medium-Scale Integration), and LSI (Large-Scale Integration)*. SSI ICs contain 12 or fewer devices, such as logic gates or transistors. MSI ICs contain 13 to 99 devices and LSI ICs contain 100 or more devices. The typical CPU, such as a Pentium (Intel trademark) microprocessor contains hundreds of thousands of devices. For a photo of a VLSI device, see *Microchip*.

LSP (Link-State Packet)   Usually called a *link-state advertisement*. It is a broadcasted data unit used by routers in a link-state network routing environment. These data units provide address information to neigh-boring routers about paths and costs. LSPs are the resource within a link-state routing protocol that gathers information used to build and maintain routing tables. See also *Link-State Routing Algorithm*.

LSU (Link State Update)   In OSPF routing, a packet sent to an ABR (Area Borer Router) or DR (Designated Router) when a link status changes. In response to this packet, the ABR or DR sends a link state ACK message and then stops forwarding packets. It recalculates its routing database with the new information and sends the link state database change to the other routers within its area. Then, during a short delay, it recalculates its routing database and begins forwarding packets again. This same process happens for the rest of the routers within its area.

LTA (Line Turnaround) The time required (in milliseconds) for a two-wire telecommunications circuit to reverse its transmission direction. A typical LTA is 1.5 ms. See also *Network Transit Time Delay*.

Lug More frequently called a *binding post*. A lug or binding post is a small threaded bolt with a nut used to attach wires. The binding post usually has a number, which is a reference used to identify where twisted copper pairs are terminated in access points, cross boxes, and terminals. When a technician looks for a specific pair in a cable (called a *cable pair*), they refer to documents that list the pairs and which binding posts they are spliced to.

M

M Plane   One of the three entities of frame-relay network management. The three planes are: The *User Plane* (the U Plane defines the transfer of information), the *Management Plane* (the M Plane defines the LMI, Local Management Interface), and the *Control Plane* (the C Plane is delegated for signaling and switched virtual circuits). The M Plane (also called the *Local Management Interface, LMI*) reports the status and configuration of connections to the *FRAD (Frame-Relay Access Device)*. It provides notification of the addition, deletion, availability, and presence of all *Data-Link Connections (DLCs)*. It also provides the exchange sequence that maintains the data connections when they are not in use (also called *keep-alive data*).

M1   A reference to the Meridian 1 *PBX (Private Branch Exchange)* switching system manufactured by Nortel Networks.

Ma Bell   A reference to AT&T—the company that is said to have given birth to the Baby Bells, better known as the RBOCs.

MAC (Media-Access Control)   The protocol (there are several types, e.g., Ethernet MAC) that determines the transmission of information on a local-area network. The MAC is a part of the *OSI (Open Systems Interconnect model)* data-link layer that interfaces with the physical layer. The MAC is referred to as a *sublayer* of the data-link layer. Its purpose is to manage the transfer of data to the "wire" or "fiber optic," as required by the used protocol. See also *Data-Link Layer* and *LLC*.

MAC Address (Media-Access Control Address)   A standardized *OSI (Open Systems Architecture model)* data-link layer identifier that is manufactured (burned into ROM) into every port or device that connects to a LAN. Network-control devices use MAC addresses to create and update routing tables and network data structures. MAC addresses are six bytes long (48 bits). The first 24 bits of the address are a vendor/manufacturer code, and the next 24 bits are the interface serial number (Fig. M.1). The IEEE delegates MAC address number ranges to manufacturing companies worldwide. MAC addresses are also called *hardware addresses, MAC-layer addresses,* or *physical addresses.* See also *Network Address.*

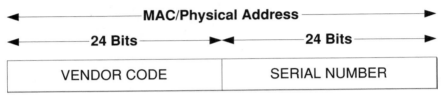

Figure M.1 Mac Address (Also Called a Physical Address)

MAC Address Learning   Service that characterizes a learning bridge, in which the source MAC address of each received packet is stored so that future packets destined for that address can be forwarded only to the bridge interface on which that address is located. Packets destined for unrecognized addresses are forwarded out every bridge interface. This scheme helps to minimize traffic on the attached LANs. MAC address learning is defined in the IEEE 802.1 standard. See also *Learning Bridge* and *MAC Address.*

Machine Language   The lowest-level programming language. PROMs are the key to machine language. They contain instructions that assist the microprocessor in decoding the 8-, 16-, or 32-bit instructions into functions that the microprocessor executes. The instructions (machine-language scripts) are burned into the PROM when it is programmed. Typical machine-language instructions include: MOV A,M, which is actually entered as an OP-Code of 167, instructs a microprocessor to move the contents of memory address A to memory address M. Another machine instruction is OUT, entered as an OP-Code of 323, which instructs a microprocessor to move the contents of the previous memory address to a port that is identified in the next instruction.

MacIP   Network layer protocol that encapsulates IP packets in DDP packets for transmission over AppleTalk. MacIP also provides proxy ARP services. See also *DDP* and *Proxy ARP.*

Magnetic Ink   Ink that is used to print information that will be read electronically. Magnetic ink is made with ferrous compounds. The banking industry uses magnetic ink on some of their printed materials and typical bank-account checks have the account number printed in magnetic ink so that they can be electronically processed.

Magnetic Storage   A method of storing data by magnetizing a tiny section of a tape or disk for each bit (Fig. M.2).Hard-disk drives and floppy-disk drives utilize disks coated with ferromagnetic materials. Data cartridges contain thin plastic tape coated with ferromagnetic materials that are recorded and read in a similar fashion. Analog information can also be stored magnetically; cassette tapes are a common example.

**Figure M.2 A Magnetic Tape Back Up for a
Lucent Technologies 5ESS Central Office Voice Switch**

Magnetic Stripe   The stripe on the back of a credit card or other device. Magnetic stripes are usually used to store information, such as a name and account number, in a binary bar-code format.

Mail Bridge   In an e-mail environment, a part of a mail gateway that enforces an administrative policy with regard to what mail it forwards.

Mail Gateway   In an e-mail operating environment, a computer (usually a server) loaded with specialized software that connects two or more dissimilar electronic-mail systems and transfers mail messages between them.

Main Distribution Frame   Also called a *distribution frame*. The place where all the wire, fiber optic, or coax for a network is terminated. The distribution frame is usually placed as close to the central-office switch or PBX as possible. For a photo, see *Distribution Frame*.

Main Feeder   An F1 (first facility) cable from a central office. The feeder cable runs to cross connect points in the telephone network where F2 (second facility) cable feeds are connected/cross connected (Fig. M.3). For a photo, see *Distribution Cable*.

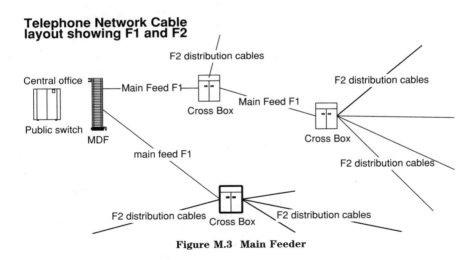

Figure M.3 Main Feeder

Main Frame   A large computer capable of retrieving information from mass-storage units and calculating/processing the data in a very short time in comparison to a client-server computing process. Main-frame computers have been regaining favoritism in large data-processing environments because of their outstanding reliability and processing power.

Main PBX   A primary PBX that interfaces with the public telephone network via CO (central office) trunk lines (Fig. M.4). The other type of PBX is an off-premises (remote) PBX, which is connected to the outside world or public network by switching through a main PBX.

A Main PBX connected to a remote PBX / Key system with analog trunks

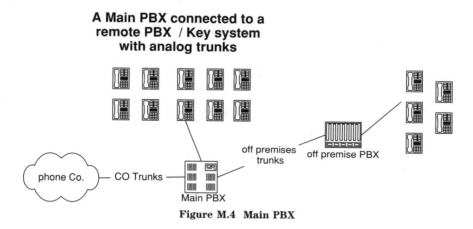

Figure M.4   Main PBX

Make Busy   1. A reference to the activation of the "busy out" feature of an ACD telephone. If the telephone is busied out, the ACD system will not transfer calls to that telephone. This is useful when an agent is on break or their shift is over.   2. A temporary fix or condition of a phone service. To "busy a line out of a hunt sequence." If a business phone line becomes defective and it is in a hunt or roll-over sequence, calls will not hunt or roll past this line. For example, four lines come into your business. The first line is the main number; if that first line is busy, then calls come in on the second line, etc. If line one goes bad, it can't be called, so it can't be busy. Because it is not busy, calls will not hunt or rotate to the next three lines. When you call the phone company repair service, they busy out the bad line, which makes it look busy to the network. Your calls then start coming in on the other three lines. When a repair technician finishes repairing the problem on the bad line, he has it unbusied. Another temporary fix is to call forward the line from the central office. The phone company can do this at the customer's request.

Malicious Call Trace   A feature offered by local telephone companies. Even though all calls are kept in an archived database, it is sometimes difficult to locate a single call—even if an accurate time is given. The malicious call trace or annoyance call trace "flags" a call when a customer hangs up and dials the call-trace feature code after receiving an annoyance call. Usually, the charge is $2.00 per trace. After the trace or "flag" has been made, telephone company security officials investigate the source of the call. If there are multiple occurrences, then the telephone company will press charges against the malicious caller. The person receiving the malicious calls will never find out who the caller is unless they are summoned to a court hearing.

MAN (Metropolitan Area Network)   See *Metropolitan Area Network*.

Managed Object   In LAN network management, a network device (such as a workstation or server) that can be managed by a network administrator through a management protocol.

Management Information Base (MIB)   A data file within a TCP/IP-loaded network device, such as a PC or router that contains information about that device. The information that is contained in the MIB include hardware addresses (MAC address), counters, statistics, and routing tables. Each software/information category is referred to as an *object*. MIB works in conjunction with *SNMP (Simple Network-Management Protocol)*.

Management Plane   One of the three entities of frame-relay network management. The three planes are: The *User Plane* (the U Plane defines the transfer of information), the *Management Plane* (the M Plane defines the LMI, Local Management Interface), and the *Control Plane* (the C Plane is delegated for signaling and switched virtual circuits). The M Plane (also called the *Local Management Interface, LMI*) reports the status and configuration of connections to the *FRAD (Frame-Relay Access Device)*. It provides notification of the addition, deletion, availability, and presence of all *Data-Link Connections (DLCs)*. It also provides the exchange sequence that maintains the data connections when they are not in use (also called *keep-alive data*).

Manchester Encoding   Also called *differential encoding*. A physical media transmission format used in IEEE 802.5 and token-ring *LANs (Local-Area Networks)*. Differential encoding combines a clocking signal with the data stream. A binary 1 is denoted by a voltage increase and a binary 0 is denoted by a voltage decrease. The voltage reset between voltage increases and decreases represents the clocking or timing source. By design, differential encoding is less prone to attenuation, but more sensitive to *RFI (Radio Frequency Interference)* than *PCM (Pulse-Code Modulation)* formats.

Manual Ring-Down Line   Not really a phone line, but two phones connected together via a pair of wires and a talk battery (9 V to 24 V). Signaling, such as ringing, is performed manually, by flipping a switch on and off rapidly, which disconnects and connects the battery. The changing voltage imitates a weak ring voltage. Rescue teams use manual ringdown lines in cave and mine-shaft rescue operations because their radio range is very limited in underground tunnels.

Map   A reference to "mapping"—a virtual tributary though an *Optical Carrier Circuit (OC-3)* over a SONET ring or end-to-end path. Mapping the tributary involves telling the SONET equipment which and how much bandwidth within the OC (Optical Carrier) will be designated a channel. The choices are DS0, DS1, DS3, and STS-1. STS-1 is an electrical version of an OC-1.

Mapping   See *Map*.

Marine Telephone   A radio telephone that is designated specific operating frequencies by the FCC. It is not cellular (cellular is a short-distance radio application), it broadcasts with more power and a much greater distance. Tellabs manufactures marine telephone equipment.

Marker Tape   A plastic label built into cables that have cable ID and specification information printed on them by the manufacturer. Marker tapes are uncommon in newer polyethylene cables, because it is much easier and less expensive to print the cable designation on the outside of the cable.

MARS (Multicast Address Resolution Server)   A mechanism for supporting IP multicast. A MARS serves a group of nodes (known as a *cluster*); each node in the cluster is configured with the ATM address of the MARS. The MARS supports multicast through multicast messages of overlaid point-to-multipoint connections or through multicast servers.

Martian (Mis-Addressed/Routed Transmission In A Network or Mis-Addressed/Routed Telepacket In A Network)   A data message that ends up in the wrong part of a network. Martians are caused by routing-table inaccuracies. Those inaccuracies are caused by the lack of maintaining static routing table entries or workstations that have been given bogus network addresses.

MARTians   *Misaddressed or Routed Telepacket* on a LAN or WAN.

Mask   See *Address Mask* and *Subnet Mask*.

Mast Clamp   A device used to attach a ram hook to a power mast (Fig. M.5). The ram hook (also called a ram horn) is used to attach an aerial service wire via a drop clamp.

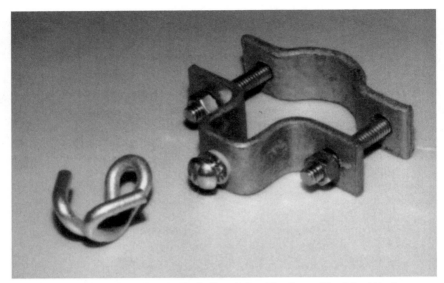

**Figure M.5 Mast Clamp, Including Other Hardware Used to Attach
Aerial Service Wires (Drop Wires)**

Master Clock   A reference to a BITSs clock. A central timing device for
synchronous networks, such as SONET networks. Bits clocks can be rack
mounted, just like other telecommunications equipment.

Master of Network Science (MNS)   A well-known industry certifica-
tion/training program offered by 3COM. This program is actually referred
to as a *credential,* rather than a *certification* by 3COM. Several hardware and administration intensive programs are available under the MNS
family of certifications. The first is the *MNS LAN Solutions track,* which
includes training on high-performance 3COM LAN switching equipment,
seven exams, and a lab test. Another is the *MNS LAN Solutions Plus
track,* which requires completion of the LAN Solutions requirements,
and extends the trainee's knowledge into the ATM products offered by
3COM. Another example track is the *MNS Network Architecture,* which
provides a rounded study of enterprise networking. Other MNS tracks
include the *MNS Network Management, MNS WAN Solutions,* and *MNS
Remote-Access Solutions.* Information regarding the 3COM MNS training programs can be found at *http://www.3com.com.*

MAT (Meridian Administration Tools)   A Northern Telecom CTI application that allows a Meridian PBX system to be managed through a
GUI environment over a LAN or single PC. It enables a user to make administrative changes to the system by clicking on the picture of an item

to be changed (such as a feature button on a telephone or the name display) and typing in the change. It also provides excellent traffic and core analysis tools, which graph the busy hours by network group. Call accounting is also a feature of MAT.

Matrix   The part of a switch that carries and routes calls. The matrix is a virtual part of the core that commands which channels to connect with what. As multiplexed bit streams run through a digital switch, they are separated and recombined from inputs (interface cards) to specified outputs (interface cards) by the core (CPU) of the switch. The matrix is not a tangible object; it is a combination of the CPU and interface equipment.

MAU (Media Attachment Unit)   Generically referred to as a *transceiver*, this device connects to a computer or other device's BNC network-interface card port and permits a convenient connection to an RJ45 twisted-pair media. Transceivers have also been called *AUIs (Access Unit Interface)* and *Media-Access Units* (with the same acronym, *MAU*).

Maximum Burst   An ATM parameter that specifies the largest burst of data above the insured and maximum rates that will be allowed temporarily on an ATM PVC, but will not be dropped at the edge by the traffic-policing function. On average, the traffic needs to be within the maximum rate. This parameter is specified in bytes or cells. See also *Maximum Rate* and *Insured Rate*.

Maximum Rate   On an *ATM (Asynchronous Transfer Mode)* connection, a parameter that defines the total bandwidth that will be permitted to traverse from point to point under any circumstances. The maximum rate is equal to the insured rate and the excess rate combined. When the ATM network that serves the connection has a low level of traffic, the connection is permitted to transmit at the excess-rate value of bandwidth. When the network is congested or very busy, only the insured rate of bandwidth is allocated to the connection. An example of an insured rate for a connection is 256 Kbps, and an example of the maximum rate for that same connection could be 512 Kbps. See also *Excess Rate*.

Mbps (Megabits Per Second)   Equivalent to one million bits per second. Memory or data transferred per unit of time is measured in bits. Memory storage is measured in bytes. The difference in abbreviations is that bits are lowercase (b) and bytes are uppercase (B).

MCLR (Maximum Cell-Loss Ratio)   In an *ATM (Asynchronous Transfer Mode)* network, the maximum ratio of cells that do not successfully transit a link or node, compared with the total number of cells that arrive at the link or node. MCLR is one of four link signals exchanged between network nodes to determine the available resources of an ATM network. The MCLR applies to cells in the *CBR (Constant Bit Rate)* and *VBR (Variable Bit Rate)* traffic classes whose *CLP (Cell-Loss Priority)* bit is set to zero. See also *CBR, CLP, UBR,* and *VBR*.

MCP (Microsoft-Certified Professional)   A well-known industry certification/training program offered by Microsoft and its training partners. The MCP certification is an in-depth learning track that incorporates Microsoft Networking products, Microsoft Applications, and general applications thereof. Up-to-date information regarding the MCP training program can be found through *http://www. microsoft.com*.

MCR (Minimum Cell Rate)   A parameter defined by the ATM Forum for ATM Traffic Management. *Minimum cell rate* is defined only for *Available Bit Rate (ABR)* transmissions and it specifies the minimum value for the *Allowed Cell-Rate (ACR)* parameter. A minimum cell rate is necessary when timing-sensitive transmissions are made, such as voice or video.

MCSE (Microsoft-Certified Systems Engineer)   A well-known industry certification/training program offered by Microsoft and its training partners. The MCSE certification is an in-depth study program that incorporates Microsoft Networking products, such as Windows NT and Microsoft Exchange Server. The learning track for this program is extensive in Microsoft data-base applications and, in general, enterprise networking. Information regarding the MCP training program can be found through *http://www.microsoft.com*.

MCU (Multipoint Control Unit)   A hardware and software device that in its respective application, whether it is voice or video, provides a point of management and control for multiple voice or video data streams that could possibly be conferenced together. MCUs are generally proprietary devices that incorporate their own features for managing the flow of a video or voice application and which users can see or hear each other, and also in what fashion the users are seen. For example, if the MCU is designed for videoconferencing, it could make four users appear in small boxes in each corner of a video monitor, or make the user that is speaking be the one viewed by all other users.   2. *Media Conference Unit* performs the same function as an MCU.

MD5 (Message Digest 5)   MD5 is the algorithm used for message authentication in *SNMP (Simple Network-Management Protocol)*. MD5 verifies the integrity of the communication, authenticates the origin, and checks for timeliness.

MDF (Main Distribution Frame)   Also called a *distribution frame*. The place where all the wire, fiber optic, or coax for a network is terminated. The distribution frame is usually placed as close to the central-office switch or PBX as possible. For a photo, see *Distribution Frame*.

Mean Opinion Score (MOS)   A voice quality measurement of Codec operation (such as G.711) over IP telephony networks. Mean opinion score rates voice quality on a scale of 1 to 5, 1 being bad and 5 being excellent. The MOS score is made by a wide range of listeners hearing a transmission and then rating it. G.711 PCM rates at an MOS of 4.1; G.729 CS A CELP rates at a 3.9; G.729A rates at a 3.7. The major quality issue to the listeners that are able to notice the difference is a slight delay and nearly undetectable voice distortions. See also *G.729*.

Measured Rate Service   Abbreviated *1MR* for residential and *1MB* for business, this type of telephone service is offered by local telephone companies. Measured-rate service means that a line is billed on a "per call basis." Telephone companies in the Southern and Western United States have tried to abolish measured service by encouraging customers to subscribe to flat-rate services, abbreviated *1FR* for residential and *1FB* for business use.

Mechanical Splice   An alternative fiber-optic splice to fusion splicing. Fusion splicing equipment is very expensive ($40,000 is typical for a fusion splicer). Mechanical splices come as a kit, which connectorizes the ends of the fibers (Fig. M.6). A tool kit is required for mechanical splicing. It consists of a microscope, polishing puck, cleavers, epoxy, and polishing compound. They cost about $1,200. An oven used to "hot cure" the epoxy is also available. With a mechanical splice, you cleave or cut the end of the fiber as square and smooth as possible, then epoxy the fiber end into a connector. The epoxy takes about 12 hours to cure without an oven and about 20 minutes with an oven. After the epoxy has cured, the tip of the connector (which should be flush with the end of the fiber optic) is polished by holding it with a device called a *puck* (it is shaped like a hockey puck). The puck holds the fiber connector while it is gently rubbed against a pad coated with polishing compound. When the polishing is done, the connector is ready to be mated with another connector and the splice is complete. Mechanical splice kits cost about $15.00 per splice and are available in SC- and ST-style connectors.

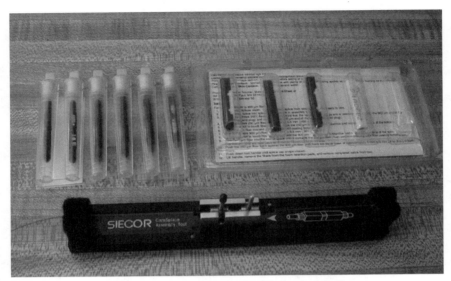

Figure M.6 Fiber-Optic Kit for Mechanical Splices

Mechanized Line Testing (MLT)   Also called *DATU (Direct Access Test Unit)*. MLT and DATU equipment is either added on or built into a central-office switch. DATU allows a technician or customer-service agent to dial the phone number of the DATU or MLT equipment and execute a test for shorts, opens and grounds remotely. In response to a digital voice, the technician enters a password and a choice of options. The results of the test can be read back to the technician by a digital recording or sent to them via an alpha-numeric pager. DATU units can also send a locating tone on the technician's choice of TIP, RING, or both TIP and RING. The test unit can also short lines and remove battery voltage for testing.

Media-Access Control (MAC)   The protocol (there are several types, e.g., Ethernet MAC) that determines the transmission of information on a local-area network. The MAC is a part of the *OSI (Open Systems Interconnect model)* data-link layer that interfaces with the physical layer. The MAC is referred to as a *sublayer* of the data-link layer. Its purpose is to manage the transfer of data to the "wire" or "fiber optic," as required by the used protocol. See also *Data-Link Layer* and *LLC*.

Media-Access Control Address (MAC Address)   A standardized *OSI (Open Systems Architecture model)* data-link layer identifier that is manufactured (burned into ROM) into every port or device that connects to a LAN. Network-control devices use MAC addresses to create and update routing tables and network data structures. MAC addresses are six bytes long (48 bits). The first 24 bits of the address are a vendor/manufacturer code, and the next 24 bits are the interface serial number. The IEEE delegates MAC address number ranges to manufacturing companies worldwide. MAC addresses are also called *hardware addresses, MAC-layer addresses,* or *physical addresses.* See also *Network Address.*

Media Attachment Unit (MAU)   Generically referred to as a *transceiver,* this device connects to a computer or other device's BNC network-interface card port and permits a convenient connection to an RJ45 twisted-pair media. Transceivers have also been called *AUIs (Access Unit Interface)* and *Media-Access Units* (with the same acronym, *MAU*). In token-ring networks, a MAU is known as a *multistation access unit* and is abbreviated *MSAU* to avoid confusion.

Media Conference Unit   (Also called a Multipoint Control Unit, MCU.) A hardware and software device that in its respective application, whether it is voice or video, provides a point of management and control for multiple voice or video data streams that could possibly be conferenced together. MCUs are generally proprietary devices that incorporate their own features for managing the flow of a video or voice application and which users can see or hear each other, and also in what fashion the users are seen. For example, if the MCU is designed for videoconferencing, it could make four users appear in small boxes in each corner of a video monitor, or make the user that is speaking be the one viewed by all other users.

Media Gateway Control Protocol (MGCP)   In IP telephony, the procedure for converting Ethernet/connectionless communication to voice/connection-oriented communication. MGCP is a feature incorporated into routers that gives them the ability to transfer/translate signaling information, such as off-hook/on-hook and dialed digits, to an IP telephony server, which can communicate the same types of signaling to IP telephones. MGCP supports common channel signaling (CCS, out-of-band signaling) but not channel-associated signaling (CAS, in band signaling).

Media Interface Connector   A fiber-optic connector.

Media Rate Adoption   A feature of LAN switches that enables them to compensate for speed differences when communications between ports/segments are different. For example, if a server that is connected to a 100 Mbps Ethernet port is communicating between a computer connected to a 10 Mbps port, media rate adoption ensures that the 100 Mbps port does not overwhelm the 10 Mbps port. Media rate adoption also allows a faster port/device to communicate with multiple slower ports/devices. See also *Backpressure* and *Aggressive Back-Off Algorithm.*

Media Termination Point (MTP) In IP telephony, a reference to a point where an RTP Ethernet packet stream is terminated while system features are executed. H.323 signaling invokes the MTP function when it is needed; however, MTP resources must be made available. For example, if a call is placed on hold or parked, the call that is on hold must temporarily terminate at some hardware point. This hardware point is known as an MTP. The hardware that makes up MTPs is usually a gateway or router port. This is inherent in hardware for calls connecting to the PSTN; however, for *internal calls* put on hold that would otherwise not route through a gateway, an MTP must be allocated. If there are no MTP resources available when a feature is executed, the feature will not work. Some MTP functions are handled through software in smaller IP telephony deployments.

Medium Frequency The range of radio frequencies from 300 to 3000 kHz.

Medium-Independent Interface (MII) In Ethernet, the MII layer defines the electrical and mechanical interfaces between the 100BaseT MAC and the physical sublayer. This setup enables users to connect media-independent products to the physical cabling using Media Attachment Units (MAUs).

Mega (M) The prefix for million. Sixteen megabytes is equal to 16,000,000 bytes, and would be abbreviated 16 MB.

Megabyte One million bytes. Mega is abbreviated "M" and bytes are abbreviated "B." Sixteen megabytes is equal to 16,000,000 bytes and would be abbreviated 16 MB.

Megacop (Slang). Short for MGCP. See, *MGCP.*

Megahertz One million hertz. Mega is abbreviated "M" and hertz is abbreviated "H." Sixteen MHz is equal to 16,000,000 Hz (*hertz* is another word for cycles in radio frequency).

Megohm One million ohm. Mega is abbreviated "M," and the symbol for ohms is "Ω." Sixteen megabytes is equal to 16,000,000 ohms (ohms are a measure of resistance to electricity) and would be abbreviated 16 MΩ.

Member Nortel's name for a trunk. For example, a T1 would contain 24 members. See also *Route.*

Memory   Electronic memory comes in two families, *ROM (Read-Only Memory)* and *RAM (Random-Access Memory)*. Memory devices are made from two different technologies, bipolar (TTL) and *MOS (Metal-Oxide Semiconductor)*. Memory is stored by a technique called *writing* and retrieved by a technique called *reading* (Fig. M.7). ROM devices can only be read and are programmed during manufacture. *PROM (Programmable Read-Only Memory)* devices can be programmed at a later date by an electronics reseller or electronic assembler for a special application using special equipment. Special ROM devices called *EPROMs (Erasable Programmable Read-Only Memory)* can be electronically erased and reused. RAM has read and write capability. The term *random* access means that any memory address can be read in any order at any time. The two types of RAM are static and dynamic. *Static RAM* can hold its memory even when power is removed. *Dynamic RAM* needs constant power to refresh its memory. The following family diagram illustrates the memory types and the technology with which they are made. See also *SD RAM* and *EDO RAM*.

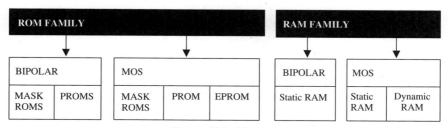

Figure M.7   Memory

Meridian 1   A large-scale *PBX (Private Branch Exchange System)* manufactured by Northern Telecom. For a photo, see *Private Branch Exchange*.

Meridian Administration Tools (MAT)   See *MAT*.

Mesh   A *WAN (Wide-Area Network)* physical or logical network topology in which devices are organized in a manageable, segmented manner with many, often redundant, interconnections strategically placed between network nodes (Fig. M.8). Mesh topologies provide efficient data transport. Many mesh applications are created through virtual private networks. Having many connections through a VPN is cost effective in comparison to a private-line network. A *full mesh* is when all nodes in the network have a connection to all other nodes. A *partial mesh topology* is when some nodes are connected to all other nodes.

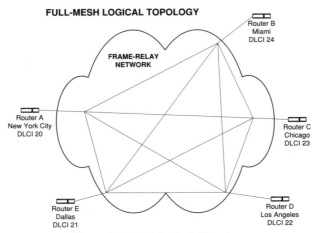

Figure M.8A Full-Mesh Topology

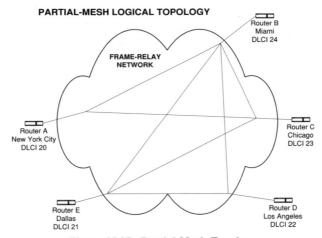

Figure M.8B Partial-Mesh Topology

Message Transfer Part The function in SS7 networking that packetizes and depacketizes signal data.

Message Waiting Usually a light on a telephone that indicates that the user of that phone has a voice mail or a written message left with a hotel clerk or administrative person.

Meters-to-Feet Conversion One meter equals 3.28 feet. One kilometer equals 3280 feet.

Metric (Routing Measure)   A method by which a routing algorithm/protocol determines that one route is better than another. This information is stored in routing tables. Metrics include bandwidth, communication cost, delay, hop count, load, best packet size, path cost, and reliability.

Metropolitan-Area Network (MAN)   A computer network that incorporates the local telephone company's facilities to communicate (Fig. M.9). MAN networks connect other LANs or computers in a city together. T1 private lines are popular for MAN applications.

A MAN NETWORK USING TELEPHONE COMPANY PRIVATE LINES

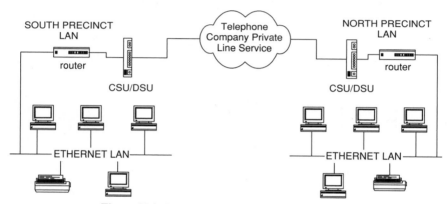

Figure M.9   Metropolitan-Area Network (MAN)

Metropolitan Statistical Area (MSA)   A term that refers to the 306 metropolitan areas that the FCC manages cellular and *PCS (Personal Communications Service)* communications in. There are also *RSA (Rural Statistical Area)* markets that the FCC determined as separate from each other. 428 RSA markets are in the United States. Each statistical area, 734 in all, has at least two licensed service providers.

MFD   A less-common abbreviation for microfarad (μF). See *Microfarad.*

MFJ (Modified Final Judgment)   The judgment that ruled against AT&T—a telecommunications monopoly. AT&T was divided into long-distance, and local telephone service. The local telephone service part of the company (actually 22 companies) were formed into seven *RBOCs (Regional Bell Operating Companies)*. The MFJ also included rules governing the business of the telephone companies involved and set tariffs/pricing limitations on telecommunications services nationwide.

MGCP (Media Gateway Control Protocol) In IP telephony, the procedure for converting Ethernet/connectionless communication to voice/connection-oriented communication. MGCP is a feature incorporated into routers that gives them the ability to transfer/translate signaling information, such as off-hook/on-hook and dialed digits, to an IP telephony server, which can communicate the same types of signaling to IP telephones. MGCP supports common channel signaling (CCS, out-of-band signaling) but not channel-associated signaling (CAS, in-band signaling).

Mho Slang for Siemens, conductance. Conductance is often confused with being the opposite of resistance, which is not the case. Conductance is the reciprocal of resistance. To get the reciprocal of a resistance, simply take one divided by that number, or the resistance. If the resistance of a circuit or component is 500 ohms, then the conductance is equal to 1/500, which is 0.002, 2 millisiemens (2 mS). The higher the number in siemens, the higher the conductance and the lower the resistance.

MHz (Megahertz) See *Megahertz*.

MIB (Management Information Base) A data file within a TCP/IP-loaded network device, such as a PC or router that contains information about that device. The information that is contained in the MIB include hardware addresses (MAC address), counters, statistics, and routing tables. Each software/information category is referred to as an *object*. MIB works in conjunction with *SNMP (Simple Network-Management Protocol)*.

Micro The prefix for one trillionth. Abbreviated "μ," the Greek letter mu. One microfarad is equal to one trillionth of one farad and would be written 0.000,000,000,001 Farad, and abbreviated 1μF.

Microchip A reference to a *VLSI (Very Large-Scale Integration)* electronic device (Fig. M.10). For more information on microchips, see *Very Large-Scale Integration*.

Microfarad Usually represented as μF. Farad is the standard unit of capacitance. A capacitor is an electronic device that has two special properties. It only allows alternating current to pass through it, and it can store an electric charge. One of the many applications of capacitors is to filter alternating current (AC) out of DC power supplies and rectifiers. This is done by placing a capacitor from the DC output to ground. The capacitor appears as an easier path to voltage fluctuations and RFI, and as an impossible path to direct current (DC). Physically, a capacitor is two plates of metal, separated by an insulator (mylar is common). The physical size of a 1-F capacitor would be two sheets of tin foil the size of a football field, insulated (or separated) by a thin sheet of mylar. The farad is a huge unit

Figure M.10 Microchip

of capacitance. This is why most capacitors are microfarads (μF) in value. For a schematic symbol of a capacitor, see *Capacitance.*

Micron   A standard unit of measurement that is equal to $\frac{1}{1000}$ of one millimeter or $\frac{1}{25,000}$ of an inch. The core and cladding of fiber optic is measured in microns.

Microprocessor   Also called a *CPU (Central Processing Unit).* The device within a computer (or switch or other machine that performs complex tasks) that controls the transfer of the individual instructions from one device connected to its bus (the data or I/O bus) to another, such as ROM, RAM, subcontrollers, decoders, and I/O ports. Some communications equipment manufacturers actually call a certain card or portion of the system the CPU. That is because they include all of the RAM, subprocessors, buffers, clocking circuitry, and ROM as a part of the CPU.

Microsoft Certified Professional (MCP)   A well-known industry certification/training program offered by Microsoft and its training partners. The MCP certification is an in-depth learning track that incorporates Microsoft Networking products, Microsoft Applications, and general applications thereof. Up-to-date information regarding the MCP training program can be found through *http://www.microsoft.com.*

Microsoft Certified Systems Engineer (MCSE)   A well-known industry certification/training program offered by Microsoft and its training partners. The MCSE certification is an in-depth study program that incorporates Microsoft Networking products, such as Windows NT and

Microsoft Exchange Server. The learning track for this program is extensive in Microsoft data-base applications and, in general, enterprise networking. Information regarding the MCP training program can be found through *http://www.microsoft.com.*

Microsoft Cluster Server (MSCS)   A feature of Windows NT that enables two NT servers that are connected on a network to logically function as one. Microsoft Cluster Server software also provides failure detection and the ability of one server to take over all functions if the other stops operating.

Microwave   In telecommunications, this is usually a reference to a terrestrial microwave link. The link is made by two radio transceivers equipped with parabolic dish antennas pointed directly at each other (Fig. M.11).

Figure M.11 Microwave

Radio can carry point-to-point transmissions of many bandwidths, including DS1, DS2, DS3, STS1, and OC1. Their range can vary, depending on the size of the antenna (dish), weather in the region, and the amount of power emitted. Including all of the previous factors, a link can range from 0 to 50 miles. For a diagram of a microwave system, see *Terrestrial Microwave*.

Mid-Span   A telephone service wire that runs from a pole to a hook attached to a cable strand, then to a house or building.

Midspan Power   In Ethernet networks, DC power provided to desktop devices on ethernet networks via spare wires within the Ethernet cable. This is not as favorable as inline power (which delivers the same DC power component over the same conductors that the transmit and receive are sent over). See also *Inline Power* and *CDP*.

Mileage of Circuit   The mileage of a private-line circuit is calculated using V and H coordinates. For a table of V&H coordinates, see *Airline Mileage*. AT&T developed a grid-coordinate system that gives every telephone central office in the United States a vertical and horizontal grid number. To calculate the mileage between two cities, the Pythagorean theorem is used. For an example of calculating airline mileage, see *Airline Mileage*.

Milli   Milli is the prefix for one-thousandth, abbreviated "m." Five mA is equal to five thousandths of an amp and is written as 0.005 A or 5 mA.

Million Packets per Second (Mpps)   In LAN and WAN architectures, a standard measure of switched traffic throughput based on a 64-byte packet. This is not an official standard as of this writing (early 2001). The size of the packet varies by several bytes, depending on the manufacturer and type of equipment. Some use packets as small as 60 bytes on networks specifically configured to carry such traffic. The minimum size for an Ethernet packet is 64 bytes. See also *Runt* and *Giant*.

MIME (Multipurpose Internet Mail Extensions)   This standard allows Internet mail users to attach nontext files, such as graphics (JPG), spreadsheets (XLS), and formatted documents (DOC) to their e-mail messages. The files can be binary, text, or multimedia. This standard is an enhancement to *SMTP (Simple Mail-Transfer Protocol)*. The other important protocol with regard to Internet e-mail is *POP (Post Office Protocol)*, which enables users to retrieve mail from outside networks. See also *S/MIME*.

Mini Connector   A physical connector used for PC applications and telecommunication hardware-interface applications (Fig. M.12).

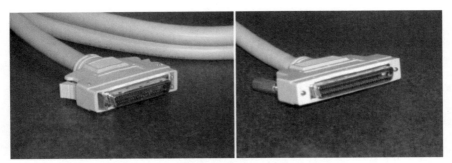

Figure M.12 Mini 50-Pin and Mini 68-Pin Connectors

Mini-T1   Also called a *Mini-T*. A means of connecting a T1 circuit from one side of a building, to another. A T1 cross-connect that runs between two *DSX (Digital Cross-Connect Panels)* within a central office or other telecommunications environment. In central telecommunications environments, there are so many circuits and connections that it is necessary to create a circuit-tracking system just to get a connection made. These Mini-Ts commonly incorporate a building's horizontal and vertical wiring system as well.

Minimum Cell Rate (MCR)   A parameter defined by the ATM forum for ATM traffic management. Minimum cell rate is defined only for *Available Bit-Rate (ABR)* transmissions and specifies the minimum value for the *Allowed Cell Rate (ACR)* parameter.

MIPS (Millions of Instructions Per Second)   A measurement of how fast a microprocessor or central processor can execute program istructions.

MIS (Management Information System)   Also called *IS (Information Systems)*. The part of a company that cares for data and voice communications/processing. The two have been merging together over the past decade and are becoming one entity as communications technology advances. The latest craze in MIS is *CTI (Computer Telephone Integration)*, which enables users to track telecommunications events and operate telecommunications equipment on a computer, with a *GUI (Graphical User Interface,* such as Windows). *IVR (Integrated Voice Response)* is a form of CTI.

MLT (Mechanized Line Testing)   Also called *DATU (Direct-Access Test Unit)*. MLT and DATU is equipment that is either added on or built in to a central-office switch. DATU allows a technician or customer-service agent to dial the phone number of the DATU or MLT equipment and execute a test for shorts, opens, and grounds remotely. In response to a digital voice, the technician enters a password and a choice of options. The results of the test can be read back to the technician by a digital recording or sent to them via an alphanumeric pager. DATU units can also send a locating tone on the technicians choice of TIP, RING, or both TIP and RING. The test unit can also short lines and remove battery voltage for testing.

Mm-Band   The band of frequencies designated by the IEEE between 110 GHz and 300 GHz. For a table, see *IEEE Radar Band Designation*.

Mnemonic   A computer programming command that is an abbreviation or shortened version of what the command does. PRT is a mnemonic in Nortel Networks applications software that makes a switch "print" a specified list of information. LOGI is a mnemonic for "log in."

MNS (Master of Network Science)   A well-known industry certification/training program offered by 3COM. This program is actually referred to as a *credential,* rather than a *certification* by 3COM. Several hardware and administration intensive programs are available under the MNS family of certifications. The first is the *MNS LAN Solutions track,* which includes training on high-performance 3COM LAN switching equipment, seven exams, and a lab test. Another is the *MNS LAN Solutions Plus track,* which requires completion of the LAN Solutions requirements, and extends the trainee's knowledge into the ATM products offered by 3COM. Another example track is the *MNS Network Architecture,* which provides a rounded study of enterprise networking. Other MNS tracks include the *MNS Network Management, MNS WAN Solutions,* and *MNS Remote-Access Solutions.* Information regarding the 3COM MNS traiing programs can be found at *http://www.3com.com.*

Mobile   A communications link made by portable radio.

Modal Dispersion   As light travels down a fiber optic, each individual light ray/particle takes a different path (Fig. M.13). Imagine that a bunch of small rubber balls are shot down a long tube at the same time. Each ball will bounce differently as they make their way around curves. At the

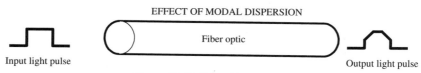

EFFECT OF MODAL DISPERSION

Fiber optic

Input light pulse

Output light pulse

Figure M.13   Modal Dispersion

end of the tube, the balls will come out at different times. Light behaves the same way. A sudden pulse of light on one side of a fiber optic will disperse itself as it traverses down the fiber, causing the pulse of light at the far end to be more of a "blip."

Modal Loss The attenuation of a light signal as it travels through a fiber optic because of tight bends. See *Modal Dispersion*.

Modem (Modulate/Demodulate) A device that transmits digital information over a telephone line (standard POTS line) or a private circuit (56 K line). Modems modulate the digital information before transmitting it. One standard of modulation is *FSK (Frequency-Shift Keying)*. Each positive (1) bit is sent as a frequency or "pitch" of sound and each (0) bit is transmitted as a different frequency or "pitch" of sound.

Modem Eliminator A device similar to the baseband modem in that it does not do any actual modulation, and allows connection of two DTE devices. It is intended for very short distances, for example, from one device to another right next to it. Modem eliminators connect at the ITU/T V.24 and V.28 (or EIA 232-D, where only the names of the pins are different) interface-level standard. Some modem eliminators are simply a null/modem cable. Some have their own clock source added for conditioning or buffering the transmission.

Modem Standards

Standard	XFR rate	Modulation	Duplex
V.21 / Bell 103	300	FSK	Full duplex
V.22	1200	DPSK	Full duplex
V.22 bis	2400	QAM	Full duplex
V.23 / Bell 202	1200/75	FSK	Half duplex
Bell 212A	1200	DPSK	Full duplex
V.32	9600	QAM	Full duplex EC
V.32 bis	14,400	TCM	Full duplex EC
V.32 ter	19,200	TCM	Full duplex EC
V.34	28,800	TCM	FULL DUPLEX EC
V.90	56,600	TCM	EC = Error Correction

Modified Cut-Through Switching There are three ways that frames/packets transverse through a LAN switch, bridge or router. The first is *store and forward*, where the entire frame and its contents are accepted and stored in the switch. Error detection is calculated (CRC), and if the frame is good, the address is looked up in the routing table. When the associated destination port/segment is found, the frame is sent on to its destination. This is a good method for routing traffic because damaged frames, runt frames, and giant frames are discarded before they are transmitted. This method is used where the network infrastructure or media is prone to damaging frames, such as RFI environments, or a poor WAN network service. The disadvantage of the store and forward method is *latency*. Storing the entire frame while the destination port is retrieved causes a delay, and in multiple-hop networks, this can cause slow network performance even when there is very little traffic.

The second way that frames/packets transverse through a LAN switch/router/bridge is the *cut-through switching* method, where only the address of an incoming frame is processed by memory. The address is associated with its destination port/segment in the routing table, and the entire frame is sent directly through. This process happens if the frame is good or not, as long as there is a nondamaged address in the frame. Cut-through switching greatly reduces latency delays through a network, but still transmits bad frames. If an NIC card or host device begins sending lots of bad erroneous frames, the network performance could be slowed greatly. Some LAN switches have safeguards in place to detect and suppress error storms from defective equipment.

The third method of forwarding frames is *modified cut-through,* which works similar to store and forward, except that it uses a limited number of bytes to check for errors rather than the entire frame. This method helps prevent the retransmission of defective frames and also provides an acceptable level of latency delay through the network.

Modular   A reference to equipment that is equipped with plug-in type interfaces, rather than being hard wired.

Modular Adapter   A device used to interconnect one wire/cable type with another, without the use of termination blocks (Fig. M.14). Also called *harmonica adapters*. See also *258A Adapter*.

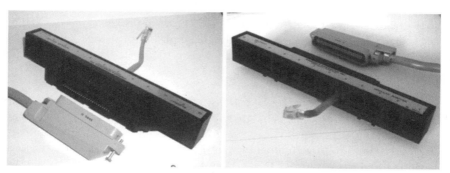

Figure M.14 Modular Adapter: 50-Pin AMP to RJ11

Modular Jack   A jack that is equipped with a plug so that devices can be easily attached and detached. Old jacks, still found in old homes, are hard wired, which means that the telephone cord had to be permanently affixed to the terminals inside the jack with screws. The same went for nonmodular or hard-wired telephones. If you wanted to have a longer cord, you couldn't buy one at the store and just plug it in. You had to call the phone company and they would send a telephone technician out to install a longer cord for you.

Modulation   A method of varying a radio carrier frequency so that a signal (the variations) can ride on it. After the carrier signal has the variations imposed on it, it is amplified and transmitted. The variations in the signal are then detected by the receiver. The variations in the carrier signal are actually voices, music, or whatever is to be transmitted. The different methods of modulating a carrier frequency are *AM (Amplitude Modulation), FM (Frequency Modulation),* and *PM (Phase Modulation).*

MOH (Music On Hold)   A feature of PBX and key systems that allows an audio signal from a tape recorder, radio, or other audio device to be fed to callers that are on hold. The PBX or key system user manual designates how to cross connect the audio signal into the system.

Monopole Antenna   An antenna mast of one pole extending from the ground. These are popular with cellular and PCS wireless services (Fig. M.15).

MOP (Method Of Procedure)   When network engineers design an addition to a network (a new node in a SONET ring is a good example), they write a MOP. The MOP instructs technicians step by step which circuits to reroute, which circuit cards to swap and when, and when to activate the new node. MOPs are a crucial communication tool between engineers and technicians.

Morse Code   In 1836, Samuel F.B. Morse built the first working telegraph. He also derived a code that enabled people to exchange information (Fig. M.16). The Morse Code is still in use today. It is used by amateur radio operators, ships at sea, etc.

MOS   Another reference to *CMOS (Complementary Metal-Oxide Semiconductor).* The reason why many computer and other high-speed components are static sensitive. CMOS' largest advantage over TTL is their low power consumption (less than $\frac{1}{10}$ of TTL); they switch on without drawing very much current in contrast to TTL technology.

Figure M.15 Monopole Antenna With PCS/Cellular Antenna Array

Because very little current is drawn, very little power is consumed and very little heat is given off. This allows the devices to be much smaller.

MOS (Mean Opinion Score) A voice quality measurement of Codec operation (such as G.711) over IP telephony networks. Mean opinion score rates voice quality on a scale of 1 to 5, 1 being bad and 5 being excellent. The MOS score is made by a wide range of listeners hearing a transmission and then rating it. G.711 PCM rates at an MOS of 4.1; G.729 CS A CELP rates at a 3.9; and G.729A rates at a 3.7. The major quality issue to the listeners that are able to notice the difference is a slight delay and nearly undetectable voice distortions. See also *G.729*.

MOSPF (Multicast OSPF) A multicast addressing and transmission method that uses OSPF (Open Shortest Path First) to develop a multicast tree, then MOSPF to implement a broadcast. See also *PIM Dense Mode* and *PIM Sparse Mode*.

MORSE CODE

A • —	N • —	1 • — — — —
B — • • •	O — — —	2 • • — — —
C — • — •	P • — — •	3 • • • — —
D — • •	Q — — • —	4 • • • • —
E •	R • — •	5 • • • • •
F • • — •	S • • •	6 — • • • •
G — — •	T —	7 — — • • •
H • • • •	U • • —	8 — — — • •
I • •	V • • • —	9 — — — — •
J • — — —	W • — —	0 — — — — —
K — • —	X — • • —	. • — • — • —
L • — • •	Y — • — —	? • • — — • •
M — —	Z — — • •	- — • • • • —

Figure M.16 Morse Code

Mouse   A serial-bus I/O device used to point and select objects on a computer monitor screen. There are many manufacturers of mouse hardware and many variations thereof. Some mice have fixed roller balls, some are used with the thumb, others with the forefinger. The three major types of mice interfaces are: USB, serial, and PS2. They are distinguishable by their connector interface.

MPEG (Motion Picture Experts Group)   A widely used menu of standards for compressing video. MPEG1 is a bit-stream standard for compressed video and audio optimized to fit into a maximum bandwidth of 1.5 Mbps and with less resolution, MPEG2 is intended for higher-quality video-on-demand applications and runs at data rates between four and nine Mbps. MPEG4 is a low-bit-rate compression algorithm intended for 64 Kbps connections.

MPLS (MultiProtocol Label Switching)   A method used to carry IP over WAN ATM. MPLS evolved from the Cisco Systems' proprietary communications method called *tag switching,* which was submitted to the IETF in 1997. This packet-to-cell loading and unloading method enables telecommunications service providers to furnish IP WAN transport services to customers. This method is favorable by service providers because customer traffic can be less expensively inserted directly into backbone architecture without having to use an intermediate service, such as frame relay.

MPOA (Multiprotocol Over ATM)   An ATM Forum standardization effort that specifies how existing and future network-layer protocols, such as IP, IPv6, AppleTalk, and IPX run over an ATM network with directly attached hosts, routers, and multilayer LAN switches. See also *Local-Area Network Emulation.*

MPPS (Million Packets per Second)   In LAN and WAN architectures, a standard measure of switched traffic throughput based on a 64-byte packet. This is not an official standard as of this writing (early 2001). The size of the packet varies by several bytes, depending on the manufacturer and type of equipment. Some use packets as small as 60 bytes on networks specifically configured to carry such traffic. The minimum size for an Ethernet packet is 64 bytes. See also *Runt* and *Giant.*

MQA (Multiple Queue Assignment)   A reference to the ability of an ACD system to allow agents to log into multiple queues. A Northern Telecom Meridian PBX with MQA software is capable of allowing agents to log into and receive calls from five separate queues. When properly used, MQA makes call centers more efficient by allowing agents to share the incoming call load more effectively. See also *Skills Based Routing.*

MRP (Multi-Service Route Processor)   In Cisco Systems IP telephony, a card that fits into an ICS7750 cabinet. The MRP gives the system layer 3 capability and has two WIC/VIC slots. This enables the MRP card to act as an end station interface for a FAX machine, 2500 telephone, or modem. With a WIC card on board, the MRP can act as an interface to a telephone company through T1, or single loop start, E&M, or ground start trunk. The ICS7750 is a midsized IP telephony system.

MSA (Metropolitan Statistical Area)   A term that refers to the 306 metropolitan areas where the FCC manages cellular and PCS communications in the United States. The FCC has determined 428 *RSA (Rural Statistical Area)* markets as being separate from each other. Each statistical area, 734 in all, has at least two licensed service providers.

MSB (Most Significant Bit)   The bit in an octet that carries the most value. You can better understand this by comparing it to our base 10 numbering system. Imagine a "most significant number." If you are 43 years old, the 4 is the most significant number and the three is the least significant number. If the 4 were lost, then you would only be three (a very significant difference). If the 3 were lost, you would still be 40 (not so significant).

MSC (Mobile Switching Center)   A place where cellular telephone call traffic is controlled. A cellular switch is used to perform the functions of the MSC. Bandwidth and cells are switched between users, and trunks that interface to landlines are also managed here.

MSCS (Microsoft Cluster Server)   A feature of Windows NT that enables two NT servers that are connected on a network to logically function as one. Microsoft Cluster Server software also provides failure detection and the ability of one server to take over all functions if the other stops operating.

MTP (Media Termination Point)   In IP telephony, a reference to a point where an RTP Ethernet packet stream is terminated while system features are executed. H.323 signaling invokes the MTP function when it is needed; however, MTP resources must be made available. For example, if a call is placed on hold or parked, the call that is on hold must temporarily terminate at some hardware point. This hardware point is known as an MTP. The hardware that makes up MTPs is usually a gateway or router port. This is inherent in hardware for calls connecting to the PSTN; however, for *internal calls* put on hold that would otherwise not route through a gateway, an MTP must be allocated. If there are no MTP resources available when a feature is executed, the feature will not work. Some MTP functions are handled through software in smaller IP telephony deployments.

MTSO (Mobile Telephone Switching Office)   This is where all control is done for a cellular switching network within a LATA. Smaller MSCs hand off calls to the MTSO. The MTSO is where the billing, trafficking, maintenance monitoring, and hand-offs to long distance and local land-based carriers happens.

Mu Law   The ITU-T companding standard used in conversion between analog and digital signals in cellular radio systems. The Mu-Law standard is used in North America; the majority of the world utilizes the European A-Law standard. Also see *Companding*.

Muldem (Multiplexer Demultiplexer)   Another name for a multiplexer.

Multicast   In Ethernet, packets are classified by routers into three categories: unicast, broadcast and multicast. A unicast address is unique to a single host (a computer, IP phone, or other network appliance). A broadcast packet floods a network, which means it is sent to every host, and every host processes the packet. Broadcast packets are not forwarded by routers, so they are limited to the domain that they are sent in. To meet the needs of business applications such as videoconferencing, where more than one user must receive the data yet not all hosts are forced to receive and process the packets and packets must be forwarded beyond a router, multicast techniques were developed. Multicast enables select groups of users to receive the same data transmission on a network. Multicast is made possible by PIM (Protocol Independent Multicast). A network can be configured to utilize PIM as one of two features: PIM Dense Mode and PIM Sparse Mode. PIM Dense Mode floods a network, sending the packets that make up the transmission to every device on a network. Routers on the network then identify which users respond to the Multicast flood, and then prune (discontinue) packet transmissions from devices that do not respond. The advantage of PIM Dense Mode is that its operation is simple to end users. There is no need for an advance request to start receiving the transmission. It appears to be automatic to the end user because it is. PIM Sparse Mode incorporates IGMP, which is a signal for end users to request a multicast packet stream. The advantage of this method is that in large networks, the initial flood in PIM Dense Mode never happens. The transmission is sent to a single router called a *rendezvous point*. The rendezvous point router then duplicates the packets to the necessary segments that have requested them. PIM Sparse Mode is a more efficient use of a network's resources. The drawback is the need for requests to refresh the transmission.

There are many multicast protocols that add features to PIM; however, they are almost always proprietary to one vendor and designed to suit a very specific traffic solution.

Multicast Address   A single address that transmits to multiple network devices. IP multicast addresses range from 224.0.0.0 to 239.255.255.255. Multicast addresses are also called *group addresses*. See also *Multicast* and *Broadcast*.

Multicast, Protocol Independent (PIM Dense Mode/Sparse Mode)
In Ethernet-type networks such as 802.3, multicast is made possible by a set of communications procedures called PIM (Protocol Independent Multicast). A network can be configured to utilize PIM as one of two features: PIM Dense Mode and PIM Sparse Mode. PIM Dense Mode floods a network, sending the packets that make up the transmission to every

device on a network. Routers on the network then identify which users respond to the Multicast flood, and then prune (discontinue) packet transmissions from devices that do not respond. The advantage of PIM Dense Mode is that its operation is simple to end users. There is no need for an advance request to start receiving the transmission. It appears to be automatic to the end user because it is. PIM Sparse Mode incorporates IGMP, which is a signal for end users to request a multicast packet stream. The advantage of this method is that in large networks, the initial flood in PIM Dense Mode never happens. The transmission is sent to a single router called a *rendezvous point*. The rendezvous point router then duplicates the packets to the necessary segments that have requested them. PIM Sparse Mode is a more efficient use of a network's resources. The drawback is the need for requests to refresh the transmission.

Multihop   A reference to microwave links that require two or more links to get to a destination. Multihop links can extend distance and enable a more flexible path to go around buildings or mountains.

Multilayer LAN Switch   A LAN switch that is capable of bridging frames at the OSI layer 2 (MAC address) level and/or forward (route) packets at the OSI layer 3 (i.e., IP address) level. Some multilayer switches are capable of switching or routing traffic at layer 4 and above, identifying the application of the data inside the packet so it can be prioritized as voice, video, and data. See also *LAN Switch* and *Layer 3 Switch*.

Multi-Layered Switching   (Also known as hierarchical switching.) A method of LAN network design where there are three layers of switches: a core layer, a distribution layer, and an access layer. This model is used when designing LAN networks because it allows for methodical scalability and easy deployment of redundancy where needed.

Multimeter   An electronic test device used to measure voltage levels, electric current, and circuit resistance. Some multimeters are analog and some are digital. For a photo of a digital multimeter, see *Voltmeter*. For a photo of an analog cable test meter, see *145A*.

Multimode   The alternative to *Single-Mode* fiber optic. Multimode has a larger core (50 to 100 micron). Therefore, it accepts more light and more frequencies of light. Multimode is used for shorter-distance applications, such as LANs. Singlemode fiber optic has a smaller core (5 to 15 micron), but is capable of longer-distance transmissions. It is used in the public network more often and is the choice for SONET applications. Multimode fiber optic is made with an orange-colored tube or insulation, and single mode is made with yellow.

MultiNAM   A cellular phone that is programmed to have multiple phone numbers, usually two. MultiNAM cell phones can have numbers that are subscribed to from different cellular companies.

Multiple Domain Network   SNA network with multiple SSCPs. See also *SSCP*.

Multiple Queue Assignment (MQA)   A reference to the ability of an ACD system to allow agents to log into multiple queues. A Northern Telecom Meridian PBX with MQA software is capable of allowing agents to log into and receive calls from five separate queues. When properly used, MQA makes call centers more efficient by allowing agents to share the incoming call load more effectively. See also *Skills Based Routing*.

Multiplex   Multiplexing is the process of encoding two or more digital signals or channels on to one. Channels are multiplexed together to save money. When we use all of the wires in a cable and need more, it costs less to add electronics on the ends of a cable than to install a new one (imagine the expense from LA to NY). A T1 encodes 24 channels into 1 by using frequency-division multiplexing. In a simpler explanation, a T1 makes it possible to place 24 lines that once needed 24 pairs on only 2 pairs. When a group of signals are multiplexed together, they are all sampled at a high rate of speed, faster than the combined speed of all the channels being multiplexed. For a diagram on the multiplexing process, see *Time-Division Multiplexing*.

Multiplexer   An electronic device that encodes several digital signals into a single digital signal for transmission on a single medium (such as a pair of wires). For a diagram on the multiplexing process, see *Time-Division Multiplexing*. Figure M.17 shows an Alcatel DS3 to DS1 multiplexer. For another photo, see *Mux*.

Multipoint Control Unit (MCU)   1. A hardware and software device that in it's respective application, be it voice or video, provides a point of management and control for multiple voice or video data streams that could possibly be conferenced together. MCUs are generally proprietary devices that incorporate their own features for managing the flow of a video or voice application and which users can see or hear each other, and also in what fashion the users are seen. For example, if the MCU is designed for video conferencing, it could make four users appear in small boxes in each corner of a video monitor, or make the user that is speaking be the one viewed by all other users. 2. *Media Conference Unit* Performs the same function as an MCU.

Figure M.17 Multiplexer: DS3 to DS1

Multipoint Line   An older name for a bus physical topology. In Ethernet, a bus is a communications line, twisted pair or coax, with multiple points of connection. In bus physical topology, all points of connection must be terminated to a workstation (host) or a line terminator. See also *Bus Topology*.

MultiProtocol Label Switching (MPLS)   A method used to carry IP over WAN ATM. MPLS evolved from the Cisco Systems' proprietary communications method called *tag switching*, which was submitted to the IETF in 1997. This packet-to-cell loading and unloading method enables telecommunications service providers to furnish IP WAN transport services to customers. This method is favorable by service providers because customer traffic can be less expensively inserted directly into backbone architecture without having to use an intermediate service, such as frame relay.

MultiProtocol Over ATM (MPOA)   An ATM Forum standardization effort specifying how existing and future network-layer protocols, such as IP, IPv6, AppleTalk, and IPX, run over an ATM network with directly attached hosts, routers, and multilayer LAN switches. See also *Local-Area Network Emulation*.

Mushroom Board   Also called a *white board* or *peg board*. It is placed between termination blocks (such as 66M150 blocks) to provide a means of support for routing cross-connect wire. For a photo, see *White Board*.

Music on Hold (MOH)   See *MOH.*

Mute   A feature of PBX and key telephones that turns off the microphone. Mute is also used in conjunction with handsfree to prevent the other party's voice from cutting in and out during a call.

Mux   A shortened name for multiplexer (Fig. M.18). For a diagram of the multiplexing concept, see *Time-Division Multiplexing.* For an additional photo, see *Fig. M.17.*

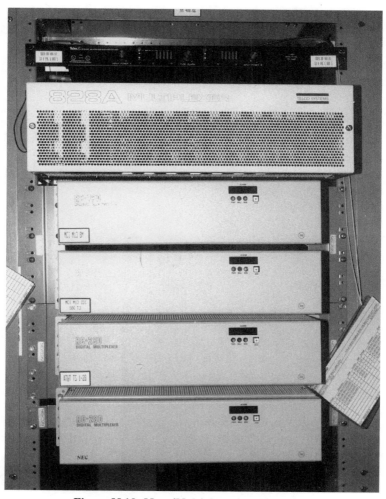

Figure M.18 Mux (Multiplexer): DS3 to DS1

N

N-Type Connector   A connector used to fit RG-8 coax. N-type connectors are used in microwave radio applications where the indoor radio unit connects to an outdoor dish unit via a coaxial cable feed. These connectors are also very popular in cellular/PCS/paging radio applications (Fig. N.1).

Figure N.1   N-Type Connector Used in Cellular and Microwave Radio Applications

N2   A parameter setting of mechanical cycles (usually 10 or 20) in the X.25 protocol that determines how many times a sending device (DCE device) will poll a receiving device (DTE device). The poll is sent because the far equipment is not responding. After the N2 is reached, a *Set Asynchronous Balanced Mode (SABM)* will be transmitted by the polling device, which will ultimately reset the entire link.

NADN (Nearest Active Downstream Neighbor) In token-ring or IEEE 802.5 networks, the closest downstream network device from any given device that is still active.

NAK The ASCII control-code abbreviation for negative acknowledge. The binary code is 0101001 and the hex is 51.

Naked Call An incoming call that receives no greeting message and no call menus or flexible routing before it is routed into an ACD queue.

NANP (North American Number Plan) See *Area Code.*

NAT (Network Address Translation) An early network protocol/ feature that advertises address changes. This has been replaced with newer methods. See *Layer 2,3,4 Switching.*

National Access Fee A Federal tax placed on telecommunications services provided by telephone companies.

National Television Standards Committee (NTSC) The pre-HDTV broadcast standard in Canada, Japan, the United States, and Central America. NTSC defines 525 vertical scan lines per frame and yields 30 frames per second. The scan lines refer to the number of lines from top to bottom on the television screen. The frames per second refer to the number of complete images that are displayed per second. See also *Television Broadcast Standards.*

Native Protocol The format of the LAN that is being transmitted over a WAN. For instance, if a LAN is Ethernet and it connects to a remote LAN via a frame-relay network, the native protocol is Ethernet, and the connection protocol is frame relay.

NAUN (Nearest Active Upstream Neighbor) In token-ring or IEEE 802.5 networks, the closest upstream network device from any given device that is operational.

NCP (Netware Core Protocol) The heart of the Novell NetWare operating system. It is a set of programmed instructions that enables communication to occur between software programs on a workstation and a file server's operating system. It manages the high-level aspects of communication, including: user account authorization, file retrievals, and remote printing services. Furthermore, NCP is a connection-oriented service. It acknowledges packets that have been received and requests retransmissions of lost or discarded packets.

NDIS (Network Driver Interface Specification)   A standard data-link layer (OSI level 2) protocol within the TCP/IP family that enables multiple protocols and multiple network (physical layer) adapters to exist on the same computer. NDIS allows all upper layer protocols to use the same *Network Interface Card (NIC)*.

NDT (No Dial Tone)   An abbreviation frequently used on telephone company repair orders and by service personnel.

NE (Network Element)   A device attached to a network via hardware or software that performs a service or function to the network. A network element can be a router, a host, a workstation, a hub, a central office switch, a private branch exchange switch, a voice-mail system, a firewall/security program, or any other network-servicing entity.

Near-End Cross Talk (NEXT)   The uncommon phenomena of signals sent over twisted copper pairs bleeding onto each other via magnetic fields produced at cross connections, or within defective electronic equipment. The *near end* refers to the problem occurring between a switch and a DSLAM or other device within a central office/node. For a diagram, see *Far-End Cross Talk*.

Neighboring Routers   A reference to routers that are adjacently connected within a network. In multiple-protocol networks, neighbors are dynamically discovered by the OSPF Hello protocol.

NetBIOS (Network Basic Input/Output System)   A set of instructions within the Novell NetWare protocol stack that extend a PC's BIOS instructions to include those that enable communicating beyond its own hardware and into a network.

NetWare   A trademark of Novell. Netware is a widely utilized network operating system software that was developed by Novell from the *XNS (Xerox Network Systems)* architecture, which was originally released in 1981. The Netware protocol suite is defined in the top five layers of the OSI, and can be made to run on virtually any data-link and physical layer system (Fig. N.2).

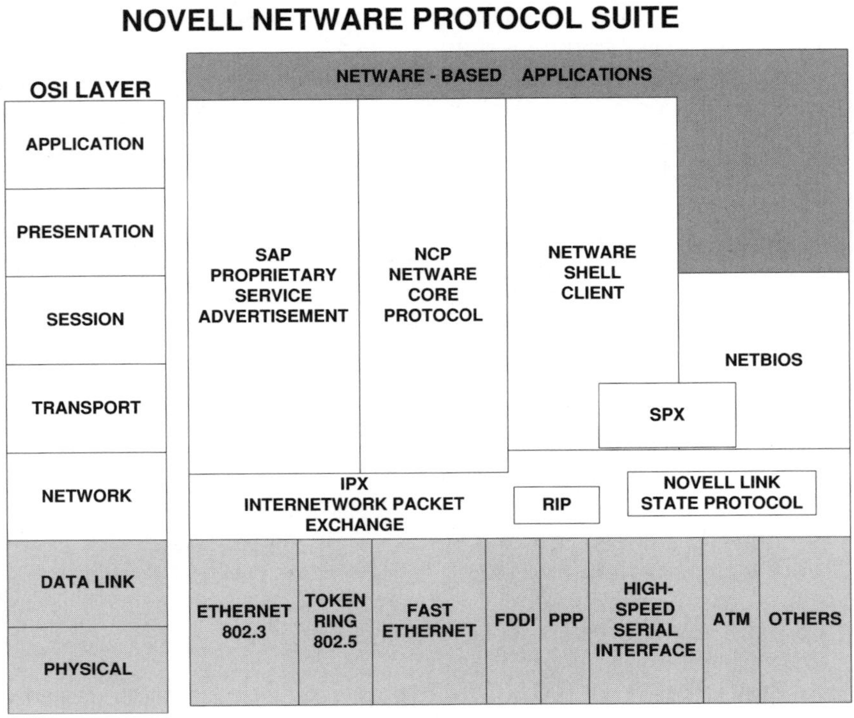

NOVELL NETWARE PROTOCOL SUITE

Figure N.2   Novell NetWare Protocol Suite

NetWare Core Protocol (NCP)   The heart of the Novell NetWare operating system. It is a set of programmed instructions that enables communication to occur between software programs on a workstation and a file server's operating system. It manages the high-level aspects of communication, including: user account authorization, file retrievals, and remote printing services. Furthermore, NCP is a connection-oriented service. It acknowledges packets that have been received and requests retransmissions of lost or discarded packets.

NetWare Loadable Module (NLM)   A Novell trademark. An individual program or application that can be loaded into memory and function as part of the Novell NetWare *NOS (Network Operating System).*

NetWare Shell   A function in the Novell Netware Protocol Stack. The NetWare Shell or "requestor shell" stays resident in a workstation or server's memory. It decides whether or not to send data/instructions entered by the user (or application) to the network. Each time the workstation user executes a command, the NetWare Shell software program determines whether the call/instruction is for the user's PC or for a remote server on the network.

Network   A group of devices that communicate back and forth using a set of rules or a set of protocols (called a *protocol stack* in data communications). The medium that the devices communicate through can be copper wire (UTP), fiber optic, coax, fiber optic, air/vacuum (radio), or light (infrared).

Network Architecture   The combination of software and hardware type of a network. Each network architecture can have one or more protocols within it.

Network Element (NE)   A device attached to a network via hardware or software that performs a service or function to the network. A network element can be a router, a host, a workstation, a hub, a central office switch, a private branch exchange switch, a voice-mail system, a firewall/security program, or any other network-servicing entity.

Network File System (NFS)   As commonly used, a distributed file-system protocol suite developed by Sun Microsystems that allows remote file access across a network. In actuality, NFS is simply one protocol in the suite. NFS protocols include NFS, RPC, XDR, and others. These protocols are part of a larger architecture that Sun refers to as *ONC*. See also *ONC*.

Network Forwarding Rate   In bridges or routers, the amount of data in packets per second that a device such as a switch/router/bridge can transfer traffic in on one port and out on another.

Network Interface (NI)   Also called a *Standard Network Interface (SNI)*, demarcation point, or lightning protector. The device that contains carbons to protect a phone line from being overloaded by lightning and acts as the separation point between the telephone company's wire and the customer's wire, which is also called the *IW (Inside Wire)*. For a photo, see *Standard Network Interface*.

Network Interface Card (NIC)   An expansion board that plugs into a motherboard via an ISA or PCI expansion socket/slot. The network interface card provides the electronic and the physical interface for the network of its type. Network types include, but are not limited to, Ethernet and token ring. For a photo of a PCI Ethernet NIC, see *NIC*.

Network Layer   A layer in a communications protocol model. In general, the network layer does the job of switching and routing of the data being transmitted within the protocol. A central-office switch would be a good example of a network layer function. The latest model (guideline) for communications protocols is the *OSI (Open Systems Interconnect)*. It is the best model so far because all of the layers or functions work independently of each other. For a diagram of the OSI model and its layers, see *Open Systems Interconnection*.

Network Node Interface (NNI) In the ATM world, this type of connection provides 4096 virtual paths and 65,536 virtual connections within each path for a total of 268,435,456 channels between two ATM networks, switches, or users that are physically connected over a *UNI (User Network Interface)*.

Network Operating System (NOS) A software that manages communications of devices. An administrator sets access and security privileges to users of a network and monitors network performance via this software. In Novell networks, the NOS provides sockets at OSI level 4, which provide the link between software applications and lower network layers. Network operating systems are distributed file systems. Examples include Novell NetWare, *NFS (Network File System* for UNIX), Windows NT, and Banyan VINES.

Network Service Access Point (NSAP) The logical, software, or virtual interface between the *OSI (Open Systems Interconnect)* network layer (3) and transport layer (4). This is known as a *socket* in Novell networks.

Network Service Provider (NSP) A company that provides telecommunications services such as frame relay, ADSL, HDSL, ATM, Internet Access or other services via their own switching equipment, or leased switching equipment. NSP's usually provide services by utilizing a local telco or CAP for the last mile (local loop). They also link their nodes via high capacity circuits, and have interconnection agreements with other telecommunications companies.

Network Termination Equipment (NTE) The generic term for *DCE* and *DTE (Data Communication Equipment* and *Data Termination Equipment)*. With newer transmission methods, such as those provided in the xDSL family, DCE and DTE are replaced with near-end NTE and far-end NTE. The near end is the central office transmission interface and the far end is the xDSL router.

Network Transit-Time Delay The amount of time (in milliseconds) that it takes for a data packet to traverse across a network. Transit time delay is often rated based on a data packet making a round trip via a ping command. A typical round-trip transit time delay across a very robust network that stretches from Tokyo to New York is 200 ms. Transit time delay is increased by the number of routing or switching devices that a packet passes through. A concern for network designers is to have enough route redundancy to ensure that alternate routes can be taken by packets when connections fail. The disadvantage to having redundancy is that more switching and routing equipment is required, which increases transit-time delay.

Neutral   Also called common or floating ground. Neutral/common is a reference point and is ungrounded. It is usually a signal return or DC reference coupling for transmission circuits.

NEXT (Near-End Cross Talk)   The uncommon phenomena of signals sent over twisted copper pairs bleeding onto each other via magnetic fields produced at cross connections, or within defective electronic equipment. The *near end* refers to the problem occurring between a switch and a DSLAM or other device within a central office/node. For a diagram, see *Far-End Cross Talk*.

Next-Hop Resolution Protocol (NHRP)   A set of instructions used by routers that enables them to automatically discover the physical-layer address of other routers and hosts connected to a *NonBroadcast Multi-Access (NBMA)* network. Data routed with NHRP can then directly address sections of the network. This allows users/devices to communicate without traffic having to use an intermediate hop, increasing performance in ATM, frame-relay, SMDS, and X.25 environments.

NFAS (Non-Facility Associated Signaling)   In ISDN PRI (Primary Rate Interface) signaling, the ability of D channels to carry signaling information for multiple 24 channel PRIs. This is an advantage for those who would like to squeeze that last (24th) B channel out of their circuits. The only drawback is that if the D channel should go down, then all PRI circuits associated to that PRI go down with it. It is a feature of NFAS signaling, and very common to configure a back-up D channel on one of the secondary PRI circuits.

NFS (Network File System)   As commonly used, a distributed file-system protocol suite developed by Sun Microsystems that allows remote file access across a network. In actuality, NFS is simply one protocol in the suite. NFS protocols include NFS, RPC, XDR, and others. These protocols are part of a larger architecture that Sun refers to as *ONC*. See also *ONC*.

NHRP (Next Hop Resolution Protocol)   A set of instructions used by routers that enables them to automatically discover the physical-layer address of other routers and hosts connected to a *NonBroadcast Multi-Access (NBMA)* network. Data routed with NHRP can then directly address sections of the network. This allows users/devices to communicate without traffic having to use an intermediate hop, increasing performance in ATM, frame-relay, SMDS, and X.25 environments.

NI (Network Interface)   See *Network Interface.*

Nibble   Four bits, or 1/2 of a byte.

Nibble Coding   It takes four binary bits to make a decimal number, so two decimal numbers can be fit into one byte. X.25 uses a method referred to as *nibble coding,* which breaks overhead bytes into nibbles to help make it easier to understand the addressing of the packets. Back when X.25 was new (1970s), dynamic memory was very costly, so compacting addresses into nibbles was more efficient and cost-effective.

NIC (Network Interface Card)   An expansion board that plugs into a motherboard via an ISA or PCI expansion socket/slot. The network interface card provides the electronic and the physical interface for the network of its type. Network types include, but are not limited to, Ethernet and token ring. A PCI Ethernet NIC is shown in Fig. N.3.

Figure N.3 PCI Slot NIC Card

Night Service   A feature of PBX and hybrid key systems that allows the lines ringing into an office to be handled differently during certain times of the day. The phone system is programmed as two different systems, usually a day system, and a night, or after-hours system. If a user would like all calls that come into the office after hours to ring to a voice-mail system, or be forwarded to security, it can be done with the night-mode feature. Some systems are equipped with software that allows the night-mode feature to activate automatically at certain times of the day.

NLM (Netware Loadable Module)   A Novell trademark. An individual program or application that can be loaded into memory and function as part of the Novell NetWare *NOS (Network Operating System)*.

NMC (Network Management Center)   A place where large or public telephone networks are managed, monitored, and maintained from a central location.

NNCDE (Nortel Networks Certified Design Expert)   A well-known industry certification/training program that is offered by Nortel Networks. This certification recognizes an advanced level of network design, planning, and optimization using Nortel Networks products. It requires industry experience as well as Nortel training. Information regarding Nortel Networks' certification programs can be found at *http://www.nortelnetworks.com.*

NNCNA (Nortel Networks Certified Network Architect)   A well-known industry certification/training program that is offered by Nortel Networks. This advanced certification not only requires learned technical knowledge, but also industry experience. Students seeking this certification are required to obtain a required number of points in certification-related areas. A specific number of points are obtained through job experience and a number are obtained by taking Nortel Networks courses. Information regarding Nortel Networks' certification programs can be found at *http://www.nortelnetworks.com.*

NNCSE (Nortel Networks Certified Support Expert)   A well-known industry certification/training program that is offered by Nortel Networks. This certification recognizes an advanced level of technical post-sales expertise required to implement, support, troubleshoot, and optimize Nortel Networks products. This certification requires industry experience points as well as Nortel Networks training. Information regarding Nortel Networks' certification programs can be found at *http://www.nortel-networks.com.*

(NNCSS) Nortel Networks Certified Support Specialist A well-known industry certification/training program that is offered by Nortel Networks. This certification recognizes a fundamental level of post-sales technical expertise required to deploy, operate, and troubleshoot Nortel Networks products. This certification requires industry experience points, as well as Nortel Networks training. Information regarding Nortel Networks' certification programs can be found at *http://www.nortel networks.com.*

NNI (Network Node Interface) In the ATM world, this type of a connection provides 4096 virtual paths and 65,536 virtual connections within each path for a total of 268,435,456 channels between two ATM networks, switches or users that are physically connected over a *UNI (User Network Interface).*

NOC (Network Operations Center) A place where large or public telephone networks are managed, monitored, and maintained.

NOD (Network Outward Dialing) See *Network Outward Dialing.*

Node 1. In local-area networking, any entity or device attached to a network that possesses a *MAC (Media-Access Control)* address. 2. In wide-area networking, a node is a router or switch that serves as a control point in a network, and those control points (routers/switches) incorporate a routing protocol to communicate network information and to route traffic through the network. 3. In voice networking, a node is a device that has switching capability, or a central office. 4. In general, it is common for a node to be used as a reference for any device that is connected to a communications network via a copper wire, fiber-optic, radio link, or infrared light. These devices include routers, bridges, terminals, computers, hubs, controllers, and switches.

Noise Noise is any kind of distortion or unwanted signal. The two main categories of noise are electromagnetic interference and ambient noise. Electromagnetic interference is caused by a radio signal or other magnetic field inducing itself onto a medium (twisted-/nontwisted-pair wire) or device (telephone or other electronics). The world we live in is full of radio waves that are emitted from electric appliances, such as blenders, automobile engines, transmitters, and even fluorescent lights. Even though we take preventative measures to avoid receiving these unwanted signals, they sometimes find their way into places that they are not wanted.
Electromagnetic interference is usually caused by one of two things. The first is when a wire connected to a device acts like an antenna and

receives the EMI, which is then passed on to the electronics inside the device and amplified. The second is when an electronic component inside a device acts like an antenna because of poor design, poor shielding, or because the component is defective. Ambient noise is noise caused by the random movement of electrons in an electronic circuit when the power is off or by the random movement of air.

Noise Canceling   Noise canceling is accomplished by filtering a sample of the noise from a preamp stage of a circuit, then inverting the signal 180 degrees and adding the inverted noise signal to the original signal containing the noise (Fig. N.4). The noise combined with the inverted sample of the noise cancel each other out (electronically add to 0 V). When the original noise signal goes positive in its cycle, the noise sample goes negative and the resultant output is 0 V. A good application of noise canceling is in the radio headsets that aircraft pilots use. The cockpit noise is sampled and fed into the radio system, inverted, re-fed into the amplification system, and the surrounding noise is canceled out.

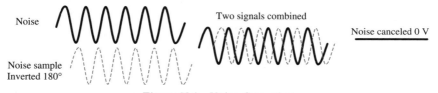

Figure N.4   Noise Canceling

Non-Canonical   (Also known as *LSB, Least Significant Bit,* or *Little Endian.*) A method of placing the least significant bit on the wire or in the header of a frame first. Ethernet is a non-canonical transmission method. Other canonical protocols such as FDDI and token ring place the MSB (Most Significant Bit) on the wire or in the packet header first.

Non-Facility Associated Signaling (NFAS)   In ISDN PRI (Primary Rate Interface) signaling, the ability of D channels to carry signaling information for multiple 24 channel PRIs. This is an advantage for those who would like to squeeze that last (24th) B channel out of their circuits. The only drawback is that if the D channel should go down, then all PRI circuits associated to that PRI go down with it. It is a feature of NFAS signaling, and very common to configure a back-up D channel on one of the secondary PRI circuits.

NonReturn To Zero Inverted (NRZI)   Signals that maintain constant voltage levels with no signal transitions (no return to a zero-voltage level), but interpret the presence of data at the beginning of a bit interval as a signal transition and the absence of data as no transition. See also *NRZ*.

Nortel Networks Certified Design Expert (NNCDE)   A well-known industry certification/training program that is offered by Nortel Networks. This certification recognizes an advanced level of network design, planning, and optimization using Nortel Networks products. It requires industry experience as well as Nortel training. Information regarding Nortel Networks' certification programs can be found at *http://www.nortelnetworks.com*.

Nortel Networks Certified Network Architect (NNCNA)   A well-known industry certification/training program that is offered by Nortel Networks. This advanced certification not only requires learned technical knowledge, but also industry experience. Students seeking this certification are required to obtain a required number of points in certification-related areas. A specific number of points are obtained through job experience and a number are obtained by taking Nortel Networks courses. Information regarding Nortel Networks' certification programs can be found at *http://www.nortelnetworks.com*.

Nortel Networks Certified Support Expert (NNCSE)   A well-known industry certification/training program that is offered by Nortel Networks. This certification recognizes an advanced level of technical post sales expertise required to implement, support, troubleshoot, and optimize Nortel Networks products. This certification requires industry experience points as well as Nortel Networks training. Information regarding Nortel Networks' certification programs can be found at *http://www.nortelnetworks.com*.

Nortel Networks Certified Support Specialist (NNCSS)   A well-known industry certification/training program that is offered by Nortel Networks. This certification recognizes a fundamental level of post-sales technical expertise required to deploy, operate, and troubleshoot Nortel Networks products. This certification requires industry experience points, as well as Nortel Networks training. Information regarding Nortel Networks' certification programs can be found at *http://www.nortelnetworks.com*.

North American Area Codes   See *Area Codes*.

North American Numbering Plan   See *Area Codes*.

NOS (Network Operating System)   A software that manages communications of devices. An administrator sets access and security privileges to users of a network, and monitors network performance via this software. In Novell Networks, the NOS provides sockets at OSI level 4, which provide the link between software applications and lower network layers. Network operating systems are distributed file systems. Examples include Novell NetWare, *NFS (Network File System* for UNIX), Windows NT, and Banyan VINES.

Notch Filter   A filter that is designed to pass or block a specific band of frequencies. The three types of filters are low pass/block, high pass/block, and notch pass block. What determines if the filter is a pass or block filter is how the filter is arranged. If the filter is set in series with a circuit, then it passes the desired frequencies down the line. If it is connected to ground, it will pass the desired frequencies to ground, thus preventing them from continuing through the circuit to block them.

NPA (Number Plan Area)   Also called an *Area Code.* Each area code contains central offices and each central office has a set of prefixes (first three digits of a seven-digit number) that identify that central office to all other central offices within the associated area code. Some people actually sit around and plan what numbers will belong to which central office.

NRZ (NonReturn to Zero)   1. In a physical medium, signals that maintain constant voltage levels with no signal transitions (no return to a zero-voltage level) during a bit interval. See also NRZI.   2. A binary encoding method used to write information to hard-disk drives in computers.

NRZI (NonReturn to Zero Inverted)   1. In a physical medium, signals that maintain constant voltage levels with no signal transitions (no return to a zero-voltage level), but interpret the presence of data at the beginning of a bit interval as a signal transition and the absence of data as no transition. See also NRZ.   2. A binary encoding scheme used to write information to hard-disk drives in computers.

NSAP (Network Service Access Point)   The logical, software, or virtual interface between the *OSI (Open Systems Interconnect)* network layer (3) and transport layer (4). This is known as a *socket* in Novell networks.

NSP (Network Service Provider)   A company that provides telecommunications services such as frame relay, ADSL, HDSL, ATM, Internet

Access or other services via their own switching equipment, or leased switching equipment. NSP's usually provide services by utilizing a local telco or CAP for the last mile (local loop). They also link their nodes via high capacity circuits, and have interconnection agreements with other telecommunications companies.

NTE (Network Termination Equipment)   The generic term for *DCE* and *DTE (Data Communication Equipment* and *Data Termination Equipment)*. With newer transmission methods, such as those provided in the xDSL family, DCE and DTE are replaced with near-end NTE and far-end NTE. The near end is the central office transmission interface and the far end is the xDSL router.

NT1 (Network Terminal 1)   Another reference for an ISDN terminal adapter.

NTSC (National Television Standards Committee)   The pre-HDTV broadcast standard in Canada, Japan, the United States, and Central America. NTSC defines 525 vertical scan lines per frame and yields 30 frames per second. The scan lines refer to the number of lines from top to bottom on the television screen. The frames per second refer to the number of complete images that are displayed per second. See also *Television Broadcast Standards.*

NUL   The ASCII control-code abbreviation for null. The binary code is 0000000 and the hex is 00.

Null Modem   A communications cable, such as an RS-232 cable, that has the transmit and receive wires switch places in pin-out from one end to the other. These cables are used to connect *DCE (Data Communications Equipment)* with *DTE (Data Termination Equipment)*, so the transmit of one reaches the receive of the other (Fig. N.5).

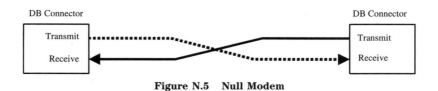

Figure N.5   Null Modem

Null-Modem Adapter   An adapter that is used to convert a straight-through cable into a null-modem cable. Null-modem adapters are available in many different pin-outs and connector types (Fig. N.6).

Figure N.6   Null Modem Adapter

Number Crunching   A reference to data processing or data manipulation done by a PC, server, or mainframe.

Numbering Plan   A plan of what numbers will be used where. In a local phone company, each central office has its own numbering plan or range of numbers. In a PBX or key system, numbering plans are implemented to ease the complexity of accounting, and sometimes they aid in remembering what a person's extension number is. For example, sales can be extensions that range from 3000 to 3999, manufacturing can be extensions that range from 4000 to 4999, off-premises extensions can range from 5000 to 5999, etc. Numbering plans can be formed any way that a user/administrator likes with the following exceptions: Usually no extension on a PBX starts with 9 because 9 as a first digit is used to access outside lines, so 9000 to 9999 is not used in a numbering plan. Zero (0) is also restricted from a numbering plan because it is often used to dial the attendant or operator. A good numbering plan will make call accounting much easier. Call reports can be sorted by department if every department has its own unique numbering plan.

Number Portability   Number portability is still in the legal, financial, and architectural planning process (Fig. N.7). When it is completed (different

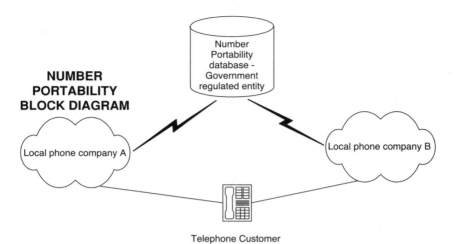

Figure N.7   Number Portability

places will be implemented at different times, the goal for starting was 1998), there will be a national data base that stores every phone number subscribed to by every user. The ultimate goal is to automate the ability of a customer to switch telephone companies and take their phone number to whichever exchange area (within an area code) they wish. If a customer decides to switch companies, their number must first be disconnected by the old company, then reconnected/reactivated by the new. If both numbers are active at the same time, the telephone network will become confused and most likely not complete calls. With the planning and implementation of *CTI (Computer Telephone Integration)*, it will be possible for a company to enter into a data base (operated by a third party company), the order to disconnect, and when another company enters the connect, a computer will make the actual switch with no outage of service. This will eliminate the possibility of one telephone company interfering with a customer's change-over by failing to disconnect service at the proper time.

Competition in the telecommunications industry is hampered by the fact that no customer can switch local telephone companies and take their phone number with them (1-800/888 long-distance service is a different story. Customers can transfer those numbers.). The cost to re-advertise phone numbers for business is too costly and too inconvenient for the patrons of businesses. The estimated cost to implement number portability is 100 million dollars per LATA (area code). The legal argument at the writing of this definition is that the new phone companies do not want to pay for the number portability upgrade because the cost would outweigh the profit. The RBOCs don't want to pay for it because they are regulated by the government, which means that any increase in costs of the phone network are passed on to the subscribers/rate payers. It seems unfair that telephone customers would have to pay for huge corporate investments—especially if they are not clients/patrons of the particular company that is receiving the benefit of the investment.

NVRAM (NonVolatile RAM)   Random Access Memory that retains its contents when a unit is powered off.

NYNEX (New York New England Exchange)   One of the original seven regional Bell Operating Companies that was divested from AT&T.

Nyquist Theorem   A theory that states that any analog signal to be converted to a digital signal (ADC conversion) must be sampled at twice the frequency of the top end of the bandwidth of the signal to be converted. If you would like to convert a high-fidelity recording to a compact disc, you would need to sample the audio at a minimum of 36 kHz (36,000 times per second) because the bandwidth of high-fidelity music is 18 kHz (18,000 cycles per second). This sample rate would give two samples per cycle at the highest frequency of human hearing, which is 18 kHz. A DSO channel in a channel bank samples a voice at 8000 times per second (8 kHz). This gives a Nyquist standard sample up to 4 kHz, which is sufficient to sample all sounds in the voice range.

OAI (Open Application Interface) A means for a computer system and a PBX to exchange information. It is an older name for *CTI (Computer Telephony Integration)*. It allows a person in the workplace to enter information into a computer by using their telephone. Some of the information is time-reporting information, inventory information, etc.

OC (Optical Carrier) A prefix for SONET carrier hierarchies, which is followed by a number, such as OC-1, OC-3, etc.

OC-1 (Optical Carrier 1) The beginning of the SONET-level transmission speeds. An OC-1 is capable of carrying one DS-3 within its payload. Its transmission carrier speed is 51.840 Mbps. OC-1 can be converted into an electrical signal, which is called an *STS-1 (Synchronous Transport Signal-1)*. For more information on OC speeds, see *OC-N*.

OC-12 (Optical Carrier 12) A SONET level of transmission speed. It is capable of transporting three DS-3 signals, which is equal to 622.080 Mbps (Figs. O.1 and O.2). For more information on OC speeds, see *OC-N*.

Figure O.1   OC-12 FLM SONET

Figure O.2   OC-12 TBM SONET Shelf

OC-192 (Optical Carrier 192)   A SONET level of transmission speed. It is capable of transporting three DS-3 signals, which is equal to 9.953 Gbps. For more information on OC speeds, see *OC-N*.

OC-3 (Optical Carrier 3)   A SONET level of transmission speed. It is capable of transporting three DS-3 signals, which is equal to 255.520 Mbps (Fig. O.3). For more information on OC speeds, see *OC-N*.

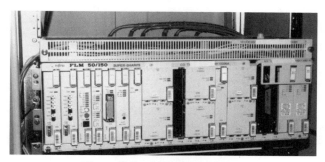

Figure O.3   OC-3 FLM SONET

OC-48 (Optical Carrier 48)   A SONET level of transmission speed. It is capable of transporting three DS-3 signals, which is equal to 2.488 Gbps (Fig. O.4). For more information on OC speeds, see *OC-N*.

Figure O.4   OC-48 FLM SONET

OC-N (Optical Carrier N)   The N denotes a number in the SONET optical carrier hierarchy, which now extends from OC-1 to OC-192 (Fig. O.5).

Name/Acronym	Bandwidth	Equivalent DS0	Equivalent DS1	Equivalent DS3	Comments
DS0	64 Kb/s	1	*	*	one phone line
DS1/T1	1.544 Mb/s	24	1	*	popular service
DS1C	3.152 Mb/s	48	2	*	equipment
E1/CEPT1	2.048 Mb/s	32	1	*	European
DS2	6.312 Mb/s	96	4	*	equipment
E2	8.448 Mb/s	96	4	*	European
DS3/T3	44.736 Mb/s	672	28	1	popular service
E3	34.368 Mb/s	512	16	1	European
DS4	274.176 Mb/s	4,032	168	6	long haul radio
STS-1	51.84 Mb/s	672	28	1	electrical OC1
OC-1	51.84 Mb/s	672	28	1	SONET
OC-3	255.520 Mb/s	2016	84	3	SONET
OC-12	622.080 Mb/s	8,064	336	12	SONET
OC-48	2.488 Gb/s	32,256	1,344	48	SONET
OC-192	9.953 Gb/s	129,024	5,376	192	SONET

Figure O.5   OC-N Hierarchy

Octal   A numbering system. Base 8. The Base 10 system has 10 different characters to represent numbers, 0 through 9. Base eight uses only eight of those characters, 0 through 7.

Octel   A company that manufactures stand-alone voice-mail equipment. For a photo of an Octel voice-mail system, see *Voice Mail.* Octel is now a part of Lucent Technologies.

Octet   Another term for *byte,* a string of eight bits.

Octopus Cable   Also called a *Y* or *three-way cable.* An octopus cable is used to break a larger connector (usually from a bus) to two or more smaller connectors. The Nortel Meridian utilizes a cable with 50 pins on one connector that breaks out in two different RS-232 connectors. Having the single large connector plug into the back plane uses less space than three smaller connectors.

ODBC (Open Data Base Connectivity)   A standard software interface for accessing data in both relational and nonrelational data-base management systems. Using this application programming interface, data-base applications can access data stored in data-base management systems on a variety of computers—even if each data-base management system uses a different data-storage format and programming interface. ODBC is based on the call-level interface specification of the X/Open SQL access group and was developed by Digital Equipment

Corporation, Lotus, Microsoft, and Sybase. See also *Java Database Connectivity.*

Odd Parity   A method of bit-stream checking. Parity is used in error correction. The number of logic "ones" is counted in a bit stream. There is "odd parity" and there is "even parity." Which is used depends on if you like odd or even numbers, or if the modem you are trying to connect with likes odd or even numbers. Parity is a part of error-checking protocol. It is simply the part of the protocol where the two devices are told if they are counting odd number bits or even number bits. In odd parity, if the number of ones is odd, then a parity bit is set to "one" at the end of the bit stream. This is odd parity because the parity bit is set to one when the number of "ones" is odd. In even parity, the parity bit is set to "one" when the number of "one" bits is even. See *Parity.*

ODI (Open Data-Link Interface)   A Novell specification that provides a standardized interface for *NICs (Network Interface Cards)* that allows multiple protocols to use a single NIC. See also *NIC.*

OE (Office Equipment)   Also referred to as *line equipment.* Line equipment is the actual interface port (from a circuit card) on the central-office switch in a telephone-company central office. It is the equivalent to a station or *IPE (Intelligent Peripheral Equipment)* card in a PBX system (*TN* for Nortel Specifics). Each telephone line has an associated line equipment or office-equipment interface. That particular port is what defines the telephone service provided to the customer connected to it via the OSP network. The CPU (core) of the central-office switch associates a phone number with a line-equipment port (OE). When a customer of the phone company calls and requests that their phone number be changed, the service order eventually finds its way to a central-office technician or service translator that reprograms the line equipment with a new phone number.

Off-Premises Extension (OPX)   Off-premises extension adapter, also called a *loop extender.* An OPX adapter is an add-on device for a PBX switch or central-office switch that allows operation over an abnormally long loop or twisted pair, usually more than 12,000 feet for a central-office switch and 1500 feet for a PBX. The PBX or key-system manufacturer usually offers special equipment for a long-reach application. Some OPX adapters for PBX systems can be programmed to dial digits into an outgoing trunk, which will automatically ring a telephone somewhere else, such as the CEO of a company's home office.

Off-Sight Night Answer   A feature of PBX and some key systems that allows a main line to be forwarded to a telephone number programmed in by the user/administrator when the system is put into night mode.

OHD   Optical hard drive.

Ohm   The unit of resistance, represented by the Greek letter Omega, Ω. Resistance is just what its name depicts, resistance to electric current flow. A 100-W, 120-V household light bulb has about one ohm of resistance. The more resistance is in a circuit, the less current flows through it.

Ohm's Law   A series of mathematical relationships for electronics. The relationships are based on voltage, resistance, power, and amperage. The two basic Ohm's law formulas are:

P (power in watts) $= I$ (current in amps) $\times E$ (voltage in volts)

E (voltage in volts) $= I$ (current in amps) $\times R$ (resistance in ohms)

Omnidirectional   A reference to a microphone that receives sound from all directions.

On-Board Modem   A term that refers to an internal modem.

ONA (Open Network Architecture)   The architecture of the public telephone network. Under FCC rulings, the Bell operating companies must allow other companies that offer "value-added services" to connect to and offer services through the local telephone companies network. Value-added services, under open network architecture, are voice mail, operator services, and IVR telephone-shopping applications. You don't have to use the Bell companies' voice mail if another voice-mail service provider is available. If you order the other company's voice mail, all of your voice-mail connections will go through the alternative value-added service provider, on their separate equipment. The problem with ONA is that from a technology sense, access is equal, but in competition for market share, it is not.

One-Ten Termination Block   A twisted-pair connectivity method. See *110 Termination Block.*

One-Way Trunk   A reference to a *DID (Direct Inward Dial)* or *DOD (Direct Outward Dial)* trunk used in PBX applications.

Ones Density   A reference to the maximum number of consecutive "zero" bits can be transmitted in a row using specific transmission equipment without losing the timing of the carrier (T1). To eliminate successive zeros in T1 transmissions line/protocols, such as B8ZS, have been implemented.

ONI   Optical Network Interface.

ONU (Optical Network Unit)   A reference to an access node that converts optical signals transmitted via fiber to electrical signals. ONUs are a part of hybrid fiber/coax/twisted-pair networks that are emerging in local markets. The ONU enables mass amounts of bandwidth to be delivered to areas that are beyond electrical transmission range of a central office or head end. ONUs can be pole mounted or placed in cable vaults. See also *Remote Mini Fiber Mode*.

OOF (Out Of Frame)   A fault condition of a T1 carrier circuit. If an OOF condition exists, the circuit is down and not operational. Many T1 carrier equipment manufacturers implement recovery measures in the operating system to help systems come back on line automatically.

Open   See *Open Circuit*.

Open Application Interface (OAI)   See *OAI*.

Open Architecture   The ability of different systems to integrate with each other, such as a PBX system and a Novell LAN. The newer term for open network architecture is *CTI (Computer Telephony Integration)*.

Open Circuit   A circuit fault. Many confuse an open with a short. An open is literally an open, a "disconnection" in a circuit. A short is a "crossed circuit," an easier path to ground caused by a bad component, water, or other means for electricity to get to where it is not wanted (Fig. O.6).

Open and Short circuit faults

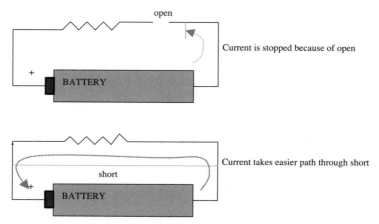

Figure O.6   Open Circuit/Short Circuit

Open Data-Link Interface (ODI)   A Novell specification that provides a standardized interface for *NICs (Network Interface Cards)* that allows multiple protocols to use a single NIC. See also *NIC*.

Open Data Base Connectivity (ODBC)   A standard software interface for accessing data in both relational and nonrelational data-base management systems. Using this application programming interface, database applications can access data stored in data-base management systems on a variety of computers—even if each data-base management system uses a different data-storage format and programming interface. ODBC is based on the call-level interface specification of the X/Open SQL access group and was developed by Digital Equipment Corporation, Lotus, Microsoft, and Sybase. See also *Java Database Connectivity*.

Open-Ended Access   A term that refers to a switched telephone line that is not restricted from any calling prefixes or area codes. Most residential customers have subscribed to service that has open-ended access, not knowing what it is called.

Open Network Architecture (ONA)   See *ONA*.

Open Shortest Path First (OSPF)   1. A standard traffic-control program used by routers. OSPF was developed by the Internet Engineering Task Force during the late 1970s. Because OSPF is one of the original routing protocols, it is supported by virtually every router manufacturer. OSPF is a link-state hierarchical routing protocol.   2. (Also known as *Link State Routing Protocol, Distributed Routing Protocol, Shortest Path First,* and *Interior Gateway Routing Protocol.*) OSPF is a layer 3 routing protocol that enables routers to identify other routers to use as a path to forward packets across a network. A router that has OSPF enabled sends a layer 3 broadcast packet called a *Link State Advertisement* (LSA) every n seconds (by default in Cisco Systems' IOS). The LSA is a list of known routers and routes to destinations within a network. Only routers that are operating with the OSPF feature enabled send and recognize the LSAs, and only routers that send LSAs are considered by other routers as alternate route providers if there is a failure. In addition to LSAs, the OSPF features a "hello" message that is generated by all routers every 5 seconds to let other routers know that all is well and operating OK. In enterprise LAN environments, OSPF is often run as a layer 3 alternate routing mechanism in conjunction with STP (Spanning Tree Protocol), which provides alternate layer 2 routing for switches and VLANs. What makes OSPF more complicated yet far more expandable than RIP is its design based on multiple areas. Breaking a large network area into smaller and more manageable pieces makes a network more efficient. What makes OSPF better than IS-IS is that it recovers from link failures almost 20 times faster. The significant drawback to OSPF is that the larger the network gets, the more areas there are. This inherently makes the system complicated to manage for network administrators. Another consideration that may be a drawback is that in large implementations, OSPF can push CPU utilization to its limit. This means that OSPF requires higher-quality routers. Alternative protocols to OSPF include IS-IS and EIGRP (Cisco Proprietary). See also *Autonomous System, Area Border Router,* and *Autonomous System Area Border Router.*

Open Shortest Path First Area (OSPF Area)   In OSPF (Open Shortest Path First) routing, a group of routers that share identical link state databases provided by a designated router. It is recommended that no more than 40 routers exist within one OSPF area.

Open Systems Interconnection (OSI)   The latest model, or guideline for communications protocols is the *OSI (Open Systems Interconnect).* It is the best model so far because all of the layers (functions) work independently of each other. Older proprietary communications models are shown in the figure. The OSI model is a seven-layer or "step" process for communications. The different functions are:

- *Application layer*   The seventh and highest layer of the OSI communications protocol model. The applications layer is the function of connecting an application file or program to a communications protocol.

- *Presentation layer*   The sixth layer in the OSI. In general, the presentation layer performs the function of encoding and decoding the data to be transmitted within the communications protocol.

- *Session layer*   The fifth layer of the OSI model. In general, the session layer establishes and maintains connection to the communications process of the lower layers. It also controls the direction of the data transfer.

- *Transport layer*   The fourth layer or function in a communications protocol model. In general, the transport layer performs the function of error correction and the direction of data flow (transmit/receive).

- *Network layer*   The third layer in the OSI model. In general, the network layer does the job of switching and routing of the data being transmitted within the protocol.

- *Data-link layer*   The second layer or function in the OSI model. In general, the data-link layer receives and transmits data over the physical layer media (twisted pair, fiber optic, etc.).

- *Physical layer*   The first layer in the OSI model. In general, the physical layer is the actual media of the communications transmission (twisted-pair wire, coax, air, fiber optic, etc.). It is also the types of connectors used and the pin-outs of those connectors. The 568B wiring scheme for CAT 5 wire is a physical-layer function. For a basic conceptual diagram of the OSI model, see *OSI Standards*.

Open Wire   Also called "C" wire. Wire that is steel strengthened for long-span aerial-plant applications. Some C wire is uninsulated, so it is also called *open wire*. Open wire better fits the application because it is used in wide open or very rural areas. The old telegraph system was an open-wire system.

Operating System   The computer software program that controls the functions of computer hardware. Examples of operating systems are Microsoft Windows 2000, MS-DOS, Pick, UNIX, and OS/2.

Operator   An attendant that assists callers. Operators can work for telephone companies or private telecommunications service companies.

Operator Console   The huge telephone used by a PBX attendant. The console is distinguishable by its large *BLF (Busy Lamp Field)* and many feature keys.

Optical Attenuator   A fiber-optic attenuator works like your sunglasses, it reduces the level of light entering your eyes so that you can see more effectively. They come in various connector types (Fig. O.7). Typical fiber-optic attenuator values are 5 dB, 10 dB, and 20 dB.

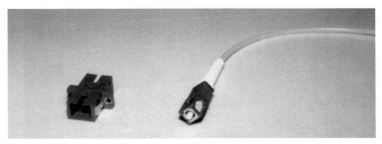

Figure O.7   Optical Attenuators: SC Type (Left) and Fiber-Optic SC Connector (Right)

Optical Drive   A term used to refer to hardware that reads and/or writes to compact discs, in general, including CD-ROM drives, DVD drives, and CD-RW drives.

Optical Fiber Patch Panel   A means of terminating fiber-optic cable. Fiber patch panels contain a fiber splice tray equipped with pigtails. The pigtails are simply fiber connectors with a piece of fiber optic connected to them so that a fiber from within a cable can be easily spliced to them. The connectors are spaced on the front of the fiber patch panel.

Optical Fiber Splice   The two types of fiber-optic splices are fusion (heat) and mechanical.

Optical Time-Domain Reflectometer (OTDR)   A testing device that measures the loss over a fiber optic and the distance from the tester (Fig. O.8). OTDRs look similar to oscilloscopes with a CRT display tube. OTDRs are a specialized optical version of a *TDR (Time-Domain Reflectometer)* used to test copper pairs. The way a TDR works is that it transmits a signal down a media (copper or glass), then waits for a reflection to come back. When the reflection returns to the device, the time difference is used to calculate the distance that the signal traveled. The size or power of the return signal is used to calculate loss.

Figure O.8   OTDR (Optical Time-Domain Reflectometer)

Optoelectric Transducer   A class of electronic components that converts light energy into electrical energy and electrical energy into light energy.

OPX Adapter (Off-Premises Extension Adapter)   Also called a *loop extender.* An OPX adapter is an add-on device for a PBX switch or central-office switch that allows operation over an abnormally long loop or twisted pair, usually over 12,000 feet for a central-office switch and 1,500 feet for a PBX. The PBX or key system manufacturer usually offers special equipment for a long-reach application. Some OPX adapters for PBX systems can be programmed to dial digits into an outgoing trunk, which will automatically ring a telephone somewhere else, such as the CEO of a company's home office.

Oscillator   An electronic circuit that produces an AC cycle from a DC power source. Oscillators are used as carrier references for transmitters and for the timing signal in clock circuits for digital instruments, such as PCs and telephone systems. Quartz crystal oscillators are the most reliable and inexpensive. For a photo, see *Crystal Oscillator.*

Oscilloscope   A testing device that allows a user to view a waveform on a screen (CRT). The screen is graduated to show different frequencies and voltage levels. The value of each graduation (or division) is determined by the setting of the frequency/division knob. The voltage level of each division is determined by the voltage-level selector knob. Oscilloscopes range in price from about $400 to more than $8000. The features that make an oscilloscope increase in price are the ability to read/display very fast frequencies and the ability to view more than two waveforms at a time. Some oscilloscopes are capable of being connected to plotters, which gives the user the ability to print a waveform displayed on the screen.

OSI (Open Systems Interconnect)   See *Open Systems Interconnect* (Fig. O.9).

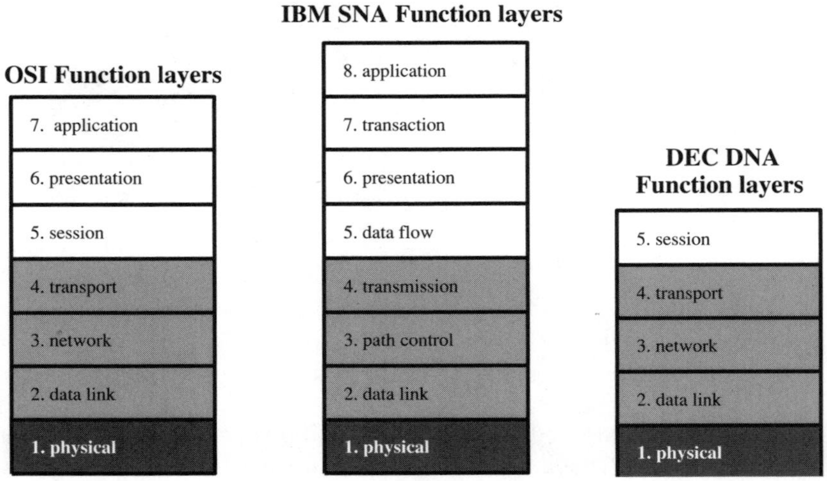

IBM SNA Function layers

OSI Function layers

8. application
7. application

OSI Function layers	IBM SNA	DEC DNA

Figure O.9   OSI Model Individual Layers Compared to SNA and DNA Models

OSI Model (Open Systems Interconnect Model)   See *Open Systems Interconnect.*

OSI Standards   An architecture set up by the *ISO (International Standards Organization)* that sets some broad standards for communications. The purpose of the standards is to help manufacturers make equipment that is universally compatible. The OSI is not perfectly followed by the telecommunications and data communications industry. It is used as a model in the design of communications protocols. The basic idea of the OSI is that seven functions, steps, or layers are in the successful completion of a communication transmission. The goal of the OSI is to make all of these layers separate and individual "entities" in hardware and software so that different manufacturers can integrate at different levels. Data communications is modeled after voice communications. Even though humans speak many different languages (protocols), there is still a common architecture of human communications. See *Open Systems Interconnect.*

OSPF (Open Shortest Path First)   1. A standard traffic-control program used by routers. OSPF was developed by the Internet Engineering Task Force during the late 1970s. Because OSPF is one of the original routing protocols, it is supported by virtually every router manufacturer. OSPF is a link-state hierarchical routing protocol.   2. (Also known as *Link State Routing Protocol, Distributed Routing Protocol, Shortest Path First,* and *Interior Gateway Routing Protocol.*) OSPF is a layer 3 routing protocol that enables routers to identify other routers to use as a path to forward packets across a network. A router that has OSPF enabled sends a layer 3 broadcast packet called a *Link State Advertisement* (LSA) every *n* seconds (by default in Cisco Systems' IOS). The LSA is a list of known routers and routes to destinations within a network. Only routers that are operating with the OSPF feature enabled send and recognize the LSAs, and only routers that send LSAs are considered by other routers as alternate route providers if there is a failure. In addition to LSAs, the OSPF features a "hello" message that is generated by all routers every 5 seconds to let other routers know that all is well and operating OK. In enterprise LAN environments, OSPF is often run as a layer 3 alternate routing mechanism in conjunction with STP (Spanning Tree Protocol), which provides alternate layer 2 routing for switches and VLANs. What makes OSPF more complicated yet far more expandable than RIP is its design based on multiple areas. Breaking a large network area into smaller and more manageable pieces makes a network more efficient. What makes OSPF better than IS-IS is that it recovers from link failures almost 20 times faster. The significant drawback to OSPF is that the larger the network gets, the more areas there are. This inherently makes the system complicated to manage for network administrators. Another consideration that may be a drawback is that in large implementations OSPF can push CPU utilization to its limit. This means that OSPF requires higher-quality routers. Alternative protocols to OSPF include IS-IS and EIGRP (Cisco Proprietary). See also *Autonomous System, Area Border Router,* and *Autonomous System Area Border Router.*

OSPF Area   In OSPF (Open Shortest Path First) routing, a group of routers that share identical link state databases provided by a designated router. It is recommended that no more than 40 routers exist within one OSPF area.

OTDR (Optical Time-Domain Reflectometer)   See *Optical Time-Domain Reflectometer.*

Out-of-Band Signaling   In telephone circuits (DS1 to be specific), the two different ways to send signals are in-band and out-of-band. Signals are digits that you dial, dial tone, the phone being off-hook, ringing, etc. An in-band telephone line is like the one in your home; the digits that you dial and the ringing are carried within the channel that you talk on. Out-of-band signaling is a method that telephone companies and businesses use for larger PBX applications and data-transfer applications. In an out-of-band signaled DS1, there are 24 multiplexed channels. The 24th channel carries the signaling for the other 23 channels or phone lines. The advantage of out-of-band signaling is that each channel has an increased capacity to carry data (8 Kb/s more) and the 23 channels are not used to find out if a line is busy (both directions, in and out). The off-hook sensing and busy signaling are performed in the 24th channel. If you have a system that receives thousands of calls per day, this can reduce traffic.

Outdoor Jack Closure   Closures are available that help protect telephone and other jacks from moisture and other outdoor weather conditions (Fig. O.10).

Figure O.10   Outdoor Jack Closure

Outside Plant A term that refers to a communications utility's twisted-pair and/or coax network that winds through towns and neighborhoods. It includes terminals, pedestals, cross boxes, and vaults.

Outsource To subcontract work to other companies, usually for their construction or technical expertise in the installation of specific electronics or software.

Outward Restriction A feature of *PBX (Private Branch Exchange)* telephone systems that prevents selected telephone extensions from dialing outside the office/building. When a user of one of these extensions dials "9" for an outside dial tone, they will just get a fast busy signal.

Overhead The part of a transmission that contains the information/signal that controls the operation of the transmission. If you are transporting yourself across town in your car, you are the payload and your car is the overhead.

P

P Connector   A 25-pair male amp connector. For a photo of the female version, called a *C connector,* see *25-Pair Connector.*

P.530   A standard for radio link planning. The International Telecommunications Union (ITU-T) publishes a reference for terrestrial radio link planning, which is available at www.itu.ch. ITU-R. It also contains recommendations for link quality objectives. ITU-R Recommendation P.530 contains information on how to plan for high-reliability in clear, line-of-sight links.

P1024B   A mainframe environment protocol used between host computers and user terminals/workstations.

P1024C   A mainframe environment protocol used between host computers and user terminals/workstations.

PA System   See *Public Address System.*

PABX (Private Automatic Branch Exchange)   The old name for *PBX, Private Branch Exchange.*

Pac Bell   The RBOC that operates the public telephone network in the state of California, owned by Pacific Telesis, who was recently purchased by Southern Bell.

Pacific Telesis   The RBOC that owns PAC Bell and Nevada Bell, which was bought out by Southern Bell.

Pacing   See *flow control.*

Packet   A unit of data at the network layer of the *OSI (Open Systems Interconnect)* model. Packets have a header that contains control information and a payload with user data. The terms, datagram, frame, packet data unit, message, and segment are also used to describe logical information groupings at various layers of the OSI reference model and in various arms of the networking industry.

Packet Assembler Dissembler (PAD)   See *PAD.*

Packet Buffer   Memory allocated or dedicated to the temporary storage of a copy of a data packet until the original has reached its destination.

Packet Controller   Another name for a packet switch. A packet switch is the central controlling device in a packet-switched network, such as switched Ethernet, switched token ring, or ISDN packet switching.

Packet Data Unit   A unit of data at the application, presentation, and session layers of the *OSI (Open Systems Interconnect)* model. Packet data units might have a header containing control information. The terms *datagram, frame, packet, message,* and *segment* are also used to describe logical information groupings at various layers of the OSI reference model and in various networking circles.

Packet Interleaving   To place many data packets from many data packet sources on one transmission channel.

Packet Internet Groper (PING)   A command followed by an IP address that sends an *ICMP (Internet Control Message Protocol)* echo request to the host specified by the IP address. If the targeted host has an active IP network connection, it will return the message. The length of time this process takes is listed by the local user system as an indication of network speed.

Packet Layer   The layer in the X.25 protocol that is equivalent to the network layer in the OSI model. The packet layer does the same functions as the network layer, but it is simply called something different. Specifically to X.25, the packet layer handles the multiplexing of data. It does this by using an address system that is embedded in its control signaling with the packet layer on the other end.

Packet Level   The part in the telecommunications process of the X.25 protocol where network-layer functions are performed, such as addressing, multiplexing, and demultiplexing.

Packet Loss   The total number of frames transmitted at wire speed, less the number received at the final destination node or host.

Packet-Over-SONET   The packet-over-SONET specification is primarily concerned with the use of the Point-to-Point Protocol (PPP) encapsulation over SONET/SDH links. Because SONET/SDH is by definition a point-to-point circuit, PPP is well suited for use over these links. PPP was designed as a standard method of communicating over point-to-point links, and packet-over-SONET is a convenient way for broadband service providers to carry guaranteed rate Internet traffic.

Packet Switching Exchange (PSE)   Part of an X.25 packet-switching network that receives packets of data from a *PAD (Packet Assembler/Dissembler)* via a modem. The PSE makes and holds copies of each packet, then transmits the packets one at a time to the PSE that they are addressed to. The local PSE then discards the copies as the far-end PSE acknowledges the safe receipt of the original.

Packet Switch   Also called a *data packet switch.* A device that routes segmented transmissions between end users, using a connectionless protocol, such as X.25, Ethernet, frame relay, token ring, ATM, or TCP/IP. Data packet switching is often performed in levels, with one protocol carrying another. For example, an Ethernet-based transmission from a LAN could be routed over a frame-relay or ATM connection via a data packet switch. Packet switches make up *PDN (Public Data Networks)* or *PSN (Packet Switching Networks),* which are the basis of the *frame-relay, ATM,* and X.25 services that public telecommunications companies offer.

Packet-Switching Network (PSN)   A name sometimes used in place of *Public Data Network (PDN).* PSNs connect to users via X.25 or frame relay. The Internet is also a type of PSN using TCP/IP packets.

Packets per Second (PPS)   In LAN and WAN architectures, a standard measure of transported traffic over a period of time based on a 64-byte packet. This is not an official standard as of 2001. The size of the packet varies by several bytes depending on the manufacturer and type of equipment. Some use packets as small as 60 bytes on networks specifically configured to carry such traffic. The minimum size for an Ethernet packet is 64 bytes. See also *Runt* and *Giant.*

PAD (Packet Assembler/Dissembler)   The device or software program in an X.25 network packet-switching network that takes a large file to be transmitted (or small) and breaks it down into smaller pieces. It gives each piece an identification number in relation to the rest of the pieces (e.g., 387 of 8954) and an address, along with error-checking information (usually CRC) and other *HDLC (High-Level Data Link Control)* information. The PAD can be a part of an end users computer or a separate device. The PAD sends the packets to a *PSE (Packet Switching Exchange)* via a modem, where the packets are individually copied and transmitted. The copies are made by the PSE in case a packet needs to be retransmitted because it was lost or corrupted.

PAD Parameters (Packet Assembler/Disassembler Parameters)

Listed are some common configuration parameters for PADs. Packet assembler/disassemblers are based on the X.3 standard.

- *Recall*   A parameter for X.25 pads that allows an administrator to set individual channels to 1 (Yes) or 0 (No) to access command mode. If the device connected to the channel is a terminal, the parameter is usually set to Yes. If the device connected to the channel is a printer, then the parameter is usually set to No. In the PAD configuration, the keystroke used to activate this parameter is optional, and is set to ^P (Control-P) by most administrators.

- *Echo*   A setting that determines what device (DCE or DTE) is responsible for displaying typed data on the screen of a terminal. If the echo setting of all of the terminals connected to a PAD (or host) is On, then the setting on the PAD (or host) should be Off, or vice versa. If possible, echo should be performed by a host on an X.25 network because echo also controls which device controls editing functions. It is an unruly task to set all of the host-emulation parameters when echo is left to the PAD. In most cases, the PAD cannot emulate them anyway.

- *Data Forward Signal (FWD <CR>)*   The parameter that determines which keystroke will send a data packet. It is most commonly set as the carriage return key (<CR>). If the parameter is set to 0 (Off), then no keystroke will initiate a packet transfer. The PAD will only ship out a packet when the packet is completely full or when a time out is reached (see *Idle Time Parameter 4*). Values on this parameter are limited to the PAD manufacturer's specifications, and the one used depends on the host application programming.

- *Idle Timer (Idle Time 1)*   The parameter that determines the time out for a packet to be shipped out. If a carriage return is not entered within this time in 20ths of seconds (see *FWD Parameter 3*), then the PAD ships out the packet automatically. Common settings are 1 (for 1/20 second) and 3 (for 3/20 second) seconds. Setting this parameter to 0 typically sets the PAD to send full packets only, or to send only on a specific keystroke. Some hosts are programmed to automatically set this parameter to the application that they are running.

- *Ancillary Device Control (Flow Ctrl XON/XOFF)*   A parameter that deals with flow control between a terminal and a PAD. This parameter gives the PAD the ability to pause or stop the terminal from communicating for a moment. A value of 0 usually is a value of No and 1 is usually a Yes.

- *PAD Service Signals (Svc Sigs YES/NO)*   This parameter sets the PAD to deliver X.28 service signals to the terminal, such as call connected, cleared, error, etc.

- *Procedure On <Break> (Break/Reset)*   A parameter used in flow control between the PAD and host. This parameter is used to tell the PAD how the terminal will stop the PAD's data flow (between the PAD and the host) and wait for further instructions. There are several ways to do this. Which way is chosen depends on the PAD, the host, and the application. Some examples of valid parameter settings are one or a combination of the following: send interrupt packet, reset the call, escape to PAD command mode, and send special predefined command for break. The signaling for this is done via an interrupt packet sent through the overhead. Whether the data in transit is discarded or saved is determined by the discard/save parameter setting (Parameter 8).

- *Discard Output (Discard/Save)*   When a PAD is in break/reset mode (break/reset is Parameter 7), the discard save parameter determines what will be done with the data that is still in transit.

- *<CR> Padding (<CR> Pad x)*   The setting that determines the number of "blank" bits that will be sent while the head of a printer returns to home after a carriage return. This parameter is set to a valid value between 0 and 7. The 0 setting is used for terminals, and 1 through 7 are used for printers.

- *Line Folding (ln fold 80)*   A printer setting for the maximum number of bytes (or characters) per printed line. If this parameter is set to high, the printer (or terminal) continues to print additional characters on top of each other at the end of each line that does not have a carriage return (<CR>). In many cases, the host is capable of reading this parameter and format the sent data to match the setting.

- *Terminal Speed (Speed 300)*   Terminal-to-PAD communication speed. Some device combinations (host, PAD, and terminal) will set this automatically (referred to as *autobaud*). The speed ranges from 50 bps to 64 Kbps.

- *Flow Control By Terminal (Pad Flow XON/XOFF)*   A PAD parameter that deals with flow control between a terminal and a PAD. This parameter sets the ability of the terminal to stop or pause communication with the PAD for a moment.

- *Line-Feed Insertion (lf CR/LF)*   A PAD parameter that is used to emulate host editing functions when the PAD is responsible (set to Yes) for the echo parameter.

- *Line Feed Padding (lf pad 10)*   A PAD parameter that is used to emulate host editing functions when the PAD is responsible (set to Yes) for the echo parameter. This particular function is to set how many blank data bits will be sent while the printer advances the paper feed one line.

- *Editing (edit YES/NO)*   A PAD parameter that is used to emulate host editing functions when the PAD is responsible (set to Yes) for the echo parameter, rather than the host. This is only set when delay times for screen display are crucial or exaggerated (satellite transmission delay, for instance).

- *Character Delete (char del ^H)*   A PAD parameter that is used to emulate host editing functions when the PAD is responsible (set to Yes) for the echo parameter, rather than the host. This is only set when delay times for screen display are crucial or exaggerated (satellite transmission delay, for instance).

- *Line Delete (ln del ^U)*   A PAD parameter that is used to emulate host editing functions when the PAD is responsible (set to Yes) for the echo parameter, rather than the host. This is only set when delay times for screen display are crucial or exaggerated (take satellite transmission delay for instance).

- *Line Redisplay (ln rdisp ^R)*   A PAD parameter that is used to emulate host editing functions when the PAD is responsible (set to Yes) for the echo parameter, rather than the host. This is only set when delay times for screen display are crucial or exaggerated (satellite transmission delay, for instance).

PAD Service Signals   A packet assembler/disassembler feature that is activated in the PADs parameter settings. This parameter sets the PAD to deliver X.28 service signals to the terminal, such as call connected, cleared, error, etc.

Page Zone   When attendants use a page feature on a telephone system, they are prompted to input a choice of zone or "area" they want to page. This area is called the *page zone.* Most PBX systems have three separate page zone options and "all zones" is an option to the attendant as well.

Pager   A small, portable device that receives simplex messages. Pagers are small enough to be worn on a belt or wrist (as a wristwatch-type pager). Pagers come in three types: numeric, alphanumeric, and PCS. The numeric are the older type, receiving numbers that a person desiring a return call inputs to the page signal. The alphanumeric pagers have a larger LCD display and have the ability to receive text messages as well as numeric messages. PCS paging is offered as a service with PCS cellular telephones by many cellular telephone companies.

PagP (Port Aggregation Protocol) An operating feature of LAN switches that enables 2 or 4 Ethernet ports (more in some switches) to be concatenated into one high-capacity link. The PagP subroutine automatically identifies ports that are configured as pairs and reports them to the Spanning Tree Protocol, which manages traffic over the combined links as one large channel. This is referred to as *Etherchannel* in marketing Cisco Systems Catalyst LAN switches.

Pair Two copper wires, or two optical fibers.

Pair Gain Usually a reference to a Lucent SLC96 or SLC2000 system. A system in the public network that multiplexes many conversations or phone lines into one or two copper pairs. T1 is a pair-gain system used by public telephone-service providers, such as USWest, PAC Bell, Brooks, ELI, and virtually every other local facilities-based phone company. The photograph shows a pair-gain system outdoor closure/cabinet (Fig. P.1).

Figure P.1   Pair-Gain System (SLC 5)

PAL (Phase Alternation Line) The television broadcast standard in Europe, the Middle East, parts of Africa, and parts of South America. PAL defines 625 vertical scan lines and refreshes the screen 25 times per second. See also *Television Broadcast Standards*.

PAM (Pulse Amplitude Modulation) See *Pulse Amplitude Modulation*.

Parabola The curve that all projected objects travel when acted on by a force of gravity (Fig. P.2). If you watch a baseball that is hit or thrown through the air, it curves during its fall. This oblong curve is called a *parabola,* and it is very special. Ancient mathematicians discovered that this curve can be duplicated mathematically with trigonometry. Radio communications engineers later used it to focus and guide radio signals because all radio waves from a single point were reflected in exactly one direction. For a diagram, see *Parabolic Dish Antenna.*

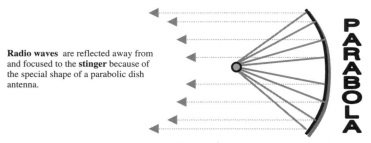

Radio waves are reflected away from and focused to the **stinger** because of the special shape of a parabolic dish antenna.

Figure P.2   Parabola

Parabolic Dish Antenna A directional antenna. This name results from its parabolic shape, which means all radians from a single point are reflected into one direction (Fig. P.3). For a photo of a parabolic microwave antenna, see *Microwave.*

Figure P.3   Parabolic Dish Antenna: 3.7 Meter With C-Band LNB

Parallel Circuit   A circuit that has more than one path for current through multiple loads or devices (Fig. P.4). The other type of circuit is a series circuit, which has only one path for current through multiple loads.

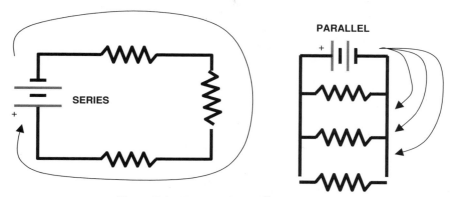

Figure P.4   Parallel Circuit Current Flow

Parallel Data   The transmission of data over a media with multiple bits being transferred at one time, such as a whole byte (Fig. P.5). The other type of transmission is serial data, which sends data one bit at a time. The illustration shows an example of 8 bits being sent in series and in parallel.

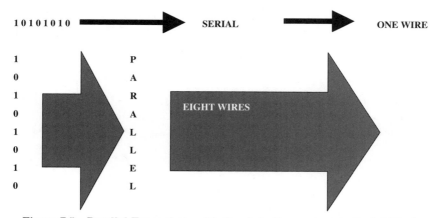

Figure P.5   Parallel Transmission (Bottom), in Comparison to Serial (Top)

Parallel Port   A standard DB25 connector used to connect computer printers. The communication speed of parallel ports is about 1.5 Mbps.

Parallel Printer Cable   A cable that is designed for transmitting multiple bits at one time, used on a parallel-printer port. See *Parallel Data.*

Parasite   A device that gets its power to operate from the telephone line. The telephone line has a -52-V battery voltage when it is idle and -12 V when it is in use. Standard telephone sets (2500 type) are parasitic telephones. Other devices, such as RFI filters and tapping devices, are parasitic.

Parity   A method of bit-stream checking. Parity is used in error correction. The number of logic "ones" is counted in a bit stream. There is "odd parity" and "even parity." Which is used depends on if you like odd or even numbers, or if the modem you are trying to connect with likes odd or even numbers. Parity is a part of error-checking protocol. It is simply the part of the protocol where the two devices are told if they are counting odd number bits or even number bits. In odd parity, if the number of ones is an odd number, then a parity bit is set to "one" at the end of the bit stream. This is *odd parity* because the parity bit is set to one when the number of "ones" is odd. In *even parity,* the parity bit is set to "one" when the number of "one" bits is even.

Odd-parity bit stream: 0 0 1 0 1 0 1 Subsequent parity bit would be 1 because the number of "ones" is odd.
Odd-parity bit stream: 1 0 1 0 1 1 0 Subsequent parity bit would be 0 because the number of "ones" is not odd.
Even-parity bit stream: 0 0 1 0 1 0 1 Subsequent parity bit would be 0 because the number of "ones" is not even.
Even-parity bit stream: 1 0 1 0 1 1 0 Subsequent parity bit would be 1 because the number of "ones" is even.

Parity Bit   A parity bit is a bit added into a bit stream, usually after every seven bits. See *Parity.*

Parity Check   An older and original method of error correction. Newer methods of error correction include *CRC (Cyclic Redundancy Checking),* which are more efficient and far more accurate (99% CRC, 50% parity).

Park   A PBX system feature that allows a user to transfer a call to a "ghost" extension. When the call is transferred to the "ghost" extension, it can then be retrieved from any telephone by dialing that extension. The "ghost" extension is thought of as a parking spot in the network.

Part 64   A reference to the part of the *MFJ (Modified Final Judgment)* handed to the *RBOCs (Regional Bell Operating Companies)* by Judge Harold Greene. It specifies the separation of customer-owned equipment *(Customer Premises Equipment, CPE)* and the telephone company-owned equipment, and telephone company demarcation.

Part 68   See *Part 64.*

Part X   Usually a reference to Part 64 or Part 68 of the MFJ (Modified Final Judgment) handed to the RBOCs (Regional Bell Operating Companies) by Judge Harold Greene. It specifies the separation of customer-owned equipment (Customer Premises Equipment, CPE) and the telephone company-owned equipment, and telephone company demarcation.

Party Line   A telephone line that is shared by multiple residences. A party line is a one-pair circuit that can have as many as eight individual residences, (each with a separate phone number) share that same pair for service (Fig. P.6). If one residence is using the line, the others can't. Each residence can have its own phone number, with the use of a *SRM (Selective Ringing Module).* The SRM is installed in the NI of each residence and contains electronics that can be configured to recognize different ringing formats using DIP switches. Some different ringing formats that an SRM would differentiate are ring voltage on the ring side, ring voltage on the tip side, ring voltage on the ring side with the tip side grounded, and ring on the tip side with the ring side grounded. The selective ringing modules are wired to recognize a certain ring. The central office sends a specific ring to reach a specific number.

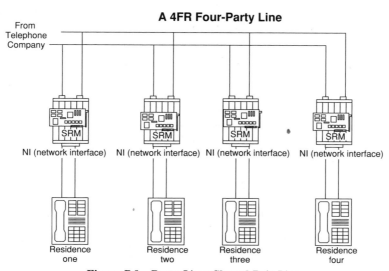

Figure P.6   Party Line: Shared Pair Line

Pass-Band Filter   Another name for a band-pass filter. A band-pass filter is used in frequency-division multiplexing as well as the equalizer in your stereo. It is usually a capacitor/resistor/inductor network that has a resonant frequency and a rating of how well it passes one frequency (or a bandwidth of frequencies) and blocks out others (called the Q, quality) of the circuit. The resonant frequency of the circuit is the frequency that the circuit will pass.

Passive Hub   As opposed to an active hub, a passive hub has no ability to amplify (extend signal transmission range) a signal on an Ethernet network, so it needs to be in close proximity to the computers it is connected to. Simply stated, a hub makes a star wiring configuration look like a bus configuration to all the devices connected to it. Hubs are utilized extensively in Ethernet networks. For a diagram of a hub application, see *Hub*.

Passive Optical Network (PON)   A fiber-optic-based transmission network that contains no electronic devices that require external power. Passive optical networks use the physical characteristics of light to separate different carriers or colors of light. These types of optical networks are relatively inexpensive to implement and maintain, compared to their active counterparts.

Passive Matrix Display   A type of laptop computer display technology where rows of liquid crystal elements are connected as a grid. Those elements are activated by their coordinate (horizontal and vertical) transistor. Passive matrix displays have a slightly fuzzy image during screen movement and screen refresh, although they are less expensive and consume less power.

Patch Panel   A panel equipped with plugs, rather than terminals, for connecting wires or fiber optics. A patch panel can be used to terminate installed wire or be used as a "plug-in" test access point for communications circuits. DS0 and DS3 patch panels are very popular in central offices for testing purposes. Cat 5 patch panels are popular in computer LAN environments for the easy connection of computers to a network of pre-installed wire.

Path   1. The process of aligning a microwave radio link. Two technicians point the dishes at each other while taking AGC readings from the transmission equipment, which is often located in the dish (also called an ODU, Outdoor Unit).   2. The space between two microwave dishes that make a microwave radio link.

Path Cost   A predetermined way to prioritize trunks for traffic and failover (as in Spantree). The path cost number is used by a switch to determine which network link is favorable and is a function of bandwidth. Common path costs are shown in the Fig. P.7.

COMMON PATH COSTS FOR SPANTREE			
TRUNK TYPE	NAME	BANDWIDTH	COST
10BaseT	Single Ethernet	10 Mbps	100
100BaseT	Fast Ethernet	100 Mbps	19
100BaseT X2	2 Fast Ethernet*	200 Mbps	12
100BaseT X4	4 Fast Ethernet*	400 Mbps	8
1000Base FX	Gigabit Ethernet	1 Gbps	4
*Based on Cisco Systems' Fast Etherchannel™ configuration.			

Figure P.7   Path Costing (Routing)

Pause   A feature incorporated with the speed-dial feature of telephones. When speed-dial numbers are programmed, a 1.5-second pause can be inserted by pressing the # key. If a user wants to program a speed dial that rings into a PBX system where an extension needs to be input after an auto attendant answers, the user can input several pauses before the extension number in the speed-dial string that they program on their phone. When activated, the speed-call feature will then dial the number, pause while the auto attendant answers, then dial the extension.

Payload   A transmission signal or packet has two components, the payload and the overhead. The payload carries the customer information, like a B Channel in an ISDN circuit. The overhead carries operational, maintenance, and synchronization information that make the protocol work. An ISDN D channel is an overhead component of an ISDN circuit.

Payphone   A coin operated telephone. Many payphones are owned and operated by local telephone companies, but there are private payphone companies too. Pay or coin-operated telephones can be purchased at telecommunications equipment distributors, such as Graybar and Anixter.

PBX (Private Branch Exchange)   See *Private Branch Exchange.*

PBX, IP-Based   See *IP Telephony.*

PC (Personal Computer)   A reference to an IBM PC or an IBM clone PC, which is a PC manufactured by some company other than IBM (such as Dell, Hewlett Packard, or Compaq). The older architecture for PCs is the *AT (Advanced Technology)* architecture. The newer is the ATX architecture developed by Intel. Illustrated is the rear-panel comparison of an AT machine (Fig. P.8) and an ATX machine (Fig. P.9). Notice that the ATX architecture blocks the serial and I/O connectors (i.e., printer port) together on a backplate. See also *AT, ATX,* and *Personal Computer.*

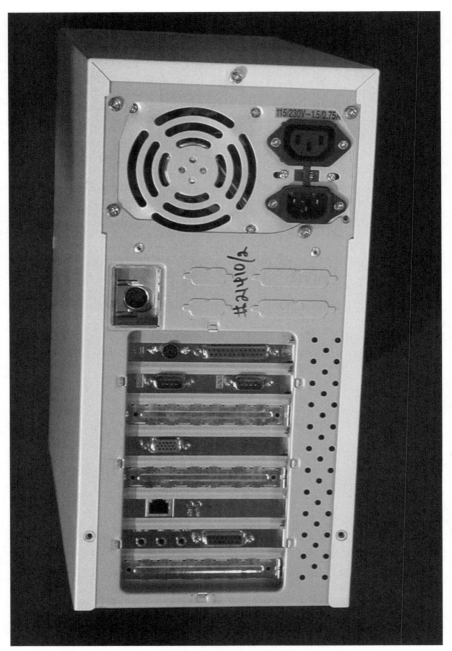

Figure P.8   Rear of AT PC

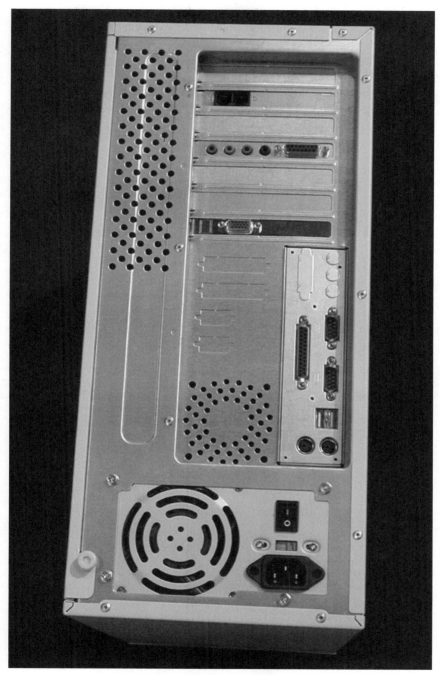

Figure P.9   Rear of ATX PC

PCB (Printed Circuit Board)   See *PC Board.*

PC Card   A short name for a PCMCIA card most commonly made for PC laptops. See *PCMCIA.*

PCI (Peripheral Component Interconnect)   A 64-bit bus standard developed by Intel. *PCI* is a term that is commonly referred to by PC users that are searching for an expansion board for their PC. PCI is an improvement over the *ISA (Industry Standard Architecture)* bus. As PCI matures in the industry, it will exceed a throughput rate of 200 Mbps. PCI slots on a personal computer's motherboard are recognizable by their white or brown color, they are relatively shorter than other sockets/slots, and they have a smaller pin size. Figure P.10 shows a PCI video card. See also *ISA* and *AGP.*

Figure P.10   PCI (Peripheral Component Interconnect) Video Card

PCL (Printer Control Language)   A printer-interface communication method developed by Hewlett-Packard, and used widely in the computer-printer industry. PCL translates page data into instructions for a printer.

PCM (Pulse-Code Modulation)   A concept that is similar to that of Morse code, digital signals are sent over a media (twisted copper pairs, radio, fiber optic, coax, etc.) one bit at a time, each bit being represented as a pulse or the absence of a pulse. A typical digital transmission is PCM.

PCMCIA (Personal-Computer Memory-Card International Association)   An organization made up of about 500 companies that developed a standard for small, credit card-sized PC devices. These devices are designed especially for laptop or other portable computing devices, and they are designed to be exchanged without rebooting the PC (although this is not always the case). The first PCMCIA cards were made for additional memory. Subsequent designs included LAN interface cards and modems (Fig. P.11). There are three types of PCMCIA interfaces. *Type-I cards* are for memory, and are the thinnest of the three types (3.3-mm thickness). *Type-II cards* are generally for LAN interface and modem applications (5.5-mm thickness). *Type-III cards* are the largest in thickness at 10.5 mm thick and are for portable disk drives. All three have the same rectangular dimensions (85.6 by 54 millimeters). Because the rectangular dimensions of the cards are the same, they all fit into the same slots as follows: A Type-III slot can hold one Type-III card or one each of a Type II and Type I. A Type-II slot can hold one Type-II or two Type-I cards. A Type-I slot can hold one Type-I card.

Figure P.11   PCMCIA Cards

PCR (Peak Cell Rate)   A parameter defined by the ATM forum for ATM traffic management. In *Constant Bit Rate (CBR)* transmissions, the peak cell-rate parameter determines how often data samples are sent. In

Available Bit Rate (ABR) transmissions, the peak cell-rate parameter determines the maximum value of the available cell rate. See also *ACR, MCR,* and *CBR.*

PCS (Personal Communications Services)   In wireless/cellular telecommunications, an allocation of frequency bands in the 1900 MHz range that were designated for the use of public wireless communication by the FCC. In the United States, the majority of telephone companies that provide PCS services do so utilizing CDMA (Carrier Division Multiple Access) transmission formats, which are a variation of spread spectrum radio. Specifically, CDMA is deployed in the PCS-1900 MHz band (called PCS-1900). Using CDMA technology, PCS services are able to provide longer battery life, variable speech compression, data services, soft-call handoff, and better reception when multipath radio reflections occur. CDMA is a second-generation wireless technology and is the precursor to CDMA2000, which is an operating system for 3G (third-generation) wireless technology. See also *3G* and *CDMA.*

PDN (Primary Directory Number)   An ISDN telephone number.

PE (Peripheral Equipment)   Devices that are not a part of a system, but work with it, such as a printer.

Peak Cell Rate (PCR)   A parameter defined by the ATM forum for ATM traffic management. In *Constant Bit Rate (CBR)* transmissions, the peak cell-rate parameter determines how often data samples are sent. In *Available Bit Rate (ABR)* transmissions, the peak cell-rate parameter determines the maximum value of the available cell rate. See also *ACR, MCR,* and *CBR.*

Peak Power   A method of calculating the power consumption or power output of an electronic/electrical device. Other methods of calculating power include true power, transparent power, and RMS (Root-Mean-Square) power. Most audio applications use either peak or RMS power. A great example to demonstrate the difference between peak power and RMS power is home and car stereo amplifiers. Many people ask which is better, peak power or RMS power? The answer is both. Some stereo manufacturers put peak-power ratings on their products because it sounds better. Some put RMS power on their products because it is closer to the true power of the device. To convert from peak power to RMS power, multiply the peak-power rating by 0.707. The result is RMS power. To convert RMS power to peak power, divide the RMS power rating by 0.707. The result is the peak-power rating.

Ped (Pedestal) See *Pedestal.*

Pedestal (Ped) Usually a small green box that houses telephone or cable-TV cable splices or terminals.

Peer Communication The communication of individual layers in a protocol with each other. For instance, the network-layer control signals from the near end are only used by the network layer on the far end. They are not used by any other layer, and are definitely not seen by the user. These types of signals or "control frames" are referred to as *peer communications.* For an example of a peer communications signal, see *SABM* and *FRMR.*

Peer-to-Peer Networking A local-area network scheme that does not use a server or host. Individual PCs are linked together via network cards and CAT-5 wire or coax. Windows 95 has its own peer-to-peer networking utility built in.

Peg Board Also called a *white board* or *mushroom board.* It is placed between termination blocks (such as 66M150 blocks) to provide a means of support for routing cross-connect wire. For a photo, see *White Board.*

Performance Management One of five categories of network management defined by ISO for management of OSI networks. Performance-management subsystems are responsible for analyzing and controlling network performance, including network throughput and error rates. See also *Accounting Management, Configuration Management, Fault Management,* and *Security Management.*

Peripheral Component Interconnect (PCI) A 64-bit bus standard developed by Intel. *PCI* is a term that is commonly referred to by PC users that are searching for an expansion board for their PC. PCI is an improvement over the *ISA (Industry Standard Architecture)* bus. As PCI matures in the industry, it will exceed a throughput rate of 200 Mbps. PCI slots on a personal computer's motherboard are recognizable by their white or brown color, they are relatively shorter than other sockets/slots, and they have a smaller pin size. Illustrated is a PCI video card. See also *ISA* and *AGP.*

Peripheral Equipment Devices that are not a part of a system, but work with it, such as a printer.

Peripheral Node A terminal, printer, or other I/O device on an *SNA (IBM System Network Architecture)* network (Fig. P.12).

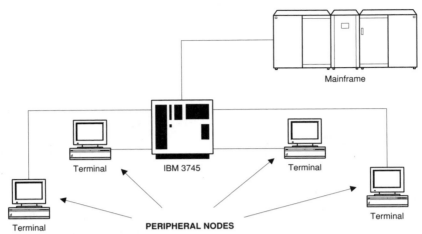

Figure P.12 Peripheral Node in an SNA Environment

Permanent Virtual Circuit   A dedicated (private line) channel in a multiplexed transmission or packet network used by telephone companies. Permanent virtual circuits are very common in T3 and SONET carrier networks. A virtual circuit is a switched circuit, like a plain telephone line. A permanent circuit is a dedicated twisted copper pair with a carrier, such as T1 for a private line.

Personal Communications Services (PCS)   In wireless/cellular telecommunications, an allocation of frequency bands in the 1900 MHz range that were designated for the use of public wireless communication by the FCC. In the United States, the majority of telephone companies that provide PCS services do so utilizing CDMA (Carrier Division Multiple Access) transmission formats, which are a variation of spread spectrum radio. Specifically, CDMA is deployed in the PCS-1900 MHz band (called PCS-1900). Using CDMA technology, PCS services are able to provide longer battery life, variable speech compression, data services, soft-call handoff, and better reception when multipath radio reflections occur. CDMA is a second-generation wireless technology and is the precursor to CDMA2000, which is an operating system for 3G (third-generation) wireless technology. See also *3G* and *CDMA*.

Personal Computer   A small and relatively inexpensive data-processing machine that is designed for one user. Early computers had many terminals and were shared among many users. Personal computers are used for word processing, data-base management, desktop publishing, Internet access, games, composing music, and many other applications. The two major types of personal computers are the Apple Macintosh and the "PC," which is based on the IBM PC architecture, and frequently referred to as an *IBM clone,* if manufactured and distributed by one other than IBM. Apple Macintosh computers are frequently referred to as *Macs.* IBM PCs and IBM clones are referred to as *PCs.*

Personal Computer Memory Card International Association (PCM-CIA)   An organization made up of about 500 companies that developed a standard for small, credit card-sized PC devices. These devices are designed especially for laptop or other portable computing devices, and they are designed to be exchanged without rebooting the PC (although this is not always the case). The first PCMCIA cards were made for additional memory. Subsequent designs included LAN interface cards and modems. There are three types of PCMCIA interfaces. *Type-I cards* are for memory, and are the thinnest of the three types (3.3-mm thickness). *Type-II cards* are generally for LAN interface and modem applications (5.5-mm thickness). *Type-III cards* are the largest in thickness at 10.5 mm thick and are for portable disk drives. All three have the same rectangular dimensions (85.6 by 54 millimeters). Because the rectangular dimensions of the cards are the same, they all fit into the same slots as follows: A Type-III slot can hold one Type-III card or one each of a Type II and Type I. A Type-II slot can hold one Type-II or two Type-I cards. A Type-I slot can hold one Type-I card.

Peta   The prefix for 1,000,000,000,000,000. It would take five million 1GB (one gigabyte) hard drives to have the capacity of one 5PB hard drive. I don't think we will see hard drives in the PB range any time soon.

Phantom DN (Phantom Directory Number)   Also called a Virtual DN. A directory number or extension on a PBX system that is used to attach a voice mailbox. The phantom DN does not really have a telephone set, but the PBX system thinks it does, so it transfers calls to that DN, which are configured to be forwarded to a voice-mail system. A user of that DN can then dial into the voice-mail system, enter their extension, and receive their messages.

Phase   A reference to a sine wave and its relative cycle to another sine wave or time source. Phase is measured in degrees (0 to 360) or radians (Fig. P.13).

A sine wave, and a cosine wave (dashed), which is 180° out of phase with the sine wave. The resultant output from combining these signals would be 0 V.

Figure P.13   Phase

Phase Alternation Line (PAL)   The television broadcast standard in Europe the Middle East, parts of Africa, and parts of South America. PAL defines 625 vertical scan lines and refreshes the screen 25 times per second. See also *Television Broadcast Standards.*

Phased Locked Loop (PLL)   A very important electronic circuit in the world of *FM (Frequency Modulation)* and *PM (Phase Modulation).* Phase-locked loops are used as the detector circuits in FM receivers and to create stable RF references for all types of transmitters and timing circuits.

Phase Modulation (PM)   A method of varying a radio carrier frequency so that a signal (the variations) can ride on it. After the carrier signal has the variations imposed on it, it is amplified and transmitted. The variations in the signal are then detected by the receiver. The variations in the carrier signal are actually voices, music, or whatever is to be transmitted. The other methods of modulating a carrier frequency are *AM (Amplitude Modulation)* and *FM (Frequency Modulation).* Phase modulation makes the phase of a carrier frequency change in conjunction with a signal that it is to carry. The color on broadcast television is sent in a PM format. A simple representation is depicted in Fig. P.14.

PHASE CHANGES OF A CARRIER SIGNAL FOR A TRANSMITTED SIGNAL (DASHED)

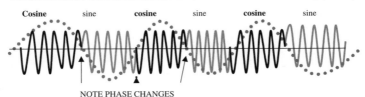

NOTE PHASE CHANGES

Figure P.14   Phase Modulation

Phase-Shift Keying (PSK)   A method of modulating a carrier frequency by making the carrier signal phase shift in conjunction with the digital input signal. In Fig. P.15, a cosine phase indicates a "1" value and a sine phase indicates a "0."

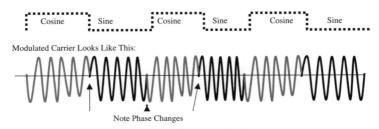

Note Phase Changes

Figure P.15   Phase-Shift Keying

Phased Array Antenna   A group of small antennas placed a multiple of a wavelength in distance from each other to create one larger antenna.

Phonetic Alphabet   A set of audibly distinct words that were chosen by the U.S. Army to identify spoken letters and numbers (Fig. P.16).

A	ALPHA	N	NOVEMBER	0	ZERO
B	BRAVO	O	OSCAR	1	ONE
C	CHARLIE	P	PAPA	2	TWO
D	DELTA	Q	QUEBEC	3	THREE
E	ECHO	R	ROMEO	4	FOUR
F	FOXTROT	S	SIERRA	5	FIVE
G	GOLF	T	TANGO	6	SIX
H	HOTEL	U	UNIFORM	7	SEVEN
I	INDIA	V	VICTOR	8	EIGHT
J	JULIET	W	WHISKEY	9	NINER
K	KILO	X	XRAY		
L	LIMA	Y	YANKEE		
M	MIKE	Z	ZULU		

Figure P.16   Phonetic Alphabet

Photoconductive Cell   Also called a *photoresistor* or *photosensitive cell* (Fig. P.17). An electronic device that conducts electricity better when it is exposed to light. Photoconductive cells are made from Cadmium Sulfide (CdS) and Cadmium Selenide (CdSe). They are most responsive to green-colored light (5500 angstrom). They react to almost the entire spectrum of light that is visible to the human eye.

Figure P.17   The Schematic Symbol for a Photoconductive Cell

Photo-Conductor   A reference to a photoconductive cell.

Photodetector   A photo-sensitive circuit whose main component is usually a photo-diode or photo-transistor. A photo-detector converts pulses of light into pulses of electricity.

Photodiode   An electronic device that acts as a light-activated switch (Fig. P.18). It operates similar to a zener diode, except that the reversecurrent effect is controlled by light.

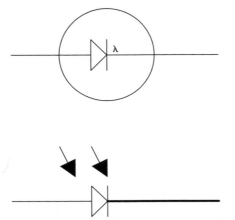

Figure P.18   Schematic Symbols for Photodiodes

Phototransistor   A transistor that is forward biased (conducts electricity as a switch) when exposed to light (Fig. P.19). Phototransistors are used the same way as switching transistors, except the base is the photosensitive part of the device.

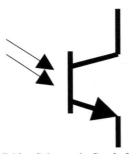

Figure P.19   Schematic Symbol for Phototransistor

Photonic Layer   Fiber optic. A reference to the physical layer in the SONET protocol architecture, which is where the type of fiber optic is defined (multimode/single mode).

Physical Address   Another name for MAC address. See *MAC Address*.

Physical Colocation   A *colocation* is an interconnection agreement and a physical place where telephone companies hand-off calls and services to each other. This is usually performed between a CLEC and an RBOC. The CLEC installs and maintains interconnection equipment usually consisting of optical carrier (SONET) equipment and a digital cross-connect system. There are other types of colocations. Alarm companies like to have their alarm-signaling equipment located in the local central office for security and convenience of connecting alarm circuits. Long-distance companies colocate with local telephone companies as well.

Physical Layer   A layer in a communications protocol model. In general, the physical layer is the actual media of the communications transmission (twisted-pair wire, coax, air, fiber optic, etc.) It is also the types of connectors used and the pin-outs of those connectors. The 568B wiring scheme for CAT 5 wire is a physical-layer function. The latest guideline for communications protocols is the *OSI (Open Systems Interconnect)*. It is the best model so far because all of the layers (functions) work independently of each other. For a diagram of the OSI, SNA, and DNA function layers, see *Open Systems Interconnection*. For a conceptual diagram of the OSI model layers, see *OSI Standards*.

Physical Medium-Dependent Sublayer (PMD)   The ATM physical layer is divided into two parts (sublayers), the *TCS (Transmission Convergence Sublayer)* and the PMD. The PMD is the lowest physical layer and it determines the types of connectors and the medium used for transmission. It also interfaces the line-coding technique, such as SONET OC-1 or STS-3c/STM1, with the TCS sublayer. See also *TCS*.

Physical Topology   A physical topology refers to the way a *Local Area - Network (LAN)* of computers is connected for communication. The three different types of physical topologies are: ring, star, and bus (Fig. P.20). The star and bus topologies work very much the same. The ring topology is also called a *token passing topology*.

PIC   PIC refers to color-coded cable. Icky-PIC is cable that is color coded and jelly filled to help protect the copper pairs inside from water.

Pico   Prefix for 1×10^{-9}. One picofarad (a capacitor) is equal to 0.000,000,001 farads. It is abbreviated as 1 pF.

Picofarad   A unit of measurement for capacitors. Pico is the prefix for 1×10^{-9}. One picofarad (a capacitor) is equal to 0.000,000,001 farads. It is abbreviated as 1 pF.

Piggybacking   Process of carrying acknowledgments within a data packet to save network bandwidth.

**LAN PHYSICAL
TOPOLOGIES**

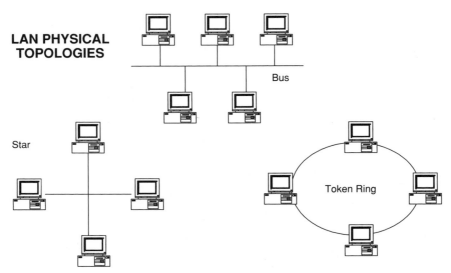

Figure P.20   Physical Topologies

PIM Dense Mode/Sparse Mode   In Ethernet-type networks such as 802.3, multicast is made possible by a set of communications procedures called PIM (Protocol Independent Multicast). A network can be configured to utilize PIM as one of two features: PIM Dense Mode and PIM Sparse Mode. PIM Dense Mode floods a network, sending the packets that make up the transmission to every device on a network. Routers on the network then identify which users respond to the multicast flood, and then prune (discontinue) packet transmissions from devices that do not respond. The advantage of PIM Dense Mode is that its operation is simple to end users. There is no need for an advance request to start receiving the transmission. It appears to be automatic to the end user because it is. PIM Sparse Mode incorporates IGMP, which is a signal for end users to request a multicast packet stream. The advantage of this method is that in large networks, the initial flood in PIM Dense Mode never happens. The transmission is sent to a single router called a *rendezvous point*. The rendezvous point router then duplicates the packets to the necessary segments that have requested them. PIM Sparse Mode is a more efficient use of a network's resources. The drawback is the need for requests to refresh the transmission.

PING   1. A DOS command that sends a broadcast packet over a network that asks for an acknowledgment from the device that is configured with the IP address entered with the PING command. PING was named so because it is a good mnemonic reference that is associated with submarines. The PING command behaves similar to submarine radar emitting audible ping sounds in attempts to locate other vessels or objects in the sea.   2. In some networking circles, it is believed that PING is a mnemonic abbreviation for Packet Internet Groper, and a reference for the same DOS command listed previously.

Plain-B Wire Connector   Most commonly referred to as *beans*. A splicing connector used to splice twisted-pair telephone wire (Fig. P.21). The connectors are crimped onto the wires to be spliced. Inside the connector are teeth, which pierce the vinyl insulation of the wire to make a good connection.

Figure P.21   Plain-B Wire Connector (Beans)

Plant   A reference to telephone company equipment, poles, cable vaults, cable, central offices, and transmission equipment.

Plant Test Numbers   Telephone numbers that when dialed provide a test tone or access to other testing resources, such as a quiet-line or automatic-number identification. Plant test numbers are used by telephone company personnel and are not given to the public.

Plenum   A reference to telephone, communications, or electrical wire that is insulated with polyvinylidene diflouride. It is made with this substance because it does not emit poison gasses when it burns, like PVC (polyvinyl chloride) does (PVC produces chlorine gas when burned). It gets its nickname because it is permitted to be placed in air ducts or plenum spaces in buildings. Plenum wiring or cable is typically three times the cost of PVC jacketed/insulated types.

Plenum Cable   See *Plenum.*

Plesiochronous   "Almost in time." Plesiochronous networks are those that the telephone companies use to synchronize T1 and T3 carrier signals. The electronic transport equipment at each end of the transmission does not get timing from the same source (thus being synchronous), but the timing of each individual device is very close. Stratum One clocks are used in this type of communications equipment, which provide a steady timing for each end to transmit and receive signals by. Transmissions between SONET networks over a long-haul circuit could be considered plesiochronous.

PMD (Physical Medium Dependent Sublayer)   The ATM physical layer is divided into two parts (sublayers), the *TCS (Transmission Convergence Sublayer)* and the PMD. The PMD is the lowest physical layer, and it determines the types of connectors and the medium used for transmission. It also interfaces the line-coding technique, such as SONET OC-1 or STS-3c/STM1 with the TCS sublayer. See also *TCS.*

POH (Path Overhead)   The overhead that is added to a signal to allow a transport network to carry it.

Point of Interface   See *Point of Presence.*

Point of Presence   In Internet terms, an access node equipped with modems that allow customers to dial into their ISP. National Internet service providers try to have a point of presence within every area code that a customer could try to access their network through.

Point to Point   Usually a reference to a private-line circuit that is leased from the telephone company. It can also be a reference to a switched service, like a plain telephone line where communications links are switched from one point to another, depending on the number dialed.

Pointing Stick   An alternative to the use of a mouse on portable laptop computers. The pointing stick is usually located in the middle of the laptop's keyboard and resembles a pencil eraser. Other laptop mouse alternatives include the roller ball and the touch/track pad.

Point-to-Point Protocol (PPP)   Also known as *dial-up IP.* PPP is the successor to SLIP that provides router-to-router and host-to-network connections over synchronous and asynchronous circuits. Whereas SLIP was designed to work with IP, PPP was designed to work with several network-layer protocols, such as IP, IPX, and ARA. PPP also has built-in security mechanisms, such as CHAP and PAP. PPP relies on two protocols: LCP and NCP. See also *CHAP, LCP, NCP, PAP,* and *SLIP.*

Poisson Distribution   A mathematical formula that used to be used in traffic engineering for calculating the probability of blocked calls in a telephone network. Now we have computer programs (CTI) that provide graphs of trunk groups and their usage. The graphs are much easier to use. When customers don't have expensive software to manage their telephone networks, they simply add more lines when they get complaints of busy signals.

Poison Reverse Updates   Routing updates that explicitly indicate that a network or subnet is unreachable, rather than implying that a network is unreachable by not including it in updates. Poison reverse updates are sent to defeat large routing loops.

Polarity   A reference to positive or negative voltage potential, or to the polarization of an antenna or dish-type antenna.

Polarization   The pointing of a microwave dish antenna so that the transmission dispersion is in a vertical or horizontal pattern. The headlights on cars are polarized in a horizontal manner so that the light dispersion is spread across the horizontal surface of the road. The other kind of polarization is vertical, where the transmission dispersion is in an up-and-down pattern. The two antennas or dishes used in a point-to-point application need to be polarized the same way.

Pole Attachment   A lease from a utility company (usually power) that permits a telecommunications company to attach their cable facilities to power poles (Fig. P.22). For more photos of different aerial-attachment hardware, see *B Washer, Strand Clamp, Guy Thimble,* and *Johnny Ball.*

Figure P.22   Pole-Attachment Hardware

Policy-Based Routing   A routing scheme that forwards packets to specific interfaces based on user-configured policies. Such policies might specify that traffic sent from a particular network should be forwarded out one interface, while all other traffic should be forwarded out another interface.

Poll/Final   A bit in the control byte of an X.25 frame. When the two devices on the ends of an X.25 link lose track of where the other is in receiving the transmission, a poll bit will be sent inside the control byte of the next frame. The device on the other end will send a frame with the poll bit set to one that is followed by the number of the next frame desired (binary one through seven).

Polling   A LAN line-sharing method where a primary network device "asks" all other devices attached to a network if they have data to transmit. This gives each device a fair chance to use the network media. Token ring is a polling-type of network protocol. In contrast, Ethernet is not a polling protocol.

PON (Passive Optical Network)   A fiber-optic-based transmission network that contains no electronic devices that require external power. Passive optical networks use the physical characteristics of light to separate different carriers or colors of light. These types of optical networks are relatively inexpensive to implement and maintain, compared to their active counterparts.

POP   1. *Post Office Protocol.* POP is a daemon program that runs on *SMTP (Simple Mail Transfer Protocol)* servers acting as if it were a mail server in itself. It processes SMTP client's mail-retrieval requests. Without POP, clients cannot receive mail unless they are online at the time it is sent to them. POP and SMTP have been further enhanced with the *MIME (Multipurpose Internet Mail Extensions)* standard, which allows users to attach files to their text messages.   2. *Point of Presence.* See *Point of Presence.*

Port Aggregation Protocol (PagP)   An operating feature of LAN switches that enables 2 or 4 Ethernet ports (more in some switches) to be concatenated into one high-capacity link. The PagP subroutine automatically identifies ports that are configured as pairs and reports them to the Spanning Tree Protocol, which manages traffic over the combined links as one large channel. This is referred to as *Etherchannel* in marketing Cisco Systems Catalyst LAN switches.

Port Number   An identification/reference system that TCP uses to identify applications that it will packetize and send. The port numbering system allows TCP to know how to handle the application for which it is performing communications. A field of the TCP header is dedicated for the port number being utilized in the current transmission. All applications have standard port numbers assigned to them. Standard port numbers are called *well-known port numbers*. Some examples of well-known port numbers are: 20 = FTP-data, 21 = FTP-control, 23 = Telnet, 25 = SMTP, 53 = Domain Name, 161 = SNMP agent, and 162 = SNMP manager. Further, port numbers can be altered. Network administrators do not always utilize the standard application's port numbers within intranets as a security design measure. See also *Socket*.

Portability   A reference to the ability to change telephone companies and take your phone numbers with you.

POST (Power On Self Test)   A set of hardware diagnostic instructions (usually retrieved from ROM) that operates within routers and other traffic-management devices when they are first turned on.

Post Office Protocol   POP is a daemon program that runs on *SMTP (Simple Mail-Transfer Protocol)* servers acting as if it were a mail server in itself. It processes SMTP client's mail-retrieval requests. Without it, clients cannot receive mail unless they are online at the time it is sent to them. POP and SMTP have been further enhanced with the *MIME (Multipurpose Internet Mail Extensions)* standard, which allows users to attach files to their text messages.

Post, Telephone, and Telegraph (PTT)   The general name for the communications service providers in countries that have not yet allowed this service to be controlled by a nongovernment agency or corporation. PTTs are often extensions of the postal service, where telecommunications service is a branch from the original letter-mail and telegraphy service. South American and African countries commonly have a PTT agency, rather than multiple telephone companies.

PostScript   A printer interface communication method developed by Adobe. PostScript incorporates ASCII code and translates it to instructions for printer devices. *PostScript* is also a reference to an Adobe standard in text fonts.

POT   1. Abbreviation for potentiometer. Also known as a *variable resistor.* Many electronic control knobs are connected to variable resistors. Variable resistors are usually made from carbon film. For a photo and schematic symbol of a potentiometer, see *Variable Resistor.*   2. *Plain-old telephone,* a reference to standard switched residential and business telephone lines.

Potato   (slang) Another name for an aerial service-wire splice. Also called a *football*. For a photo, see *Aerial Service-Wire Splice*.

Potential   A voltage difference. Potential is a voltage from one point to another. The voltage potential of a POTS telephone line is −52 volts from ring to tip.

Potentiometer (Pot)   Also known as a *variable resistor*. Many control knobs are connected to variable resistors. Most volume controls are variable resistors. For a photo and schematic symbol of a potentiometer, see *Variable Resistor*.

POTS (Plain Old Telephone Service)   A telephone line, with a telephone number, like the standard ones subscribed to by residences and many small businesses.

Pound Key   The button on a telephone dial pad with the # on it.

Power   Current multiplied by voltage. Power is measured in watts. If you use a certain amount of wattage over a certain period of time, then you have used energy. Energy is equal to watts multiplied by time, and the unit is joules.

Power On Self Test (POST)   A set of hardware diagnostic instructions (usually retrieved from ROM) that operates within routers and other traffic-management devices when they are first turned on.

Power Supply   A device that converts 120-V or 220-V standard AC power to a voltage that can be useful for an electronic system.

PPP (Point-to-Point Protocol)   Also known as *dial-up IP*. PPP is the successor to SLIP that provides router-to-router and host-to-network connections over synchronous and asynchronous circuits. Whereas SLIP was designed to work with IP, PPP was designed to work with several network-layer protocols, such as IP, IPX, and ARA. PPP also has built-in security mechanisms, such as CHAP and PAP. PPP relies on two protocols: LCP and NCP. See also *CHAP, LCP, NCP, PAP,* and *SLIP*.

PPS (Packets per Second)   In LAN and WAN architectures, a standard measure of transported traffic over a period of time based on a 64-byte packet. This is not an official standard as of the year 2001. The size of the packet varies by several bytes depending on the manufacturer and type of equipment. Some use packets as small as 60 bytes on networks specifically configured to carry such traffic. The minimum size for an Ethernet packet is 64 bytes. See also *Runt* and *Giant.*

PPSN (Public Packet-Switched Network) A reference to *Frame Relay.*

PPSS (Public Packet-Switched Service) A reference to *Frame Relay.*

PRAM (Parameter RAM) A small battery-powered unit of RAM used in computers to store basic user settings, such as time and date.

Preamplifier An amplifier designed to amplify the voltage level of a very small signal so that it can be fed to a power amplifier, which amplifies the current aspect of the signal so that it is powerful enough to drive the signal current through a loudspeaker or other device.

Predictive Dialing Another term for auto dialing or progressive dialing. Instead of telemarketers dialing digits through a list or phone book all day long, the numbers are entered into a predictive dialer system. The system then dials the numbers; when a call is answered the predictive dialer transfers the call and the associated information to the computer screen of the appropriate telemarketer.

Premises Equipment Also called *CPE (Customer Premises Equipment).* Telephones, wiring, answering machines, CSU/DSUs, and anything else you might find on the customer side of the network interface.

Premises Wire The wiring on the customer side of the communications company's demarcation point (NI, Network Interface). The premises wire is owned by the customer and is the customer's responsibility to maintain (Fig. P.23). Many communications companies sell maintenance contracts, which enable them to troubleshoot and repair the telephone wire within your home or business, at no extra charge. Typical maintenance contracts are about $2.00 per month.

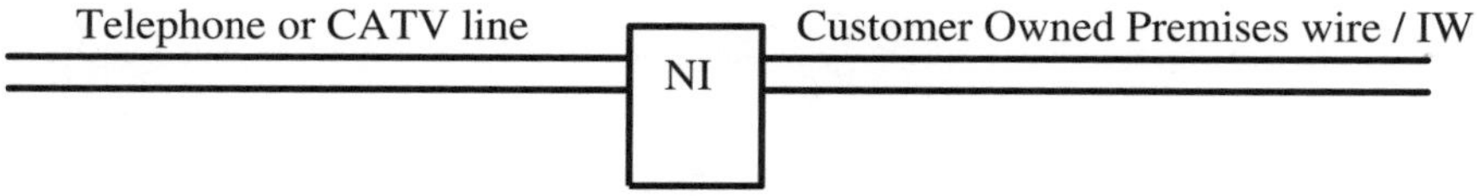

Telephone or CATV line   Customer Owned Premises wire / IW

NI

Figure P.23   Premises Wiring is Separated from Communications Company Wiring by a Network Interface

Prepaid Phone Card   A card that comes with an 800/888 number that the card owner dials to reach a network that allows them to dial anywhere they like (Fig. P.24). The service is good for the amount of time that the prepaid phone card says on its face. Prepaid calling cards are becoming bigger and bigger, especially because they don't cost much more than third-party billing to your home number when you are out of town. Typical prepaid calling cards have a rate of 30 to 35 cents per minute, flat rate, no matter when or where you call within the continental U.S. The way the system works is that a calling card company sets up a data base with card numbers in it and connects it to a calling-card platform. A calling-card platform is a computer that receives a phone call and prompts a caller to enter their calling-card number and the telephone number that they wish to dial. The calling-card platform then checks the card number to see how many minutes it has left on it (it sometimes tells the customer with a recorded message). If time left on the card, the system then dials the number on an outgoing trunk to connect the call. In reality, two long-distance calls are made, one to the calling card platform and one to the number being dialed by the customer.

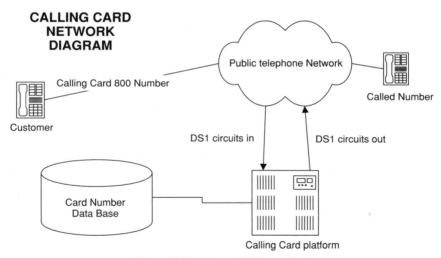

Figure P.24   Prepaid Phone Card

Prepay   A reference to a coin-operated telephone that requires a coin to be inserted before a number is dialed.

Presentation Layer   A layer in a communications protocol model. In general, the presentation layer performs the function of encoding and decoding the data to be transmitted within the communications protocol. The latest guideline for communications protocols is the *OSI (Open Systems Interconnect)*. It is the best model so far because all of the functions work independently of each other. For diagrams relating to the OSI, see *Open Systems Interconnection* and *OSI Standards*.

Pressure Cable   Telephone cable that is equipped with air-pressure equipment. In many cables, nitrogen is used instead of air because it is noncorrosive (air contains humidity, which corrodes copper pairs). Nitrogen is pumped into the cable and the pressure is monitored. If the cable is cut, the pressure drop notifies the telephone company of the cable problem and the nitrogen rushing out of the cable helps prevent any water from entering.

Presubscription   When a customer calls a local telephone company and orders a new phone line, they are asked which long-distance company they would like to subscribe to. When the customer tells them, then the telephone company sets the customers line up in translations so that when the customer dials 1 as a first digit, they are connected directly to the long-distance company that they selected.

Prewire   To install standard wiring into a building or space while it is being constructed. Standard building wiring is that all wiring from each jack terminates to a common location, usually called the *telephone closet*. Prewiring of buildings is common for telephone and CAT 5 computer LAN wiring.

PRI (Primary Rate Interface)   One of two *ISDN (Integrated Services Digital Network) circuit sizes*. ISDN first evolved in 1979. It brings the features of PBX systems and high-speed data-transfer capability to the telephone network. The only thing that complicates ISDN is the many available features. The two kinds of ISDN lines are Primary Rate Interface (PRI) and Basic Rate Interface (BRI). ISDN has two types of channels within an ISDN circuit. The B (bearer) channel carries the customer's communications and the D (data) channel provides control and signaling for the B channels. The BRI ISDN line has two B channels and one D channel. A PRI has 23 B channels and one D channel.

The separate control of the ISDN line over the D channel is what enables the broad flexibility and features available with ISDN. When you are talking or sending a data transmission over an ISDN line, the voice and/or data is carried by the B channels. While you are talking on your ISDN line, you can still dial digits (signal the central office) to change or alter the state of your service because of the separate D channel. For example, imagine you want to arrange a meeting with a client. You dial the client's telephone number on your ISDN telephone to reach the client. While you are speaking with the client, you can dial up an Internet access on your computer and put two baseball tickets in at the ticket counter while on the same BRI line. Then you can fax your client directions by downloading a map provided by the baseball ticket office, disconnect and redial your client's fax number. All of this occurs while talking to your client the entire time. Through the advanced convenience and flexibility of ISDN, you can send different types of data and messages to different places at the stroke of a few buttons, and at a much faster speed than a regular telephone line. If you are interested in ISDN, call your local phone company. They can help you decide on what kind of terminal adapter (equipment that connects your computer and phone equipment to the ISDN line) to buy and what kind of features to subscribe to. ISDN is not yet available everywhere. For a diagram that compares an ISDN BRI and ISDN PRI circuit, see *Integrated Services Digital Network.*

Primary Rate Interface (PRI)   See *PRI.*

Prime Line   A key telephone system and hybrid key telephone system feature. The feature enables a user to select the line that a key system connects to a telephone set to when the receiver is lifted. If you don't want people in the office using the main telephone line in the office to make outgoing calls, then don't select that line as a prime line for any of the telephone extensions.

Primitives   The IP layer in the TCP/IP layer operates with the smallest actions or instructions that can be initiated. These small-size actions are called *primitives.* Some examples of standard IP primitives are: Receive Datagram, Send Datagram, Select Source Address, Find Max Datagram Size, Advise on Delivery Success, Send ICMP (Internet Control Message Protocol) message, and Receive ICMP Message.

Print Server   A computer dedicated to fielding, managing, and executing (or sending for execution) print requests from other devices (such as servers or workstations) on its network.

Printed Circuit Board (PC Board)   The green- or brown-colored board that has copper-conductive tracks etched onto its surface. Electronic components are soldered onto these boards by hand or by a method called *flow soldering.* Some PC boards are layered or sandwiched, with conductive tracks inside them and on both sides.

Priority Ethernet Switching In converged (voice, video, and data) Ethernet environments, a reference to the IEEE 802.1p standard for prioritization of LAN traffic among Ethernet switches based on the switch port, MAC address, or IP address associated with the communicating end appliance (be it an IP phone, video monitor, host PC, printer, or server). Packets are tagged as belonging to a queue, which determines the priority of the packet. By the 802.1p standard, queues 0–3 are normal and 4–7 are high priority. 802.1p functions hand-in-hand with 802.1Q or VLANs.

Private Branch Exchange (PBX) A telephone system used to maximize use of telecommunications services purchased from a telecommunications company. A PBX simply takes telephone lines from the outside world and makes them accessible to extensions within a certain building, home, or office. PBX systems are available in many sizes, with many software and feature options (Figs. P.25 to P.27). PBX features include call forwarding, speed dial, internal/external paging, and call-detail recording (call accounting). The larger PBX manufacturers are AT&T, Northern Telecom, Siemons, Toshiba, Iwatsu, NEC, and Rolm. PBX systems have six main parts: the cabinet-backplane (also called a *KSU, Key Service Unit),* the station/telephone connectivity, the trunk/telco connectivity, the power supply, the telephones/extensions, and the administrative access.

Figure P.25   A Definity G3 Private Branch Exchange, Manufactured by Lucent Technologies

Figure P.26   A Nortel Networks Meridian Option 81 Private Branch Exchange

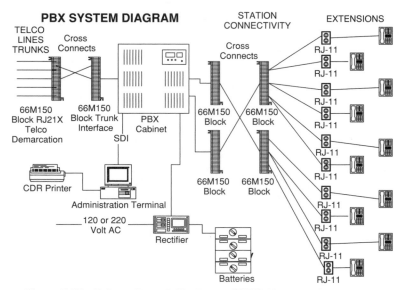

Figure P.27   Private Branch Exchange (PBX) Network Components

- *Cabinet/KSU* The cabinet of the system contains the electronics that make the PBX system work. The backplane (for a photo of a backplane see, *Backplane*) that interface cards plug into is located here. The CPU or core processor (for a photo, see *CPU*) is located in here as well. Many PBX cabinets are designed to allow for additional circuit cards (trunk interfaces/trunk cards and telephone interfaces/station cards) to be added or plugged in later on as the system grows. These spaces are called *expansion slots*.

- *Station-telephone connectivity* This wiring runs from each office or telephone location to the location of the PBX cabinet. Four-pair wiring is most popular because it is inexpensive and contains enough wire to add additional lines or telephones in the future (or additional wire if one or two should go bad). This wiring is installed in a "home run" method, which means that every wire installed runs directly from a jack (usually an RJ-11) directly to the location of the PBX cabinet. Next or near to the PBX cabinet, the individual pairs are neatly terminated and labeled on 66M150 or AT&T 110 (one-ten) blocks.

- *Trunk-telco connectivity* This is similar to the station connectivity, but it needs to be separately labeled from the station connectivity. This is the point where cross connects will be run from the telephone-company demarcation (or *NI, Network Interface*) to your PBX system.

- *Power Supply* The power source for the phone system is a very important consideration. If the power is interrupted, the PBX system will cease to function unless its power supply is incorporated with a UPS system or rectifier/battery system. The best way to go for power is the rectifier with battery back-up (a heavy-duty UPS system especially designed for telephone equipment). Different PBX systems can be ordered to run on 120-V AC or −48-V DC. The −48-V DC system is designed to be powered by a rectifier. The 120-V AC system is designed to run on standard outlet power or a UPS system.

- *Telephones* The telephones for each individual PBX system will work only with that system. They will not work if they are plugged into a regular telephone line. Each phone will determine what features can be implemented. The features are enabled or disabled by the programming or administration done on the PBX system. Some systems have an interface *(SDI, Serial Data Interface)* for a computer or terminal and some are simply programmed by using the telephone stations.

- *Administrative Access* The administrative function of a PBX system can be performed by the user or a telephone-equipment service company. The administrative responsibilities of a PBX system include changing extension numbers, moving phones, changing name displays, and other programming of the system. It also includes maintaining the *Call-Detail Reports (CDR)* of the system. The call-detail reports are reports output by a call-accounting system, which is offered as an extra by virtually every PBX manufacturer. Call-detail reports summarize numbers dialed, length of calls, and incoming calls, caller ID, and their duration.

Private Carrier   A telecommunications company not regulated by the rulings of the PUC; however, they are regulated by the Telecommunications Act of 1996.

Private Line   Also called a *leased line* or *leased circuit.* A leased line is a telephone service that is permanently connected from one point to another (Fig. P.28). Leased circuits include 56 K analog and DS1. A leased circuit acts like a pipeline that carries data from one point to another. If you put a bit in one side, the same bit pops out on the other side. It can carry data across town, across the country, or around the world. Leased lines are relatively expensive. Because leased lines have been offered, new services, such as frame-relay and switched 56 K services have evolved. Frame relay does the same job as a private line, except that it is not isochronous (real time), and you need a private line to put your frame-relay service on. Frame relay is a cost-effective solution for long-haul/long-distance data-transfer applications.

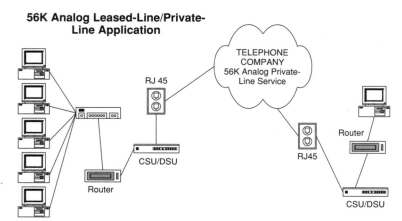

Figure P.28   Private Line

Private Signaling System 1 (PSS1)   Better known as *Q.SIG* in circuit and packet switched telephony circles. The Q.SIG protocol is a variant of ISDN D-channel voice signaling. It is based on the ISDN Q.921 and Q.931 standards and is a worldwide standard for PBX interconnection (under ETSI and ISO), although not all PBX systems and IP telephony gateways support Q.SIG services. PBX systems that are connected via Q.SIG are able to share only the basics in telephone operability. Depending on the PBX systems that are linked with Q.SIG, some advanced features that are proprietary to the individual PBXs are lost when the call connection departs that particular PBX, just the same as PBX features are not supported across the PSTN. Some features can be translated and retained, such as call forwarding or call transfer. Q.SIG is most commonly used to integrate PBX systems across WAN links. Routers that are Q.SIG enabled connect to the PBX on either end, performing IP, frame relay, T1, among other required translation services. Q.SIG has been a precursor to the newer voice over IP and IP telephony methods, and the gateways used within these networks. See also *IP Telephony* and *Gateway.*

Probe   A MAC address finding protocol developed by Hewlett-Packard that works similar to ARP. Probe is a feature that is compatible with and incorporated into many routers and layer 2 switches (LAN switches) including those that run CISCO IOS. See also *ARP* and *RARP.*

Programmable Read-Only Memory (PROM)   Electronic memory comes in two families, *ROM (Read-Only Memory)* and *RAM (Random-Access Memory)*. Memory devices are made from two different technologies: *Bipolar (TTL)* and *MOS (Metal-Oxide Semiconductor)*. Memory is stored by a technique called "writing" and is retrieved by a technique called "reading." ROM devices can only be read and are programmed during manufacture. PROM devices can be programmed at a later date by an electronics reseller or electronic assembler for a special application using special equipment. Special ROM devices called *EPROMs (Erasable Programmable Read Only Memory)* can be electronically erased and re-used. RAM has read and write capability.

The term *random access* means that any memory address can be read in any order at any time. The two types of RAM are static and dynamic. *Static RAM* can hold its memory even when power is removed. *Dynamic RAM* needs constant power to refresh its memory. For a diagram of the different types of dynamic memory, see *Memory.*

PROM (Programmable Read-Only Memory)   See *Programmable Read-Only Memory.*

Prompt   A message from a computer or interactive device that indicates that it is time for a user to input a decision, choice, or other response. Many PBX systems have software that is a "prompt response" style of programming. For example, when the user inputs extension 255, the system responds with "Hands Free?" The user then responds with "YES" or "NO."

Prompt, Response IO   See *Programming PBX.*

Propagation Time   The time for an electrical, optical, or radio signal to travel from one point to another.

Propagation Velocity   The speed that a communications signal travels from one point to another. Electromagnetic waves (radio), electricity, and light approach 300,000,000 meters per second, which is about 160,000 miles per second.

Proprietary   Specially made. All PBX equipment and other premises telephone equipment is proprietary. Northern Telecom telephones will only work with Northern Telecom PBX systems. The same goes for Lucent, Mitel, and other specialized telephone equipment manufacturers.

Protector Block   A block that has many lightning protectors, used to terminate telephone cables (Fig. P.29). A protector is a device used in telephone company network interfaces that provides an easier path for lightning to travel to ground, compared to a telephone user or inside

Figure P.29   Protector Block   2 Pair (Top) 1 Pair (Bottom)

wiring. Before lightning protectors, houses sometimes burned down because of lightning striking the telephone lines. The two types of lightning protectors are carbon and gas. The carbon protectors are simply a piece of carbon that connects tip and ring to ground. The gas protectors are the same, only they are a gas instead of solid carbon. The good thing about gas lightning protectors is that after they are hit by lightning, they do not need to be replaced.

Protocol   The organized processes and rules that communications equipment use to transfer bits and bytes (data). The many communications protocols and layers of protocols that carry other protocols (called protocol stacks), include ISDN, Ethernet, token ring, POTS signaling, DS1, ATM, frame relay, and SONET.

Protocol Analyzer   A test device that can plug into a hub or communications port on a LAN and monitor any address on that LAN at any protocol level (Figs. P.30 and P.31). Protocol analyzers are useful for verifying that an address is good through a network. Most networks are not so complex as to need a protocol analyzer to troubleshoot them.

Figure P.30   A Firebird 500 Protocol Analyzer, Manufactured by TTC

Figure P.31   A Network General Protocol Analyzer

Protocol Converter   A network device or software that converts packeted or framed data from one format to another. This is accomplished by recognizing the initial packet or frame format, removing the data from the packet or frame, and adding new frame or packet headers to the data that are conforming to the new protocol. In older network architectures, protocol converters were an individual entity on a network that existed as a device or software. In newer network architectures, routers and data switches perform the protocol-conversion function.

Protocol Data Unit   Another variation of the term *Packet Data Unit (PDU)*. See *Packet Data Unit*.

Protocol Independent Multicast (PIM Dense Mode/Sparse Mode)   In Ethernet-type networks such as 802.3, multicast is made possible by a set of communications procedures called PIM (Protocol Independent Multicast). A network can be configured to utilize PIM as one of two features: PIM Dense Mode and PIM Sparse Mode. PIM Dense Mode floods a network, sending the packets that make up the transmission to every device on a network. Routers on the network then identify which users respond to the multicast flood, and then prune (discontinue) packet transmissions from devices that do not respond. The advantage of PIM Dense Mode is that its operation is simple to end users. There is no need for an advance request to start receiving the transmission. It appears to be automatic to the end user because it is. PIM Sparse Mode incorporates IGMP, which is a signal for end users to request a multicast packet stream. The advantage of this method is that in large networks, the initial flood in PIM Dense Mode never happens. The transmission is sent to a single router called a *rendezvous point*. The rendezvous point router then duplicates the packets to the necessary segments that have requested them. PIM Sparse Mode is a more efficient use of a network's resources. The drawback is the need for requests to refresh the transmission.

Protocol Stack   A set of related communications-control programs (software) that work together and as a group. Each individual software program is called a *protocol*. A protocol stack can control communication processes at some or all of the seven layers of the OSI reference model. Not every protocol stack covers all layers of the model. Often, a single protocol in the stack will control a number of layers at once. Some protocols within a stack exist only to provide services or specific support functions for other protocols. TCP/IP is a typical protocol stack and *ARP (Address-Resolution Protocol)* is one of the protocols within that stack.

Protocol Translator   A network device or software that converts packeted or framed data from one format to another. This is accomplished by recognizing the initial packet or frame format, removing the data from the packet or frame, and adding new frame or packet headers to the data that are conforming to the new protocol. In older network architectures, protocol translators were an individual entity on a network that existed as a device or software. In newer network architectures, routers and data switches perform this function.

Provisioning   A term that refers to the process of allocating copper pairs, central-office ports/equipment, and programming of central-office equipment. This is what happens before a telephone company network technician installs a telephone service, such as a POTS line or a high-capacity digital service line.

Proxy   An entity or device that, in the interest of efficiency, essentially stands in for another entity.

Proxy Address-Resolution Protocol   A variation of the *Address-Resolution Protocol (ARP)* in which an intermediate device (for example, a router) sends an ARP response on behalf of an end node to the requesting host. Proxy ARP can lessen bandwidth use on slow-speed WAN links. See also *Address-Resolution Protocol*.

Proxy Server   A network server that is loaded with software and equipped with hardware to interface a LAN, MAN, or WAN to the Internet. Proxy servers make up the hardware part of a firewall, which is software that protects the LAN's interworkings from being accessed by strangers/unwanteds/hackers on the outside. Although firewalls are expensive and abound everywhere, hackers still manage to get through them.

PSC (Public Service Commission)   See *Public Service Commission.*

PSE (Packet Switch Exchange)   Essentially, a data packet network access or relay device in an X.25 network, such as the Nortel DPN100.

PSI   1. *Pounds Per Square Inch,* a unit of air pressure. Telephone cables (pulp-insulated cables) that are pressurized with nitrogen are kept at a pressure of 10 to 15 PSI near the central office.   2. *Packet-Switching Interface* gives a customer a means to connect with a packet switching network, such as frame relay.

PSK (Phase-Shift Keying)   See *Phase-Shift Keying.*

PSN (Packet Switching Network)   A name sometimes used in place of *Public Data Network (PDN).* PSNs connect to users via X.25 or frame relay. The Internet is also a type of PSN using TCP/IP packets.

PSS1 (Private Signaling System 1)   Better known as *Q.SIG* in circuit and packet switched telephony circles. The Q.SIG protocol is a variant of ISDN D-channel voice signaling. It is based on the ISDN Q.921 and Q.931 standards and is a worldwide standard for PBX interconnection originally pioneered by Siemens (under ETSI and ISO). Not all PBX systems and IP telephony gateways support Q.SIG services. PBX systems that are connected via Q.SIG are only able to share the basics in telephone operability. Depending on the PBX systems that are linked with Q.SIG, some advanced features that are proprietary to the individual PBXs are lost when the call connection departs that particular PBX, just the same as PBX features are not supported across the PSTN. Some features can be translated and retained, such as call forwarding, or call transfer. Q.SIG is most commonly used to integrate PBX systems across WAN links. Routers that are Q.SIG enabled connect to the PBX on either end, performing IP, frame relay, T1, among other required translation services. Q.SIG has been a precursor to the newer voice over IP and IP telephony methods and the gateways used within these networks. See also *IP Telephony* and *Gateway.*

PTN (Public Telephone Network)   Also called *PSTN (Public Switched Telephone Network)* and *PSN (Public Switched Network)*. The telephone network that we know today provides us with an open-ended dial tone, the ability to dial a telephone anywhere we wish.

PTT (Post, Telephone, and Telegraph)   The general name for the communications service providers in countries that have not yet allowed this service to be controlled by a nongovernment agency or corporation. PTTs are often extensions of the postal service, where telecommunications service is a branch from the original letter-mail and telegraphy service. South American and African countries commonly have a PTT agency, rather than multiple telephone companies.

Public Address System (PA System)   There are different types of PA systems. High-fidelity PA systems are used in studio recording and concert productions and simple systems are used for paging/intercom and loudspeaker systems. The two main components of a PA system are the amplifier and the speakers. Different components can be attached to the input of a PA system. The PA amplifier input is a high-impedance circuit (this means that it does not draw a lot of electrical current from the source, thus transferring maximum voltage). Common source (signal input) devices include microphones, musical instruments (electric), and the paging output of telephone systems (Fig. P.32). If an amplifier is used to drive external speakers (rather than the ones inside telephones), then it is called an *external paging amplifier* or *PA amplifier.*

The question that most people have about PA amplifiers is which one to buy. The answer is that the majority of the cost in a paging system is usually the wiring and the speakers. Most paging amplifiers are equipped with multiple inputs so that different areas or "zones" can be paged

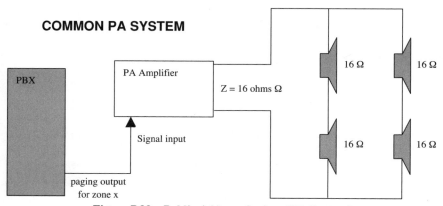

Figure P.32   Public-Address System (PA System)

individually. The factor that affects the price of a PA amplifier the most is the power output rating. The more power that an amplifier is capable of pushing through the speaker network, the more expensive it is. A good general rule is to allot 5 watts of RMS power per speaker in an office environment and 10 to 25 watts of RMS power per speaker in an industrial environment. The crucial factor in designing the speaker network is that the impedance (Z) of all the combined speakers matches (or is equal to) the output impedance of the PA amplifier, which is usually 8 or 16 Ω. If the impedance is not matched, there is a possibility of over working the amplifier and causing it to fail or having a poor performance and sound quality. Quality PA amplifiers/paging amplifiers have instructions on how to wire and arrange the connections of speakers. See the drawing for an example of how a 16-Ω output amplifier is matched with four 16 Ω speakers.

Public Data Network (PDN)   Also called *PSDN (Packet-Switched Data Network)* or *PSN (Packet-Switching Network),* a reference to public-network X.25 services. This type of service, when available, eliminates the need for a private line connection beyond the local telephone company, which is cost efficient for users who send very little data.

Public Service Commission (PSC)   The watchdog for the Public Utilities Commission. The Public Utilities Commission regulates the telecommunications companies under federal judgments (which change from time to time), and other utility companies. For a telecommunications company to be regulated, it must have a minimum number of customers. All the RBOCs are regulated by the PUCs of their area.

Public-Switched Digital Service   A general name for switched 56 K service from a local or long-distance telephone company.

Public Switched Network (PSN)   Also called *PTN (Public Telephone Network)* and *PSTN (Public Switched Telephone Network).* The telephone network that provides open-ended dial tone, the ability to dial a telephone anywhere we wish.

Public Utilities Commission (PUC)   The governing body of regulated public utility service companies and the *Public Service Commission (PSC)* that watches over them. The Public Utilities Commission regulates the telecommunications companies under federal judgments (which change from time to time) and other utility companies. For a telecommunications company to be regulated, it must have a minimum number of customers. All the RBOCs are regulated by the PUCs of their area.

PUC (Public Utilities Commission)   See *Public Utilities Commission.*

Pulling Strength   A cable specification. The maximum pulling force that can be applied to a strength member of a cable without voiding the warranty.

Pulp Cable   Telephone cable used in outside plant applications that uses paper insulation on the twisted copper pairs. The other kind of widely used cable is *pick cable,* which has color-coded plastic-insulated pairs. For a photo of pulp-insulated cable, see *Lead Jacket.*

Pulse Amplitude Modulation (PAM)   See Fig. P.33.

THREE SIGNALS SAMPLED AND PLACED ON ONE CHANNEL WITH PAM

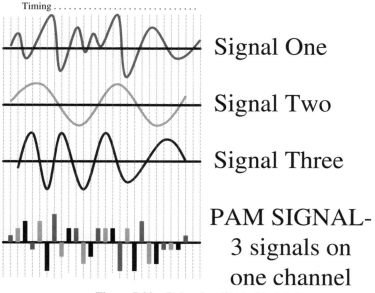

Figure P.33   Pulse Amplitude Modulation

Pulse-Code Modulation (PCM)   See *PCM.*

Pulse Density   See *Ones Density.*

Punch-Down Block   A 66M150 block, AT&T 110 (one ten) block, crone block, or other wire-terminating device. A punch-down block provides connections to neatly connect and label wires.

Punch-Down Tool   A tool that is used to terminate telephone wires onto punch-down blocks (Fig. P.34).

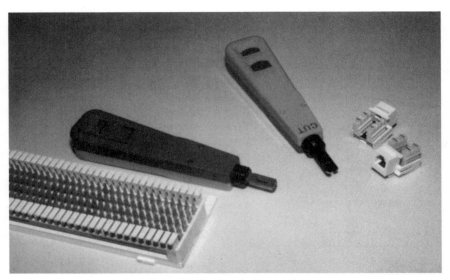

Figure P.34   Punch-Down Tool

PVC   1. *Permanent Virtual Circuit.* A logical connection made between two end-communicating devices on a packet network. PVCs are set up by using *Committed Information Rates (CIRs),* as in frame relay.   2. The substance with which common telephone wire is insulated. PVC wire is available in many colors. The other more expensive option for telephone wiring is Plenum. Plenum wiring is required in many newer buildings because when it burns, it does not emit poison gasses (PVC produces chlorine gas when burned). Plenum wiring is made from polyvinylidene diflouride, and costs about three times as much as PVC does.

PVDF (Polyvinyl Diflouride)   Better known as plenum wire.

PVDM (Packet Voice Data Module)   In Cisco Systems IP telephony, a reference to an add-on card that gives MRP (Multi-service Route Processor) cards more processing power for carrying voice, fax, and data. One PVDM allows for four simultaneous channels. The MRP is an ICS7750 specific card/module. The ICS 7750 is a compact, modular, midsized IP telephony cabinet.

Q.920/Q.921 ITU-T specifications for the ISDN UNI data-link layer. See also *UNI*.

Q.922A ITU-T specification for frame-relay encapsulation.

Q.93B ITU-T specification for signaling to establish, maintain, and clear BISDN network connections. An evolution of ITU-T recommendation Q.931. See also *Q.931*.

Q.931 ITU-T specification for signaling to establish, maintain, and clear ISDN network connections. See also *Q.93B*.

Q.SIG In circuit-based PBX telephony, Q.SIG is also known as Private Signaling System 1 (PSS1). The Q.SIG protocol is a variant of ISDN D-channel voice signaling. It is based on the ISDN Q.921 and Q.931 standards and is a worldwide standard for PBX interconnection (under ETSI and ISO). Not all PBX systems and IP telephony gateways support Q.SIG services. PBX systems that are connected via Q.SIG are able to share only the basics in telephone operability. Depending on the PBX systems that are linked with Q.SIG, some advanced features that are proprietary to the individual PBXs are lost when the call connection departs that particular PBX, just the same as PBX features are not supported across the PSTN. Some features can be translated and retained, such as call forwarding or call transfer. Q.SIG is most commonly used to integrate PBX systems across WAN links. Routers that are Q.SIG enabled connect to the PBX on either end, performing IP, frame relay, T1, among other required translation services. Q.SIG has been a precursor to the newer voice over IP and IP telephony methods, and the gateways used within these networks. See also *IP Telephony* and *Gateway*.

Q Bit (Qualified Data Bit)   A bit in the X.25 protocol that resides in the data packet header (within the X.25 frame). It is the first bit in the three-byte header that indicates whether or not data is intended for the end DTE (in which case, the bit is set to 0) or X.29 instructions for a PAD device (in which case, the bit is set to 1).

QAM (Qaudrature Amplitude Modulation)   A modulation technique used for digital signals. The line format or transmission format (for wireless applications) is analog (Fig. Q.1). QAM is a breeding of phase-modulation and amplitude-modulation. The many different variations of QAM allow for extended constellations and formats thereof. Two simple QAM formats have 16 different signals by using four different phase shifts and four different amplitudes, or eight different phase shifts and two different voltage levels. Both methods provide four bits per baud. To convert a digital bit stream to a QAM signal, the bit stream is accepted four bits at a time. Those bits are converted to a symbol, which represents a constellation, or, more technically, a voltage level and phase shift. With four bits, there are 16 possible constellations. This enables four bits to be sent with one baud cycle. Some versions of QAM incorporate 256 constellations, which enables 16 bits to be transmitted in a single baud cycle.

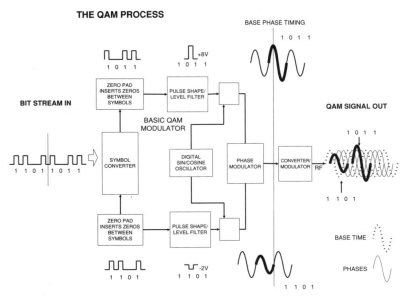

Figure Q.1 QAM (Quadrature Amplitude Modulation)

QCIF (Quarter Common Intermediate Format)   A video playback format that specifies 144 lines of luminance (with 180 pixels per line) and 72 lines of chrominance information (with 90 pixels per line). QCIF provides a better video quality than CIF (Common Intermediate Format) when the number of frames per second is less than 3. This format falls under the operation of H.261. See also *H.261.*

QLLC (Qualified Logical Link Control)   A data-link layer protocol defined by IBM that allows SNA data to be transported across X.25 networks.

QoS Enabled Switch   (Also called a Voice Enabled Switch.) A reference to an Ethernet switching device that is 802.1Q as well as 802.1p compliant that enables the device to carry IP telephony traffic. IP telephony and VoIP are not compatible with token ring networks.

QoS/Voice Enabled Router   A router that is QoS, MGCP, and H.323 enabled at minimum that enables it to carry voice traffic in an IP network and/or translate it to the PSTN (Public Telephone Network). See also *MGCP, Gateway, H.323, WIC,* and *RTP.*

Quad IW (Quad Inside Wire)   Older standard telephone wire used by telephone companies. Quad has four wires, the colors are red, green, black, and yellow. Some quad wiring is not twisted, so it is susceptible to RFI. The colors for line one are green and red, and the colors for line two are yellow and black.

Quad Lock Conduit   Conduit that is designed to be direct buried (Fig. Q.2). The four individual conduits allow communications companies

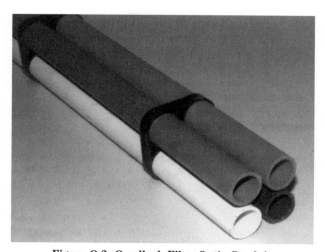

Figure Q.2 Quadlock Fiber-Optic Conduit

to lease conduit space to each other in a way that is easy to track for fiber-optic cable installers/splicers, etc.

Quad Word   In computer memory, a word is 16 bits, which is one data unit processed by the bus. Newer computers and other processing systems are built with 32- and 64-bit busses, which gives them the ability to process double words (32 bits) and quad words (64 bits).

Quadrature Amplitude Modulation (QAM)   A line-code method primarily used in twisted-pair telephony applications that increases the number of bits sent per baud (line state) by representing multiple bits as a combination of phase and amplitude levels (Fig. Q.3). See also *QAM*.

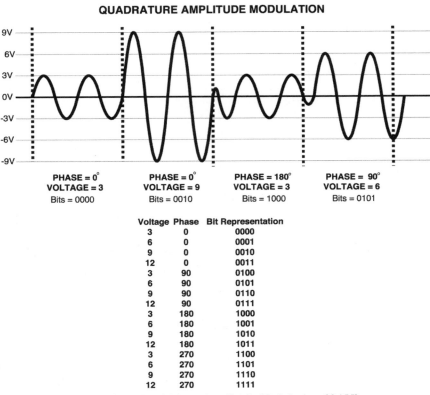

Figure Q.3 Quadrature Amplitude Modulation (QAM)

Quantizing Distortion   A form of distortion that can occur in telecommunication circuitry at the physical, data-link, or network level. It is a result of a digital signal being corrupted to some degree, thus causing an improper reproduction of the original audio in the digital-to-analog conversion process. Quantizing distortion can cause a person's voice to sound "robotic," delayed, or choppy during a wireless phone conversation.

Quarter Common Intermediate Format (QCIF)   A video playback format that specifies 144 lines of luminance (with 180 pixels per line) and 72 lines of chrominance information (with 90 pixels per line). QCIF provides a better video quality than CIF (Common Intermediate Format) when the number of frames per second is less than 3. This format falls under the operation of H.261. See also *H.261.*

Query   A name given to a programming string that asks a question about data in a relational data base. Queries are common in call-accounting applications.

Query Language   A programming language designed for manipulating data in data bases.

Queue   1. Queue is a way of saying *waiting line* in telephony. The two types of queues are line queues (very uncommon) and *ACD (Automatic Call Distribution System)* call queues. Some telephone systems have a feature called *line queuing.* If you try to dial out and you cannot get an outside line, you are put in queue, or in a waiting line for the next available trunk. Some systems can provide music, as if you were on hold for the line and some can ring your phone back. The other queuing is ACD specific. ACD systems place incoming calls in queue for the next available agent and evenly distribute calls among the agents so that the workload is not unbalanced and sales opportunities are fair.   2. Generally, an ordered list of elements waiting to be processed.   3. In LAN and WAN routing, a backlog of packets waiting to be forwarded over a router interface.   4. In a call center environment, a virtual extension on a PBX with conditional treatments for calls routed to it.   5. In Ethernet switching, a level of priority set for packets. See also *Layer 4 Switching* and *802.1p.*

Queuing Delay   In LAN and WAN data transfers over X.25 and frame relay, the amount of time that data must wait before it can be transmitted onto a statistically multiplexed physical circuit.

Quick Connect   A name given to 66M150 block and AT&T 110 (one-ten) block connectivity.

RA (Return Authorization)   Also known as *RMA (Return Material Authorization)* or *RAN (Return Authorization Number)*. A reference number in the advance-replacement process. If you receive a shipment from a distributor or manufacturer and a part is defective, you call the distributor/manufacturer and they give you an RA or RMA number to place on the package when you send it back to them. They, in turn, send you a replacement immediately.

Race Server   A method of layer 7 networking. In WAN networking or Internet applications, a router that is equipped with special software that enables it to measure the network distance between two network servers that provide the same application. To provide a user with the quickest response time while using an application, it makes sense to have them connect with the closest server. The race server receives the initial request for connection and then creates a "race" request among multiple servers. The server that responds the quickest is the closest by factual network performance, and gets the request packet forwarded to it. The transaction between end user and appropriate server then takes place. The race server is also called a boomerang server in many networking circles because the packet request initially starts out from the user in one direction, but comes back through the network from another direction.

Raceway   A trough designated for wiring. Raceways can be in ceilings, attached to walls, or built into floors.

Rack   Also called *relay rack*. The two standard dimensions of racks used in telephony and rack-mountable computer equipment are 19″ and 22″ wide. The height ranges from one to seven feet. Some racks can be attached to walls (wall mount). Most racks are rated as zone 4, which means that they are designed to withstand earthquakes to a certain degree.

Radar (Radio Detection and Ranging)   Radar is a means of detecting objects within the vicinity of a radio signal. Different types of objects can be detected, depending on the frequency of radio used. Radar works by sending a pulse from a transmitter: the pulse travels outward, bounces off objects, and is returned to a receiver. The time difference between the pulses departure and arrival determines the distance of the object. Any Doppler effect on the pulse determines the speed of the object toward (or away from) the transmitter.

Radar Detector   Radar detectors are famous for their use in speed-limit enforcement in the United States. Radar detectors use radar technology to measure the Doppler effect of radio signals that are sent from a transmitter, bounced off of a moving object, then returned at a different frequency. The movement of the object compresses the radio signal as the two come into contact, thus increasing the frequency.

Radio   The emission of electromagnetic radiation into the air, then picking it up with a receiver. Electromagnetic radiation occurs when a magnetic field changes at the rate of a carrier frequency. The magnetic field then traverses through the farthest reach of the magnetic field, which could be many miles. One determining factor in the distance that the signal will reach is the transmitting power. Broadcast radio has a typical output power of 15 to 100 kW (15,000 to 100,000 watts) and CB radios have an output power of 4 w. What makes the radio signal carry a voice or music is called *modulation*. Modulation is the act of varying a carrier signal in a way that can be sensed or "detected" by a radio receiver. Those variations are then amplified and run through a speaker so that people can hear them.

Radio Common Carrier (RCC)   A cellular/PCS service provider or paging company. A company that provides one-way (paging) or two-way (mobile phone) radio services to individuals, rather than communities. Broadcast TV or radio stations are not radio common carriers.

Radio Frequency (RF)   Any electromagnetic frequency that is above the range of human hearing. Most licensed radio transmissions range from 500 kHz (500,000 Hz) to 300 GHz (300,000,000,000 Hz).

Radio Frequency Interference (RFI)   Also called *EMI (Electromagnetic Interference)*. Interference caused by a radio signal or other magnetic field inducing itself onto a medium (twisted/nontwisted pair wire) or device (telephone or other electronics). The world we live in is full of radio waves that are emitted from electric appliances, such as blenders, automobile engines, transmitters, and even fluorescent lights. Even though we take preventative measures to avoid receiving these unwanted signals, they sometimes get into places they are not wanted. Electromagnetic interference is usually caused by one of two things. The first is when a wire connected to a device acts like an antenna and picks up the EMI, which is then passed on to the electronics inside the device and amplified. The second is when an electronic component inside a device acts like an antenna because of poor design, poor shielding, or the component is defective.

Radio Modem   A type of data-transmission device that is used where leased telephone lines are of either poor quality, are unavailable, or are very expensive. They operate at frequencies that range from 400 Hz to 2.4 GHz. They have a point-to-point range of three to 30 miles, depending on the weather of the region, the operating frequency, and the design of the radio circuitry. These types of modems have a net throughput that ranges from 300 bps to more than 2 Mbps. More recently, this communications solution has evolved to terrestrial microwave radio, which also requires licensing to operate. For more info and an illustration, see *Terrestrial Microwave*.

Radome   A cover for a radio antenna, typically used in public broadcast applications.

RADSL (Rate-Adaptive DSL)   A version of asymmetrical digital subscriber loop, where the modems test the line at start up and adapt their operating speed to the fastest the line can handle. RADSL has a maximum downstream transfer rate of 9 Mbps and a maximum upstream rate of 1 Mbps.

RAID (Redundant Array of Inexpensive Disks)   A hard-drive control technology that is intended for servers. RAID links individual drives together, enabling them to act as one storage device, or back each other up via several different storage schemes. One storage scheme is disk striping, which shares data among disks, but does not provide disk-failure protection. RAID provides failure and back-up through disk mirroring, and multiple striping duplicate data to two of many disks. The four most popular types of RAID configurations are: RAID Level 0, which provides data striping only; RAID Level 1, which provides disk Mirroring; RAID Level 3, which stripes data and uses one disk for error correction should one of the others fail; and RAID Level 5, which provides for data striping and stripe error correction. Thus, RAID Level 5 is the best-performing RAID configuration.

Rain Attenuation   1. The degradation of a radio signal (particularly in the microwave region) because of rain. The rainfall average and density of the rainfall is the determining factor (along with fog, which attenuates radio much more severely) in the distance that a radio (microwave link) can send a signal. Typical ranges for the dry climate regions of the United States are as much as 6 miles, and as little as one mile for the wetter regions (for a 7-watt transmitter).   2. A reference to the amount a signal is diminished or distorted by rain. Except in extreme conditions, attenuation (weakening of the signal) due to rain does not require serious consideration for frequencies up to the range of 6 or 8 GHz. When microwave frequencies are at 9 GHz or higher, attenuation due to rain becomes much more of a concern, especially in areas where rainfall is of high density and long duration. In cases where higher ranges of the GHz spectrum is implemented, shorter paths may be required. Wireless LAN equipment designed for 802.11b point-to-point operates at frequencies lower than 6 GHz, so rain is not a concern. In this frequency range, fog is generally considered to be the same as rain. However, fog can adversely effect the radio link when it is accompanied by atmospheric conditions such as temperature inversion (i.e., refraction). It is always a good idea to gain the most clearance possible from the ground and from other radio paths.

Raised Floor   Many computer and telecommunications rooms have a raised floor. The raised floor is a very sturdy framework of iron, with heavy 1 in. tiles placed into the framework. The tiles are easily removed and replaced with a suction cup. The raised floor is used as a giant "duct" to move and run connecting cables through, and it is also used as an airway to pump cool air through the equipment. Instead of cooling a room, cool air is blown under the floor, where it finds its way into the equipment through holes in the floor. The holes are cut into the floor when the equipment is installed.

RAM (Random-Access Memory)   See *Random-Access Memory.*

Ram Hook/Ram Horn   A hardware attachment used to hold *ASW (Aerial Service Wire)* drop clamps in aerial-span applications (Fig. R.1) Ram Horn (left) Mast Clamp (right).

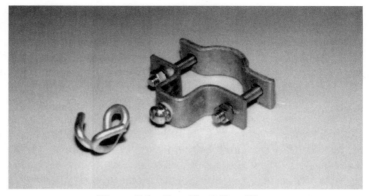

Figure R.1   Ram Hook/Ram Horn

Rambus "D" RAM (RDRAM)   A newer dynamic random-access memory technology that allows for far greater access speeds than the previously popular SDRAM. At 600 MHz, RDRAM is about six times faster than SDRAM. The package that RDRAM comes in is a small PC board like SDRAM, only it is called a *RIMM (Rambus In-Line Memory Module)*.

Rambus In-Line Memory Module (RIMM)   The package that RDRAM comes in. See *Rambus "D" RAM*.

RAMDAC (Random-Access Memory Digital-to-Analog Converter)   An integrated circuit that is commonly incorporated into video cards for personal computers. The RAMDAC converts digital images into analog video signals that a monitor can display.

RAN   1. *Recorded Announcement,* a term used in IVR and ACD call-flow analysis. If you like, you can call the recorded greeting on your answering machine a RAN. For a photo of a digital announcer that stores RAN messages, see *Digital Announcer.*   2. *Return Authorization Number,* also called *RA (Return Authorization),* or *RMA (Return Material Authorization).* A reference number in the advance-replacement process. If you receive a shipment from a distributor or manufacturer and a part is defective, you call the distributor/manufacturer and they give you an RA or RMA number to place on the package when you send it back to them. They, in turn, send you a replacement immediately.

Random-Access Memory (RAM)   Electronic memory is available in two families, *ROM (Read-Only Memory)* and *RAM (Random-Access Memory).* Memory devices are made from two different technologies, *Bipolar (TTL)* and *MOS (Metal-Oxide Semiconductor).* Memory is stored by a technique called *writing* and retrieved by a technique called *reading.* ROM devices can only be read and are programmed during manufacture. *PROM (Programmable Read-Only Memory)* devices can be programmed at a later date by an electronics reseller or electronic assembler for a special application using special equipment. Special ROM devices called *EPROMs (Erasable Programmable Read-Only Memory)* can be electronically erased and reused. RAM has read and write capability. The term *random access* means that any memory address can be read in any order at any time. The two types of RAM are static and dynamic. Static RAM can hold its memory even when power is removed. Dynamic RAM needs constant power to refresh its memory. For a diagram of the different types of dynamic memory, see *Memory.*

RAS (Registration Admission and Status Protocol)   In IP telephony, the protocol used between end points and the gatekeeper to perform network management functions, particularly in bandwidth management. The RAS signaling function performs registration, admissions, bandwidth changes, status, and disengage procedures between the VoIP gateway and the gatekeeper, which is often a router dedicated to gathering status information from other routers in a network.

Rate Adaptive   A type of data protocol that is capable of "testing" the telephone circuit for the fastest possible transmission rate, then transmitting at that rate. This test is done using a "ping" packet similar to that used in DSL (Digital Subscriber Loop). DSL is a rate adaptive protocol.

Rate-Adaptive ADSL (RADSL)   A version of asymmetrical digital subscriber loop, where the modems test the line at start up and adapt their operating speed to the fastest the line can handle. RADSL has a maximum downstream transfer rate of 9 Mbps and a maximum upstream rate of 1 Mbps.

Rate Design   A term that refers to the way utility companies figure a way to charge money for their services. Rates are designed to be affordable for everyone (PUC/PSC requirement). A good example of rate design is the way that the telephone companies charge extra money for business lines to offset the costs of residential lines. How far the offset is and how much one rate is subsidized for the other is the rate design.

Rate Elements   The individual charges and fees for a service. For instance, all of the rate elements are listed on your phone bill: the local service charge, dial tone, 911 service, etc. All of these parts of the telephone company have been separated by the FCC and billed for separately by law.

Rate Queue   In ATM, a value associated with one or more virtual circuits that defines the speed at which an individual virtual circuit transmits data to the remote end. Each rate queue represents a portion of the overall bandwidth available on an ATM link. The combined bandwidth of all configured rate queues should not exceed the total available bandwidth.

Rayleigh Fading   Rayleigh Fading is a form of signal reduction or loss because of a receiver picking up the same signal from multiple directions. The signal commonly arrives from multiple directions because of reflections from buildings when there is no line-of-site path (receiving reflections of the same signal is also referred to as *multipath reception*). When the signals meet, they add or subtract each other, causing an irregular signal strength. Rayleigh fading is a common reason for geographical "dead spots" in cellular service networks.

Rayleigh Scattering   The scattering of light in a fiber-optic cable because of impurities in the glass of the fiber. It has a similar effect to what a lamp shade has on a light bulb, just not as drastic.

RBOC (Regional Bell Operating Company)   At the time of divestiture, there were 22 BOCs, grouped into seven *Regional Bell Operating Companies (RBOCs)*. For a listing of the BOCs and RBOCs, see *Regional Bell Operating Company*.

RCA (Regional Calling Area)   The geographical area that a telephone company serves.

RCA Connector   A plug first developed and used by *RCA (Radio Corporation of America)*. These plugs are very common in audio- and video-patch applications. If you have a CD player and a separate tuner/amplifier, the cord that connects the two most likely has RCA connectors (Fig. R.2).

Figure R.2   RCA Connector

RCC (Radio Common Carrier)   A cellular/PCS service provider or paging company. A company that is in the business of providing one-way (paging) or two-way (mobile phone) radio services to individuals, rather than communities. Broadcast TV or radio stations are not radio common carriers.

RCDD (Registered Communications Distribution Designer)   A well-known industry certification/training program offered by *BICSI (Building Industry Consulting Service International)*. The RCDD certification is designed to educate professionals in the area of physical network distribution, including twisted pair and optical media. The RCDD is sometimes referred to as a *BICSI* (pronounced "bik-see") *certification*. More information can be found regarding BICSI certifications at *http://www.bicsi.org*.

RCP (Remote Copy Protocol)   A protocol that allows users to copy files to and from a file system residing on a remote host or server on the network. The RCP protocol uses TCP to ensure the reliable delivery of data.

RDRAM (Rambus "D" RAM)   A newer dynamic random-access memory technology that allows for far greater access speeds than the previously popular SDRAM. At 600 MHz, RDRAM is about six times faster than SDRAM. The package that RDRAM comes in is a small PC board like SDRAM, only it is called a *RIMM (Rambus In-Line Memory Module)*.

Reactance   Reactance is the resistance that a component gives to an AC or fluctuating DC current. The two components that cause reactance are inductors (coils) and capacitors. (Reactance is also caused by other electronic conditions where it is not useful. All wire and electronic components possess a small amount of reactive properties, e.g., twisted-pair wire causes signal attenuation because of the inductance of the copper wire and the capacitance of the two adjacent wires.) The difference between resistance and reactance is that resistance is always the same, regardless of the voltage amplitude or frequency applied to the resistive device. The reactance of a component changes along with frequency changes, the speed at which an AC current changes direction. The higher the frequency applied to an inductor, the higher the reactance or resistance to that frequency. The reason that coils of wire cause reactance is that as electricity flows through them, they force the electricity to create a magnetic field every time it changes direction. A perfect inductor has zero reactance to a DC current, and has a specific reactance or resistance to every frequency of AC current. Each coil (inductor) has a value in henries. The higher the number of henries, the more it will resist AC or fluctuating DC. Coils are used to filter out ("choke" out) DC fluctuations in power supplies. They are also used to help tune in radio or other frequencies.

Read-Only Memory (ROM)   See *Random-Access Memory*.

Real Time   A reference to the relationship of events in a communications channel, machine, or PC. *Real time* means that the inside of the machine is synchronized with real-world time that you and I live in. Another word for this is *isochronous*, which means "in time." Newer technology has changed this.

Real-Time Streaming Protocol (RTSP)   An application-level protocol designed to utilize TCP/IP to enable real-time communications over the Internet, such as voice over IP.

Rebiller   A telephone company that buys a telephone service from a facilities based telephone company and resells it. A rebiller attempts to add value to the original long-distance company's service by providing better customer service and customized technical expertise. The rebiller gets a discount from the original long-distance company, typically about 10%. Rebilling is also known as *Type-III service*, where all circuits (telephone lines) are type III.

Rebooting   To restart a computer by turning it off and turning it on again. The two ways to re-boot a computer are a hard boot and a soft boot. Hard booting is manually turning off the computer to force the microprocessor to reset. Soft booting is done by pressing Ctrl-Alt-Del at the same time. This direct code sends a positive pulse to the reset of the computer. However, it will sometimes not work if the computer's keyboard is locked up with the rest of the components.

Receiver   1. The part of a telephone handset that you talk into. The receiver has a microphone inside it.   2. A radio device that is connected to an antenna and filters and detects carrier frequencies and signals modulated on them. For more information, see *Modulation* and *AGC*.

Receiver Off Hook (ROH)   The condition of a telephone set being left off the hook, with no numbers dialed, or left off the hook after a conversation has been completed. This causes the central office to disconnect the voltage from the telephone line, which saves electricity. When the receiver is placed back on the hook, the telephone line does not become activated instantaneously. The dial tone can take one to two minutes to return. ROH is a common test result in telephone company mechanized loop testing.

Recorded Announcement (RAN)   1. *Recorded Announcement,* a term used in IVR and ACD call-flow analysis. If you like, you can call the recorded greeting on your answering machine a *RAN.*   2. *Return Authorization Number,* also called *RMA (Return Material Authorization)* or *RA (Return Authorization).* A reference number in the advance-replacement process. If you receive a shipment from a distributor or manufacturer and a part is defective, call the distributor/manufacturer and they will give you an RA or RMA number to place on the package when you send it back. They, in turn, send you a replacement immediately.

Rectifier   1. A device to convert AC power to DC power, also called a *diode.* An electronic semiconductor device that, simply put, only conducts electricity in one direction. Whether or not the device conducts depends on which direction the device is "biased." Diodes are used to change *Alternating Current (AC)* to *Direct Current (DC).* If a more positive voltage is applied to the anode lead of the diode, then the diode simply acts like a wire. If the more positive voltage is applied to the cathode lead, then it acts like there is no connection. The following illustration shows the schematic symbols of the first diode, which was a vacuum tube, and a solid-state silicon diode. Figure R.3 is a pair of diodes converting AC to DC.   2. A DC power source (Fig. R.4).

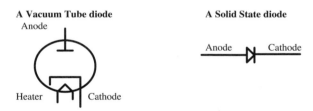

A simple one diode rectifier circuit, with a filter capacitor to eliminate DC fluctuations.

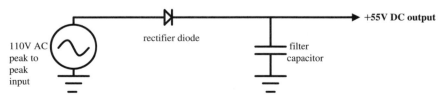

Figure R.3   Rectifier Schematic Diagrams

Figure R.4   Rectifiers

Redundancy   To have one main and one back-up. SONET equipment is
capable of being configured in a redundant manner, with two fiber-optic
routes (in a ring) and duplicates of the electronic cards that control the
communications transmission. Many PBX systems are capable of being
configured with redundant CPU and memory cards. The idea behind redundancy is that if one device fails, the other will take over, without a
loss of service.

Redundant Array of Inexpensive Disks (RAID)   A hard-drive control
technology that is intended for servers. RAID links individual drives together, enabling them to act as one storage device, or back each other
up via several different storage schemes. One storage scheme is disk
striping, which shares data among disks, but does not provide disk-failure

protection. RAID provides failure and back-up through disk mirroring, and multiple striping duplicate data to two of many disks. The four most popular types of RAID configurations are: RAID Level 0, which provides data striping only; RAID Level 1, which provides disk Mirroring; RAID Level 3, which stripes data and uses one disk for error correction should one of the others fail; and RAID Level 5, which provides for data striping and stripe error correction. Thus, RAID Level 5 is the best-performing RAID configuration.

Reference Clock   Also called a *bits clock*. A device that provides a timing pulse in the form of a 1-0-1-0-1-0-1-0 bit stream. Bits clocks are used extensively in SONET networks. The bits clock provides the timing pulse that everything in the network synchronizes itself to.

Reflexive Access List   An access list designed to allow return traffic from the Internet to pass a firewall to a server. This prevents multiple security log-ins for every user data transaction made in a session.

Refraction   *Refraction* refers to the wavelike nature of light. When light travels from one media, such as air, into another media, such as water, it bends. This is why when you look into a swimming pool, the bottom looks very distorted. Fiber-optic technology is based on the fact that light refracts (or bends) as it travels from one media to the next. A single fiber-optic strand consists of many different kinds of glass. The core is one kind and the cladding consists of many layers of glass that have different "levels" of refraction and cause light to gradually refract or bend back to the center as it travels down the fiber. A ray of light refracting as it passes through materials of different refractive indexes is illustrated in Fig. R.5.

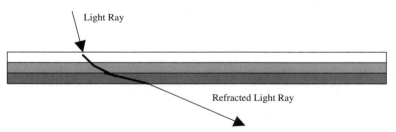

Light Ray

Refracted Light Ray

Figure R.5   A Light Beam Refracted by Materials of Different Refractive Indexes

Refractive Index   The refractive index of a material refers to how much light refracts (bends) when it travels from a vacuum into the material at an angle. For a diagram and more details, see *Refraction*.

Refurbished   Electronic equipment that has been repaired and cleaned, or remanufactured.

Regenerator   Also known as a *repeater*. A device that is used to take a signal that has traveled a long distance and make it new again. Repeaters can be coils of wire, which are used in the public telephone network for voice (POTS) lines, or they can be electronic, taking an electronic signal that has been attenuated over a long distance, reproducing it, then retransmitting it.

Regional Bell Operating Company (RBOC)   At the time of divestiture, the 22 BOCs were grouped into seven *Regional Bell Operating Companies (RBOCs):*
   BOCs: Bell Telephone Company of Nevada, Illinois Bell Telephone Company, Indiana Bell Telephone Company, Michigan Bell Telephone Company, New England Telephone and Telegraph Company, US West Communications Company, South Central Bell Telephone Company, Southern Bell Telephone and Telegraph Company, Cincinnati Bell Company, Mountain Bell Telephone Company, Mountain States Telephone and Telegraph Company, Southwestern Bell Telephone Company, The Chesapeake and Potomac Telephone Company of Maryland, The Bell Telephone Company of Pennsylvania, The Chesapeake and Potomac Telephone Company of Virginia, The Chesapeake and Potomac Telephone Company of West Virginia, The Diamond State Telephone Company, The Ohio Bell Telephone Company, The Pacific Telephone and Telegraph Company, New Jersey Bell Telephone Company, Wisconsin Telephone Company
   RBOCs: Ameritech, Bell Atlantic, Bell South, NYNEX, Pacific Telesis, Southwestern Bell, US West

Register   Also called a *shift register*. An electronic circuit used for temporarily storing memory in a serial format (Fig. R.6). Shift registers are commonly used in the serial-to-parallel conversion for data transmission. Bits are clocked into the register one at a time, then clocked out to their destination when they are needed. Each memory segment of a register is typically an RS (Reset-Set) flip-flop.

Registered Communications Distribution Designer (RCDD)   A well-known industry certification/training program offered by *BICSI (Building Industry Consulting Service International)*. The RCDD certification is designed to educate professionals in the area of physical network distribution, including twisted pair and optical media. The RCDD is sometimes referred to as a *BICSI* (pronounced "bik-see") *certification*. More information can be found regarding BICSI certifications at *http://www.bicsi.org.*

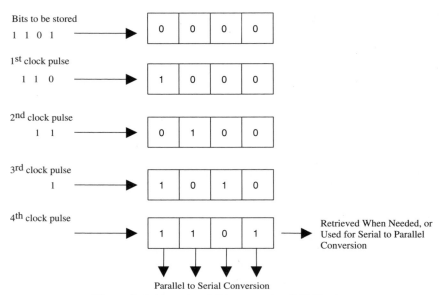

Figure R.6 **Functional Diagram of a Shift Register**

Registered Terminal Equipment   Any line or telephone service that is installed by a telephone company is terminated to registered terminal equipment. The different levels of registered equipment are as simple as an RJ45 (Registered Jack 45) or as complicated as a DSX hand-off point in a colocation.

Registration Admission and Status Protocol (RAS)   In IP telephony, the protocol used between end points and the gatekeeper to perform network management functions, particularly in bandwidth management. The RAS signaling function performs registration, admissions, bandwidth changes, status, and disengage procedures between the VoIP gateway and the gatekeeper, which is often a router dedicated to gathering status information from other routers in a network.

Registration Jack (RJ)   The prefix to many telephone company connection and interface standards.

Regulation, Power   A device that takes an unstable power source, such as public utility power, and reproduces the voltage/amperage with electronics. The electronics in a power regulator are a controlled environment that produces the desired power signal. The street power provides the energy for the device (power regulator) to do this.

Regulation, Telco   The RBOCs, CLECs, long-distance companies, and competitive-access providers are all regulated by the *Public Utilities Commissions (PUC)* of their respective areas in some fashion. The Bell companies are heavily regulated to a disadvantage to enable new telephone companies to become established. One of those disadvantages is that they are forced to charge higher rates for their services in competitive areas. This allows the new companies to attract customers with a price advantage. Many new companies do not take advantage of the price regulation, because their networking equipment is of latest technology (SONET 100%, in many cases), which therefore carries a higher value to customers. *Competitive Local Exchange Carriers (CLECs)* become regulated for pricing when they reach a certain number of customers or percentage of market share. All communications companies, new and old, must demonstrate to the PUC that they are capable and willing to provide acceptable service to the public.

Relational Data Base   A data-base application (software program) that tracks data, based on relationships. It works very similar to a manual paper-filing system, with different categories and cross references. A good example is: There are many houses. Each house has many (one or more) residents. If a data base were created for this, each house would be listed and people would be related to the house. People make many telephone calls. The calls people make can be related to the people that make them. The way that a computer program would think, if queried, to find every person that dialed the phone number 1-602-555-1212, is as follows. It would scan through the data base and, as it found the phone number, it would list the relational items to it. More simply stated, it would list every person and house that dialed that phone number.

Relay   An electromechanical switch (Fig. R.7). Relays are used in electronic and electrical circuits as switches. A relay consists of a coil of wire wrapped around a thin cylindrical-shaped piece of iron, called a *core*. When electricity flows through the coil, it magnetizes the core, which is close to a pair of electrical contacts. The magnetic field attracts one of the contacts to move and make a connection. A popular application for relays is between a small voltage to run a switch that connects a very large voltage. This is why you can start the motor in a large truck with a small keyed switch. The key applies voltage to a relay, which, in turn, connects a large contact from the battery to the starter of the motor. Relays were used in old central offices (some still in service), called *stepper switches*. By dialing a rotary phone, you manipulated a vast network of relays and logic circuits to connect your call. Today, relays have been replaced with transistors; in large or heavier applications, they have been replaced with semiconductor devices called *SCRs (Silicon-Controlled Rectifiers)*. However, electromechanical relays still have the special ability to physically isolate one voltage or device from another. They are still used in higher-end home-audio equipment, where if you turn down the volume, then turn on your stereo, after a short pause, you will hear a "click." That "click" is a relay connecting the output signal voltage to your speakers.

Figure R.7   Mechanical Relay Switch

Relay Rack   Large racks that got their names from a time when they were used as a mounting platform for electromechanical relay circuits in telephone central offices. Their standard size is 7 feet tall by 22 inches wide (Fig. R.8). Relay racks are available in a 19-inch wide size as well.

Remote Call Forward   A feature of PBX systems that enables users/subscribers to make calls dialed to their telephone ring to a different telephone of their choice and activate the feature from a different phone. When a user wants to activate the feature, they can dial a feature code, dial their extension, then dial the extension that they would like their calls to ring on. The great thing about this feature is that if you are in a meeting, you

Figure R.8 Relay Rack (22 Inches by 7 Feet)

can pick up a phone and forward the phone in your office to another co-worker or anywhere else within the PBX system. Some systems allow you to forward your extension to an off-premises telephone number, such as your home. In this case, you can work out of your home, and not miss any calls coming to your extension because they will ring directly to you.

Remote Mini-Fiber Node   A device used in the outside plant portion of cable telephony networks to convert light signals to RF electrical signals (Fig. R.9). Mini-fiber nodes can be made to be pole mounted or can be placed in hand holes/cable vaults. Because cable-TV networks are far spread and rural, fiber optic is a necessity in extending the bandwidth required to areas distant from head ends. The mini-fiber node provides the means for this extension.

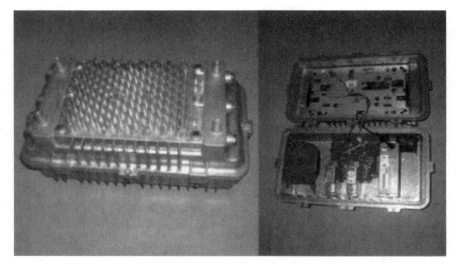

Figure R.9 Remote Mini-Fiber Node

Remote Order Wire   A telephone line that is used to monitor an electronic system. A dial-up maintenance line for a server or mainframe is an order wire.

Remote Shell (RSH)   1. A UNIX operating system level of access. See also *C-Shell* and *Root*. 2. *Remote Shell Protocol*. It is referred to as *R-shell* from the similarities in the UNIX command set. An application or subprotocol that allows a user to execute commands on a remote system without having to log into the system. For example, *rsh,* along with a password (if the network OS supports passwords), can be used to remotely examine the status of network devices without connecting to each communication server, executing the command, and then disconnecting from the communication server.

REPACCS (Remote Cable-Pair Cross-Connect System)   A remote-controlled/automated cross box. A less-sophisticated *DCS (Digital Cross-Connect System)*. In certain areas of cities that are hazardous for telco workers, REPACCS systems are implemented so that F1 and F2 cable pairs can be cross connected remotely.

Repeater   Also known as a regenerator. A device that is used to take a signal that has traveled a long distance and make it new again. Repeaters can be coils of wire, which are used in the public telephone network for voice (POTS) lines or they can be electronic, taking an electronic signal that has been attenuated over a long distance, reproducing it, then retransmitting it. The repeater closure illustrated in Fig. R.10 is an electronic (active) repeater (right).

Figure R.10 Repeater Closure (Left) and Splice Pedestal (Right)

Repeater Coil A radio-type transformer that is used to amplify voice signals on copper twisted-pair telephone wires. Repeater coils have a typical inductance value of 33 mH and are placed every 3000 to 5000 feet.

Request To Send (RTS) 1. After a modem receives a *CD (Carrier Detect)* signal from another modem, the next step is to send some data. Before it sends data, it sends an *RTS (Request To Send)*. After it receives a *CTS (Clear To Send)* from the far-end modem, it begins sending data. 2. A control signal that has a dedicated wire in the RS-232 protocol. When the far device places a logic "one" or 5-V voltage on this wire, it enables the near modem to initiate a transmission.

Rerouting To change the physical path or medium of a communications signal. Rerouting is a part of *SHARP (Self-Healing Alternate-Route Protection)* service from telephone companies over their SONET networks. If a cable is cut, the electronic equipment reroutes the transmission with very little or no interruption in service. If you are talking on a voice line while a fiber is cut on a SONET ring network that is very busy and fully utilized, you might hear a very light click sound.

Resale Carrier A long-distance company that leases long-distance facilities and sells service on them. Sprint and MCI are resellers in some areas; in some areas, they have their own switches, fiber-optic lines, and microwave equipment. In those areas, they are facilities-based carriers.

Reseller Also called an *aggregator*. A long-distance or cellular/PCS reseller. They sign up with a long-distance company as a reseller and all their customers are "aggregated" together for a bulk discount. The long-distance or cellular company provides the service and does the billing. The advantage to the long-distance company is that they have more people pushing their long distance. The advantage to the customer is the value-added service and personal consulting of the aggregator.

Reset See *Rebooting*.

Resistance The unit of resistance is the ohm, abbreviated/represented by the Greek letter omega (Ω). Resistance is just what its name depicts, resistance to electric current flow. A 100-W 120-V household light bulb has about one ohm of resistance. The more resistance in a circuit, the less current is allowed to pass through it.

Resistor An electronic component/semiconductor usually made from carbon (Fig. R.11). Resistors are usually used to limit current flow through a circuit or create RC/RL (resistor-capacitor/resistor-inductor) frequency filters.

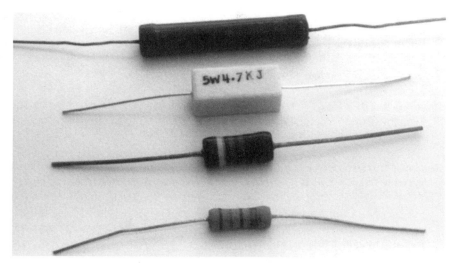

Figure R.11   Carbon Resistors

Resistor Color Code   See *Appendix F*. Resistors have four color bands. They are regarded to as the first, second, third, and fourth bands. The first band is the closest to one side of the resistor and the following bands count to the inside. The first band indicates the first integer of the value of resistance. The second band indicates the second integer of the value of resistance. The third band indicates a multiplier or number of zeros to be placed after the first two band numbers.

Resonance   A circuit is resonant if the inductive reactance and capacitive reactance are equal. This condition occurs for all inductor/capacitor circuits. The frequency at which the resonance happens is determined by the value in mH of the coil and the value of the capacitor in μF.

Resource Reservation Protocol (RSVP)   1. A transport-layer protocol that is intended to provide quality-of-service transmission levels in conjunction with TCP/IP over the Internet. The RSVP protocol makes the sender of data responsible for notifying the receiver that a call is to be made (or data to be sent) and what *QOS (Quality of Service)* will be needed. The responsibility of selecting the resources or path by which the transmission will take is given to the receiver or called party. RSVP is modeled to work with IPv6 and IPv4.   2. In IP telephony and IP video, a protocol that enables QoS features to provide a reserved amount of throughput to meet the requirements of a real-time application, such as a live voice or video stream.

Retrofit   To make older equipment work with newer equipment. *Retrofit* is a term commonly used among telephone company network technicians in reference to upgrading telephone equipment.

Return Authorization Number (RAN)   See *RAN*.

Return to Zero (RZ)   A transmission format where each positive bit returns or drops to a zero value during its timing period. The drop-to-zero format assists in timing/synchronizing of the transmission signal.

Reverse Battery Supervision   A form of answer supervision. An in-band signaling method, if a telephone call goes from one central office to another (or PBX), the originating central office needs to know when the call has been answered so that a billing cycle can begin. The terminating central office briefly reverses the voltage on the connecting trunk line as a signal.

Reverse Channel   Also called a *backward channel*. The channel that flows upstream in an asymmetrical (uneven) transmission. An asymmetrical communications transmission that is characterized by one direction being very fast, compared to the other. Cable-TV is an example of asymmetrical communication. The cable-TV head end sends massive amounts of video and audio information down a coax one way and the cable-TV set-top decoder boxes send small amounts of ID and status information the other way back to the head end over the same coaxial connection. Sometimes asymmetrical channels are referred to as "upstream" for the slow channel and "downstream" for the fast channel, or "forward" for the fast channel and "backward" for the slow channel.

RF (Radio Frequency)   Any electromagnetic frequency that is above the range of human hearing. Most licensed radio frequency transmissions range from 500 kHz (500,000 Hz) to 300 GHz (300,000,000,000 Hz).

RF Choke   A coil of wire that filters out high frequencies (Fig. R.12).

RF Splitter   Used to make a junction point or split a signal so that it will travel down multiple paths over coax. Also called a *splitter* and *UHF/VHF splitter*. Figure R.13 shows a four-way splitter.

RFI (Radio Frequency Interference)   See *Radio Frequency Interference*.

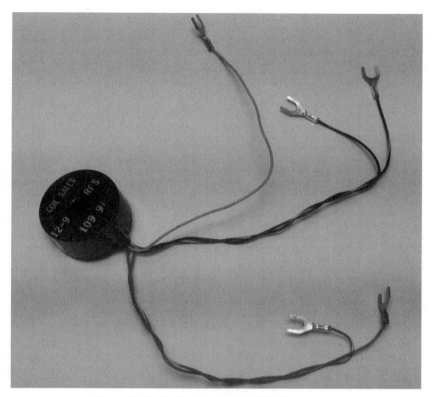

Figure R.12   Radio Frequency (RF) Suppressor

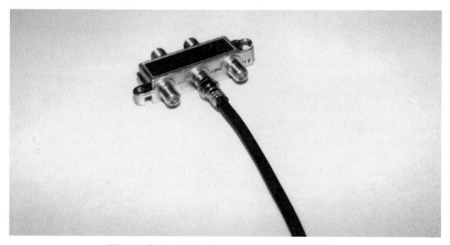

Figure R.13   UHF/VHF Four-Way RF Splitter

RFP (Request For Proposal)   Also called *RFQ (Request For Quotation)*. A formal invitation that a company or individual gives to other individuals or companies, to bid or price a service.

RFQ (Request For Quotation)   See *RFP.*

RG-8   A type of coaxial cable that has a transmission impedance of 50 ohms. For more information on different types of coaxial cable, see *Coax* and *Characteristic Impedance*. For a photo of RG-8, see *DIN Connector*.

RG-58   A type of coaxial cable that has a transmission impedance of 50 Ω. It is used primarily in LAN applications and wired in a bus physical topology. For more information on different types of coaxial cable, see *Coax* and *Characteristic Impedance*.

RG-59   A type of coaxial cable designed for television antenna use that has an impedance of 75 ohms. For more information on different types of coaxial cable, see *Coax* and *Characteristic Impedance*.

RG-62   A type of coaxial cable with a transmission impedance of 93 ohms. It is primarily used in LAN applications and wired in a bus physical topology. ARCnet utilizes RG-62 as its transmission media. For more information on different types of coaxial cable, see *Coax* and *Characteristic Impedance*.

RG-U   The military designation for general-use coaxial cable.

RI (Ring Indicator)   An indicator on a modem that indicates that a ring voltage (90 volts AC) is present on the telephone line to which it is connected.

RIF (Routing Information Field)   In token ring LAN bridging, a part of the token ring frame header (IEEE 802.5) that contains ring number and bridge/router number.

RIMM (Rambus In-Line Memory Module)   The package that RDRAM comes in. See *Rambus "D" RAM*.

Ring   One wire in a POTS telephone line. The ring side of the line is usually marked red when terminated and carries the 90V AC ring-voltage signal that makes the telephone ring.

Ring Banding   Some pic cable comes with no ring banding, which means that the color code is determined by two wires twisted together (e.g., a white and a blue). Ring-banded cable comes with color rings painted around each wire, and the same twisted pair listed before would be white/blue bands and blue/white bands.

Ring Cycle   The ring cycle for a North American POTS telephone line is two seconds of ringing, then four of quiet. Ringing cycles vary throughout the world.

Ring-Down Box   Used in building ring-down circuits. A device that you put on each end of a copper twisted pair that provides battery and ring voltages that a central office would. However, no dialing is involved. When you pick up one phone, the one on the other end automatically rings. When the ringing phone is picked up, the lines are connected together with a talk battery and people can talk on both ends, just like a normal telephone call. Tellabs manufactures a wide variety of ring-down devices.

Ring-Down Circuit   A simple telephone line that is made using ring down boxes. See *Ring-Down Box*.

Ring Generator   A ring generator is the part of a PBX or central-office switch that provides the source of the ring voltage that rings telephones. Ring generators are an individual circuit card in many PBX systems. In some systems, the ring-generation capability is built into the station/telephone interface cards.

Ring Latency   In a token-ring network, ring latency is the time required for a transmission packet to go all the way around the ring.

Ring Topology   A LAN topology (a MAN topology in SONET) that connects all devices on a network in a ring configuration (Fig. R.14). The data transmitted through the network goes through each device. As the devices receive the data, they check to see if it is intended for them. If it is, then they keep it; if it is not, then they pass it along. Unlike Ethernet star and bus topologies, the ring topology is not contention based; each device gets a specified turn in sending and receiving data.

Ring Voltage   Ring voltage on a POTS telephone line is 90V AC.

Ringer Equivalency Number (REN)   A number that references a device's load on a telephone line when the line rings. Telephones, modems, answering machines, and other devices connected to telephone lines are required to have this number printed or stamped somewhere on the device, or placed in the device's literature. A telephone line in North America is capable of driving 5 bells, or 5 devices with a ringer equivalency

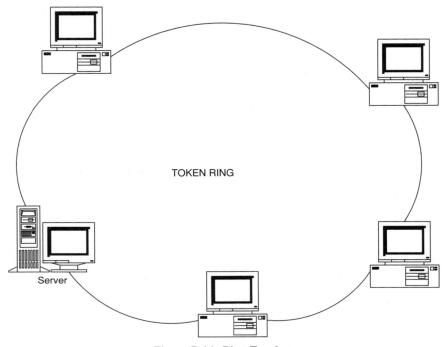

TOKEN RING

Server

Figure R.14   Ring Topology

number of 1. If a telephone line has more than 5 such devices plugged into it, it may fail to ring when called.

RIP (Routing Information Protocol)   A traffic-control method used by routers. RIP was originally developed by Xerox Corporation in the early 1980s for use in *Xerox Network Systems (XNS)* networks. Cisco Systems has developed their own version of RIP that works well as a standard with other manufacturers' routers and has additional proprietary features that work only on Cisco products. In general, RIP works well in small environments, but has serious limitations when used in larger internetworks. For example, RIP limits the number of router hops between any two hosts in an Internet to 16. RIP is also slow to converge, meaning that it takes a relatively long time for network changes to become known to all routers. RIP determines the best path through an Internet by looking only at the number of hops between the two end nodes. This technique ignores other routing metrics, such as differences in line speed, line utilization, and other metrics, many of which can be important factors in choosing the best path between two nodes. See also *Routing Protocol*.

Rip Cord   An aid in stripping the jacket off of bundled pair cable. It is a nylon string that is put in telephone wire and cables when it is manufactured. The string is used by installers to rip the jacket or insulation when it is being installed.

RIPv2 (Router Information Protocol Version 2)   An enhanced version of the original RIP that allows the ability to incorporate subnetting ability into networks. The original RIP was developed before the Interent protocol standard had the subnet mask provision, which gives a network address user options on how a local router reads network addresses. RIPv2 allows for 256 router hops rather than 16, and converges slightly faster than the original RIP. See also the more complex and feature-rich protocols *EIGRP, BGP,* and *OSPF.*

Riser   A telephone cable feed inside a building that runs vertically from floor to floor. It is called a *riser* because it is usually placed in a place that architects call *risers.* We call them *elevator shafts, plenums,* or *airways,* whichever the riser is used for. Typical copper pair riser cables are in the hundreds of pairs in size (100, 200, 300 pair, etc.).

Riser Cable   A twisted-pair cable (usually several hundred pairs) distribution system that progresses from the telephone company Demarc or point of entrance in a building to each floor of that building.

RJ (Registration Jack)   The prefix to many telephone company connection and interface standards.

RJ11   The telephone jack that most of us have come to know. It has a 6× plug with four conductors. Handset cords are a smaller plug, a 4× plug with four conductors. If you look at the two of them side by side, you will notice the difference. For a photo, see *Jack.*

RJ21   Also known as RJ21X. See *RJ21X.*

RJ21X   An RJ21X is a 66M150 block that is designated as the demarcation point for telephone company-provided communications lines (Fig. R.15). Most RJ21X blocks have an orange cover where the telephone numbers of the lines are written.

RJ45   An 8-position, 8-conductor modular jack. RJ45 is used in many computer LAN applications.

RJ48   An 8-position, 8-conductor modular jack. Used to terminate T1 service.

Figure R.15 RJ21X Network Interface

RJ48X   An 8-position, 8-conductor modular jack. They are used to terminate T1 service and they have a shorting bar that is built-in for making manual loop-backs.

RMA (Return Material Authorization)   See *Return Material Authorization*.

RMON (Remote Monitor)   A network management protocol that has an *MIB (Management Information Base)* extended beyond what *SNMP (Simple Network-Management Protocol)* provided. RMON enables a network manager to monitor, configure, and troubleshoot a network. RMON2 is a newer version of the original RMON that enables visibility for a system administrator to view the network and physical layers by protocol. See also *SNMP*.

RMS   See *Root Mean Square*.

Roaming   When a cellular/PCS telephone travels outside of its calling area, it is roaming. When a cellular/PCS telephone roams, it continues to send a signal out that tells any cellular site it can reach (regardless of company) that "I am here." If it can communicate with a cellular site, then the roam indicator is displayed on your phone. If you have a roaming service with your cellular/PCS company, then you can still receive calls even though you are out of your calling area. If you don't pay extra for roaming service, you can still make outgoing calls; however, they cost extra.

Robbed-Bit Signaling   Also called *bit robbing,* but usually known as in-band signaling. The practice of taking a bit here and there in the beginning and end of a digital transmission for use in the overhead of the transmission equipment. Bit robbing is bad news when the signals being multiplexed into the transmission are data. Robbing a bit from a data stream severely corrupts it. Bit robbing is reserved for multiplexing multiple voice circuits onto a T1. Circuits intended to transmit data use out-of-band signaling or clear-channel signaling.

Robust   A term that is synonymous with fast, flexible, and reliable.

ROH (Ringer Off Hook)   See *Receiver Off Hook*.

Rolled Cable   A reference to a null modem cable, where the transmit and receive pairs "roll" within the cable, thus reversing the connection to match transmit to receive and vice versa in relation to connected devices.

Rolm (Siemens-Rolm)   A telephone equipment manufacturer.

ROM (Read-Only Memory)   See *Read-Only Memory*.

Root Account   A type of log-on account that allows access to UNIX subsystems software used exclusively by network or system administrators. Root-account logins commonly give the user a "#" as a prompt, similar to the way that MS-DOS would give a C>. Programming changes are generally unrestricted when logged into a system as a root user. See also *C Shell* and *V Shell*.

Root Mean Square (RMS)   A method of calculating the power consumption or power output of an electronic/electrical device. RMS power is ultimately the average of an AC waveform, which is the peak voltage, multiplied by 0.707. The other methods of calculating power include true power, peak power, and transparent power. Most electronic/audio applications use either peak or RMS power. To convert from peak power to RMS power, multiply the peak power rating by 0.707. The result is RMS power. To convert RMS power to peak power, divide the RMS power rating by 0.707. The result is the peak power rating.

Round-Trip Time (RTT)   The time (usually measured in milliseconds) required for a network communication to travel from the source to the destination and back. RTT includes the time required for the destination to process the message from the source and generate a reply. RTT is used by some routing algorithms to aid in calculating optimal routes. See also *Network Transit Time Delay*.

Route   Nortel's name for a trunk group. See also *Member*.

Route Map   How routes are selected in BGP.

Routed Protocol   A network communications language that is supported or "encapsulated and transmitted" by routers. Examples of routed protocols are Ethernet, Appletalk, TCP/IP, frame relay, and X.25. Routed protocols are encapsulated by routing protocols, such as *Intermediate System to Intermediate System (IS-IS),* Cisco System's *Enhanced IGRP (Enhanced Interior Gateway Routing Protocol),* and *RIP (Routing Information Protocol).* See also *Routing Protocol*.

Router   A network layer device that uses one or more measures (cost, number of hops, etc.) to determine the optimal path along which communications traffic should be forwarded (Figs. R.16 and R.17). Routers forward packets from one network to another, based on address information. They are also capable of translating (or repackaging) data into dissimilar routed protocols, such as Ethernet, token ring, RS-232, etc. Routers were once called *gateways*. See also *Routing Protocol*.

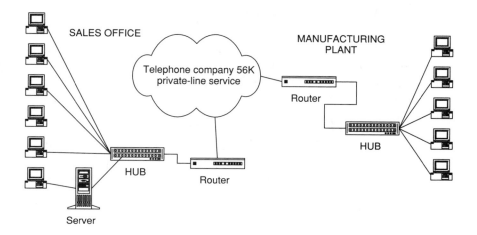

SALES OFFICE

MANUFACTURING
PLANT

Telephone company 56K
private-line service

Router

Router

HUB

HUB

Server

Figure R.16 Router Application

Figure R.17 Router

Router on a Stick   A router connected to a switch that provides connectivity among different Virtual LAN (VLAN) segments within a single Ethernet switch. More advanced LAN switches have "on-board" routers built into them so that the traffic does not actually have to leave the switch via a link. See also *Layer 3 Switching* and *VLAN*.

Router, Voice Enabled   A router that is QoS, MGCP, and H.323 enabled at minimum that enables it to carry voice traffic in an IP network and/or translate it to the PSTN (Public Telephone Network). See also *MGCP, Gateway, H.323, WIC,* and *RTP*.

Routing Domain   A group of routers and hosts operating under the same set of routing protocols. Within each routing domain is one or more areas. Each is uniquely identified by an area address.

Routing Information Protocol (RIP)   A traffic-control method used by routers. RIP was originally developed by Xerox Corporation in the early 1980s for use in *Xerox Network Systems (XNS)* networks. Cisco Systems has developed their own version of RIP that works well as a standard with other manufacturers' routers and has additional proprietary features that work only on Cisco products. In general, RIP works well in small environments, but has serious limitations when used in larger internetworks. For example, RIP limits the number of router hops between any two hosts in an Internet to 16. RIP is also slow to converge, meaning that it takes a relatively long time for network changes to become known to all routers. RIP determines the best path through an Internet by looking only at the number of hops between the two end nodes. This technique ignores other routing metrics, such as differences in line speed, line utilization, and other metrics, many of which can be important factors in choosing the best path between two nodes. See also *Routing Protocol*.

Routing Metric (Routing Measure)   A method by which a routing algorithm determines that one route is better than another. This information is stored in routing tables. Metrics include bandwidth, communication cost, delay, hop count, load, best packet size, path cost, and reliability.

Routing Protocol   The functional part of a router's operating system that works with other routers within a network to transport data from one location to another. The three classes of routing protocols are: distance vector, link-state, and hybrid. Routing protocols work by incorporating algorithms that help the router know which port to relay data from and to. This is accomplished through routing tables that can be created statically (by a user) or dynamically (automatically by the routing protocol). Routing protocols carry data of other protocols (such as Ethernet, X.25, or TCP/IP) by placing an address in front of the packet to be routed. This address is translated by the routing protocol in routing tables to correspond to a *MAC (Media-Access Control)* address and/or a router port. Each time a packet passes through a router, the routing protocol header is read, stripped off, and a new header is added. Examples of routing protocols include Cisco System's *Enhanced IGRP (Enhanced Interior Gateway Routing Protocol), IS-IS (Intermediate System to Intermediate System),* and *RIP (Routing Information Protocol)*. See also *Routed Protocol*.

Routing Table   A reference to call-handling instructions input to an ACD (Automatic Call Distribution) system. The routing table lists each incoming trunk and the steps that the call goes through. The "steps" the call goes through are also called a "call treatment or a script." A routing table could list trunk number 1 and the treatment that calls receive when they come in on that trunk.   2. An incoming digit translation file in a PBX system's memory that instructs the system on which extension to route a call based on DNISI or D10 digits.   3. A list of paths or connections through a network that is kept in a router's memory. See also *OSPF*.   4. A data file stored in a router or other internetworking device that keeps track of routes to particular network destination addresses and their associated circuit interface port on the router. Many routing protocols store metric data within routing tables as well. See also *Routing Protocol* and *Routing Metric*.

Sample Meridian 1 CCR Script
GOTO CLOSED IF LOGGED AGENTS QUEUE 1234=0
QUEUE TO 1234
GIVE RAN 105
GIVE MUSIC 100

SECTION LOOP
WAIT 60
GIVE RAN 106
GOTO DIFFICULTIES IF LOGGED AGENTS QUEUE 1234=0
GOTO LOOP

SECTION DIFFICULTIES
GIVE RAN 107
FORCE DISCONNECT

SECTION CLOSED
GIVE RAN 108
FORCE DISCONNECT

–RAN 105 Thank you for calling abc company, your call will be answered by the next available agent.

–RAN 106 Please continue to hold

–RAN 107 We are currently experiencing technical difficulties, please call alternate number xxx-xxxx

–RAN 108 We are closed, our business hours are . . . please call back during these times.

RP   *Redirecting Forward Party.* Another name for a phone that is call forwarded.

RS   1. The ASCII control-code abbreviation for record separator. The binary code is 1110001 and the hex is E1.   2. RS Connectors, *Recommended Standard,* the prefix in RS-232, RS-328, etc.

RS-232   A popular physical-layer interface now known as *EIA/TIA-232.* See *EIA/TIA-232.*

RS-232C   A communications protocol that was developed by the EIA so that data devices could communicate. The standard includes the different functions and signals for two devices to communicate. The signals are physically interfaced to a cable via a 25-pin or 9-pin connector. Each pin having a signal function. In order for an RS-232 connection to work, the cable and pin-outs must match (Fig. R.18). Even though

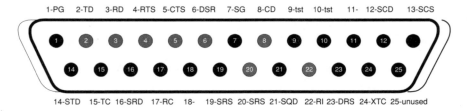

1-PG   2-TD   3-RD   4-RTS   5-CTS   6-DSR   7-SG   8-CD   9-tst   10-tst   11-   12-SCD   13-SCS

14-STD   15-TC   16-SRD   17-RC   18-   19-SRS   20-SRS   21-SQD   22-RI   23-DRS   24-XTC   25-unused

PIN DESCRIPTIONS

1 Protective Ground
2 Transmitted Data
3 Received data
4 Request To Send
5 Clear To Send
6 Data Set ready
7 Signal Ground
8 Received Line Signal Detector
9 Reserved for Data Set Testing
10 Reserved for Data Set Testing
11 Unassigned
12 Secondary Received Line Signal Detector
13 Secondary Clear to Send

14 Secondary Transmitted Data
15 Transmit Signal Element Timing
16 Secondary Received Data
17 Receiver Signal Element Timing
18 Unassigned
19 Secondary Request to Send
20 Data Terminal ready
21 Signal Quality Detector
22 Ring Indicator
23 Data Signal Rate Selector
24 Transmit Signal Element Timing
25 Unassigned

Figure R.18   A Common RS-232C Pinout Diagram for DB25

most modem and DCE/DTE manufacturers use the RS-232 protocol, not all use the same pin-outs. To complete a connection, a cable must be a null cable (null-modem cable), which means that transmit and receive are reversed inside the cable from one end to the other.

RS-328   The first EIA facsimile standard (1966).

RS-366   EIA standard for auto-dialing.

RS-422   Balanced electrical implementation of EIA/TIA-449 for high-speed data transmission. Now referred to collectively with RS-423 as *EIA-530*. See also *EIA-530* and *RS-423*.

RS-423   Unbalanced electrical implementation of EIA/TIA-449 for EIA/TIA-232 compatibility. Now referred to collectively with RS-422 as *EIA-530*. See also *EIA-530* and *RS-422*.

RS-449   1. EIA standard that is the newer version of RS-232. RS-449 uses a 37-pin connector and each of multiple transmit and receive pairs are balanced. RS-449 is faster and able to transmit longer distances (300 feet) than RS-232 (limited to 50 feet).   2. Popular physical-layer interface. Now known as *EIA/TIA-449*. See EIA/TIA-449.

RSA (Rural Service Area)   The counterpart to *MSA (Metropolitan Service Area)*. A term that refers to the 306 metropolitan areas where the FCC manages cellular and PCS communications. There are also *RSA (Rural Statistical Area)* markets that the FCC determined as separate from each other. 428 RSA markets are in the United States. Each statistical area, 734 in all, has at least two licensed service providers.

RSC (Remote Switching Center)   A common term for a long distance carrier's central office, or relay point. Many long distance carriers have offices that are not occupied by personnel, only equipment.

RSH   1. *Remote Shell*. A UNIX operating system level of access. See also *C Shell* and *Root*.   2. *Remote Shell Protocol*. Referred to as *R Shell* from the similarities in the UNIX command set. An application or sub-protocol that allows a user to execute commands on a remote system without having to log into the system. For example, *rsh*, along with a password (if the network OS supports passwords), can be used to re-motely examine the status of network devices without connecting to each communication server, executing the command, and then disconnecting from the communication server.

RSVP (Resource Reservation Protocol)   1. A transport-layer protocol that is intended to provide quality-of-service transmission levels in conjunction with TCP/IP over the Internet. The RSVP protocol makes the sender of data responsible for notifying the receiver that a call is to be made (or data to be sent) and what *QOS (Quality of Service)* will be needed. The responsibility of selecting the resources or path by which the transmission will take is given to the receiver or called party. RSVP is modeled to work with IPv6 and IPv4. 2. (Also known as *Resource Reservation Setup Protocol.)* For the sake of IP telephony and IP video, a networking feature that supports the reservation of packet transfer resources across an IP network for "real-time" services. Applications running on IP end systems can use RSVP to indicate to other nodes the nature (bandwidth, jitter, maximum burst, etc.) of the packet streams they want to receive. RSVP depends on IPv6; however, it is backward-compatible for IPv4.

RTE (Registered Terminal Equipment)   See *Registered Terminal Equipment.*

RTMP (Routing Table Maintenance Protocol)   The Macintosh/Apple Computer's proprietary routing protocol that was derived from RIP. RTMP establishes and maintains the routing information that is required to route datagrams from any source socket (logical channel) to any destination socket (logical channel) in an AppleTalk network. Using RTMP, routers dynamically maintain routing tables to reflect changes in topology. See also *RIP (Routing Information Protocol).*

RTP   1. Realtime Transport Protocol: An IETF standard used to enable end-to-end network transport functions for applications transmitting real-time data, such as IP voice messaging; audio, video, or simulation data, over multicast or unicast network services. RTP is a layer 4 protocol and provides services such as payload type identification, sequence numbering, timestamping, and delivery monitoring to real-time applications. RTP is one of the IPv6 protocols.   2. Rapid Transport Protocol: In an IBM SNA environment network feature that provides pacing and error recovery for APPN (Advanced Peer-to-Peer Networking) data as it crosses the APPN network. With RTP, error recovery and flow control are done end-to-end rather than at every node. RTP prevents congestion rather than reacts to it.

RTS (Request To Send)   After a modem receives a *CD (Carrier Detect)* signal from another modem, the next step is to send some data. Before it sends data, it sends an *RTS (Request To Send)*. After it receives a *CTS (Clear To Send)* from the far-end modem, it begins sending data.

RTSP (Real Time Streaming Protocol) The IETF standards-based protocol for control over the delivery of data with real-time properties such as audio and video streams. It is useful for large-scale broadcasts and audio or video on-demand streaming, and is supported by a variety of layer 7 vendors of streaming audio and video multimedia, including Cisco IP/TV, RealNetworks RealAudio G2 Player, and Apple QuickTime 4 software. The RFC allows for RTSP to run over either UDP or TCP, although not all IP telephony or IP video systems support both. RTSP establishes a TCP-based (the more popular) control connection, or channel, between the multimedia client and server. RTSP uses this channel to control commands such as "play" and "pause" between the client and server. These requests and responses are text-based and are similar to HTTP. RTSP does not typically deliver continuous data streams over the control channel, usually relying on a UDP-based data transport protocol such as standard Real-Time Transport Protocol (RTP) to open separate channels for data and for RTP Control Protocol (RTCP) messages. Typically, RTP and RTCP channels occur in pairs, with RTP being an even-numbered port and RTCP channel being the next consecutive port.

RTT (Round Trip Time) The time (usually measured in milliseconds) required for a network communication to travel from the source to the destination and back. RTT includes the time required for the destination to process the message from the source and generate a reply. RTT is used by some routing algorithms to aid in calculating optimal routes. See also *Network Transit Time Delay*.

Runt In LAN networking, an invalid Ethernet frame that is less than 64 bytes long. These frames are discarded when detected by LAN switches/router devices.

RZ (Return to Zero) A transmission format where each positive bit returns or drops to a zero value during its timing period. The drop-to-zero format assists in timing/synchronizing the transmission signal.

S

S Band   The band of frequencies designated by the IEEE between 2 GHz and 4 GHz (15 cm to 7.5 cm). For a table, see *IEEE Radar Band Designation.*

S/MIME (Secure Multipurpose Internet Mail Extensions)   MIME is the standard format for attaching nontext files, such as graphics (JPG), spreadsheets (XLS), and formatted documents (DOC) to text-based e-mail messages. S/MIME enhances the MIME standard by encrypting or adding digital signatures to MIME-formatted messages. Digital signatures ensure the recipient that the message is from who it says and that it is private, thus preventing forgeries.

SAA (Supplemental Alert Adapter)   An AT&T term that is specific to the Merlin telephone system. It is the equivalent to a loud ringer.

SABM (Set Asynchronous Balanced Mode)   It is the command defined by the last three bits in the control byte of an unnumbered or "control" type frame in the X.25 protocol being 100. It is a code that an X.25 device sends when it has come into service and is ready to receive packets. It is also referred to as *set* or *reset.* This command resets all timers to zero and clears all buffers on all of the devices communicating on an X.25 link.

SABME (Set Asynchronous Balanced Mode Extended)   It is the extended version SABM command. It is defined by the last three bits in the control byte of an unnumbered or control-type frame in the X.25 protocol being "100." It is a code that a device sends when it has come into service and is ready to receive frames. It is also referred to as *set* or *reset.* This command resets all timers to zero and clears all buffers on all of the devices communicating on an X.25 link.

SAC (Single Attached Concentrator)   A FDDI or CDDI concentrator that connects to the network by being cascaded from the master port of another FDDI or CDDI concentrator.

Safety Belt   Used by communications/power/construction personnel to harness themselves to telephone/power poles or tower structures (Fig. S.1). Also called a *body belt* and *climbing belt.*

Figure S.1   A Safety Belt (Also Called a *Body Belt* or *Climbing Belt*)

Sag   If an outside plant engineer refers to *sag,* it is the amount that an aerial span dips down between telephone poles. Different cable needs to have a different sag, depending on the climate, the weight of the cable, the type of poles being used, etc.

SAP (Service Advertising Protocol)   A program within the Novell NetWare protocol stack that allows nodes on a network to notify devices on that network what services they are designed to deliver. Devices perform their advertising through SAP packets. The SAP packets communicate with files that reside in servers and routers that track which devices in a network provide required services. The routers and servers return SAP packets with requested information.

SARM (Set Asynchronous Response Mode)   An old (1970s) X.25 LAP-vintage command defined by the last three bits in the control byte of an unnumbered or control-type frame in the X.25 protocol. A device sends this command to notify that it is now ready to receive frames.

Satellite   A self-sustained electronic device/platform that orbits the earth at an altitude of about 22,000 miles. Communications satellites transmit and receive signals in the microwave range and are used for broadcast TV, telecommunications, global positioning, and many other applications.

Satellite Antenna   A reference to a parabolic dish that has an *LNB (Low-Noise Block Converter)* attached to the end of a protruding arm in front of the dish. The actual antenna element is very small and is inside the LNB. For a photo, see *LNB Converter.*

Satellite Link   A communications path that includes a satellite.

Satellite Receiver   The electronics that the output from a satellite antenna's LNB converter feeds to. The satellite receiver demodulates the communications information into audio, video, and data. It is the equivalent of a tuner in a TV set or a radio. The satellite dish receives and sends a band of RF frequencies to the satellite receiver, which, in turn, "tunes" to a desired station.

SBC Communications   One of the RBOCs, formerly (but more commonly) known as *Southwestern Bell.*

SC Connector   A square-shaped snap-on fiber-optic plastic connector. SC connectors come in single or dual. For a photo of a single SC connector, see *Fiber-Optic Connector.*

Scattering   Attenuation of light in a fiber optic because of the light changing direction in the fiber.

SCC (Specialized Common Carrier)   An old term for an *IXC (interexchange carrier)* other than AT&T.

Schematic   A diagram of an electronic circuit. Schematics can be drawn at many different levels from block diagrams to the discrete component level.

Scotchlok   A family of splicing connectors that are manufactured by 3M. Scotchloks are used in splicing copper twisted pairs in outside and inside plant applications (Fig. S.2).

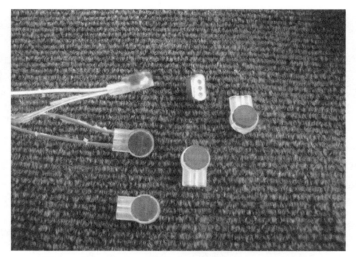

Figure S.2   A Scotchlok Twisted-Pair Crimp Splice

SCR (Silicon Controlled Rectifier) An electronic component that works like a transistor, but much better in power switching applications. An SCR has three leads, one for the anode, gate, and cathode (Fig. S.3). When a negative bias pulse is applied to the gate, the SCR acts like a switch, turns on and stays on, even though the pulse is gone.

Schematic Symbol for an SCR

Figure S.3   SCR (Silicon-Controlled Rectifier)

SCSI (Small Computer System Interface) Pronounced as "scuzzy." A parallel interface standard for linking internal computer components (such as hard drives) and peripheral devices (such as printers). The good thing about SCSI interfaces is that they are faster than others (such as IDE/ATA). The drawbacks are the higher cost and that many of the SCSI formats that are on the market are not compatible with each other (Fig. S.4). If you upgrade your PC, you might end up replacing the SCSI interface in your printer as well. Some examples of the SCSI formats and their transfer speeds are: Fast SCSI, 8-bit bus 10 Mbps; Fast Wide SCSI, 16-bit bus 20 Mbps; Ultra Wide SCSI, 16-bit bus 40 Mbps; and Wide Ultra2 SCSI, 16-bit bus 80 Mbps. See also *IDE*.

Figure S.4   SCSI Host Adapter Card Interface

Scuzzy   See *SCSI.*

SDH (Synchronous Digital Hierarchy)   A European family of digital carrier rates. SDH defines a set of rate and format standards that are transmitted using optical signals over fiber. SDH is the term used by the ITU to refer to SONET OC rates referred to in the United States. Its basic building block is a rate of 155.52 Mbps, designated at STM-1 (OC-3). See also *SONET* and *STM-1.*

SDI (Serial Data Interface)   A connection designed for transmission of data over a media one bit at a time. The other type of transmission is parallel, which sends multiple bits at a time, over multiple wires. Many printers run in a parallel manner, modems work in a serial manner. For a diagram, see *Serial Interface.*

SDRAM (Synchronous Dynamic Random-Access Memory)   SDRAM is a newer type of RAM used in personal computers. It is capable of running at speeds greater than 100 MHz, and it synchronizes itself with the CPU clock. SDRAM was made obsolete by *RDRAM (Rambus D RAM),* which can run at speeds of 600 MHz. Figure S.5 shows a 128-pin 32-MB SDRAM dual-inline memory module.

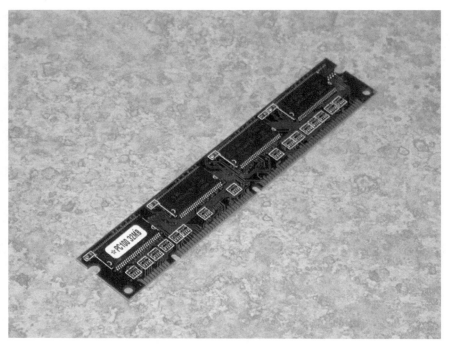

Figure S.5   SDRAM

SDLC (Synchronous Data-Link Control)   A revised revision of the IBM Bisync protocol that was submitted by its creator (IBM) to the ISO and ANSI in hopes that it would become a standard. Instead, it was evolved by the ISO into *HDLC (High-level Data-Link Control)* and evolved by the ANSI into *ADCCP (Advanced Data Communication Control Procedure)*. The CCITT reviewed and modified HDLC and called it *LAP (Link-Access Procedure)*, which became the basis for the frame or network layer of the X.25 standard in 1976. The LAP protocol was further modified by the CCITT to become *LAPB (Link-Access Procedure Balanced mode)* in 1978.

SDSL (Symmetrical Digital Subscriber Line)   A physical-layer telecommunications protocol that delivers high-speed data networking over a single pair of copper phone lines. SDSL is symmetrical, which means that upstream and downstream data-transfer rates are the same. Speeds range from 160 Kbps to 1.544 Mbps. The base transmission distance is 24,000 feet (about 4.5 miles), and it can be extended to greater than 30,000 feet with repeaters. SDSL can be extended to any distance over fiber optic. SDSL is ideal for business applications that require identical downstream and upstream speeds, such as video conferencing or collaborative computing, as well as similar applications that are appropriate for ADSL technology. SDSL uses the same kind of line-modulation technique used in ISDN, known as *2B1Q*.

SECAM (Système Electronique pour Couleur Avec Mémoire)   The analog television broadcast standard in France, Russia, and regions of Africa. SECAM is a variant of PAL, but it delivers the same number of vertical scan lines as PAL and uses the same refresh rate. See also *Television Broadcast Standards*.

Second Dial Tone   The dial tone that you get after dialing 9 on a PBX system. When you first pick up the handset, you hear a dial tone, which is the PBX internal dial tone, then you dial 9 to get an outside dial tone.

Secondary Ring   The redundant and nontraffic carrying of the two rings making up an FDDI or CDDI ring. The secondary ring is usually reserved for use in the event of a failure of the primary ring, which is active and carries traffic.

Secondary Winding   A reference to the output of an electronic transformer. Transformers have at least one primary and at least one secondary winding (Fig. S.6). Transformers are made to work with AC voltages. If you connect a transformer to a DC voltage with no filtering electronics, the transformer will overheat and be destroyed. Transformers are used to "step-up" or "step-down" AC voltage levels. Transformers are rated with a ratio of the primary to secondary winding. A common transformer is a 10:1. This means that for every 10 windings of wire on the primary side of the transformer, only one winding is on the secondary side. Transformers have the same ratio to voltage as they do windings, so if a transformer with a 10:1 ratio that has 120 volts is applied to the primary winding, then 12 volts will be the output on the secondary winding.

Schematic Symbol for an iron core Transformer
(iron core is designated by two lines between coils)

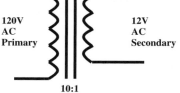

Figure S.6   Secondary Winding

Seed Router   A master router within an AppleTalk network that has the network number or cable range entered into its port descriptor. The seed router automatically responds to configuration queries from nonseed routers on its connected AppleTalk network, allowing those routers to confirm or modify their configurations accordingly. All AppleTalk networks need to have at least one seed router.

Segment   1. A physical link in a network between two devices.   2. In data-encapsulation terminology, the name of a *PDU (Packet Data Unit)* when it reaches the transport layer in the *OSI (Open Systems Interconnect)*. After becoming a data segment, the PDU becomes a packet at the network layer, and then a frame at the data-link layer.

Selective Ringing Module (SRM)   A device that is attached to each individual network interface for customers that are sharing a party line. The SRM contains electronics that can be configured to recognize different ringing formats using DIP switches. Some different ringing formats that an SRM would differentiate are ring voltage on the ring side, ring voltage on the tip side, ring voltage on the ring side with the tip side grounded, and ring on the tip side with the ring side grounded. See also *Party Line.*

Selectivity   The measure in dB of a radio receiver or tuner to select or pass a wanted signal carrier and reject all others. The higher the selectivity in dB, the better the receiver/transmitter/tuner. Selectivity is sometimes called the Q (quality) of a tuner, but the Q rating is usually used in reference to specific fixed-frequency filters.

Self Diagnostics   A feature of many telecommunications testing and transmitting equipment. PBX switches, microwave radio equipment, SONET equipment, and central-office switches are equipped with troubleshooting aids that indicate where trouble is. They are not always accurate, but they are of great assistance in the diagnosis of faulty equipment or transmission paths.

Self Test   The ability of telecommunications test, transmission equipment, or switching equipment to run a test on its hardware components and software. Most telecommunications equipment, as well as personal computer equipment, run a self test when they are first turned on. If there is a problem, then an error of some kind is displayed, which can be cross referenced in a user's or administrator's manual.

Semiconductor   Germanium, silicon, and carbon are all semiconductors. They are not great conductors (like copper wire) and they are not insulators (like plastic or rubber). They have special properties that allow a controlled amount of current to flow through them, given a certain amount of voltage, under certain conditions. Transistors, diodes, and other "active" components (devices that require power to do their job) are made from silicon or germanium. Passive devices (such as resistors) are made from carbon.

Sensitivity   A reference to the ability of a radio receiver or tuner to receive a tiny electronic signal from an antenna and amplify it. A sensitivity rating is given in microvolts (μV). If a sensitivity rating is better, the sensitivity for the same given output is lower.

SEPT (Signaling End-Point Translator)   The part of the SS7 (Signaling System 7) network that receives coded signals from another central office and translates those codes into a number plan or set of codes used in the central-office exchange. For more information, see *Translations.*

Sequenced Packet Exchange (SPX)   A part of the Novell NetWare protocol stack. SPX is used for very specific applications. SPX ensures complete delivery of messages. Applications that use SPX usually include interserver or other device communications, such as printing reports done by servers. An application that would use SPX is for the remote control/administration (RCONSOLE) of servers.

Serial Bus   A bus that transmits one bit at a time. Serial busses are usually one pair of wires: one used for transmit/receive and the other is ground (or common, for a balanced line). A modem line to your computer can be thought of as a kind of serial bus. See *Parallel Bus.*

Serial Data Interface   A connection designed for transmission of data over a media one bit at a time. The other type of transmission is parallel, which sends multiple bits at a time, over multiple wires. Many printers run in a parallel manner; modems work in a serial manner. For a diagram, see *Serial Interface.*

Serial Data Transmission   The transmission of data over a media one bit at a time. The other type of transmission is parallel, which sends multiple bits at a time, over multiple wires. Many printers run in a parallel manner; modems work in a serial manner. For a diagram, see *Serial Interface.*

Serial Interface   The transmission of data over a media, one bit at a time (Fig. S.7). The other type of transmission is parallel, which sends multiple bits at a time, over multiple wires. Many printers run in a parallel manner; modems work in a serial manner.

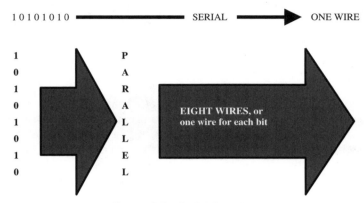

Figure S.7 Serial Interface

Serial Line Internet Protocol (SLIP)   A protocol that enables IP datagrams to be transmitted over dial-up telephone lines. The successor to SLIP is PPP, which provides error detection and automatic configuration.

Serial Port   A computer interface that is used to connect such devices as mice or modems. Serial port interfaces are commonly RS-232 (for modems), USB, or DIN (for keyboards).

Series Circuit   A circuit that has only one path for current through multiple loads. The other type of circuit is a parallel circuit, which has more than one path for current through multiple loads, or devices. For a diagram of a parallel and series circuit, see *Parallel Circuit.*

Server   A computer that is dedicated to providing services to other computers. File servers store common data for other computers to access via a *LAN (Local-Area Network),* and there are application servicers, which other computers access to run large or complicated tasks. The whole idea behind a LAN, or any other network, is to share information and/or processing power.

Server-Based PBX   See *IP Telephony.*

Service Access Code (SAC)   Service access codes are three-digit numbers that are used like an area code, but they are not an area code. These codes are used for special services, such as 800/888 or 900 numbers. Five SACs are in use at the time of this writing: 600, 700, 800, 888, and 900.

Service Advertising Protocol (SAP)   A program within the Novell NetWare protocol stack that allows nodes on a network to notify devices on that network what services they are designed to deliver. Devices perform their advertising through SAP packets. The SAP packets communicate with files that reside in servers and routers that track which devices in a network provide required services. The routers and servers return SAP packets with requested information.

Service Affecting   A reference to a problem that is critical and is interfering with the operation or ability of a network to meet its transmission objectives.

Service Area   The geographic area of a telecommunications service provider. The local service area for USWest is Washington, Oregon, Idaho, Montana, Wyoming, Utah, Arizona, New Mexico, Colorado, North Dakota, South Dakota, Nebraska, Iowa, and Minnesota.

Service Code   A three-digit code or shortened phone number that has a specific purpose, such as 911.

Service Entrance   Also called a building entrance. The place where communications cables enter a building.

Service Profile Identifier (SPID)   An *ISDN (Integrated Services Digital Network)* telephone number. The number that some telephone companies use to define the services to which an ISDN device subscribes. ISDN devices use SPID numbers when accessing the telephone company's switch to identify the device it would like to be connected to. In the United States, SPID numbers can look like ordinary public-service telephone numbers with an extension, such as 972-555-1212 4455. There is one SPID for each B Channel. See also *ISDN*.

Session   A related set of communications transactions between two or more network devices.

Session Layer   A layer in a communications protocol model. In general, the session layer does the job of establishing and maintaining connection to the communications process of the lower layers. It also controls the direction of the data transfer. The latest model or guideline for communications protocols is the OSI (Open Systems Interconnect). It is the best model so far because all of the layers or functions work independently of each other. For a diagram of the OSI and older proprietary communications models, see *Open Systems Interconnection*.

Set Asynchronous Balanced Mode (SABM)   It is the command defined by the last three bits in the control byte of an unnumbered or "control" type frame in the X.25 protocol being 100. It is a code that an X.25 device sends when it has come into service and is ready to receive packets. It is also referred to as *set* or *reset*. This command resets all timers to zero and clears all buffers on all of the devices communicating on an X.25 link.

Set Asynchronous Balanced Mode Extended (SABME) It is the extended version SABM command. It is defined by the last three bits in the control byte of an unnumbered or control-type frame in the X.25 protocol being "100." It is a code that a device sends when it has come into service and is ready to receive frames. It is also referred to as *set* or *reset*. This command resets all timers to zero and clears all buffers on all of the devices communicating on an X.25 link.

Set Asynchronous Response Mode (SARM) An old (1970s) X.25 LAP-vintage command defined by the last three bits in the control byte of an unnumbered or control-type frame in the X.25 protocol. A device sends this command to notify that it is now ready to receive frames.

SF 1. *Single frequency,* a method of in-band signaling in switched telephone networks. A single 2600-Hz frequency tone is used for signaling. 2. *Superframe.* A common framing type used on T1 circuits. SF consists of 12 frames of 192 bits each, with the 193rd bit providing clocking. SF is superseded by ESF, but is still widely used. SF is also called *D4 framing.* See also *ESF.*

SFC (Switch Fabric Controller) An interface between the CPU/core and multiple networks of telecommunications switches.

SGRAM (Synchronous Graphics RAM) A type of dynamic random-access memory used as a buffer/temporary storage in computers that enhances the performance of the graphics accelerator and video adapters.

SHARP (Self-Healing Alternate Route Protection) A service offered from Local Telephone Companies over SONET networks. Sharp service is made possible by the SONET ring technology, which incorporates its network on a ring of fiber-optic cable. If the fiber is cut, all traffic is rerouted the other way around the ring. If you are talking on a telephone over a SONET sharp-based service and a fiber is cut, you might hear a very faint "click" sound. Other than that, you would never know there was a problem.

Shift Register An electronic circuit used to temporarily store memory in a serial format. Shift registers are commonly used in the serial-to-parallel conversion for data transmission. Bits are clocked into the register one at a time, then clocked out to their destination when they are needed. Each memory segment of a register is typically an RS flip-flop. For a functional diagram, see *Register.*

Shiner   A common defect found in twisted-pair copper plant, where the plastic insulation on the copper conductors in terminals literally turns to dust and exposes the sheen of the copper conductor to other copper conductors and moisture.

Short Circuit   A circuit fault. A short is a "short circuit," or an easier path to ground caused by a bad component, water, or other means for electricity to get to where it is not wanted. Many confuse an open with a short. An open is literally a "disconnection" in a circuit. For a diagram of open- and short-circuit faults, see *Open Circuit.*

Short-Haul Modem   Also called a "limited-distance modem" or "line driver." Short-haul modems are commonly used to extend the distance of a printer or other *DTE (Data-Termination Equipment)* device from its host. One example is to extend the printer dedicated to printing call-accounting records from a PBX to an accountant's office. For a network diagram and photo of a limited-distance modem, see *Limited-Distance Modem.*

Short-Tone DTMF   A reference to a telephone or other dialing equipment that sends a short pulse of touch-tone DTMF (100 to 300 ms), regardless of how long you hold the button down. Some telephony equipment cannot "hear" a tone that is that short, so equipment manufacturers have implemented adjustable short-tone lengths, and options to remove the short tone altogether.

Shortest Path First (SPF)   (Also known as *Link State Routing Protocol, Distributed Routing Protocol,* and *Interior Gateway Routing Protocol.*) An SPF routing protocol is a methodology used in router protocol design that enables routers within an autonomous network (i.e., corporate LAN) to identify each other and the status of their port connections. SPF protocols create three databases within a router's memory: a neighboring router database, a link database, and a routing table. The routing table is created by applying Dykstra's algorithm to the first two databases. The most widely used SPF Protocol is Open Shortest Path First (OSPF). See also *OSPF.*

Shortest-Path-First Algorithm (SPF)   SPF is also referred to as a *link-state* or *Dijkstra's Algorithm.* A class of router-operating software that enables routers to build their own complex address routing tables that detail every router and node within their network. The routing table-building process is accomplished through information multicasts. The routing-table multicasts are referred to as *LSPs (Link-State Packets)* and they consume payload bandwidth to transmit this information. The process of sending and receiving LSPs is called the *discovery process*. Multicasts are only sent when there is a change in the network, such as a circuit connection going down, or a new router or connection being added. Link-state algorithms use tremendous amounts of router system memory (20 MB to 30 MB in a 30-node network), and consume significant processor resources within a router's circuitry. During the startup of a link-start network, the discovery process can take hours. The great advantage to this complex operating method is that routing loops are not created. See also *Distance Vector Routing Algorithm* and *Hybrid Routing Algorithm*.

SI   The ASCII control-code abbreviation for shift in. The binary code is 1111000 and the hex is F0.

Sideband   A sideband is a harmonic radio frequency that is a result of modulation on an AM carrier, and is a transmission characteristic of an FM carrier. In AM radio transmissions, one sideband above the carrier frequency and one sideband below the carrier frequency are created. In FM, the modulation of the carrier itself is an infinite number of sidebands.

Side Tone   When you talk on a telephone, you can hear a little bit of your voice being sent back into the earpiece. This is called a *sidetone* and it lets you know that the line is live.

Signal Strength   Signal strength is measured in dB, dBrn, or DB.

Signal-to-Noise Ratio   Signal-to-noise ratio is the amount of desired signal, in comparison to the amount of unwanted signal, expressed as a ratio.

Signaling System 7   A method of out-of-band inter-office signaling for telephone circuits. Simply stated, *out of band* means that a special separate line is used to carry signaling, such as dialed touch tones, ringing signals, busy tones, (everything, but the actual voices/conversation), etc.

Remember that the two different ways to send signals in telephone transmissions are: in-band and out-of-band. Signals are digits that you dial, dial tone, the phone being off-hook, ringing, etc. An in-band telephone line is like the one in your home; the digits that you dial and the ringing are carried within the channel you talk on. Out-of-band signaling is a method that telephone companies and businesses use for larger PBX applications and data-transfer applications. An out-of-band signaled DS1 has 24 multiplexed channels. The 24th channel carries the signaling for the other 23 channels (phone lines). The advantage of out-of-band signaling is that each channel has an increased capacity to carry data (8 Kb/s more) and the 23 channels are not used to find out if a line is busy (both directions, in and out). The off-hook sensing busy signaling and other signaling previously mentioned is done in the 24th channel. If your system receives thousands of calls per day, this can reduce traffic. SS7 makes it easy for long-distance companies to let us dial a phone number, get a busy signal, and not be billed for it because we are not really using a call channel to do this.

Silicon   The important thing to know about silicon is that it is an element used to make electronic components. It is used because it has special atomic properties that enable it to conduct or not conduct electricity, depending on the way it is *doped*. Doping is the implantation of impurities into the silicon, additional electrons to be specific. When a transistor, diode, or any other silicon active device is made, at least two types (two types of silicon doped differently) of silicon are used to form a *junction*. The first type of silicon is called a *P-type* (for positive) and the second type is called an *N-type* (for negative). The two (very small) pieces of silicon are placed together to form a P-N junction. A single PN junction is used to make a diode and three pieces of silicon are used to make a PNP or NPN junction transistor. Another element that is not as frequently used, but used in the same manner, is Germanium.

SIM (Single Interface Module)   An NEC trademark, this is a smaller PBX in the NEC telephone equipment family. Larger PBXs include the IMG and the MMG.

SIMM (Single-Inline Memory Module)   A small circuit board, about 1" by 3", that contains memory components for PCs and other memory-using devices. The edge of the SIMM circuit board has a single row of contacts so that it can be plugged into a socket or slot. The SIMM gets its name from the type of socket it plugs into. See also *SDRAM*.

Simple Mail-Transfer Protocol (SMTP)   A standard interchange format used by e-mail applications to exchange messages with each other. SMTP does not provide a user interface or method for a user to create a message. It only provides a way for the message to be transferred. For example, Lotus Notes can be configured to use SMTP when it sends an outgoing message to an Internet e-mail address. SMTP is used in conjunction with *POP (Post Office Protocol)* and *MIME (Multipurpose Internet Mail Extensions)* to provide TCP/IP Internet e-mail.

Simple Network Management Protocol (SNMP)   A standard protocol for managing diverse internetworks. SNMP is an application-layer protocol standard designed to facilitate the exchange of management information between network devices and is used almost exclusively in Transmission Control Protocol/Internet Protocol (TCP/IP) networks. SNMP is a relatively simple protocol, yet its feature set is sufficiently powerful to handle the difficult problems presented in trying to manage today's heterogeneous networks. By using SNMP-transported data (such as packets per second and network error rates), network administrators can more easily manage network performance, find and solve network problems, and plan for network growth. SNMP has two versions: Version 1 and Version 2. Most of the changes introduced in Version 2 increase the security capabilities of SNMP. Other changes increase interoperability by more rigorously defining the specifications for SNMP implementation. The creators of SNMP believe that after a relatively brief period of coexistence, SNMP Version 2 (SNMPv2) will largely replace SNMP Version 1 (SNMPv1).

Simplex   Communications in one direction. FM radio and broadcast TV are forms of simplex communication. Other methods are half duplex and full duplex.

Single-Frequency Signaling (SF)   Mostly referred to as *single frequency,* a method of in-band signaling in switched telephone networks. A single 2600-Hz frequency tone is used for signaling.

Single-Mode Fiber   The alternative to multi-mode fiber optic. Single-mode fiber optic has a smaller core, but is capable of longer-distance transmissions. It is used in the public network more often and is the choice for SONET applications. Multi-mode has a larger core, and therefore accepts more light and more frequencies of light. Multi-mode is used for shorter-distance applications, such as LANs. Multi-mode fiber optic is made with an orange-colored tube or jacket, and single-mode fiber is made with yellow.

Six Ones   Simple terminology for a "flag" in the X.25 protocol. Flags are bytes that are inserted and deleted by the frame layer (or network layer, if you refer to OSI terminology) in between each frame. The rules for the X.25 frame layer are defined by the *LAPB (Link-Access Procedure Balanced mode)* protocol. An X.25 flag bit is as follows: 01111110. See also *Bit Stuffing.*

Six-Pack Coax   A type of bundled coax that is used in STS/SONET environments. Bundled coax comes in many sizes, including 6 pack, 10 pack, 12 pack, and 24 pack. Bundled coax is generally used to interconnect SONET equipment, digital cross-connect systems, and routers. See also *12-Pack Coax.*

Six-Pair Can   See *6-Pair Can* (Fig. S.8).

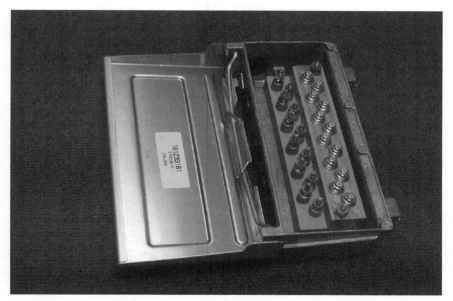

Figure S.8   A Six-Pair Can Outdoor Enclosure With Lightning Protection

Skinny Client Control Protocol (SCCP)   In IP telephony- or server-based PBX environments, an application layer program that replaces legacy circuit based PBX and Public Telephone Network signaling such as off-hook, on-hook, and dialed digits for Common Channel Signaled (CCS or out-of-band signaling) T1/E1 communications circuits (Fig. S.9). Although a proprietary Cisco Systems protocol initially, SCCP is offered via OEM to any telecommunications manufacturer that would like to implement it, and has been submitted to standards bodies to become a standard. There is a need for skinny protocol because without it or another signaling protocol like it, all IP telephones would require an "on board" PC to digest the protocol soup serviced in IP and 7 layer OSI communication. Skinny protocol enables a device (such as an IP telephone) with minimal intelligence to be managed by a central server or cluster of servers. This is similar to the IBM SNA architecture where there is one central computer and many user terminals attached to it, only in this case there are telephones instead of dumb terminals. Within IP telephony environments, multiple vendors (such as voice mail, IVR, ACD, and other telephone manufacturers) can build SCCP support into their products to manage signaling between an IP telephony server, IP telephones, and IP telephony gateways. The Skinny Gateway Program provides signal translation information from IP telephones to layer two IP telephony gateways. I say layer 2 because gateways do not provide layer 3 (IP network) layer services; therefore, Skinny Protocol does not service routers that function as gateways. (Currently, routers currently use MGCP and H.323v2 to perform PSTN to IP Network translation.) The signal translation information is a go-between from dialed digits and off-hook information to IP addressing through translation tables in a server-based PBX (or an IP telephony server). Simply stated, Skinny Protocol is to the IP telephony environment what a 24th out-of-band signaled T1 and the ISDN D-channel found is to a circuit-based PBX TDM environment.

Slamming   The illegal practice of changing a customer's telecommunications carrier (long-distance or local) without their consent. To identify the primary long-distance carrier for a telephone circuit in the United States, dial 1-700-555-4141. To identify the local toll carrier for a circuit (line), dial 1 (the circuit area code) 700-4141.

SLC96   Also known as *Slick 96*. A Lucent Technologies "pair-gain" system that multiplexes 96 telephone lines onto eight pairs of twisted-pair wire. It is used extensively in the public telephone network to provide telephone service to areas that do not have enough twisted pairs to meet customer needs. The SLC 96 actually uses four T1 circuits (24 lines per T1) to achieve the 96-line transport. The SLC 96 is configured in a cabinet, one for inside rack-mount central-office use and the other (far end) as an outdoor cabinet. The circuit cards that are incorporated into the SLC 96 design are separate and redundant power cards, battery back-up for the remote end, common equipment (control) cards, and a separate card for every two lines that are multiplexed (48-line cards for a full system).

SLIP (Serial Line Internet Protocol)   A protocol that enables IP datagrams to be transmitted over dial-up telephone lines. The successor to SLIP is PPP, which provides error detection and automatic configuration.

Sloppy Floppy Copy   To copy data to a floppy disk, then load it onto another computer's hard drive when you can't get the LAN to work right or if you don't have a LAN. Also called *Sneaker Net*.

Slot 1 A "package" that CPUs for personal computers come in (Fig. S.9). Slot-1 packages provide users with easier interchangeability and a lower production cost to manufacturers, as opposed to the older Socket-7 packages. Slot-1 CPUs can only be installed in Slot-1 motherboards, which are equipped with the proper slot interface, which the CPU plugs into.

Figure S.9   Slot-1 Package 500-MHz CPU

Slots A reference to expansion capability of a PBX system, PC, or other electronic equipment. PBX manufacturers make extra slots for electronic circuit cards to plug into the backplane of a KSU or cabinet for future network expansion.

Slotted Ring A LAN topology that is, for all practical purposes, a switched token ring.

Small Computer System Interface (SCSI) Pronounced as "scuzzy." A parallel interface standard for linking internal computer components (such as hard drives) and peripheral devices (such as printers). The good thing about SCSI interfaces is that they are faster than others (such as IDE/ATA). The drawbacks are the higher cost and that many of the SCSI formats that are on the market are not compatible with each other (Fig. S.10). If you upgrade your PC, you might end up replacing the SCSI interface in your printer as well. Some examples of the SCSI formats and their transfer speeds are: Fast SCSI, 8-bit bus 10 Mbps; Fast Wide SCSI, 16-bit bus 20 Mbps; Ultra Wide SCSI, 16-bit bus 40 Mbps; and Wide Ultra2 SCSI, 16-bit bus 80 Mbps. See also *IDE*.

Smart Card A credit card that not only has a magnetic strip (ROM) like all traditional credit cards, but also has a RAM component. Smart cards are being implemented in places where it is inconvenient to carry cash,

Figure S.10   SCSI 58-Pin Adapter

like on battleships at sea or in amusement parks. To buy something, you simply place your card into a machine, the machine deducts the balance from your account (which is directly on the card) and the transaction is completed.

Smart Jack   Also known as an *RJ68*. A smart jack is an RJ45 (8-pin modular jack) that has some simple electronic components inside it that enables it to be remotely placed in a loop-back mode for testing purposes.

SMDR (Station Message Detail Reporting)   Another term for *call accounting*. A call-accounting system is a computer (usually a dedicated PC) that connects to a PBX switch via a serial data port and monitors the details of every phone call made through that switch. The call details are stored as call records; with the appropriate software, they can be retrieved, sorted, processed, and queried to almost any specific nature that the call-accounting system administrator desires. These systems are used by hotels to track all the calls you make from your room so that they can bill you for them. They are also used by companies to do bill back reports for individual departments within the company.

SMDS (Switched Multimegabit Data Service)   A service offered by local telephone companies that is intended for the transport of large amounts of data at high speed from point to point over a switched type of network. Enter in the address or number that you would like your data to be sent and the SMDSU (SMDS Unit) packetizes the data and the SMDS network transports it. SMDS is a packet- or frame-type technology that is available in five transmission rates. Class 1 is 4 Mb/s, Class 2 is 10 Mb/s, Class 3 is 16 Mb/s, Class 4 is 25 Mb/s, and Class 5 is 44.7 Mb/s.

SMF   Single Mode Fiber.

SMSA (Standard Metropolitan Statistical Area)   An area that the FCC manages rights to provide cellular service. Most SMSAs have two cellular service providers.

SMTP (Simple Mail-Transfer Protocol)   A standard interchange format used by e-mail applications to exchange messages with each other. SMTP does not provide a user interface or method for a user to create a message. It only provides a way for the message to be transferred. For example, Lotus Notes can be configured to use SMTP when it sends an outgoing message to an Internet e-mail address. SMTP is used in conjunction with *POP (Post Office Protocol)* and *MIME (Multipurpose Internet Mail Extensions)* to provide TCP/IP Internet e-mail.

S/N Ratio (Signal-to-Noise Ratio)   The amount of desired transmission received in comparison to the amount of interference or distortion received with it. Expressed as a ratio.

SNA (Systems Network Architecture)   IBM protocol and architecture for mainframe/terminal computing environments. See *Fig. S.11.*

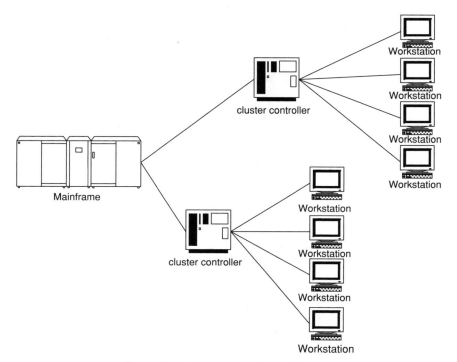

Figure S.11   SNA Network Architecture

SNADS (SNA, System Network Architecture, Distribution Services)
A data-transaction protocol that is capable of being encapsulated within frame relay and X.25. A software set of SNA programs that work together to provide asynchronous information transactions between end users. SNADS is one of three SNA transaction services. See also *Distributed Data Management* and *Document Interchange Architecture.*

Sneaker Net   See *Sloppy Floppy Copy.*

SNI (Standard Network Interface)   Also called *TNI (Telephone Network Interface)* and *NI (Network Interface).* It is a device used to terminate telephone service at the customer's location and provide lightning protection. One side of the SNI is for telephone company use only and the other side provides a place for customers to access their telephone lines (Fig. S.12). For other photos of SNIs, see *Standard Network Interface* and *Lightning Protector.*

Figure S.12   SNI (Standard Network Interface)

Snips   Scissors that telephone cable splicers and cable installation technicians use when installing/splicing cable. Snips have serrated blades, are very sturdy, and are capable of cutting copper that is as thick as a penny. For a photo, see *Cable Knife.*

SNMP (Simple Network Management Protocol)   A status and monitor subprogram that is used almost exclusively in TCP/IP networks. SNMP provides a means to monitor and control network devices, such as hosts and routers, on a LAN or WAN. It enables the user to troubleshoot, manage configurations, change/assign port addresses, collect statistics, performance, and security. In regard to SNMP, each device with an IP address on a TCP/IP network is called an *agent*. The station, host or terminal that an administrator uses to monitor the agents on the network is called the *SNMP manager*. Information and statistics gathered by SNMP include: the number of IP packets that have been sent and received, the number of errors encountered at the MAC layer, and the number of TCP retransmits.

Snowshoe   A device that is used to maintain a minimum bend radius for installed fiber-optic cable (Fig. S.13). The cable shown in the photo has a slack length to allow for future splicing. The slack is run along the strand and looped around snowshoes.

Figure S.13   Snowshoe

SO   The ASCII control-code abbreviation for shift out. The binary code is 1110000 and the hex is E0.

Socket   1. In TCP/IP, a socket is a combination of a TCP port number and an IP address (Fig. S.14). Sockets uniquely identify all connections within an internet.   2. An AppleTalk socket is similar in concept to a TCP/IP port.

Socket 7   A 321-pin *ULSI (Ultra Large-Scale Integration)* flat package that 586 CPUs for personal computers are manufactured in. Only Socket-7 CPUs can be used on Socket-7 motherboards (Fig. S.15).

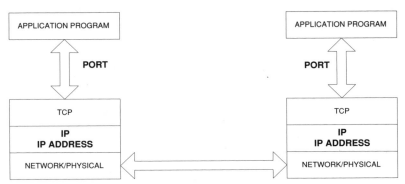

socket   **socket**
CONNECTION = (PORT NUMBER, IP ADDRESS) - (PORT NUMBER, IP ADDRESS)

Figure S.14   Socket for TCP Logical Connections to Higher OSI Layers

Figure S.15   Socket-7 CPU

Socket 8   Also called a *zero insertion-force socket*. A combination 387-pin package and matching socket for advanced 586 CPUs for personal computers. Socket-8 plugs (or sockets) have a lever that holds the CPU in place. Only Socket-8 CPUs can be used with Socket-8 motherboards. The Socket-8 architecture was outdated by the Slot-1 physical format.

Soft Call Forward   A type of call transfer where an incoming call is sent to another location after it has not been answered. A soft call forward is usually set by a voice-mail administrator or by a voice-mail user. In contrast, a *hard call forward* is a type of call transfer where an incoming call is immediately routed to an alternate destination with no ringing. A hard call forward is usually set on a phone or phone line by a user.

Software   Also referred to as *application*. Software is the programming instructions that are loaded into a computer's memory that tells it how to function. Microsoft Word is a software application and so is Lotus Notes.

SOH   The ASCII control-code abbreviation for start of heading. The binary code is 0001000 and the hex is 10.

Solid State   The term *solid state* came about when the transistor was invented. Before the transistor, vacuum-tube amplifiers were in wide use. For a long time, many appliances you bought said "solid state" on them because they had been made with transistors and other silicon devices instead of tubes.

SONET (Synchronous Optical Network)   SONET is strictly a broadband transport system. It is implemented over fiber optic and is able to be configured in a ring, which allows it to reroute traffic with no interruption of service if a fiber is cut (Fig. S.16). CLECs are implementing SONET as the mainstay of their network construction. SONET is based on a hierarchy of *STS (Synchronous Transport Signals),* which is the electrical version of an *OC-1 (Optical Carrier Level 1)*. An OC 1 has a transmission speed of 51.84 Mbs. The hierarchy of telephone communications services and their speeds is shown in the table. SONET permits a virtual tributary to be created from one node to another on a network. Virtual tributaries can be equal to a DS1, DS3, STS-1, or any of the OC levels. The important thing to know about a SONET network is that it simply replaces the older telecommunications technology copper twisted-pair outside plant with fiber optic and electronics.

Sound Card   An expansion board/card that is required by personal computers to process digital audio information and produce sound through speakers that are connected to it.

Name/Acronym	Bandwidth	Equivalent DS0	Equivalent DS1	Equivalent DS3	comments
DS0	64 Kbps	1	*	*	one phone line
DS1/T1	1.544 Mbps	24	1	*	popular service
DS1C	3.152 Mbps	48	2	*	equipment
E1/CEPT1	2.048 Mbps	32	1	*	European
DS2	6.312 Mbps	96	4	*	equipment
E2	8.448 Mbps	96	4	*	European
DS3/T3	44.736 Mbps	672	28	1	popular service
E3	34.368 Mbps	512	16	1	European
DS4	274.176 Mbps	4032	168	6	longhaul radio
STS-1	51.84 Mbps	672	28	1	electrical OC1
OC-1	51.84 Mbps	672	28	1	SONET
OC-3	255.520 Mbps	2,016	84	3	SONET
OC-12	622.080 Mbps	8064	336	12	SONET
OC-48	2.488 Gbps	32,256	1,344	48	SONET
OC-192	9.953 Gbps	129,024	5376	192	SONET

Figure S.16   SONET and Other Physical-Layer Transmission Rates

Source Route Bridging (SRB)   In token ring LAN networks, a bridge/ router operating protocol that builds a table of ring numbers with associated network connections (ports) on a bridge device. Source route bridging uses the RIFs (Routing Information Fields) from the headers of frames to route them, as opposed to MAC addresses as in transparent bridging. Bridges operating under source route bridging rules use the *All-Routes Explorer* and *Spanning Explorer* frames sent by clients or servers attached to the token network to build their routing tables.

In modern networks, routers are implemented to perform bridge functions (and frequently called *bridges* even though they are routers) with the appropriate operating protocol enabled (i.e., source route bridging) to suit the networking task at hand. Bridges are used to isolate traffic between common users on a shared bandwidth network, thus making communication between devices more efficient. In large-scale networks, LAN switches are implemented, with bridges/routers connecting or "bridging" the switches. See also *Source Route Bridging, Source Route Transparent Bridging, Spanning Tree,* and *Source Route Switching.*

Source Route Transparent Bridging (SRT)   In mixed networks using Ethernet and token ring, SRT uses RIFs (Routing Information Frames) and MAC addresses to accomplish packet transfer between separate and unlike LANs. SRT is a combination of transparent bridging, which uses MAC addresses to route frames, and Source Route Bridging, which uses RIFs (Routing Information Fields) to route frames. No standards group has adopted SRT. There are some complicated issues in accomplishing the protocol conversion from Ethernet to token ring that make it challenging for the companies that create products to suit the task. Some of those challenges include a different maximum packet size (limited to 1500 bytes by Ethernet), incompatible spanning tree algorithms, and unlike bit ordering (in Ethernet, the least significant address bit is put on the wire first; in token ring, the most significant is first). When SRT is implemented, all incompatibility issues need to be addressed on a case-by-case basis by the end users.

Source Routing Bridge   A type of network bridge that relies on routing information provided by an external sending system. See also *Bridge*.

Southwestern Bell Corp.   Merged with *SBC (Southern Bell Communications)*.

Spade Lug   A connector with two flat surfaces shaped like a two-prong fork that crimps onto a wire so that the wire can be mounted with screws.

SPAM (SPecialized Automated Mail)   The electronic version of "junk mail," unrequested e-mail messages that advertise products to users that have Internet e-mail boxes.

Span   One section of aerial wire.

Spanning Tree   A term given to routing translation tables that are created automatically by network operating software algorithms.

Spantree (STP)   In Ethernet switching, the IEEE 802.1d standard for Spanning Tree Algorithm that prevents layer 2 loops in redundantly connected LAN switches. Spantree is automatically enabled when redundant bridges are connected. If redundant bridges were connected to a network without Spantree enabled, the dual connected bridges would forward the same frames to each other in an endless loop. This condition saturates bandwidth immediately and renders all devices associated with the loop useless. The way that Spantree works is that when bridges are initialized (powered on), they send a signal to other networked devices called a Bridge Protocol Data Unit (BDPU). When bridges/switches receive these BDPUs from other devices, they become "aware" that other bridges are connected to the network and whether any are connected in redundancy to them. Using BDPU information, bridges on the network elect a "root bridge" and a "designated bridge." Depending on the way the bridges are physically connected, all ports are "blocked" or partially disabled except for "root ports" and "designated ports," which are bridge ports closest (by number of hops) to a designated or root bridge. If a link is lost, an alternate port then becomes the root port. New BDPU messages are sent to notify other bridges of the status change. Most makers of bridging hardware set the default to automatically send BDPUs and enable Spantree to On. This is so that if a network is unknowingly connected with bridges in parallel, it will not bring the network down. The 802.1d standard evolved from Digital Equipment Corporation's (DEC) Spantree algorithm. 802.1d and the original *DEC Spantree* are not interoperable (Fig. S.17). Further, when incorporated with 802.1Q (VLANs), one instance of spanning tree must be set up for each and every VLAN. See also *STP Bomb*.

COMMON PATH COSTS FOR SPANTREE			
TRUNK TYPE	NAME	BANDWIDTH	COST
10BaseT	Single Ethernet	10 Mbps	100
100BaseT	Fast Ethernet	100 Mbps	19
100BaseT X2	2 Fast Ethernet*	200 Mbps	12
100BaseT X4	4 Fast Ethernet*	400 Mbps	8
1000Base FX	Gigabit Ethernet	1 Gbps	4
*Based on Cisco Systems Fast Etherchannel ™ configuration.			

Figure S.17   Path Cost for Spantree

Spark Gap   Two wires, one hot, and one ground that are separated by a gap of air. Spark gaps are used for lightning protection. If the hot lead, or anything connected to it, is struck by lightning, the lightning will arc across the spark gap because it is engineered to be the easiest path to ground.

Spatial Redundancy   Having multiple fail-over devices in multiple locations. Often, especially in non-server-based PBX (IP telephony) environments, the entire PBX with all of its redundant electronics reside within one room. Spatial redundancy means having redundant electronics reside in different rooms or buildings. This is advantageous in areas where there is threat of disasters such as tornadoes, explosions, earthquakes, floods, and the like. Spatial redundancy is one of the inherent and easy to implement features of IP telephony.

Spatial Reuse Protocol (SRP)   A solution for reliably and economically transporting IP over fiber networks. As bandwidth requirements and demand for IP in metropolitan markets increase, service providers need equipment that is optimized for transporting and managing IP traffic over an optical infrastructure. SRP allows IP-based metropolitan networks to offer the same protection and restoration benefits as SONET-based networks while doubling bandwidth efficiency.

Speakerphone   A feature of telephones. Speaker phone allows a user to talk on the phone as if it were an open intercom system in the room, without using a handset.

Spectrum Analyzer   A type of test equipment used to evaluate the status of a broadband transmission. Spectrum analyzers are commonly used by cable-TV head-end technicians to troubleshoot and isolate noise and other troubles on cable-TV networks (Fig. S.18).

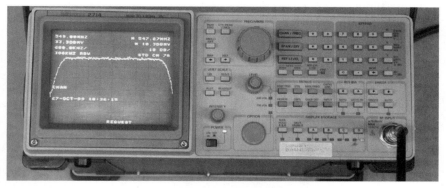

Figure S.18   Spectrum Analyzer

Spectrum, Frequency   All electromagnetic radiation is categorized by its frequency in Hz. If some of the frequencies were vibrations, rather than electromagnetic waves, people could hear them. Some electromagnetic radiation (theoretically) is visible as light. The full spectrum of electromagnetic radiation is listed in Fig. S.19, along with a brief description of its use or what types of transmissions are broadcast over those frequencies. See also *IEEE Radar Band Designation.*

Frequency Range	US designator	ITU designator	use
30 Hz to 300 Hz	ELF (extremely low freq)	2	Submarine/power
300 Hz to 3 KHz	ULF (ultra low freq)	3	Human audio
3 KHz to 30 KHz	VLF (very low freq)	4	Human audio
30 KHz to 300 KHz	LF (low freq)	5	
300 KHz to 3 MHz	MF (medium freq)	6	AM radio
3 MHz to 30 MHz	HF (high freq)	7	
30 MHz to 300 MHz	VHF (very high freq)	8	FM radio Broadcast TV
300 MHz to 3 GHz	UHF (ultra high freq)	9	Broadcast TV
3 GHz to 30 GHz	SHF (super high freq)	10	Terrestrial microwave/satellite
30 GHz to 300 GHz	EHF (extremely high freq)	11	Terrestrial microwave/satellite
300 GHz to 3 THz	THF (tremendously high freq)	12	Heat infrared
3 THz to 30 THz		13	Infra-red light
30 THz to 300 THz		14	
300 THz to 3 PHz		15	Visible light
3 PHz to 30 PHz		16	Ultra violet light
30 PHz to 300 PHz		17	
300 PHz to 3 EHz		18	
3 EHz to 30 EHz		19	
30 EHz to 300 EHz		20	
300 EHz to 3000 EHz		21	

Figure S.19   Spectrum Frequency

Speed Dial   A feature of telephone sets that enables a user to input a frequently dialed telephone number and assign that number a speed-dial code. To initiate the speed dial, the user dials the code instead of the entire number.

Speed of Light   The approximate speed of light in a vacuum is 300,000,000 m/s.

SPF (Shortest Path First Algorithm)   SPF is also referred to as a *link-state* or *Dijkstra's Algorithm*. A class of router-operating software that enables routers to build their own complex address routing tables that detail every router and node within their network. The routing table-building process is accomplished through information multicasts. The routing-table multicasts are referred to as *LSPs (Link-State Packets)* and they consume payload bandwidth to transmit this information. The process of sending and receiving LSPs is called the *discovery process*. Multicasts are only sent when there is a change in the network, such as a circuit connection going down, or a new router or connection being added. Link-state algorithms use tremendous amounts of router system memory (20 MB to 30 MB in a 30-node network), and consume significant processor resources within a router's circuitry. During the startup of a link-start network, the discovery process can take hours. The great advantage to this complex operating method is that routing loops are not created. See also *Distance Vector Routing Algorithm* and *Hybrid Routing Algorithm*.

SPF (Shortest Path First)   (Also known as *Link State Routing Protocol, Distributed Routing Protocol* and *Interior Gateway Routing Protocol.*) An SPF routing protocol is a methodology used in router protocol design. This methodology enables routers within an autonomous network (i.e., corporate LAN) to identify each other and the status of their port connections. SPF protocols create three databases within a router's memory: a neighboring router database, a link database, and a routing table. The routing table is created by applying Dykstra's algorithm to the first two databases. The most widely used SPF protocol is Open Shortest Path First (OSPF). See also *OSPF*.

SPID (Service Profile Identifier)   An *ISDN (Integrated Services Digital Network)* telephone number. The number that some telephone companies use to define the services to which an ISDN device subscribes. ISDN devices use SPID numbers when accessing the telephone company's switch to identify the device it would like to be connected to. In the United States, SPID numbers can look like ordinary public-service telephone numbers with an extension, such as 972-555-1212 4455. There is one SPID for each B Channel. See also *ISDN*.

Splice   The connecting of two wires, cables, coax cables, or cable pairs together. A splice is shown on an engineering diagram as an arrow. The actual splices of twisted-pair telephone cable are done with modular-type splices, plain B wire connectors, or 3M Scotchloks. For photos, see *Plain B Wire Connector, Modular Splice Tool,* and *Scotchlok*. Fiber-optic cable is spliced via mechanical or fusion splicing, and coaxial cable is spliced with barrel connectors. For a photo of a barrel connector, see *Barrel Connector.*

Splice Tray   A place within a fiber-optic patch panel or other fiber-optic splice closure that holds fusion splices. In Fig. S.20, the fiber optic can be seen below the patch panel. It is wrapped in a circle in the two splice trays.

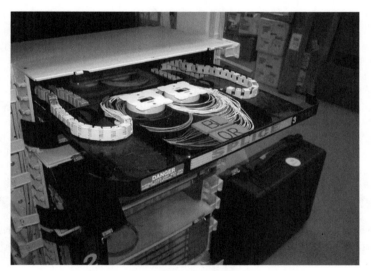

Figure S.20   Fiber-Optic Splice Tray

Split   The Lucent Technologies name for a queue. It was named so because it evenly "splits" incoming calls among agents logged into an ACD system. See *Queue.*

Split Horizon Update   A routing technique in which information about routes is prevented from exiting the router interface through which that information was received. Split-horizon updates are useful in preventing routing loops that count to infinity. They are incorporated into newer router operating systems, such as link-state, distance vector, and hybrid.

Split Pair   The use of one wire from two pairs to make a pair. Sometimes split pairs are done by mistake, but the majority of the time, it is done as a desperate measure to deliver a telephone service to a customer. If there are two bad pairs, but each pair has one good wire in it, the good wire is taken from each pair to make one good pair. Split pairs often cause inductive cross talk and pick up RFI. After RFI filters are placed on the split pair, it will work until new/more telephone wire can be installed or until the old wire can be fixed.

Splitter   Used to make a junction point or split a signal so that it will travel down multiple paths over coax. Also called an *RF splitter* and a *UHF/VHF Splitter.* For a photo, see *RF Splitter.*

Spoofing   1. A method used by routers to cause a host to believe an interface is up and supporting a communications session. The router accomplishes this by responding to keep-alive messages sent from the host. This convinces the host that the communications session is still running. Spoofing is useful to routers in dial-on-demand environments, in which a router will disconnect a call when a transmission is complete, but keep its connection to a workstation active. By doing this, the router will know when the workstation needs to send packets again and initiate a call. This provides the ability to have effective routing and save on toll charges.   2. The illegal act of sending a data packet that claims to be from an address from which it was not actually sent. Spoofing is designed to foil network security mechanisms, such as filters and access lists.

Spread Spectrum   The radio type used in PCS cellular transmissions. Spread-spectrum radio transmits and receives carrier signals over a wide spectrum of frequencies ("channels"). Several technology platforms in wireless communications are considered to be spread spectrum. A spread-spectrum system is any system that occupies more bandwidth than the minimum required for data signal transfer. Two data formats transmitted on spread-spectrum platforms include CDMA (Code-Division Multiple Access) and TDMA (Time-Division Multiple Access).

SPS (Standard Positioning Service)   The *GPS (Global-Positioning System)* service that civilians get, but does not correct signal dithering.

Spud   A hand shovel especially made for digging holes for telephone poles.

Spudger   A device that is shaped and sized like a pencil that telephone technicians use to poke their way through telephone cable when they are looking for a certain pair of wires. See *Fig. S.21.*

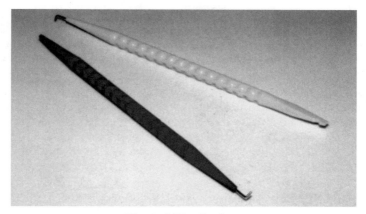

Figure S.21   Spudger

Spurs   What telecommunications and power company personnel wear to climb wooden telephone and power poles. The official name for these devices are *lineman's climbers*. They are also called *climbers, hooks,* and gaffs. They consist of a steel shank that has straps on it so that it can be strapped to a person's leg. On the inside of the shank is a spike that is used to stab into the pole. For a photo, see *Climbers*.

SPX (Sequenced Packet Exchange)   A part of the Novell NetWare protocol stack. SPX is used for very specific applications. Usually they include interserver or other device communications, such as printing reports, done by servers. An application that would use SPX is for remote control/administration (RCONSOLE) of servers.

SQL (Structured Query Language)   An international standard language for defining and accessing relational data bases.

Squelch   An electronic circuit or filter that is incorporated into microphone circuits that makes them have an adjustable sensitivity to the loudness of a sound the microphone will pick up. Squelch circuits are used in speakerphones to cut out background and transient noise.

SRAM (Static Random-Access Memory)   Electronic memory is available in two families, *ROM (Read-Only Memory)* and *RAM (Random-Access Memory)*. Memory devices are made from two different technologies: *bipolar (TTL)* and *MOS (Metal-Oxide Semi-conductor)*. Memory is stored by a technique called *writing* and retrieved by a technique called *reading.* ROM devices can only be read, and are programmed during manufacture. *PROM devices (Programmable Read-Only Memory)* can be programmed at a later date by an electronics reseller or electronic assembler for a special application using special equipment. Special ROM devices called *EPROMs (Erasable Programmable Read Only Memory)* can be electronically erased and re-used. RAM has read and write capability. The term *random access* means that any memory address can be read in any order at any time. The two types of RAM are static and dynamic. Static RAM can hold its memory even when power is removed. Dynamic RAM needs constant power to refresh its memory. For a diagram that depicts the types of memory, see *Memory*.

SRB (Source Route Bridging)   In token ring LAN networks, a bridge/router operating protocol that builds a table of ring numbers with associated network connections (ports) on a bridge device. SRB uses the RIFs (Routing Information Fields) from the headers of frames to route them, as opposed to MAC addresses as in transparent bridging. Bridges operating under Source Route Bridging rules use the *All-Routes Explorer and Spanning Explorer* frames sent by clients or servers attached to the token network to build their routing tables.

In modern networks, routers are implemented to perform bridge functions (and frequently called bridges even though they are routers) with the appropriate operating protocol enabled (i.e., Source Route Bridging) to suit the networking task at hand. Bridges are used to isolate traffic between common users on a shared bandwidth network, thus making communication between devices more efficient. In large-scale networks, LAN switches are implemented, with bridges/routers connecting or "bridging" the switches. See also *Source Route Bridging, Source Route Transparent Bridging, Spanning Tree,* and *Source Route Switching*.

SRP (Spatial Reuse Protocol)   A solution for reliably and economically transporting IP over fiber networks. As bandwidth requirements and demand for IP in metropolitan markets increase, service providers need equipment that is optimized for transporting and managing IP traffic over an optical infrastructure. SRP allows IP-based metropolitan networks to offer the same protection and restoration benefits as SONET-based networks while doubling bandwidth efficiency.

SS7 (Signaling System 7)   A method of out-of-band interoffice signaling for telephone circuits. Simply stated, *out-of-band* means that there is a special separate line used to carry signaling, such as dialed touch tones, ringing signals, busy tones, (everything but the actual voices/ conversation) etc.

The two different ways to send signals in telephone transmissions are in-band and out-of-band. Signals are digits that you dial, dial tone, the phone being off-hook, ringing, etc. An in-band telephone line is like the one in your home, the digits that you dial, and the ringing are carried within the channel you talk on. Out-of-band signaling is a method that telephone companies and businesses use for larger PBX applications and data-transfer applications. An out-of-band signaled DS1 has 24 multiplexed channels. The 24th channel carries the signaling for the other 23 channels or phone lines. The advantage of out-of-band signaling is that each channel has an increased capacity to carry data (8 Kb/s more) and the 23 channels are not used to find out if a line is busy (both directions, in and out). The off-hook sensing, busy signaling, and other signaling previously mentioned is performed in the 24th channel. If your system receives thousands of calls per day, this can reduce traffic. SS7 makes it easy for long-distance companies to let us dial a phone number, get a busy signal, and not be billed for it because we are not really using a call channel.

ST Connector (Straight Tip Connector)   An older type of fiber-optic connector. The newer is the SC connector, which is constructed of plastic instead of metal. For a photo of an ST connector, See *Fiber-Optic Connector.*

Standard Network Interface (SNI)   The device used to terminate telephone service at the customer's location and provide lightning protection (Fig. S.22). One side of the SNI is for telephone company use only, the other side provides a place for customers to access their telephone lines. For photos of other types of network interfaces, see *Two-Line Network Interface.*

Figure S.22   Standard Network Interface

Standby Processor   A second (redundant) CPU that takes over if the primary one fails.

Standing Wave   In radio transmission, a standing wave occurs when voltage and current form uneven points along a transmitter's antenna or transmission line. This is caused by a mismatch of load (antenna) impedance to transmission-line impedance, and causes an inefficient transmission. The term relating to standing waves is the *SWR (Standing-Wave Ratio),* which is the ratio of maximum current points on the line to the minimum current points on the line. To picture a standing wave, imagine you make waves in a bathtub so that waves reflecting from the sides of the tub collide in perfect timing with new waves made from your hand. The waves would appear to stand still and thus be called *standing waves.*

Star LAN Network   See *Star Topology.*

Star Topology   A topology or type of *LAN (Local-Area Network).* The star topology is used in Ethernet applications (Fig. S.23). Ethernet is one of the oldest communication protocols for personal computers. When a LAN is mentioned, the two things that should immediately come to mind are physical topology and the protocol that the LAN uses to manage communications between devices. Ethernet can be implemented in a bus or star physical topology. The alternative family of LAN protocols is the token-passing type, which is configured as a ring topology (see token ring).

   In an Ethernet LAN, computers are given a means to communicate with each other called a protocol. A protocol is a set of rules and instructions for communicating. Within the protocol is a "logical topology." Even though a network might be connected as a star, it can still "look" like a bus to the communications equipment because all of the computers/ devices are connected to the same wire (in the star diagram, the hub is a device that connects all the wires together). Ethernet works similar to the way that people talk in a group. Instead of using wire to carry the binary coded information as Ethernet does, people use air to carry sound information. When there is a silence, then one of the persons in the group is able to speak. When the people speak, they say "Johnny, do you know the answer for 5+5?" Even though all the people in the group hear this message, they know it is for Johnny because the message was "addressed" to him. Only Johnny will respond "10." Then imagine Dawn and Vicki both acknowledge a silence and try to speak at the same time. This is confusing and no one understands the information. Ethernet has the same problem and it is called a *collision.* Collision is the disadvantage of Ethernet. Because of the possibility of collisions (which happen very frequently), Ethernet is called a "contention-based" protocol because all of the connected devices are contending for use of the network. Manufacturers have

ETHERNET TYPES

PROTOCOL	PHYSICAL TOPOLOGY	WIRING USED
10 BASE 2	BUS	RG 58 COAX (50 ohm)
10 BASE 5	BUS	RG 8 COAX (50 ohm)
10 BASE T	STAR	CAT 4 or 5 UTP/STP*
100 BASE T	STAR	CAT 5 UTP/STP*

* unshielded twisted pair / shielded twisted pair

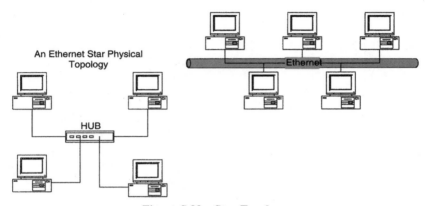

Figure S.23   Star Topology

come out with new ways to avoid collisions, called *CSMA/CD* and *CSMA/CA*. Ethernet has many different types of wiring to connect devices and many different *NICs (Network Interface Cards)* to select from that need to be installed in each computer or device on the network. The list shows Ethernet protocols and the type of wiring used for each.

Start of Heading (SOH)   The ASCII control code for start of heading. The binary code is 0001000 and the hex is 10.

Static RAM (SRAM, Static Random-Access Memory)   Electronic memory is available in two families, *ROM (Read-Only Memory)* and *RAM (Random-Access Memory)*. Memory devices are made from two different technologies: *bipolar (TTL)* and *MOS (Metal-Oxide Semiconductor)*. Memory is stored by a technique called *writing* and retrieved by a technique called *reading*. *ROM devices* can only be read, and are programmed during manufacture. *PROM devices (Programmable Read-Only Memory)* can be programmed at a later date by an electronics reseller or electronic assembler for a special application using special equipment. Special ROM devices called *EPROMs (Erasable*

Programmable Read Only Memory) can be electronically erased and re-used. RAM has read and write capability. The term *random access* means that any memory address can be read in any order at any time. The two types of RAM are static and dynamic. Static RAM can hold its memory even when power is removed. Dynamic RAM needs constant power to refresh its memory. For a diagram that depicts the types of memory, see *Memory*.

Static Route   In network routing, a path that is explicitly configured and entered into the routing table by a network administrator. Static routes take precedence over routes chosen by dynamic routing protocols.

Static VLAN   In Ethernet (802.3), there are two kinds of VLANs: static and dynamic. Static VLANs are associated with switch ports, and dynamic VLANs are associated with the MAC addresses of devices attached to the switch. Dynamic VLANs allow users to move to another office that could have a data connection installed. The switch would recognize the MAC address of the device and automatically include its traffic in the same VLAN as the previously connected switch port. See also *VLAN* and *Frame Tagging*.

Station Message-Detail Reporting (SMDR)   Another term for call accounting. A *call-accounting system* is a computer (usually a dedicated PC) that connects to a PBX switch via a serial data port and monitors the details of every phone call made through that switch. The call details are stored as call records. With the appropriate software, they can be retrieved, sorted, processed, and queried to almost any specific nature that the call-accounting system administrator desires. These systems are used by hotels to track all the calls that you make from your room so that you can be billed. They are also used by companies to bill back reports for individual departments within the company.

Statistical Time-Division Multiplexing   A multiplexing technology that gives users automatic adjustable bandwidth.

STDM (Statistical Time-Division Multiplexing)   A multiplexing technology that gives users automatic adjustable bandwidth.

Step-Down Transformer   A transformer that is wired in a fashion to receive an AC voltage on its primary winding and reduce that voltage through electromagnetic induction into the secondary winding (Fig. S.24). Reversing the way a transformer is wired changes it from a step-down to a step-up and vice versa.

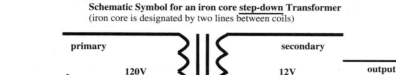

Schematic Symbol for an iron core step-down Transformer
(iron core is designated by two lines between coils)

primary

120V
AC

input

secondary

12V
AC

output

10:1

Figure S.24   Step-Down Transformer

Step-Up Transformer   See *Step-Down Transformer* (Fig. S.25).

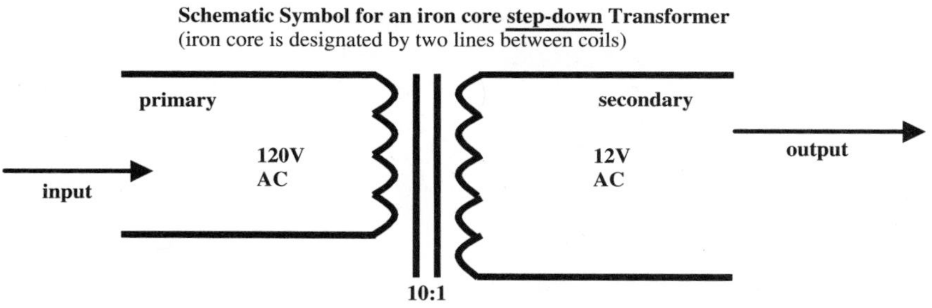

Schematic Symbol for an iron core step-down Transformer
(iron core is designated by two lines between coils)

primary

120V
AC

input

secondary

12V
AC

output

10:1

Figure S.25   Step-Up Transformer

Stepped-Index Fiber Optic   A fiber optic that has a core made of glass consisting of one refractive index. Stepped index fiber is available in multi-mode and single-mode. The alternative to stepped-index fiber optic is graded-index fiber optic (Fig. S.26). The core of graded-index fiber optic consists of many layers of glass with different refractive indexes that cause the light to gradually bend as it approaches the outside of the fiber. Graded-index fiber (like stepped index fiber) is available in multi-mode or single-mode, and it is more expensive.

LIGHT AS IT TRAVERSES THROUGH THE CORE OF A FIBER OPTIC

Stepped index fiber

Graded index fiber

Figure S.26   Stepped-Index Fiber vs. Graded-Index Fiber

Stepper Switch   Also called a *crossbar switch*. The old analog telephone switch had mechanical relays that connected telephone calls. This is where the term switch comes from. Old central-office switches contained literally thousands of mechanical switches.

STM (Synchronous Transfer Mode)   New technologies, such as SONET, *ATM (Asynchronous Transfer Mode)*, and *ISDN (Integrated-Services Digital Network)*, are leading up to STM. It is also referred to as *BISDN (Broadband Integrated-Services Digital Network)*. It will enable the user to have a DS3 private line, a POTS line, or an *ISDN BRI (Basic-Rate Interface)* automatically, depending on the device they use to access the line, a high-definition TV set, a telephone, or a PC.

STM-1   The basic bandwidth building block for *SDH (Synchronous Digital Hierarchy)*. The STM-1 signal has a bandwidth of 155.52 Mbps. Furthermore, the STM-1 format is identical to the electrical STS-1 and the optical OC-1 formats. See also *SDH* and *SONET.*

Stop Bit   In serial data transmission, the stop bit is a logical one (1) after the transmission of each character (each character is seven or eight bits long).

Store and Forward   A packet-switching process in which frames are completely received and their headers read before they are forwarded out the appropriate port. This processing includes calculating the *CRC (Cyclic Redundancy Check)* and verifying the destination address. In addition, frames might be temporarily stored in memory until an available link (or channel) is available to carry the message. This method has advantages and disadvantages compared to cut-through packet switching.

Store and Forward Switching   There are three ways that frames/packets transverse through a LAN switch, bridge or router. The first is *store and forward,* where the entire frame and its contents are accepted and stored in the switch. Error detection is calculated (CRC) and if the frame is good, the address is looked up in the routing table. When the associated destination port/segment is found, the frame is sent on to its destination. This is a good method for routing traffic because damaged frames, runt frames, and giant frames are discarded before they are transmitted. This method is used where the network infrastructure or media is prone to damaging frames, such as RFI environments or a poor WAN network service. The disadvantage of the store and forward method is *latency*. Storing the entire frame while the destination port is retrieved causes a delay, and in multiple-hop networks, this can cause slow network performance even when there is very little traffic.

The second way that frames/packets transverse through a LAN switch/router/bridge is the *cut-through switching* method, where only the address of an incoming frame is processed by memory. The address is associated with its destination port/segment in the routing table, and the entire frame sent directly through. This process happens if the frame is good or not, as long as there is a nondamaged address in the frame. Cut-through switching greatly reduces latency delays through a network, but still transmits bad frames. If a NIC card or host device begins sending lots of bad erroneous frames, the network performance could be slowed greatly. Some LAN switches have safeguards in place to detect and suppress error storms from defective equipment.

The third method of forwarding frames is *modified cut-through,* which works similar to store and forward, except that it uses a limited number of bytes to check for errors rather than the entire frame. This method helps prevent the retransmission of defective frames and also provides an acceptable level of latency delay through the network.

STP (Spantree)   In Ethernet switching, the IEEE 802.1d standard for Spanning Tree Algorithm, which prevents layer 2 loops in redundantly connected LAN switches. Spantree is automatically enabled when redundant bridges are connected. If redundant bridges were connected to a network without Spantree enabled, the dual connected bridges would forward the same frames to each other in an endless loop. This condition saturates bandwidth immediately, and renders all devices associated with the loop useless. The way that Spantree works is that when bridges are initialized (powered on), they send a signal to other networked devices called a Bridge Protocol Data Unit (BDPU). When bridges/switches receive these BDPUs from other devices, they become "aware" that other bridges are connected to the network and whether any are connected in redundancy to them. Using BDPU information, bridges on the network elect a "root bridge" and a "designated bridge." Depending on the way the bridges are physically connected, all ports are "blocked" or partially disabled except for "root ports" and "designated ports," which are bridge ports closest (by number of hops) to a designated or root bridge. If a link is lost, an alternate port then becomes the root port. New BDPU messages are sent to notify other bridges of the status change. Most makers of bridging hardware set the default to automatically send BDPUs and enable Spantree to On. This is so that if a network is unknowingly connected with bridges in parallel, it will not bring the network down. The 802.1d standard evolved from Digital Equipment Corporation's (DEC) Spantree algorithm. 802.1d and the original *DEC Spantree* are not interoperable. Further, when incorporated with 802.1Q (VLANs), one instance of spanning tree must be set up for *each and every* VLAN. See also *STP Bomb*.

Straight-Tip Connector   A type of round, metal fiber-optic connector, usually called an *ST connector.* The newer connector (and rapidly becoming more popular) is the *SC connector,* which is square in shape and made of plastic.

Strand Clamp   A pole-attachment device that is used to hold steel strand to utility poles (Fig. S.27).

Figure S.27   Strand Clamps with 14-inch Mounting Bolts

Stranded Copper   See *Stranded Wire.*

Stranded Wire   A wire that is made up of many small wires, rather than one big solid one. Stranded wire is not used in telephony applications because it doesn't stay connected to 66M150 or AT&T 110 termination blocks.

Strand, Steel   The support for telephone cable, cable-TV coax, and fiber-optic cable when installed in aerial applications. See also *Pole Attachment.*

Strap   An electrical connection (usually a wire or metal jumper) from one point to another. *Strap* is a common term used in reference to the configuration of rectifiers, power supplies, and circuit cards.

Stripe Pitch   A measurement (in millimeters) of the distance between dots along the vertical wires in a monitor that incorporates aperture grill technology. The smaller the stripe pitch, the higher the resolution capability of the monitor.

STS (Synchronous Transport Signal)   In SONET networks, the optical signal must be converted to electricity at one time so it can be de-multiplexed and further processed. This electrical version of the SONET OC-1 level signal is called a *synchronous transport signal 1 (STS-1)*, and is transported from node to node or node to digital cross-connect system, or to DSX cross-connect panels via 50-ohm coax.

STS-1 (Synchronous Transport Signal 1)   See *STS* (Fig. S.28).

Figure S.28   STS-1 BNC Interface on a Cisco Switch/Router

STS-3 (Synchronous Transport Signal 3)   Referred to by some as a level 2. The electrical equivalent of an *OC-3 (Optical Carrier Level 3)*. 155 Mbps.

STS-3c (Synchronous Transport Signal 3c)   A physical-layer transmission format that is known as an *STM1* outside of the United States and Japan. The STS-3c is the electrical equivalent of the SONET OC-3. The transfer rate is 155 Mbps, including overhead and payload. The net payload on an STS-3c is 149.76 Mbps. See *Fig. S.29.*

SONET 155 Mbps STS-3c / STM1 FRAME STRUCTURE

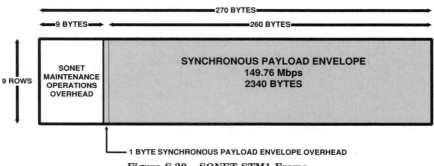

Figure S.29   SONET STM1 Frame

Stub Area   In network routing, a router at the bottom of a hierarchical design or on the end of a hub-and-spoke design. It is a router with only one exit point from its local network, so a default gateway address is used for packets to be routed out of the local network. In Cisco Systems routers, routers that are in an OSPF network can be configured to be "totally stubby," which prevents them from receiving database updates from area border routers. This feature enables inexpensive routers to be placed on the edge of OSPF networks.

Studio-Transmitter Link (STL)   A studio-transmitter link is basically a remote control for a distant transmitter. It is a separate radio channel from the broadcast radio station itself and it is used to adjust the amount of power that the transmitter emits, performance monitoring, and provide remote access for testing and troubleshooting.

Stutter Dial Tone   A dial tone that blinks on and off when a telephone line initially goes into an off-hook state. A stutter dial tone is used to signal a user that a new voice message is in their voice mailbox, which is also known as *voice-mail notification.*

STX   The ASCII control-code abbreviation for start text. The binary code is 0010000 and the hex is 20.

SUB   The ASCII control-code abbreviation for substitute. The binary code is 1010001 and the hex is A1.

Subarea Network   A group of devices that communicate through the same control node on an *SNA (IBM System Network Architecture)* network. Two types of nodes are Type-4 communications controller nodes and Type-5 host nodes. All devices in a subarea share the same subarea address and have unique element addresses.

Subnet   A method to more efficiently use Internet IP addresses. In 1985, RFC 950 was written to standardize a procedure for dividing Class A, B, and C networks into smaller, more manageable sections. Subnetting incorporates a subnet mask into an IP address as a guide to networking devices in identifying which bits to use when identifying a network. A subnet mask extends the bits processed by a router beyond the standard classful addressing scheme. Rather than use the first three bits to identify the size (or number of bits for) of the network IP address, the subnet mask is used. The additional bits beyond the classful method identify the network. Subnetworks are networks arbitrarily segmented by a network administrator in order to provide a multilevel, hierarchical routing structure while shielding the subnetwork from the addressing complexity of attached networks.

Subnet Zero   In IP addressing, a network address of zero. When an IP address is broken into subnets using a subnet mask, the highest subnet is all ones (i.e., 11111111), and the lowest is all zeros (i.e., 00000000). Incorporating the *all-zeros address* into the IP address plan is done only by network administrators as a last resort when all other possible address numbers have been used. Using the subnet zero improperly can cause data to be misrouted, which causes poor network performance and network failure. For example, if network 131.108.0.0 is subnetted as 255.255.255.0, subnet zero would be written as 131.108.0.0, which is equal to the network address. This scenario would cause a complete network failure in many cases.

Subscriber   A telecommunications customer. This includes telephone, cable TV, and cellular (PCS and wireless).

Subscriber Loop   The pair of wires that runs from a telephone company central office (or from extended transmission equipment) to the customer's network interface. A loop is a pair of wires.

Subscriber Loop Carrier (Another term for SLC 96)   See *SLC 96.*

Super Server   A server that has multiple microprocessors.

Superframe Format   A framing format for T1 that consists of 12 T1 frames, 193 bits each, transmitted in succession. The superframe format allows for maintenance and monitoring information to be sent along with the 24 DS0 channels. *ESF (Extended Super Frame)* is the newer version of T1 framing format.

Supernetting   1. The opposite of subnetting. *Subnetting* is the process of taking a network and *dividing it into smaller* subnetworks using a subnet mask. *Supernetting* is the process of taking *several discrete networks and advertising them in one routing update;* for example, if an organization had been assigned a full block of Class C addresses, say 192.10.1.0/24 to 192.10.254.0/24. Instead of advertising 254 separate networks to the Internet, the organization may advertise only the single route to 192.10.0.0 /16 to the Internet. Full connectivity is possible because any datagram destined for a 192.10.x network is bound for the same organization. When the packet gets to the organization, it is the responsibility of the routers of the organizations to get the datagram to the proper network.   2. In IP addressing, a method used to link two or more Class C addresses together so they function as one single network or Internet site.

Supervisory Signal   A way that telephone electronics communicates with each other to initiate a command. If you have call waiting on your home telephone, when you get another call (hear the beep or click indicating another call), you momentarily press the hook switch (flash button) to signal the central office to give you the other line. This is an example of a supervisory signal. It is called a supervisory signal because it relates to the connect and disconnect of a phone line. The general term for the ability for a central office or PBX switch to recognize that a telephone conversation has ended is *disconnect supervision.* When the telephone is "hung-up," the central office or PBX recognizes the decrease in current flow and disconnects the call. *PBX (Private Branch Exchange)* phone systems have disconnect supervision, which means that when a call is ended, it recognizes a "hook-flash" from the central office and disconnects the call, or vice versa. If the PBX did not have this feature, it would not release (hang up) telephone calls.

Surface Mount   1. A reference to a modular jack that is shaped like a box, and can be fastened to the surface of a wall, baseboard, or anywhere else you do not have prewired outlets. Surface-mount jacks are also called *biscuit jacks* and *baseboard jacks.* 2. A reference to an electronic component package where the device is soldered to the surface of a PC board, rather than have leads that extend through the board.

Surge Protector A power-filtering device. Most surge protectors come in the form of an extension cord with six to eight outlets box attached to the end. Not all surge protectors are created equal. Most surge protectors are more useful as an extension cord than any kind of protection from a voltage surge or spike. Some good surge protectors use fast-switching components to sense overvoltage and spikes on the power source. These good surge protectors cost about $50. If you are truly concerned about protection from all of the evil electrophysical characteristics of public power, a small UPS system is the best protection. A good *UPS (Uninterrupted Power Supply)* costs about $150.

SVC (Switched Virtual Circuit) Any circuit that can be connected for a temporary amount of time with the use of electronic circuit switching equipment. This includes a plain-old telephone call. When the phone is on the hook and no one is using it, the telephone wire runs to the central office and ends. When you pick up the receiver and dial a number, the central office makes a connection through its electronics from one phone line to another that lasts only as long as the receivers are off hook. It is a "switched" circuit because it can be switched to any telephone with a number that you dial. It is virtual because the actual path through the electronics is multiplexed, as opposed to a physical pair of wires.

SVGA (Super Video Graphics Array) The suggested monitor for the PC that is used for call-accounting applications, and other computer-telephony integration applications. The SVGA monitor is capable of resolution to 1024 by 768 pixels (dots of light on the computer screen) per inch. SVGA is the newer version of *VGA (Variable Graphics Array)*.

Switch 1. Another name for a PBX (phone system) or central office. 2. A reference to a circuit connecting device, as in the *circuit*-oriented PBX/central office switch in a telephone environment, or a *packet*-forwarding switch as in a LAN environment (Fig. S.30). In both environments, a switch is a device that forwards traffic through the utilization of layer 2 (Link/MAC layer) silicon hardware devices. See also *LAN Switch, PBX,* and *Transparent Bridging.*

Figure S.30   Ethernet Switch

Switch Clustering   Some manufacturers of Ethernet switching equipment built in the ability to link their switches together logically, making them appear as one easily manageable switch to an end user. The switches may be physically stacked on top of each other or located in different buildings. Either way, they appear as one switch to the system administrator. This enables QoS and traffic policies to be configured an a global basis. This feature is commonly referred to as *Switch Clustering*.

Switch Room   A room that is dedicated for switching equipment. Switch rooms are usually kept at a temperature of 65 degrees F and a humidity of 50% to reduce ESD.

Switched 56   A service offered by local and long-distance telephone companies that works like a regular telephone line except, that it is intended for data/modem use. Switched 56K lines have 7-digit telephone numbers (plus area code) and they are available in digital or analog. When the telephone company installs the line, they condition the copper pair (remove bridge taps and coils) and install a *DSU (Digital Service Unit)*, which is a line/signal amplifier. Connect it to your 56K modem; it will transmit at a rate of 56,000 bits per second as long as you are talking to another 56K modem on another 56K line.

Switched Ethernet   An efficiency enhancement to the original Ethernet specification. Switched Ethernet incorporates modified layer 2 (Data Link Layer) electronics in hubs and repeaters. This design provides an individual 100 Mbps (for fast Ethernet) to each far end segment (Fig. S.31).

Switched Ethernet Priority   In converged (voice, video, and data) Ethernet environments, a reference to the IEEE 802.1p standard for prioritization of LAN traffic among Ethernet switches based on the switch port, MAC address, or IP address associated with the communicating end appliance (be it an IP phone, video monitor, host PC, printer, or server). Packets are tagged as belonging to a queue, which determines

Figure S.31   Ethernet Switched Hub

the priority of the packet. By the 802.1p standard, queues 0–3 normal and 4–7 are high priority. 802.1p functions hand-in-hand with 802.1Q or VLANs.

Switched Multimegabit Digital Service (SMDS)   A service offered by local telephone companies that is intended for the transport of large amounts of data at high speed from point to point over a switched type of network. You enter in the address or number that you would like your data to be sent and the SMDSU (SMDS Unit) packetizes the data and the SMDS network transports it. SMDS is a packet- or frame-type technology that is available in five transmission rates. Class 1 is 4 Mb/s, Class 2 is 10 Mb/s, Class 3 is 16 Mb/s, Class 4 is 25 Mb/s, and Class 5 is 44.7 Mb/s.

Switched Private Line   A reference to switched 56K service.

Switched Service   Basic voice telephone lines (POTS lines) are switched services.

Switched Virtual Circuit   A telephone line is a switched virtual circuit. The circuit only exists while the conversation is happening. After the conversation is over, the circuit, which acts like a pair of wires from point A to point B, is gone. Another form of a switched virtual circuit is like those in switched Ethernet or switched token-ring technology.

Switching Center   Another name for a telecommunications company's central office. A location for switching equipment/electronics and transport equipment/electronics.

Switching Hub   Also called a concentrator. A *LAN (Local-Area Network)* element that links a device to a network with a specific amount of bandwidth or exchange rate of data, regardless of the number of users on the network (Fig. S.32). A switching hub performs the same function as a nonswitching hub, except that the nonswitching hub only connects many users into the same channel, where they all share the same bandwidth. For a photo of a switched Ethernet hub, see *Switched Ethernet.*

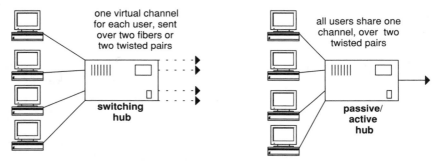

The switching hub vs. the passive/active hub

one virtual channel for each user, sent over two fibers or two twisted pairs

switching hub

all users share one channel, over two twisted pairs

passive/ active hub

Figure S.32   Switching Hub

Symmetrical Digital Subscriber Line (SDSL)   A physical-layer telecommunications protocol that delivers high-speed data networking over a single pair of copper phone lines. SDSL is symmetrical, which means that upstream and downstream data-transfer rates are the same. Speeds range from 160 Kbps to 1.544 Mbps. The base transmission distance is 24,000 feet (about 4.5 miles), and it can be extended to greater than 30,000 feet with repeaters. SDSL can be extended to any distance over fiber optic. SDSL is ideal for business applications that require identical downstream and upstream speeds, such as video conferencing or collaborative computing, as well as similar applications that are appropriate for ADSL technology. SDSL uses the same kind of line-modulation technique used in ISDN, known as *2B1Q*.

SYN   The ASCII control code abbreviation for synchronous idle. The binary code is 0110001 and the hex is 61.

Synchro Daemon   A background program or subroutine that is used for a timing reference in telecommunications platforms.

Synchronize   See *Synchronous.*

Synchronous   "With time." A reference to communications equipment that is timed from a common timing source, such as a bits clock. SONET networks are synchronous and modem transmissions are asynchronous. Asynchronous means "without timing." People talk asynchronously. Even though one person talks very fast and another very slowly, they both exchange information and process it in their brains.

Synchronous Data Link Control (SDLC)   A revised revision of the IBM Bisync protocol that was submitted by its creator (IBM) to the ISO and ANSI in hopes that it would become a standard. Instead, it was evolved by the ISO into *HDLC (High-level Data-Link Control)* and evolved by the ANSI into *ADCCP (Advanced Data Communication Control Procedure).* The CCITT reviewed and modified HDLC and called it *LAP (Link-Access Procedure),* which became the basis for the frame or network layer of the X.25 standard in 1976. The LAP protocol was further modified by the CCITT to become *LAPB (Link-Access Procedure Balanced mode)* in 1978.

Synchronous Digital Hierarchy (SDH)   A European family of digital carrier rates. SDH defines a set of rate and format standards that are transmitted using optical signals over fiber. SDH is the term used by the ITU to refer to SONET OC rates referred to in the United States. Its basic building block is a rate of 155.52 Mbps, designated at STM-1 (OC-3). See also *SONET* and *STM-1.*

Synchronous Graphics RAM (SGRAM)   A type of dynamic random-access memory used as a buffer/temporary storage in computers that enhances the performance of the graphics accelerator and video adapters.

Synchronous Transfer Mode (STM)   See *STM.*

Synchronous Transport Module Level-1 (STM-1)   The basic bandwidth building block for *SDH (Synchronous Digital Hierarchy).* The STM-1 signal has a bandwidth of 155.52 Mbps. Furthermore, the STM-1 format is identical to the electrical STS-1 and the optical OC-1 formats. See also *SDH* and *SONET.*

Synchronous Transport Signal (STS 1)   See *STS 1.*

Synchronous Transport Signal 3c (STS-3c)   A physical-layer transmission format that is known as an *STM1* outside of the United States and Japan. The STS-3c is the electrical equivalent of the SONET OC-3. The transfer rate is 155 Mbps, including overhead and payload. The net payload on an STS-3c is 149.76 Mbps. See the diagram under STS-3c.

SyncLink Dynamic RAM (SLDRAM)   A class of random-access memory used for personal computers that is half as fast as *RDRAM (Rambus Dynamic RAM)* at 400 MHz, but less costly (at initial inception to the market). It is still faster than SDRAM, which exchanges data with central computer components at a rate of 100 MHz.

System Speed Dial   A feature of *PBX (Private Branch Exchange)* and key telephone systems. Unlike standard speed dial, where each individual user programs a speed dial under a button on their own phone, system speed dial programs a speed-dial number under a code. The telephone number programmed in the system speed dial can be dialed by entering the code from any telephone on the system (that has the feature). The Northern Telecom Meridian 1 PBX system is capable of having 1000 system speed-dial numbers.

Système Electronique pour Couleur Avec Mémoire (SECAM)   The analog television broadcast standard in France, Russia, and regions of Africa. SECAM is a variant of PAL, but it delivers the same number of vertical scan lines as PAL and uses the same refresh rate. See also *Television Broadcast Standards*.

Systems Network Architecture (SNA)   See *SNA*.

T

T Carrier A reference to T1 or T3 transmission systems. A T1 has 24 voice channels (standard telephone lines) on one 4-wire circuit, and a T3 has 28 T1 channels (also called *tributaries*) on one 4-wire circuit. For a diagram of a T1 and a chart showing T1 signaling and framing applications, see *T1*.

T Connector A coax connector used in *LAN (Local-Area Networks)* that has a male BNC on one end and two female BNC connectors on the other. It's shaped like a "T."

T Interface An ISDN-compatible digital interface. For example, a T interface fits between an ISDN telephone and an ISDN line.

T PAD (Terminal Packet Assembler/Dissembler) A PAD that is specifically located on the host end of a communications link. Even though PADs are the same on each end, sometimes technicians refer to them in a specific manner. The T PAD is a device that is located at the terminal or user end (as opposed to the host end) of a virtual communications link in a frame protocol environment that reassembles and disassembles large files of data. The *HPAD (Host Packet Assembler Dissembler)* also adds and removes address, envelope, and HDLC information.

T Span Another name for a T1 or T3. See *T1*.

T.38 Phase 2 A Cisco Systems' developed fax gateway protocol that is now an ITU-T standard, ITU-T T.38. A fax gateway is a device that allows a single telephone line to receive both faxes and voice calls. The fax gateway answers all incoming calls. It listens for a fax handshake tone and if it does not hear one, it rings the telephone connected to it. T.38 is a feature that can be enabled on individual router FXS ports.

T1   A T1 ("T" one) is a standard 1.544-Mbps carrier system used to transport 24 telephone lines or various broadband services from one point to another (also called a *DS1,* but a DS1 is the service given to a customer without the –135-V carrier). T1 is the standard carrier for the United States, Canada, Japan, and Singapore. All other countries use the E1 standard (30 channels on four wires). The T1 is a four-wire circuit, two wires for

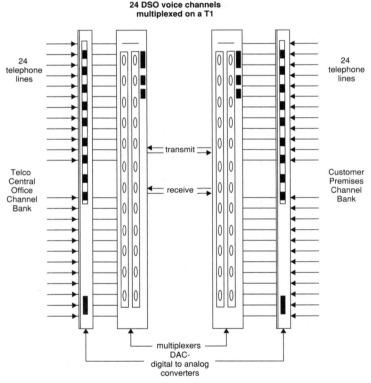

24 DS0 voice channels multiplexed on a T1

24 telephone lines

24 telephone lines

transmit

receive

Telco Central Office Channel Bank

Customer Premises Channel Bank

multiplexers
DAC-
digital to analog
converters

Figure T.1   T1/DS1 for 24 DS0 Channels

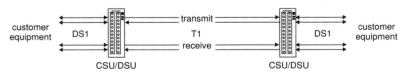

customer equipment

DS1

transmit

T1

receive

DS1

customer equipment

CSU/DSU

CSU/DSU

Figure T.2   T1/DS1 for a Private-Line Data Communications Channel

DS1/T1circuit/line types and applications

Line format/coding	framing format	signaling	Application
AMI	SF/D4	in-band	24 voice/modem channels
AMI	ESF	in-band	24 voice/modem channels
AMI	ESF	out-of-band	23 voice/modem or digital/data channels
B8ZS	SF/D4	in-band	24 voice/modem channels
B8ZS	ESF	in-band	24 voice/modem channels
B8ZS	ESF	out-of-band	23 voice/modem or digital/data channels

Figure T.3   T1/DS1 Frame and Code Types

transmit and two wires for receive. The T1 line voltage is −135 V. The T1 circuit can carry voice or data. Its use determines the variables of T1 service, framing format, and line format (Figs. T.1 to T.3). See also *DS1*.

T1 CAS (T1 Channel Associated Signaling)   A reference to the traditional circuit-switched method of signaling T1 ESF (Extended Super Frame) of 24 individual 64 Kbps channels, where 8 Kbps is "robbed" from each channel for signaling purposes, such as off-hook, ring, and so on. T1 CAS is usually a term used in IP telephony gateway consideration when there is a need to connect to the telephone company with non-ISDN PRI signaling. ISDN PRI is preferred in IP telephony implementations. See also *T1*.

T1 Test Set   Many instruments are manufactured for testing T1 transmission circuits. See *Fig. T.4*.

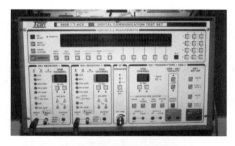

Figure T.4   T1 Test Sets

T1C   A 3.152 Mbps multiplexed signal that carries the equivalent of two T1 signals. The T1C signal is not a widely embraced form of transport; however, it does exist within the stages of some multiplexing equipment.

T568A   A standard for Ethernet wiring. The jack pinouts are shown in Fig. T.5. See Appendix G for diagrams of other LAN wiring standards.

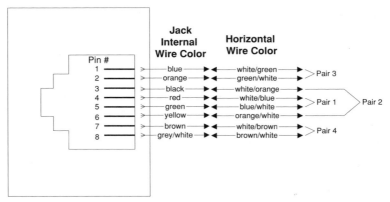

Figure T.5   568A Wiring Diagram

T568B   A standard for Ethernet wiring. The jack pin-outs are shown in Fig. T.6. See Appendix G for diagrams of other LAN wiring standards.

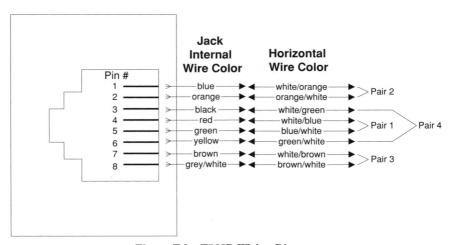

Figure T.6   T568B Wiring Diagram

TA (Terminal Adapter)   A device that converts an *ISDN (Integrated Services Digital Network)* line into regular *POT (Plain-Old Telephone)* service so that you can connect a standard telephone and/or modem to an ISDN line. For a diagram, see *Terminal Adapter.*

Tag Switching   1. A Cisco Systems proprietary protocol that was submitted to the IETF (Internet Engineering Task Force) in 1997 as a solution for integrating IP and ATM. Tag switching evolved into MPLS (MultiProtocol Label Switching). This IP over ATM standard is used worldwide by telecommunications service providers in IP network switching equipment.   2. Tag switching is also used as a reference to VLAN, or IEEE 802.1Q. It is a feature on many post-1998 LAN switches that makes selected ports behave as if they were attached to the same segment, or hub. Another good name for this feature would be *V-segment* or *virtual segment.* Devices/users that exchange a large amount of information are usually placed within the same VLAN segment. This helps make the operation of the LAN switch more efficient, keeping traffic contained within specified ports. This allows other ports on separate VLANs to carry other nonrelated traffic simultaneously. VLANs are configured by a network engineer, network analyst, or network administrator. When IP telephony is implemented over an Ethernet-switched network, the telephone devices connected to the network are best placed into their own VLAN. Most switches that are 802.1Q compatible can recognize more than 1,000 VLANs. Further, there are two kinds of VLANs: static and dynamic. Static VLANs are associated with switch ports, and dynamic VLANs are associated with the MAC addresses of devices attached to the switch. Dynamic VLANs allow users to move to another office that could have a switch port connection preinstalled. The switch would recognize the MAC address of the device and automatically include its traffic in the same VLAN as the previously connected switch port. See also *Frame Tagging.* Eventually, tag switching became the 802.1p and 802.1q standard for Ethernet VLANs.

Talk Battery   Typically –10 V on the public telephone network. When a telephone line is idle, a –52-V central-office battery is on the line. When the receiver is picked up, the voltage drops to 10 V (talk battery). The ringing voltage on a telephone line is 90V AC.

Talk Path   Two wires with a battery voltage that can be used to connect two telephones to provide a simple talk path. Sometimes when telephone company employees are working on telephone lines, they connect a battery to a pair of wires, connect a test set (butt-set/goat/test telephone) to each end, and talk to each other. They call this arrangement a *talk path.*

Tandem 1. A classification of telephone company central office or node that contains a switch in which all inter and outer area-code traffic is handled. The main LEC central office is in an area code where the hand-off for long-distance service happens. For a diagram, see *Access Tandem.* 2. A central office that carries a call, but does not connect it with the end customer, it switches ("sends") the call to the central office from which the called customer is served.

Tandem Office See *Tandem.*

Tandem PBX A *PBX (Private Branch Exchange)* switch that carries telephone calls, but does not terminate them to a telephone (Fig. T.7). It switches the calls to another PBX. See also *Tie Trunks.*

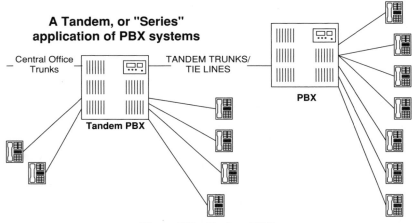

Figure T.7 Tandem PBX

Tandem Switch A central office that carries ("links") a call, but does not connect it with the end customer, it switches ("sends") the call to the central office from which the customer is fed. See also *Tandem.*

Tandem Trunks Trunk lines that connect central offices/switching equipment. See *Tandem PBX.*

Tap A device used to monitor telephone lines. With newer technology, the telephone company is capable of tapping (monitoring) a telephone line with a stroke of a few keys. They can even set the telephone line to be monitored by a different telephone line, anywhere they want. Telephone companies (especially the Bell Companies) have extremely strict security guidelines regarding the monitoring of telephone conversations. They will not set up a tap or a monitor service without legal procedures being followed according to the laws of the area where they are operating. Watch dogs are in place to ensure that telephone company employees do not monitor telephone lines when they are not supposed to. Some "spy" shops sell telephone line tapping and recording devices, but their use is not recommended because of strict laws regarding telephone privacy.

Tap Button Key   A button on some *PBX (Private Branch Exchange)* telephones. Most PBX systems are purchased with electronic/digital telephone sets. Those that are not, are purchased with standard single-line telephones that have an extra button, called a *flash* or *tap key* (most PBX systems can be configured for either). The flash/tap key sends a hook-flash signal to the PBX system to tell it that it is going to receive a command (e.g., more dialed digits) to activate a feature, such as transfer, hold, or conference.

Tape Drive   A RAM storage device that utilizes magnetic tape for data storage. The magnetic tape is similar to that of the magnetic tape used in a typical audio cassette. The disadvantage of tape-drive memory storage is the long amounts of time required to retrieve data.

TAPI (Telephone Application Program Interface)   A software program standard that allows PC support for a broad range of communications devices, including telephone switch (PBX, ACD, IVR, and central office) equipment.

Tariff   A pricing structure of telecommunications services that is offered by a communications services company and accepted by the *Public Utilities Commission (PUC)*.

TBM (Transport Bandwidth Manager)   Northern Telecom's name for their SONET-operating platform.

TBOS (Telemetry Byte-Oriented Serial)   An alarm or maintenance protocol for communicating with distant or remote equipment.

TCP (Transmission Control Protocol)   The part of the TCP/IP protocol suite that ensures that data is delivered in the proper sequence to the upper application layer without errors, missing data, or duplication. TCP is a connection-oriented protocol, similar to a voice telephone conversation in the way that a number (address) is dialed and a connection is made. The TCP layer sections data into segments and gives each segment a header and sequence number. The TCP segment is then given an IP header, which makes the handoff to lower layers and the remote host connectionless. The TCP segment and IP header together are called an *IP datagram.* All parts of the TCP/IP suite (such as ICMP, TCP, and IP) are needed to completely transfer a file or message across a network.

TCP Header   The datagram (packet) header in a TCP/IP transmission. It is of variable length and carries many elements of information. They are listed in the following section and illustrated in Fig. T.8:

- *Source and Destination Ports*   Identify the sending and receiving applications at each end of the TCP/IP connection. Both fields are 16 bits in length.

- *Sequence Number*   Allows for the receiver to identify the order of sequence for this segment.

- *Acknowledgement*   Provides an acknowledgment of receipt from the opposite end for received TCP segments. This field is only valid if the ACK bit (in the code field) is set to 1.

- *HLEN*   Header length. The 4-bit field that contains the number of 32-bit words in the TCP header, which helps the receiving device know how to read the frame.

- *Reserved*   These six bits are reserved and not used at the time of this writing.

- *Code*   A six-bit field also referred to as the *control bits field.* Each bit has its own meaning. The first is the URG/Urgent Pointer Field.

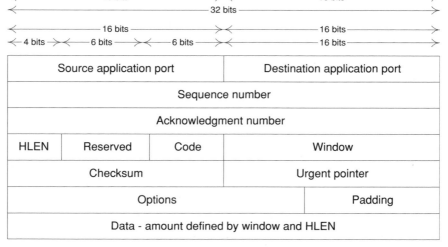

Figure T.8   TCP Header

When it is set to one, then the data in this packet is a priority. The second bit is the ACK bit. This bit being set to one is an indication to the receiving device that the acknowledgment number field is valid. The third is the PSH/Push bit. Its function is to notify the buffer to go ahead and send its data, regardless of whether it is full or not. The fourth is the RST; it is used to terminate the connection (in case a close function will not work). The fifth is the SYN/synchronize bit. When it is set to one, it is an indication to the receiver that the Sequence Number Field contains the initial sequence number to be used for the connection. The sixth is the FIN/Final bit. When it is set to one, it is an indication to the device receiving it, that the sender will not be sending any more data in this TCP connection.

- *Window Size*   Used to communicate the largest amount of data that can be received in the next packet. "Here is your data, now send me this quantity or less data back."

- *Checksum*   This 16-bit field is used to carry the checksum computation result made by the sender.

- *Urgent Pointer*   This 16-bit field contains the position in the window where urgent data ends. Urgent data is to be processed before any other data in transit.

- *Options*   A field used to carry other optional data about the transmission or the way that it is executed.

- *Padding*   Bits are added here to ensure that the header ends on a factor of 32 bits.

TCP/IP (Telecommunications Protocol Internet Protocol)   The standard communications protocol used for file exchange over the Internet (Fig. T.9). TCP/IP was developed by *ARPA/DARPA (Advanced Research Projects Agency/Defense Advanced Research Projects Agency)* of the United States Federal Government.

TCP Segment   The TCP header and its associated (or carried) application data. For a diagram of a TCP segment and its parts, see *IP Datagram.*

TCR (Transaction Confirmation Report)   The report that a fax machine prints. It details all of the faxes that were sent and received for a specified duration of time.

TCS (Transmission Convergence Sublayer)   The ATM physical layer is divided into two parts (sublayers), the TCS and the *PMD (Physical Layer Medium Dependent).* The TCS sublayer determines where cells

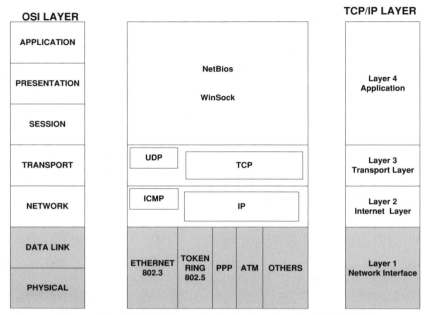

Figure T.9   TCP/IP Protocols in Contrast to the OSI Model

begin and end. It also performs blank-cell insertion functions when no data is being transmitted. See also *PMD*.

TDD (Telecommunications Device for the Deaf)   A reference to devices that attach to telephone handsets and allow deaf people to type messages to each other. Most of the devices look like Wyse terminals with an acoustic coupler attached. (An acoustic coupler is a pad that a telephone handset rests on so that data transmitted over the line can be received through the mouthpiece and earpiece. The transmission rate is a very slow 300 baud.)

TDM (Time-Division Multiplex)   For an explanation and diagram, see *Multiplex*.

TDMA (Time-Division Multiple Access)   Also called *FDMA (Frequency-Division Multiple Access)*. Another name for time-division multiplexing. Only TDMA and FDMA band can multiplex signals together over a transmission frequency. It is commonly used in spread-spectrum cellular radios (transmitters) that the cellular telephone industry uses.

TDR (Time-Domain Reflectometer)   A testing device that measures the distance of a copper twisted pair. The Dynatel 965T is a popular

TDR. A TDR works by transmitting a signal down a copper twisted pair, then it waits for a reflection to come back. When the reflection returns to the device, the time difference is used to calculate the distance that the signal traveled. For a photo of a TDR, see *965T.*

TE (Terminal Equipment)   The equipment at the end (customer side) of an ISDN line. It is classified in two categories: Type 1 (TE1) for equipment that is directly ISDN compatible and Type 2 (TE2) for equipment that requires a converter to convert the ISDN *BRI (Basic Rate Interface)* into two separate phone lines so that analog modems and telephones can be used.

TE 1 (Terminal Equipment Type 1)   Equipment that is directly ISDN compatible or is capable of plugging directly into an ISDN line.

TE 2 (Terminal Equipment Type 2)   Regular telephones and terminal adapters that convert the digital ISDN signal into an analog signal so that normal POTS (Plain Old Telephone Service) equipment can be used.

Technology Without An Interesting Name (TWAIN)   An interface standard for image scanners. Most scanners are sold with a TWAIN driver. These drivers allow a graphics program to automatically control a scanning device through its GUI interface.

Telco   Abbreviation for telephone company.

Telco Connector   A reference to a 50-pair amphenol connector. See also *Amphenol Connector, P-Cable,* and *C-Cable.*

Tele   Greek for "far."

Telecaster   The marketed name for the Cisco Systems, Inc. IP telephones (model CP-79xx). These are distinguished by their charcoal gray color and rounded style handset.

Telecommunications   To exchange information across a distance.

Telemetry   A reference to the remote monitoring of communications equipment, and its environment. Popular protocols for this are T-BOS and TL1.

Telephone   A device that consists of six major parts. A switch-hook, a dialing circuit (DTMF tone or Rotary), a ringer, a microphone, a speaker (for the ear-piece), and a hybrid coil.

Telephone Application Program Interface (TAPI)   A software program standard that allows PC support for a broad range of communications devices, including telephone switch (PBX, ACD, IVR, and central office) equipment.

Telephone Server   See *IP Telephony.*

Telephony Server   Also known as a *FEP (Front-End Processor).* A communications "front end" device that can be loaded with a "firewall" to prevent unwanted users from accessing the communications network. An FEP can also perform routing, and differentiate between different communications protocols, depending on the software that runs on it. For a diagram, see *Front End Processor.*

Teletext   Data that comes over a television transmission in the form of text at the bottom of the screen.

Teletypewriter (TTY)   A device that used to work like a telegraph (its almost as old), except that it would send typewritten messages over phone lines instead of Morse Code. It had a keyboard and a printer.

Television (TV)   The current name for video broadcasting and the receiver used to pick it up. Currently, the standard for terrestrial broadcast television in North America is NTSC, which offers a video resolution of 525×495 lines and FM stereo audio. Different standards, such as PAL and SECAM, are used in other parts of the world. The FCC is now implementing new HDTV standards, which should offer a higher-resolution signal via a digital transmission process. Figure T.10 shows the different blocks of a television receiver.

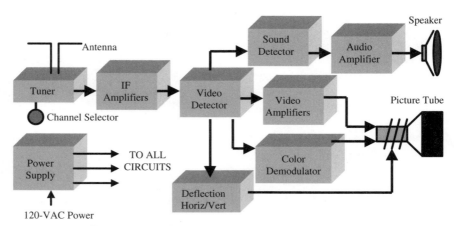

Figure T.10   Television Receiver Basic Block Diagram

Television Broadcast Standards   There are three principal standards in the world for analog television broadcast transmission for video. The first is National Television Standards Committee (NTSC), which is the broadcast standard in Canada, Japan, the United States, and Central America. NTSC defines 525 vertical scan lines per frame and yields 30 frames per second. The scan lines refer to the number of lines from top to bottom on the television screen. The frames per second refer to the number of complete images that are displayed per second. The second standard is PAL (Phase Alternation Line), the broadcast standard in Europe, the Middle East, parts of Africa, and parts of South America. PAL defines 625 vertical scan lines and refreshes the screen 25 times per second. The third is SECAM. It is the broadcast standard in France, Russia, and regions of Africa. SECAM is a variant of PAL but it delivers the same number of vertical scan lines as PAL and uses the same refresh rate.

Telex   An older international switched message service that is being replaced by fax machines and the Internet/public packet-switched network.

Telnet   A connection between an administrator's user workstation and a remote network device, such as a router, switch, hub, host, or server. Telnet enables an administrator or technician to remotely configure or maintain network equipment. Many UNIX-based operating systems have a built-in Telnet software package.

Ten/One Hundred   A reference to the newer family of Ethernet as a whole. 10BaseT is 10 Mbps and 100BaseT is 100 Mbps. It is also referred to as *802, 10/100*. Because the 10BaseT and 100BaseT can interconnect, the network as a whole is frequently called a *ten/one-hundred network*.

Tera   The prefix for Trillion. Five THz means 5,000,000,000,000 hertz.

Terabyte (Tb)   1,000,000,000,000 bytes.

Terminal   A closure where a telephone cable is terminated. It is usually a green box if it terminates buried cable or a silver box if pole mounted. 2. A video I/O device with a keyboard that is used to enter and retrieve information (data) from computers.

Terminal Adapter   A device that converts an *ISDN (Integrated Services Digital Network)* line into regular *POT (Plain Old Telephone) service* so that you can connect a standard telephone or modem to an ISDN line (Fig. T.11).

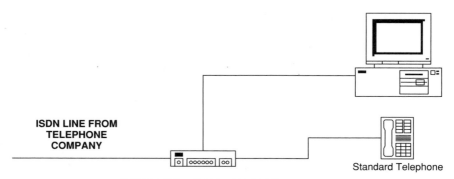

ISDN LINE FROM
TELEPHONE
COMPANY

Standard Telephone

ISDN TERMINAL ADAPTER
Figure T.11   ISDN Terminal Adapter

Terminal Block   A reference to a 66M150, AT&T 110, Crone, or some other block that is designed to terminate (permanently affix) wire.

Terminal Emulation   The use of a PC to act or communicate as a dumb I/O terminal. To use a PC in this application (such as to plug into a microwave link and boost its power), it must be equipped with terminal-emulation software. The microwave-link device has its own microprocessor and only needs a device (usually a VT100 terminal) to communicate with. Terminal emulation software allows your PC to "look like" a VT100 terminal to the microwave radio equipment.

Terminal Equipment   The equipment at the end of a communications circuit or the equipment that is on the "receive" end of the transmission, such as a printer or telephone.

Terminals   1. A workstation, as in an IBM SNA network.   2. A reference used to identify where copper twisted pairs are terminated in access points, cross boxes, and terminals. Physically, a binding post is a pair of teeth on a 66M150 block or a pair of $\frac{9}{16}''$ lugs. Each binding post has a number. When a technician looks for a specific pair in a cable (called a *cable pair*), they refer to documents that list the pairs and binding posts on which they are spliced.   3. A box on a telephone pole or on the ground where the telephone cable is terminated. From the terminal, the service wire is run to the house or building. For a photo of a buried cable terminal, see *Pedestal.*

Terminate   1. To complete a communications path.   2. To fasten wire to some form of connectivity, such as an RJ45 jack, 66M150 block, AT&T 110 block, etc.

Terminator, Ethernet   A 50-ohm BNC "dead end" connector that is used at the end of a coax run in Ethernet *LANs (Local-Area Networks)*.

Terrestrial Data Service   A reference to data signal transport service via terrestrial microwave. Terrestrial Microwave can be used for voice or data transport. See also *Terrestrial Microwave.*

Terrestrial Facilities   Communications facilities that are on the ground, such as transmitters, telephone lines and poles, central offices, etc. Anything but satellites.

Terrestrial Microwave   Microwave radio has become a very economical way to bypass construction costs of broadband private-line services. Many *CAPs (Competitive-Access Providers)* have access to microwave radio resources, such as licensing, equipment, and installation (Fig. T.12). Digital microwave is also called an *eyeball shot, 38 Gig,* or just *radio.* Most of the microwave being installed for private-line service today is in the 33- to 39-GHz frequency range. These microwave units use an FM-FSK over two sidebands for transmitting at full duplex. They are available in T1, DS3, and STS-1 (which is a DS3 formatted for SONET). 38-GHz microwave has a range that depends on the size of the antenna (dish) placed on the outdoor radio unit. The choices in antenna size are one or two feet in diameter. The one-foot antenna has a maximum range of one to three miles, depending on the regional weather conditions (rainfall, snow, and especially fog, drastically attenuate microwave transmissions). The two-foot dish has a range of two to seven miles, also

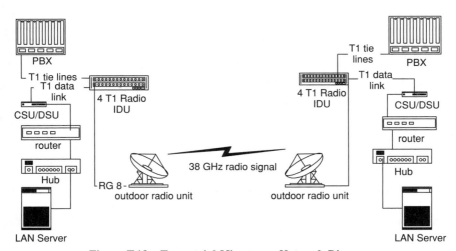

Figure T.12   Terrestrial Microwave Network Diagram

depending on the weather in the region. Below is a microwave application. For a photo of a microwave antenna, see *Microwave.* Figure T.13 illustrates an *IDU (Indoor Unit.* The dish is referred to as the *ODU, Outdoor Unit).*

Figure T.13   A 38-GHz IDU (Indoor Unit) Terrestrial Microwave System

Test Set   Also called a "goat" or "butt set." A test telephone set used by telephone installation and repair personnel. Instead of a plug on the end of the cord, it has a pair of alligator/bed-of-nails clips.

Test Shoe   A testing device aid that connects to distribution frame blocks or other cable-terminating blocks that helps prevent shorts and crosses while technicians work on or test the cable.

Test Tone   Also called an installer's tone. A small box that runs on batteries and is used to put an *RF (Radio Frequency)* tone on a pair of wires. If a telephone technician can't find a pair of wires by color or binding post, they attach a tone to one end, then go to the other end and use an inductive amplifier (also called a *banana* or *probe*) to find the beeping tone.

TFT (Thin-Film Transistor)   A transistor used in PC laptop displays. Each pixel is controlled by an individual transistor.

Thermal Noise Also referred to as *ambient current*. The result of heat that causes random movement of electrons in a circuit when the power is off. The current/movement of electrons causes ambient voltage. Ambient current/voltage is why oscillator circuits start oscillating when the power is turned on. The natural oscillations of the electrons become filtered and amplified when the power is applied to the circuit. As electronic circuits become warmer, thermal noise becomes more prevalent. If it is amplified so that you can hear it, it sounds like white noise or a radio that is not tuned to a station.

THF (Tremendously High Frequency) The American standard name for frequencies within the spectrum of 300 to 3000 GHz. For more information, see *Spectrum, Frequency.*

Thin Client A network-attached workstation that relies on a shared server to handle the bulk of application processing. Depending on the application and in a true "thin" computing environment, workstations do not need, and usually do not have, a hard drive. The thin-client computing strategy is similar to that of the mainframe/dumb terminal environment, yet is more expandable and flexible. See also *Fat Client.*

Thin Ethernet (Thinnet) A popular Ethernet *LAN (Local-Area Network)* that utilizes thin RG-58 coax (Fig. T.14). The less-popular thick Ethernet cable is slightly better for distance, but harder to work with and install. Whether you are using RG-58 thick or thin coax is not so

Ethernet Network
for thicknet and thinnet

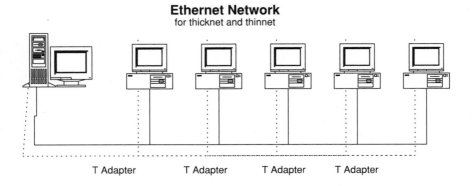

T Adapter   T Adapter   T Adapter   T Adapter

COAX:
Solid = inner
conductor
Dotted = shield

Figure T.14   Thin Ethernet Over Coax

much of an issue because the network interface cards inside the computers don't know the difference.

Third-Generation Network (3G) In wireless communications, a convergence of voice, data, and multimedia services at initial bandwidths of 144 Kbps, with a future bandwidth maturity to 1 Mbps and beyond. A simple identification of wireless communications technology evolvement is defined in generations. The first generation was AMPS, which utilized FDM technology to carry one call on each analog channel. The second generation is CDMA/TDMA/GSM, and placed multiple digital calls within PCS bandwidths as well as provided enhanced services. The third generation, 3G, is intended to unify not only voice, data, and multimedia, but also application formats. The standard application interface for whatever radio is used is IP. In the United States and Japan, among other countries that have deployed both GSM and CDMA for wireless technology, CDMA2000 is the (OSI layer 2) G3 migration path. For countries that use the European TDMA and GSM formats, GPRS (General Packet Radio Service, or W-CDMA) is the migration path. Regardless of the wireless link between end users, the applications that are accessed are done through IP. Therefore, end users are able to exchange application information via open standards. For multinational users, handset manufacturers are planning to produce devices that are compatible with both technologies by incorporating both types of radio technology in them.

Third-Party Call Not a direct call and not a collect call. A third-party telephone call is charged to a telephone number other than the one that is being dialed or used to make the call. Calling cards have been offered by many phone companies to make third-party calls easier. Without a calling card, you need an operator to dial a third-party call. Typical charges for third-party calls are about 30 cents per minute. The expensive rate has brought on competition from prepaid calling-card companies.

Three Command Set A reference to Cisco Systems' method of interacting with a router or switch, particularly the 5000 series line of switches and many router models. The three commands used to manipulate and view the IOS settings are SET, CLEAR, and SHOW.

THz (Terahertz) Tera is the prefix for trillion. For example, 5 THz means 5,000,000,000,000 Hertz.

TIA (Telecommunications Industry Association) An organization that hosts great communications trade/product shows. For more information, see *http://www.tiaonline.com.*

TIE Communications (Telephone Interconnect Equipment Communications)   An older telecommunications equipment manufacturer/ distributor that had their name stamped on *PBX (Private Branch Exchange)* and key telephone systems, like the BK2464, Delphi 6/16, and the *VDS (Visual Display System)*. Tie Communications' old products can still be purchased in catalogs and from telephone equipment companies, but it is now Nitsuko.

Tie Line   A tie trunk that is dedicated to one phone. Tie trunks are telephone lines that connect PBX systems together.

Tie Trunk   A telephone line that connects two *PBX (Private Branch Exchange)* telephone systems together so that calls can be transferred between them.

Tight Jacket Buffer   The alternative to loose tube buffer in fiber-optic cable manufacturing. Tight jacket buffer cable has insulation around the cable that is tight against the fiber optic. Loose tube buffer allows the fiber to move freely within the cable.

Time-Division Multiple Access (TDMA)   See *TDMA*.

Time-Division Multiplex (TDM)   The process of encoding two or more digital signals or channels onto one through timesharing the media (wire, air, fiber, etc.). The reason that we multiplex channels together in communications is because it saves money. When we use all of the wires in a cable and need more, it costs less to add electronics on the ends of a cable than to install a new one (imagine the expense from LA to NY). A T1 encodes 24 channels into one by using frequency-division

Individual 4 channels to multiplexed. The dotted vertical lines are timing segments. This multiplexed output signal is four times faster than a single input signal.

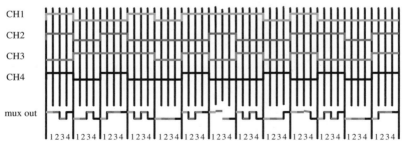

The 4 channel multiplexed output.

Figure T.15   Time-Division Multiplex

multiplexing. In a simpler explanation, a T1 makes it possible to place 24 lines that once needed 24 lines on only two pairs. When a group of signals are multiplexed together, they are all sampled at a high rate of speed, faster than the combined speed of all the channels being multiplexed. Figure T.15 illustrates the concept of *Frequency-Division Multiplexing (FDM)*.

Time-Domain Reflectometer (TDR)   A testing device that measures the distance of a copper twisted pair. The Dynatel 965T is a popular TDR. The TDR works by transmitting a signal down a copper twisted pair, then it waits for a reflection to come back. When the reflection returns to the device, the time difference is used to calculate the distance that the signal traveled. For a photo of a TDR, see *965T.*

Time-of-Day Routing   A call-forwarding feature offered by telecommunications companies and featured in large PBX switches that call forwards incoming traffic to different target parties, depending on the time of day.

Time Out   1. Refers to a "system time-out" test result on an *MLT (Mechanized Loop Test)* from a telephone company central office. When the MLT system is busy, it holds the test request in queue for several minutes until the testing system is available. If it does not become available, then the MLT system returns a "system time-out" result to the user/tester, meaning that the line was not tested.   2. A condition of a central-office line or trunk that occurs when a telephone is left off hook or the pair (facility), it is on is shorted. The central office sends a message that says "please hang-up and try your call again" followed by some harsh, loud beeping tones (also called an *off-hook indicator/warning tone*). If the line is not returned to an "on-hook" state, the central-office equipment will automatically time-out the line. When the line is timed-out, the −52-V battery is disconnected to save power. The central-office equipment will restore the −52-V battery voltage when the ringer is placed back on hook or the short is repaired.

Time Sharing   A reference to the shared use of a computer's processing power by many applications or many users on a *LAN (Local-Area Network)*.

Time Slot   A channel on a time-division multiplexed circuit (a T1, for example) is also referred to as a *time slot.* For more details, see *Multiplex.*

Time To Live   A field in an IP header that indicates how long a packet is considered valid. When the time to live expires, the packet is discarded.

Tinsel Wire   The stranded copper wire that many base cords and handset cords are made with. It is used in these applications because of its ability to withstand lots of bending without breaking.

Tip   1. The positive side of a two-wire telephone circuit, which is supposed to be terminated to black (top position). The other side of a two-wire telephone circuit (or the other wire) is called *ring* and it is designated red (bottom position) when terminated (connected to something). 2. The "tip" of a phono-type plug, which is used for signal patching and for consumer headphones. The other part of a phono plug is the "ring," which is the lower part of the phono plug.

Tip and Ring   The more electrically positive side of a *POTS (Plain Old Telephone Service)* telephone line (0 V) is tip. It is designated internationally as black, but in the U.S., it is often designated green (Fig. T.16). Its counterpart is ring (the more negative side, 52 V), which is designated red internationally and in the U.S. When tip and ring are terminated on a connecting block, tip usually goes on top (left), and ring usually goes on the right (bottom).

Figure T.16   Tip and Ring Positioning on Blocks and Terminals.
Tip is on Top (Left) and Ring is on Bottom (Right)

T-Load (Technology Load)   A reference to a test version of software or a version of software that has been modified from the original version with patches to alleviate technical problems or "bugs."

TN   In Nortel PBX administration, a Terminal Number. The terminals are hardware interfaces for telephones, and there are 16 per station (or telephone) interface card. Telephone pairs are connected to TNs. Each TN identifies a telephone set, and each button within that telephone set can have an associated feature function, or DN (extension number). See also *DN*.

TNC (Threaded Naval Connector)   A threaded version of the "twist on" BNC connector, which makes the two ends attach in a nut and bolt fashion. See also *BNC Connector.*

Token   The bit message in a token-ring network that is passed from one node (computer or other device) to the next that grants the right to transmit to the other nodes on the ring architecture.

Token Bus   The physical ring of wire or fiber optic that interconnects computers and other devices (printers) on a token-ring physical *LAN (Local Area Network)* topology. The "physical topology" is the way that the LAN is wired (ring or star) and the "logical topology" is the way that the devices communicate (the protocol).

Token Passing   A reference to a token-ring logical topology. The token-ring topology is a type of *LAN (Local-Area Network)*. What makes a token-ring network unique from other topologies is that each device on the network takes turns using the ring (the wiring, or fiber) in an organized, systematic way. The organization is achieved by having a "virtual token" that the devices hand off to each other. While a device possesses the token, it is allowed to transmit over the network. The other popular type of network is called *Ethernet,* which includes 10-BaseT, Thicknet, Thinnet, and 100-Base-T (the newer). In Ethernet protocols, there is no "token" and no taking turns. Devices simply listen to the network; if it is clear, they transmit. The problem with this is that many devices attempt to communicate at close to the same time. The transmitted data collides on the network, becomes corrupted, and must be retransmitted. Ethernet is referred to as a *contention-based protocol* because the devices (computers) are always contending for the use of the network.

Token Ring   A physical and logical *LAN (Local-Area Network)* topology. The physical topology is the way that the computers and other devices are wired together. For a token-ring physical topology, the devices are wired in a ring. The logical topology is the protocol, the rules for communication. In a ring logical topology, each device listens to everything transmitted on the network and only processes what is intended for it to receive (devices know what is meant for them by reading the address information). The devices connected to the network only transmit when they have possession of the *token.* The token is a message that each of the computers pass to each other that allows them to transmit data. This organized method of network utilization prevents more than one device from communicating at one time (a collision). If two devices try to communicate at the same time, the data that they both send becomes corrupted. Token ring networkers are free of this problem. We can't talk to two people at the same time because the words we hear become mixed up and we become confused. The same goes for computers on Local Area Networks. For a diagram, see *Ring.* For more information, see *Token Passing.*

Token Ring LAN Switch   A hardware device that provides 4 or 16 Mbps of communications bandwidth to each user. Each user's host (computer) is connected directly to the switch, rather than in the traditional daisy chain (nonswitched) topology. Although token ring does not have the bandwidth of 100BaseT Ethernet (Ethernet has a 1518-byte maximum frame size), its 4500-byte frame size makes up for the difference in carefully engineered networks. Token ring is not recommended for IP telephony networks because of its low bandwidth and large packet size.

Toll Restriction   A *PBX (Private Branch Exchange)* and key telephone system feature that enables an administrator to make a telephone extension unable to make long-distance (toll) telephone calls. Toll restriction can be made to restrict the dialing of a specific area code, a specific group of area codes, or all area codes.

Tone Generator   The part of a *PBX (Private Branch Exchange)* system that creates dial tone. The tone generator in a telephone system is usually an individual circuit card, with its own slot in the system cabinet.

Tone Probe   Also called a *banana*. Telephone line installers use test tones (also called *installer's tones*) to place an *RF (Radio Frequency)* signal on a pair of wires so that they can locate that pair on the other end of a feed. For a picture, see *Installer's Tone*.

Topology   The two types of topology are physical and logical. The logical topology defines the way that a LAN communicates. The physical topology defines the way that a LAN is physically wired. For example, even though an Ethernet network might be physically wired into the formation of a star, it really works as though it were a bus. The wire is just physically laid out and connected differently, and the electronics are a little different.

Touch Pad/Track Pad   An alternative to the use of a mouse on laptop computers. The track pad is usually located directly below the keyboard. Other mouse alternatives for laptop computers include the pointing stick and roller ball.

TouchTone (DTMF, Dual-Tone Multiple Frequency)   The tones that you hear when you dial a single-line push-button phone. The tones are a mixture of two frequencies. For a diagram of DTMF frequencies, see *Dual-Tone Multiple Frequency*.

Trace   Also called a *trap*. When the telephone company tracks the calls made to a customer's telephone number to catch a malicious caller. The standard for malicious call trace is to dial *57 after receiving a malicious call. The telephone company will charge anywhere from $1 to $4 to investigate the call. By using the service, you agree to press charges. The only disturbing thing is that you never find out who made the calls until the court hearing.

Trac Splice Closure   An aerial splice closure designed to resist condensation (Fig. T.17). Trac Splice Closure is a registered trademark of RayChem.

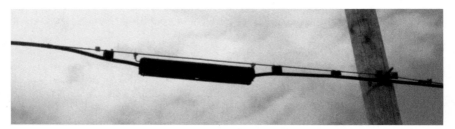

Figure T.17   Trac Splice Closure

Traffic   A measure of the amount of call attempts and active calls on a telephone switch. Traffic is measured in centum call seconds (*CCS,* one phone call for one second) or Erlangs. Many larger *PBX (Private Branch Exchange)* telephone systems and central-office switches now have *CTI (Computer Telephony Integration)* applications that will calculate traffic, CPU % utilization, busy hours, and other useful information.

Traffic Engineering   The process of figuring out how much equipment and what equipment will be needed and how to allocate the resources of that equipment to prevent call blocking, or keep call blocking to a minimum. Telephone switches and *PBX (Private Branch Exchange)* systems are engineered according to the "busy hour" of the network, which is the time when the network has the most traffic. The busy hour can be for the day, month, or year. Traffic is measured in *centum call seconds (CCS,* one phone call for one second), or Erlangs. Many larger *PBX (Private Branch Exchange)* telephone systems and central-office switches now have *CTI (Computer Telephony Integration)* applications that will calculate traffic, CPU % utilization, busy hours, and other useful information.

Transceiver   A transmitter and receiver built into the same device (Fig. T.18). A CB radio is a type of transceiver.

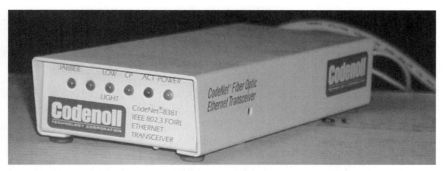

Figure T.18   Transceiver: Ethernet Category 5 UTP to Single-Mode Fiber Optic

Transcoder   A device capable of converting Codec formats such as G.711 to G.729, and vice versa. Transcoders are silicon devices that reside within IP telephony–enabled gateways/routers. In the older methods of voice communication, all calls required the same bandwidth, which was 64 Kbps. In the server-based PBX communications environment, there are many variations of devices that may be utilizing many variations of Codec compression. For example, a user that makes a telephone call from an IP telephony system that incorporates G.711 (24 Kbps) encoding to another user on an IP telephony system that incorporates G.729 (8 Kbps) encoding would need to pass through a gateway that has transcoding ability built in. This is usually the G.729-based device, which transcodes to the older Codec method of G.711. Gateways/routers that have transcoding abilities are referred to as being "voice-enabled" or having DSP (Digital Signal Processing) ability.

Transducer   An electronic component that converts one form of energy to electrical energy or vice versa. Some examples of transducers are:

- Photocell   Light to electricity
- Speaker   Electricity to sound
- Microphone   Sound to electricity
- Light bulb   Electricity to light
- Coil winding/motor   Electricity to motion
- Coil winding/generator   Motion to electricity

Transfer   A feature of *PBX (Private Branch Exchange)* telephone systems and key telephone systems. The transfer feature allows users to send a phone call to another extension by pressing a "transfer" button while on the call, entering the extension they want to transfer to, then pressing the transfer key again (this is true for Northern Telecom Systems).

Transfer Rate   How fast a data transmission can send data in *Bits Per Second (bps, b/s)*, e.g., T1 has a transfer rate of 1.544 Mb/s including payload and overhead.

Transformer   An electronic or electrical component used to "step-up" or "step-down" AC voltage. A transformer wired in a fashion to receive an AC voltage on its primary winding and increase that voltage through electromagnetic induction into the secondary winding. Increasing voltage through a transformer does the opposite for current. Reversing the way that a transformer is wired, changes it from a step-up to a step-down transformer and vice versa (Fig. T.19). The amount the voltage is stepped up or down is equal to the ratio of wire windings of the primary and secondary. Figure T.20 shows a step-down transformer that has a 10:1 winding ratio.

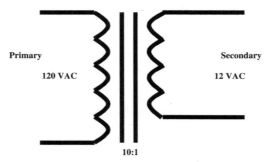

Primary

120 VAC

Secondary

12 VAC

10:1

**Figure T.19   The Schematic Symbol for an Iron-Core Transformer
(Iron Core is Designated by Two Lines Between the Cores)**

Figure T.20   An Iron-Core Transformer for Power Supply Applications

Transistor   A device that is used as a switch (for logic) or amplifier (signal). The great thing about transistors is that they use far less power than vacuum tubes, they are much smaller, they are much faster, and they cost less (Figs. T.21 and T.22). The only disadvantage of transistors to tubes is that transistors cannot amplify odd-order harmonics (natural sounds/music) and they are much more susceptible to cosmic radiation (outer-space/nuclear-fallout applications).

Bipolar NPN   DE MOSFET (CMOS) N-Channel

Figure T.21   Transistor Schematic Symbols

Figure T.22   Transistor Packages

Transit Delay   The time required for a transmission to get from one point on a network to another.

Transit Time Delay   The amount of time (in milliseconds) required for a data packet to cross a network. Transit time delay is often rated based on a data packet making a round trip. A typical round-trip transit time delay across a very robust network that stretches from Tokyo to New York is 200 ms. Transit time delay is increased by the number of routing or switching devices that a packet passes through. A concern for network designers is to have enough route redundancy to ensure that alternate routes can be taken by packets when connections fail. The disadvantage to having redundancy is that more switching and routing equipment is required, which increases transit time delay.

Transitional Voltage   On an RS-232 connection, or V.24, V.28, EIA 232-D as it is also called, the voltage on transmit and receive when it is somewhere between −3 V and +3 V (Fig. T.23). In the RS-232-C specification, any voltage between +3 and +24 volts is considered high (or On) and any voltage between −3 and −24 volts is considered low (or Off).

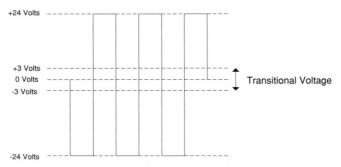

Figure T.23   Transitional Voltage

Translating Bridge   Also called a *protocol converter* or *router* (Fig. T.24). A device that is usually in the form of a PC that is loaded with

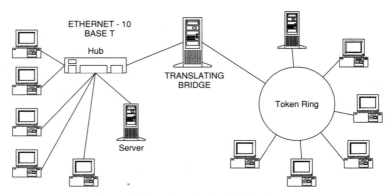

Figure T.24   Translating Bridge/Router

software that converts protocols, such as Ethernet and token ring. The PC has a network card installed for each of the different network connections.

Translational Bridging   The transfer of data frames between two networks that have dissimilar *MAC (Media-Access Control)* sublayer protocols, such as TCP/IP and Novell IPX. The MAC address information is converted into the format of the destination network at the bridge. Another bridging method is encapsulation bridging, which does not remove or modify the original data packet; it only adds additional header information. Both bridging methods are accomplished by Cisco and Nortel router/routing equipment.

Translator   A device that receives a transmission, converts it, and retransmits it on a different format or frequency. Translators are used to rebroadcast radio programming. Translators in data networks are called *translating bridges.*

Translations   The part of switching or switch set-up that converts signaling messages from other switches. Translations are very important and are probably the most crucial part of the switch set up. When a telephone call is handed from one switch to another or from one central office to another, digits (addressing information) are sent with it. Each switch has its own definitions for number set up and configuration. The switch converts these numbers for every call processed and the process is called *translating.*

Transmission   The sending of a signal, analog, digital, or light-wave, across a media.

Transmission Convergence Sublayer (TCS)   The ATM physical layer is divided into two parts (sublayers), the TCS and the *PMD (Physical Layer Medium Dependent).* The TCS sublayer determines where cells begin and end. It also performs blank-cell insertion functions when no data is being transmitted. See also *PMD.*

Transmitter, Radio   A radio transmitter emits an electromagnetic field. The electromagnetic radiation is simply an antenna being made to change its magnetic field at the rate of a carrier frequency. The magnetic field then traverses through the farthest reach of the magnetic field, which could be many miles. The distance that the signal will reach is determined by the transmitting power, weather, sunspot cycle, time of day, ionospheric conditions, and surrounding terrain. What makes the radio signal carry a voice or music is called *modulation.*

Transparent Bridge   A type of network bridge that learns which systems are on each network by listening to traffic and building its own reference tables.

Transparent Bridging   1. A routing method used in Ethernet and IEEE 802.3 networks in which routers pass frames along one hop at a time, based on tables that associate end nodes with router ports. Transparent bridging is so named because routers were once called *bridges,* and the presence of routers is transparent to network end nodes. See also *Source Route Bridging.*   2. In Ethernet LAN networks, a bridge/router operating protocol that builds a table of MAC addresses that contains all connected network devices in its memory. A "transparent" bridge builds its reference table by receiving data frames from its network connections and keeping track of the MAC addresses, sending and receiving on each connection. Transparent bridging operates on the logic that if a message or "frame" received on port 1 contains a sender MAC address of 0555.3d05.1111, then the device with MAC address 0555.3d05.1111 is connected to the network in the direction of port 1. Therefore, any frame received with this address as the destination is transmitted on that port connection *and no other.* Since Bridges use MAC addresses to forward frames, they are referred to as a *layer 2 device,* in conjunction with the OSI layer 2. Bridges that operate with transparent bridge rules are called *transparent bridges* or *learning bridges.*

Two important details of transparent bridging are that parallel (redundant) paths are not supported unless Spanning Tree Protocol is used in conjunction, and frames containing Router Information Fields (RIFs) are not forwarded (RIFs are for token ring environments).

In modern networks, routers are implemented to perform bridge functions (and frequently called *bridges* even though they are routers) with the appropriate operating protocol enabled (i.e., transparent bridging) to suit the networking task at hand. Bridges are used to isolate traffic between common users on a shared bandwidth network, thus making communication between devices more efficient. In large-scale networks, LAN switches are implemented, with bridges/routers connecting or "bridging" the switches. See also *Source Route Bridging, Source Route Transparent Bridging, Spanning Tree,* and *Source Route Switching.*

Transparent Network   A network connection that is made through more than one network, but appears to be only one network to the end user. Many *Local-Area Networks (LANs)* within companies are connected to the Internet so that users can access the Internet or Internet e-mail by clicking on a desktop icon. This is a form of transparent networking. A nontransparent version of this would be to open your communications application, dial out on your modem, enter your e-mail address, then read your e-mail.

Transparent Routing   To send data across more than one network without any additional coaxing from the user. The network does all of the protocol conversion and gives the illusion that there is only one network.

Transponder   An electronic device that receives a radio signal, then emits a response to the signal that it has received.

Transport Layer   A layer in a communications protocol model. In general, the transport layer performs the function of error correction and the direction of data flow (transmit/receive). The latest guideline for communications protocols is the *OSI (Open Systems Interconnect)*. It is the best model so far because all of the layers (functions) work independently of each other. For a diagram of the OSI, DNA, and SNA function layers, see *Open Systems Interconnection*. For a functional/conceptual diagram of the OSI layers, see *OSI Standards*.

Transport Medium   A reference to the physical component of what a transmission signal is being sent over; it can be fiber optic, coax, copper twisted pair, or air.

Transport Protocol   A reference to how data should be presented to a device or layer in a transmission process. In general telecommunications transport, protocols include SONET, T1, STS-1, and OC-N. Transport protocols can become very complicated, but if you understand the four listed, you are in good shape.

Trap   1. A diagnostic software tool used to monitor network events and report them to a system user/administrator. Traps can be used to monitor T1 lines for errors or they can be used to identify the calling parties on a subscriber's telephone line. Each manufacturer of network equipment produces its own proprietary trap software.   2. See *Trace*.

Trellis Coding   An error detection method used in high-speed modems.

Tremendously High Frequency (THF)   The American standard name for frequencies within the spectrum of 300 to 3000 GHz.

Tributary Circuit   A lower-level multiplexed channel that exists within a larger multiplexed channel. For example, a T1 channel riding inside or on an OC-3 SONET (an OC-3 can carry 84 T1 circuits) carrier is considered to be a tributary. This term is frequently used within the realm of SONET networking.

Triode   An electronic vacuum tube that works in similar applications as a transistor (Fig. T.25). Like a diode vacuum tube, it has a heater filament, anode, and cathode; the triode has a third added element, called a *grid*.

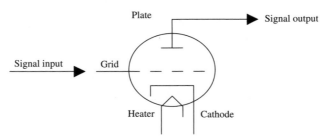

Figure T.25   Vacuum-Tube Triode

Triple DES (3DES)   Data encryption standard. A 168-bit encryption method that incorporates an algorithm developed by the United States National Bureau of Standards.

TRL (Transistor/Resistor Logic)   Logic circuits that consist of bipolar transistors, rather than *CMOS (Complimentary Metal-Oxide Semiconductor)* transistors. TRL is used in environments with lots of static electricity.

Tromboning   A method of regenerating trunk signaling, such as DNIS digits, by routing calls out of a switch on a trunk group (which provides the desired signal) and then directly back in on a different trunk group.

Trunk   There are various types of trunks. A trunk in all cases is a *link between switches* that carries traffic between various switch end ports. This is true in packet switches and circuit switches. Trunk types for circuit switches (voice) include Loop Start, Ground Start, ISDN, and E&M. For more information about LAN trunking between switches, see *STP* and *EtherChannel.*

Trunk Group   A group of telephone lines that connect a *PBX (Private Branch Exchange)* or key telephone system to the phone company, and are used for a specific application, such as incoming customer-service lines, sales lines, or information lines. A specific group of trunks can also be configured for outgoing calls only.

Truth Table   A diagram used to portray a logic statement (Fig. T.26). Logic is a mathematical process first developed by the Irish mathematician George Boole in the 1850s. The premises of logic is to tell if a

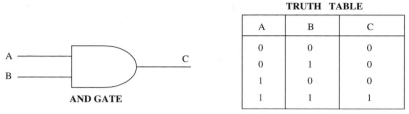

TRUTH TABLE

A	B	C
0	0	0
0	1	0
1	0	0
1	1	1

AND GATE

Figure T.26   Truth Table: For an AND Gate, If A and B Conditions Exist, then the Resultant Output is C

certain statement is true or false. An example is "the light is ON." This statement can only be true or false. Logic couples this statement with others like "the switch is ON, the power is ON; therefore, the light must be ON." If the switch is OFF and the power is ON, then the light is OFF. If the switch is ON and the power is OFF, then the light is OFF. These statements depict the truth table for an AND electronic logic gate, which is a primary building block of microprocessors. In the table depicted, the light switch would be A, the power would be B, and the light bulb would be C. Truth tables are written in ones and zeros, rather than ONs and OFFs. The science of this math is called *Boolean Algebra*. It is a book in itself and is usually covered well in textbooks that cover digital electronics.

TTL (Transistor-Transistor Logic)   1. Digital devices that consist of many bipolar transistors (the original transistor as we know it). TTL is not as static sensitive as its newer and more commonly used competitor, *CMOS (Complementary Metal-Oxide Semiconductor)* transistor, but it uses much more power.   2. *Time To Live.* A field in an IP header that indicates how long a packet is considered valid. When the time to live expires, the packet is discarded.

TTR (Touch-Tone Receiver)   An electronic device that converts *DTMF (Dual Tone Multi-Frequency)* tones dialed by telephones and other communications equipment to digits.

TTY (Teletypewriter)   See *Teletypewriter.*

Tube (Vacuum Tube)   An active electronic device that is literally a tube, with no air or other gasses in it. Tubes were the predecessors to transistors and other "solid-state" devices. Tubes have four main parts: the plate, the cathode, the filament (or heater), and the grid. The tube works when electricity flows through the tube, from the cathode to the plate. The electricity flows when the filament is heated, which is actually the source of the electron flow. The signal to be amplified is fed to the grid, which manipulates the amount of current flowing through the tube. The resultant is an amplified output on the plate. Tubes are still used in many applications. Today's vacuum tube applications include high-power applications (such as radio transmitters), high-end consumer audio equipment (where it is necessary to have odd-order harmonics accurately reproduced, transistors do not amplify odd-order harmonics, tubes do), and high radiation-risk applications (such as outer space because transistors are sensitive to cosmic radiation). For a schematic symbol of a vacuum tube, see *Triode.*

Tunneling   To encapsulate data from one application or protocol into another. Tunneling is used to transport multiple protocols over a single connection, and it is also used as a measure of network security.

Turnkey   A method of network installation. Also referred to as *EF&I (Engineer, Furnish, and Install).* Many companies have equipment or entire networks built in this manner, so that when it is ready, all they need to do is use it.

TV (Television)   See *Television.*

TWAIN (Technology Without An Interesting Name)   An interface standard for image scanners. Most scanners are sold with a TWAIN driver. These drivers allow a graphics program to automatically control a scanning device through its GUI interface.

Twisted Pair   Communications wiring consists of 19 to 26-AWG solid and insulated wires. Twisted-pair wire consists of "pairs" of color-coded wires. Common sizes of twisted-pair wire are 2 pair, 3 pair, 4 pair, 25 pair, 50 pair, and 100 pair. Twisted-pair wire is commonly used for telephone and computer networks. It comes in ratings of CAT 3 (for voice), CAT 4 (voice and 10-base-T), and CAT 5 (for 100-base-T and token ring).

Two Binary One Quarternary (2B1Q)   A type of *Pulse Amplitude Modulation (PAM),* where two bits presented at different possible voltage levels represent four bits at one voltage level. This line code is a mainstay for ISDN, and is also used in some ADSL and IDSL implementations.

Two-Wire Circuit   A circuit that utilizes two wires to work. A plain telephone line, such as the one in your house, is a two-wire circuit, and the two wires are called a *subscriber loop.*

Type-0 Supervisory Frame   An X.25 S-frame that has the control bits set to Receive Ready RR. The actual format is 00, which is zero in binary. For more details on X.25 frame-level control, see *X.25 S Frame.*

Type-1 Cable   3 IBM two-pair 22 AWG shielded data cable designed for switched token ring. Characteristic impedance is 150 ohms. It is usually terminated to genderless IBM data connectors.

Type-1 Operation   IEEE 802.2 (LLC) connectionless operation.

Type-1 Supervisory Frame   An X.25 S frame that has the control bits set to Reject REJ. The actual bit format is 01, which is 1 in binary. For more details on X.25 frame-level control, see *X.25 S Frame.*

Type-2 Cable   3 IBM two-pair 22 AWG shielded data cable designed for switched token ring. Characteristic impedance is 150 ohms. It is usually terminated to genderless IBM data connectors.

Type-2 Operation   An IEEE 802.2 (LLC) connection-oriented operation.

Type-2 Supervisory Frame   An X.25 S frame that has the control bits set to Receive Not Ready RNR. The actual bit format is 11, which is two in binary. For more details on X.25 frame-level control, see *X.25 S Frame.*

Type-3 Cable   3 IBM 24-AWG nonshielded four-pair data cable designed for switched token ring. Characteristic impedance is 150 ohms. It is usually terminated to genderless IBM data connectors.

Type-4 Cable   Presently, no type-4 cable exists.

Type-4 Node   A communications controller in an *SNA (IBM System Network Architecture)* network. See also *Subarea Network.*

Type-5 Cable   3 IBM 2 optical fiber-data cable designed for switched token ring. Characteristic impedance is 850 ohm. It is usually terminated to genderless IBM data connectors.

Type-5 Node   A host in an *SNA (IBM System Network Architecture)* network. See also *Subarea Network.*

Type-6 Cable   3 IBM two-pair 26-AWG shielded data cable designed for switched token ring. Characteristic impedance is 105 ohms. It is usually terminated to genderless IBM data connectors.

Type-7 Cable   Presently, no type-7 cable exists.

Type-8 Cable   3 IBM 2-pair flat 23-AWG shielded data cable designed for switched token ring. Characteristic impedance is 150 ohms. It is usually terminated to genderless IBM data connectors.

Type-9 Cable   3 IBM 2-pair 26-AWG shielded data cable designed for switched token ring. Characteristic impedance is 150 ohms. It is usually terminated to genderless IBM data connectors.

U Interface A two-wire (one pair) *ISDN (Integrated Services Digital Network) BRI (Basic Rate Interface)* telephone line. This popular type of digital telephone line has two B (bearer channels) and one D channel. For more information, see *ISDN*.

U Frame (Unnumbered Frame) One of three SDLC frame formats. See also *I Frame* and *S Frame*.

U-mm Band The band of frequencies designated by the IEEE between 300 GHz and 3000 GHz. For a table, see *IEEE Radar Band Designation*.

U Plane (User Plane) One of the three entities of frame-relay network management. The three planes are: *The User Plane* (the U Plane defines the transfer of information), the *Management Plane* (the M Plane defines the LMI, Local Management Interface), and the *Control Plane* (the C Plane is delegated for signaling and switched virtual circuits). The U Plane is based on the ANSI TI.602 (LAPD) core aspect standard. It provides the interface between the end user and the network, bidirectional transfer of frames, frame-order preservation, congestion avoidance, and determines the priority of *PVCs (Permanent Virtual Circuits)*.

UA (Unnumbered Acknowledgement) A command defined by the last three bits in the control byte of an unnumbered or "control" frame in the X.25 protocol being 110. The UA control frame is a response to a *Set*

Asynchronous Balanced Mode (SABM) frame, which resets the entire link. See *U Frame*.

UART (Universal Asynchronous Receiver/Transmitter)   An integrated circuit, attached to the parallel bus of a computer, used for serial communications. The UART translates between serial and parallel signals, provides transmission clocking, and buffers data sent to or from the computer.

UBR (Unspecified Bit Rate)   A *Quality of Service (QOS)* defined by the ATM Forum for ATM Networks that allows any amount of data up to a specified maximum to be sent across the network. A connection would be rightfully commissioned as an UBR connection if it carried nontime sensitive data that could be retransmitted without a large inconvenience. The UBR quality of service does not guarantee freedom from cell loss or delay. Other QOSs defined by the ATM Forum for ATM connections include *CBR (Constant Bit Rate), ABR (Available Bit Rate),* and *VBR (Variable Bit Rate)*.

UCD (Uniform Call Distributor)   A less-expensive and less-smart version of an ACD (Automatic Call Distributor). The UCD receives incoming calls and equally distributes them among agents in a call center. For more details, see *ACD*.

UDC (Universal Digital Carrier)   Also referred to as *DLC (Digital Loop Carrier)*. A method of placing two analog telephone lines onto one digital copper pair (Fig. U.1). In areas where a telephone customer has requested

Figure U.1 UDC (Universal Digital Carrier) Central Office Interface

an additional line and there are no additional copper pairs available, many telephone companies digitize the existing line to carry two. The central office routes the two telephone lines to a UDC trunk, which is then connected to the copper pair that carries the digital transmission to the customer's premises. The telephone company installs a UDC adapter at the demarcation point, which converts the digital UDC transmission into two analog lines. The UDC carrier is *DSL (Digital Subscriber Loop)* based. UDC is a *2B1Q (Two Binary One Quarternary)* line format that rides on a −137-volt battery. If the UDC carrier fails, it will go into an override/bypass mode and switch the pair to a −52-volt battery to provide the original POTs service. The telephone number that will continue to work is line one, and line two can be set to call forward to line one. For a diagram, see *DLC*.

UDC Remote Unit   The electronic equipment placed at the customer end of the UDC pair that converts the digital transmission to analog telephone service (Fig. U.2). See *UDC* for operational details.

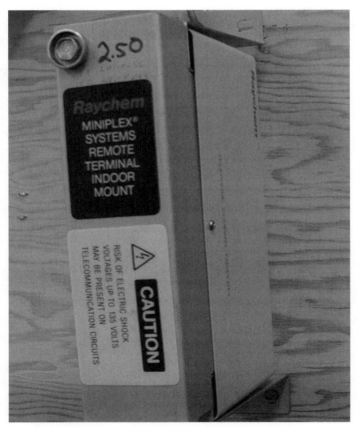

Figure U.2   UDC (Universal Digital Carrier) Remote Unit

UDP (User Datagram Protocol)   A connectionless transport-layer protocol that helps make up the TCP/IP protocol stack. UDP is a simple protocol that exchanges datagrams/packets without acknowledgments or guaranteed delivery, requiring that error processing and retransmission be handled by other protocols. UDP is defined in RFC 768. An *IP (Internet Protocol)* transport-layer protocol that is sometimes used in place of TCP when transaction-based application programs are communicating (i.e., *SNMP, Simple Network Management Protocol*).

UHF (Ultra-High Frequency)   The part of the radio-frequency spectrum that ranges from 300 MHz to 3 GHz. It is used for broadcast TV and other radio communications.

UHF/VHF Splitter   Used to make a junction point or split a signal so that it will travel down multiple paths over coax. It is also called an *RF splitter* or *splitter*. For a photo, see *RF splitter*.

UL (Underwriters Laboratories)   A private company that certifies manufacturer's products for safety.

Ultra High Frequency (UHF)   The part of the radio-frequency spectrum that ranges from 300 MHz to 3 GHz. It is used for broadcast TV and other radio communications.

Ultra Large-Scale Integration (ULSI)   A classification of microchips that have more than 100,000 devices incorporated into their circuitry.

Under-Floor Raceways   Wire ducts that are designed to be used under raised-floor systems, like those in computer rooms.

UNI (User Network Interface)   1. An ATM Forum specification that defines an interoperability standard for the interface between ATM-based products (a router or an ATM switch) located in a private network and the ATM switches located within the public carrier networks. An ATM physical-layer interface (DS3, OC-N, STS-N, or STM-N) that provides physical-layer services to an NNI. It is also used to describe similar connections in frame-relay networks. See also *NNI, Q.920/Q.921,* and *SNI (Subscriber Network Interface).*   2. *UNI (User Network Interface).* Also called a *cable voice port* or *voice port*. In cable-TV networks, a device located at the telephone subscriber's premises that modulates and demodulates the DS0 voice upstream and downstream channels. The modulated DS0 voice signal is sent by the *HDT (Host Digital Terminal)* located at the cable-TV head end. The voice port provides connection to the customer premises' telephone wiring. Channel control between the head end and the customer site voice port can be maintained remotely, from the head end or other office. Because the voice service in cable telephony is provided via radio channels, the channels can be switched when a subscriber is experiencing static or radio interference on line. This would be equivalent to a pair change in a twisted-pair-based telephone network. A pair change cannot be performed by remote control and requires a technician on site. For a photo, see *Voice Port*.

Unicast   In Ethernet-switched networks, a file/packet transfer between two entities. A unicast can be initiated by a server to a workstation, a workstation to a server, workstation to network printer, or any other single entity to single entity. Unicast transmissions are seen by all other users connected to a subnet in a hub environment or users on the same VLAN, but only at the layer 2 (datalink/MAC layer). Packets with addresses not pertinent to the receiving device are discarded by the NIC card. In contrast, packets containing multicast and broadcast addresses are passed by the NIC through layer 3 up to layer 7.

Uninterruptable Power Supply (UPS)   A battery back-up system. When the power goes out, the UPS converts the DC battery power to AC power to run the system.

Unity Gain   1. Gain (another word for amplification) in an electronic circuit that is equal to 1. In oscillator circuits, and some other radio receiver circuits that have a feed back (generate their own input), the amplification between the output and the input must be one. If the gain is not one, the signal will either diminish to nothing or be amplified beyond the saturation point of the active component (e.g., transistor) in the circuit.   2. In broadband networks, the balance between signal loss and signal gain through amplifiers.

Universal Asynchronous Receiver/Transmitter (UART)   An integrated circuit, attached to the parallel bus of a computer, used for serial communications. The UART translates between serial and parallel signals, provides transmission clocking, and buffers data sent to or from the computer.

Universal Call Distributor (UCD)   A less expensive and less smart version of an *ACD (Automatic Call Distributor)*. The UCD receives incoming calls and equally distributes them among agents in a call center. For more details, see *ACD*.

Universal Digital Carrier (UDC)   Also referred to as *DLC (Digital Loop Carrier)*. A method of placing two analog telephone lines onto one digital copper pair. In areas where a telephone customer has requested an additional line and there are no additional copper pairs available, many telephone companies digitize the existing line to carry two. The central office routes the two telephone lines to a UDC trunk, which is then connected to the copper pair that carries the digital transmission to the customer's premises. The telephone company installs a UDC adapter at the demarcation point, which converts the digital UDC transmission into two analog lines. The UDC carrier is *DSL (Digital Subscriber Loop)* based. UDC is a *2B1Q (Two Binary One Quarternary)* line format that rides on a -137-volt battery. If the UDC carrier fails, it will go into an override/bypass mode and switch the pair to a -52-volt battery to provide the original POTs service. The telephone number that will continue to work is line one, and line two can be set to call forward to line one. For a diagram, see *DLC*.

Universal Resource Locator (URL)   An Internet address. URLs consist of two parts. The first part is the protocol identifier and the second is the domain identifier. For example, in the address *http://www.mcgrawhill.com,* the "http://" specifies a request to fetch a Web page using the *HTTP (Hyper Text Transfer Protocol)* and the "mcgraw-hill.com" specifies a domain name (which converts directly into an IP address) on the *World Wide Web (WWW)*.

Universal Serial Bus (USB)   A latter external bus standard that emerged in 1996. It supports data transfer rates of 12 Mbps, and can connect as many as 127 peripheral devices (mice, printers, video equipment, modems, and keyboards) simultaneously. USB is also able to support Plug and Play and can be hot pluggable. For a photo of a USB connector, see *USB*.

UNIX   An operating system similar to MS Windows, only it is designed to operate on *RISC (Reduced Instruction Set Computers),* like those made by Sun Microsystems. (It can work on Intel Pentium and other 586-based PCs, too.) UNIX performs all the functions that MS Windows does, only the icons look different and have different names. UNIX and RISC computers are used in conjunction with *CTI (Computer Telephony Integration)* applications, such as front-end processors and integrated voice-response systems, because of their ability to execute small instructions and tasks very quickly, and on a real-time basis.

Unlisted Number   A phone number that is not listed in a telephone book, but can be found by calling directory assistance and giving the person's name. The other type of private listing is an unpublished number, which is not listed in a telephone book and cannot be found by calling directory assistance.

Unpublished Number   See *Unlisted Number.*

Unshielded Twisted Pair (UTP)   Twisted-pair wiring that is unshielded, meaning it does not have a foil wrapping around the group of conductors within the jacket. Unshielded twisted pair is the most commonly used wiring for voice and data networks. Twisted-pair wire consists of pairs of color-coded wires. Common sizes of twisted-pair wire are 2 pair, 3 pair, 4 pair, 25 pair, 50 pair, and 100 pair. Twisted-pair wire is commonly used for telephone and computer networks. It comes in ratings of CAT 3 (for voice), CAT 4 (voice and 10-base-T), and CAT 5 (for 100-base-T and token ring). See also *Plenum* and *PVC.*

Unspecified Bit Rate (UBR)   A *Quality of Service (QOS)* defined by the ATM Forum for ATM Networks that allows any amount of data up to a specified maximum to be sent across the network. A connection would be rightfully commissioned as an UBR connection if it carried nontime sensitive data that could be retransmitted without a large inconvenience. The UBR quality of service does not guarantee freedom from cell loss or delay. Other QOSs defined by the ATM Forum for ATM connections include *CBR (Constant Bit Rate), ABR (Available Bit Rate),* and *VBR (Variable Bit Rate).*

Uplink Fast™   A very popular feature of Cisco Systems' multilayer LAN switches that is a modification for Spantree protocol. If a link in a network fails, VLANs must reconfigure traffic paths between switches under spantree. This can take from 30 to 50 seconds. Uplink Fast is a method of Spantree rerouting where the entire Spantree tables are not recalculated; therefore, traffic begins to flow 1 to 5 seconds after a link failure.

UPS (Uninterruptable Power Supply)   A battery back-up system. When the power goes out, the UPS converts the DC battery power to AC power to run the system.

Upstream   In asymmetrical broadband transmissions, a reference to the bandwidth or information flow toward the service provider and away from the customer/subscriber. ADSL and cable-modem Internet services are asymmetrical. They consist of a larger downstream and smaller upstream component. Asymmetrical transmissions are best suited for end-user Internet services. See also *ADSL* and *Cable Telephony.*

URL (Uniform Resource Locator)   An Internet address. URLs consist of two parts. The first part is the protocol identifier and the second is the domain identifier. For example, in the address *http://www. mcgrawhill.com,* the "http://" specifies a request to fetch a Web page using the *HTTP (Hyper Text Transfer Protocol)* and the "mcgraw-hill.com" specifies a domain name (which converts directly into an IP address) on the *World Wide Web (WWW).*

US   The ASCII control-code abbreviation for unit separator. The binary code is 1111001 and the hex is F1.

US Security   A reference to the 3DES security standard. Called *triple DES* (Data Encryption Standard), it is a 168-bit encryption method that incorporates an algorithm developed by the United States National Bureau of Standards.

USART (Universal Synchronous/Asynchronous Receiver/Transmitter)
The part of a computer's serial communications port (most PCs have two) that converts the parallel data from the data bus into serial data to be sent to a device connected to the port, like a modem.

USB (Universal Serial Bus)   A latter external bus standard that emerged in 1996. It supports data transfer rates of 12 Mbps, and can connect as many as 127 peripheral devices (mice, printers, video equipment, modems, and keyboards) simultaneously. USB is also able to support Plug and Play and can be hot pluggable (Fig. U.3).

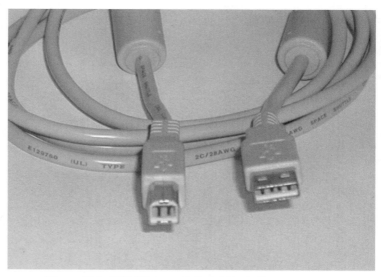

Figure U.3   USB (Universal Serial Bus) Connectors

User Datagram Protocol (UDP)   A connectionless transport-layer protocol that helps make up the TCP/IP protocol stack. UDP is a simple protocol that exchanges datagrams/packets without acknowledgments or guaranteed delivery, requiring that error processing and retransmission be handled by other protocols. UDP is defined in RFC 768. An *IP (Internet Protocol)* transport-layer protocol that is sometimes used in place of TCP when transaction-based application programs are communicating (i.e., *SNMP, Simple Network Management Protocol*). UDP usually carries noncrucial network information.

User Network Interface (UNI)   1. An ATM Forum specification that defines an interoperability standard for the interface between ATM-based products (a router or an ATM switch) located in a private network and the ATM switches located within the public carrier networks. An ATM physical-layer interface (DS3, OC-N, STS-N, or STM-N) that provides physical-layer services to an UNI (Fig. U.4). It is also used to describe similar connections in frame-relay networks. See also *NNI, Q.920/Q.921,* and *SNI (Subscriber Network Interface).*   2. *UNI (User Network Interface).* Also called a *cable voice port* or *voice port.* In cable-TV networks, a device located at the telephone subscriber's premises that modulates and demodulates the DS0 voice upstream and downstream channels. The modulated DS0 voice signal is sent by the *HDT (Host Digital Terminal)* located at the cable-TV head end. The voice port provides connection to the customer premises' telephone wiring. Channel control between the head end and the customer site voice port can be maintained remotely, from the head end or other office. Because the voice service in cable telephony is provided via radio channels, the channels can be switched when a subscriber is experiencing static or radio interference on line. This would be equivalent to a pair change in a twisted-pair-based telephone network. A pair change cannot be performed by remote control and requires a technician on site. For a photo, see *Voice Port.*

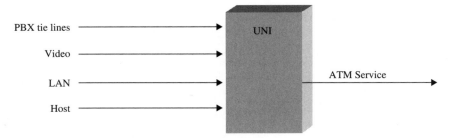

Figure U.4   User Network Interface (UNI) for ATM

User Plane   One of the three entities of frame-relay network management. The three planes are: *The User Plane* (the U Plane defines the transfer of information), the *Management Plane* (the M Plane defines the LMI, Local Management Interface), and the *Control Plane* (the C Plane is delegated for signaling and switched virtual circuits). The U Plane is based on the ANSI TI.602 (LAPD) core aspect standard. It provides the interface between the end user and the network, bi-directional transfer of frames, frame-order preservation, congestion avoidance, and determines the priority of *PVCs (Permanent Virtual Circuits)*.

USOC (Universal Service Order Code)   A code that defines different equipment and services within Regional Bell Operating Companies (RBOCs). Different RBOCs have different meanings for each code, as far as services, but the USOC codes for equipment has remained the same since the old Bell System (pre-1984).

USRT (Universal Synchronous Receiver Transmitter)   The part of a computer's serial communications port (most PCs have two) that converts the parallel data from the data bus into serial data to be sent to a device connected to the port, like a modem.

USWest   A *Regional Bell Operating Company (RBOC)*. Their territory includes the States of Washington, Oregon, Idaho, Montana, Wyoming, North Dakota, South Dakota, Minnesota, Iowa, Nebraska, Utah, Colorado, Arizona, and New Mexico.

UTP (Unshielded Twisted Pair)   Twisted-pair wiring that is unshielded, meaning it does not have a foil wrapping around the group of conductors within the jacket. Unshielded twisted pair is the most commonly used wiring for voice and data networks. Twisted-pair wire consists of "pairs" of color-coded wires. Common sizes of twisted-pair wire are 2 pair, 3 pair, 4 pair, 25 pair, 50 pair, and 100 pair. Twisted-pair wire is commonly used for telephone and computer networks. It comes in ratings of CAT 3 (for voice), CAT 4 (voice and 10-base-T), and CAT 5 (for 100-base-T and token ring). See also *Plenum* and *PVC*.

V

V (Volt) The basic unit of electric force, pressure, or *EMF (Electromotive Force)*. In Ohm's Law formulas, EMF is used as the designator for voltage, *E*. The two main components of electricity are current (amperage) and voltage. Voltage is also referred to as the potential energy between two points in a circuit.

V.13 An older modem standard that enabled full-duplex modems to act as half-duplex modems when necessary.

V.14 An older modem protocol that enables modems equipped with error correction to talk to modems that don't have error correction.

V.17 A data-transfer protocol for one-way facsimile (fax machines).

V.21 A modem protocol that is now used more outside the U.S. and Canada than within. Modems that are made in the U.S. and Canada are capable of receiving or sending transmissions to these modems.

V.22 Bell 212A. An older modem protocol for transfer rates of 1200 bp/s.

V.22bis An older modem protocol that allows for automatic increase/decrease in speed of the V.22 protocol to 2400 bp/s. The "bis" is an indicator of the ITU-T that means "second edition, or second in family."

V.23 An older modem protocol that provides up to 1200 bp/s in a forward channel and 75 bp/s in a reverse or backward channel.

V.24 An ITU-T standard for a physical-layer interface between DTE and DCE. V.24 is essentially the same as the EIA/TIA-232 standard.

V.25bis   An ITU-T specification describing procedures for call setup and tear down over the DTE-DCE interface in a packet-switched data network.

V.26   An older four-wire circuit protocol for 1200 baud modems. The two pairs (four wires) were transmit and receive.

V.27   An older modem standard that was capable of being configured for two-wire half-duplex operation or four-wire full-duplex operation.

V.29   A modem standard for transmission rates of 9.6 Kbps on private lines (no dialing, just wire).

V.32   An ITU-T standard "dial-up" serial-line protocol for bidirectional data transmissions at speeds of 4.8 or 9.6 Kbps. See also *V.32bis*.

V.32bis   An ITU-T standard that extends V.32 to speeds up to 14.4 Kbps. See also *V.32*.

V.34   An ITU-T standard that specifies a serial line "dial-up" protocol. V.34 offers improvements to the V.32 standard, including higher transmission rates (28.8 Kbps) and enhanced data compression. Compare with *V.32*.

V.34bis   A newer version of the V.34 standard that incorporates leaner compression techniques to achieve speeds that extend beyond the original 19.2 Kbps to 31.2 Kbps and 33.6 Kbps.

V.35   An ITU-T standard describing a synchronous, physical-layer protocol used for communications between a network access device, such as a router and a packet network. V.35 is recommended for speeds up to 48 Kbps, but it is commonly used for 56-/64-Kbps connections (Fig. V.1).

Figure V.1 V.35 Connector

V.42   An ITU-T family of protocols that include standards for bit transmission and error correction using *LAPM (Link-Access Procedure for Modems)*. The receiving device detects errors and requests a repeat transmission. LAPM is an older pre-X.25 protocol. Where these protocols still exist (usually in mainframe/terminal environments), they are converted by PADS before they are carried by X.25 packet networks. See also *LAPB (Link-Access Procedure level B)*.

V.54   The ITU-T standard for loop-back test capability in modems. A loopback test is when you send data to a far-end modem, and the modem simply sends your own data back to you. If you receive your own data in a loop back, then you know that your modem and the line are OK.

V.110   *ISDN (Integrated Services Digital Network)* two-wire *BRI (Basic Rate Interface)* protocol standard for data transfer over the B (Bearer) channel.

V.120   *ISDN (Integrated Services Digital Network)* two-wire *BRI (Basic Rate Interface)* protocol standard for data transfer over the B (Bearer) channel.

V Band   The band of frequencies designated by the IEEE between 40 GHz and 75 GHz. For a table, see *IEEE Radar Band Designation.*

V Fast   Also called V.34. A modem standard/compression protocol that transfers data at speeds up to 19.2 Kp/s.

V&H Coordinates   A method for calculating airline mileage between cities. For a listing of cities and their respective V&H coordinates, see *Airline Mileage.*

V Shell   One of the UNIX program operator access levels. The other levels are C Shell and Root. To access C Shell or V Shell, a password must be entered. To work in a UNIX program under the root command set, an additional password must be used. The different shells permit different operations, which have different command sets. This allows a root user to allow limited access to C-Shell and V-Shell users. Many telecommunications manufacturers incorporate the UNIX operating system into their equipment.

V. Standards   *ITU/T (International Telecommunications Union/ Telegraphy)* formerly known as *CCITT (Consultative Committee International Telephony and Telegraphy)* standards for modems, analog data communications, and methods of compression.

Vacant Code   A central-office prefix (the first three digits in a seven-digit phone number) or area code that is not in use or not assigned yet.

Vacuum Tube   An active electronic device that is literally a tube, with no air or other gasses in it. Tubes were the predecessors to transistors and other solid-state devices (Fig. V.2). Tubes have four main parts: the plate, the cathode, the filament (or heater), and the grid. The way that the tube works is that electricity flows through the tube from the cathode to the plate. The electricity flows when the filament is heated. This is actually the source of the electron flow. The signal to be amplified is fed to the grid, which manipulates the amount of current flowing through the tube. The resultant is an amplified output on the plate. Tubes are still used in many applications. Today's vacuum tube applications include: high-power applications (such as radio transmitters), high-end consumer audio equipment where it is necessary to have odd-order harmonics accurately reproduced (transistors do not amplify odd-order harmonics, but tubes do), and in applications in outer space where radiation damage is a high risk (transistors are sensitive to cosmic radiation). For a schematic symbol of a triode vacuum tube, see *Triode*.

Figure V.2 Vacuum Tubes: Triode (Left) and Pentode (Right)

Vampire Tap   A *LAN (Local Area Network)* connector that crimps onto the outside of a coax cable. The tap connector is equipped with teeth that penetrate through the jacket, shield and dielectric of the coax to make a solid connection with the inner conductor. The inner conductor of a coax is used as the bus in Ethernet networks.

Variable Bit Rate (VBR)   A *Quality of Service (QOS)* defined by the ATM Forum for ATM Networks that has two subcategories. The first is *VBR-RT (Variable Bit Rate Real Time)*. It is used for connections that are timing sensitive. A connection would be rightfully commissioned as an ABR RT connection if it carried video or voice. The second subcategory is *VBR-NRT (Variable Bit Rate NonReal Time)*. It is used for bursty or other nontime-sensitive data transmissions. Both categories of the VBR quality of service provide a guaranteed minimum cell loss and delay. Other QOSs defined by the ATM Forum for ATM Networks include *CBR (Constant Bit Rate), UBR (Unspecified Bit Rate),* and *ABR (Available Bit Rate)*.

Variable Resistor   Also called a *potentiometer (pot)*. A resistor that is usually made of carbon film and has a control knob or slide connected to it. Many electronic control knobs are connected to variable resistors (Figs. V.3 and V.4).

Figure V.3 Variable Resistors: Slide (Top) and Rotary (Bottom Left and Right)

Figure V.4 Variable Resistor Schematic

VBR (Variable Bit Rate) A *Quality of Service (QOS)* defined by the ATM Forum for ATM Networks that has two subcategories. The first is *VBR-RT (Variable Bit Rate Real Time)*. It is used for connections that are timing sensitive. A connection would be rightfully commissioned as an ABR RT connection if it carried video or voice. The second subcategory is *VBR-NRT (Variable Bit Rate NonReal Time)*. It is used for bursty or other nontime-sensitive data transmissions. Both categories of the VBR quality of service provide a guaranteed minimum cell loss and delay. Other QOSs defined by the ATM Forum for ATM Networks include *CBR (Constant Bit Rate), UBR (Unspecified Bit Rate),* and *ABR (Available Bit Rate)*.

VCI (Virtual Channel Identifier) The ATM transport method over a physical connection is divided into paths and channels. Paths and channels can be administrated by subscribers of ATM service. Over one physical connection (i.e., OC-3 or DS3) are a possible 4096 VPIs. For each VPI is a possible 65,536 VCIs (Fig. V.5).

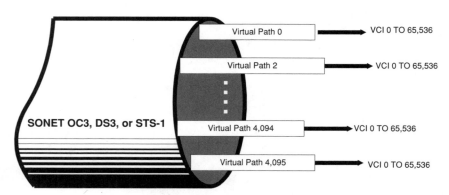

Figure V.5 VCI ATM Virtual Channel Interface

VDSL (Very High Bit-Rate Digital Subscriber Line) A physical-layer protocol that is the fastest of the DSL technologies. VDSL downloads up to 13–52 Mbps and uploads at 1.5 to 2.3 Mbps over a single pair of copper wires. VDSL is limited to a maximum range of 1000 to 4500 feet

(1.6 to 7.2 km) from the central office, depending on the speed. VDSL can be extended from central offices to remote communities using fiber optic. For a table that compares DSL formats, see *xDSL*.

Vertical and Horizontal Coordinates   See *Airline Mileage*.

Vertical Redundancy Check (VRC)   A part of *Longitudinal Redundancy Checking (LRC)*. A method of checking for errors in communications/modem transmissions by combining vertical error checking and longitudinal error checking. A transmission device sends data in bytes, which are logically stacked on top of each other. The stack forms a block. The last bit of each line is used to form a check sequence. LRC is about 85% accurate in detecting and re-transmitting blocks that contain errors. The newer method of error checking is *CRC (Cyclic Redundancy Checking)*.

Very High Bit-Rate Digital Subscriber Line   See *VDSL*.

Very High Frequency (VHF)   Radio frequencies in the range of 30 to 300 MHz. VHF is utilized for broadcast television and FM radio.

Very Low Frequency (VLF)   Radio frequencies in the range of 3 kHz to 300 kHz.

VHF (Very High Frequency)   Radio frequencies in the range of 30 MHz to 300 MHz. VHF is utilized for broadcast television and FM radio.

VIC (Voice Interface Card)   In Cisco Systems' product line, a module that provides POTS IP voice connectivity through a gateway or router. VICs come in station (FXS to plug a telephone into) or office (FXO to connect to the telephone company or PBX) options, and can be Loop Start, Ground Start, or E&M.

Video Conferencing   A video phone. A television connected to a telephone line that carries video and audio information. Some companies specialize in this service. Some even have offices that contain video conferencing equipment that can be "rented" for meetings with people at other far away video-conferencing locations. PictureTel is a popular manufacturer of video-conferencing equipment.

Video IP   There are three categories of video, be they in a public broadcast, satellite, cable TV, or IP or analog radio format. In IP, or packetized video, the same demands for video exist as in legacy technologies. The first category is broadcast video, where there is one source and many destinations. The broadcast is scheduled, and then broadcast via IP *multicast* to network users; this can be live or previously recorded. The second category is video on demand, where there is one end user that desires to view a previously stored video file as an IP unicast transmission. This is a prerecorded video that is streamed from a storage server file. The third category of video is the interactive two-or-more way video conference, where both ends are live and not prerecorded. This is the most complex type of video and requires both an IP unicast transmission and in some cases an integrated multicast, depending on each individual scenario. See also *H.323, H.320, Unicast,* and *Multicast.*

Virtual Channel Identifier (VCI)   The ATM transport method over a physical connection is divided into paths and channels. Paths and channels can be administrated by subscribers of ATM service. Over one physical connection (i.e., OC-3 or DS3) are a possible 4096 VPIs. For each VPI is a possible 65,536 VCIs.

Virtual Channel Link (VCL)   A connection between two ATM devices. A VCC consists of one or more VCLs. See also *VCC.*

Virtual Circuit Number (VCN)   A 12-bit field in an X.25 PLP header that identifies an X.25 virtual circuit. It allows DCE to determine how to route a packet through the X.25 network. See also *LCI* and *LCN.*

Virtual Colocation   The two types of colocations (also spelled *collocation*) are virtual and physical. A colocation is an interconnection agreement and a physical place where telephone companies hand-off calls and services to each other. This is usually done between a *CLEC (Competitive Local Exchange Carrier)* and an *RBOC (Regional Bell Operating Company).* A virtual colocation is when telephone company A (the CLEC) requests that their phone company's network be connected to telephone company B's (the RBOC's) network. Telephone company B charges company A lots of money. Company B owns, installs and maintains the equipment. To company A the interconnection is virtual, because they never physically do anything to it when and after it is installed. Company B likes this because company A does not get free access to their premises.

Virtual DN (Virtual Directory Number)   1. Also called a *phantom directory number.* A directory number or extension on a PBX system that is used to attach a voice mailbox to. The virtual DN does not really have a telephone set, but the PBX system thinks it does, so it transfers calls to that DN, which are configured to be forwarded to a voice-mail system. A user of that DN can then dial into the voice-mail system, enter their extension, and receive their messages.   2. DNIS numbers are also referred to as virtual DNs.

Virtual LAN (VLAN)   (Pronounced "vee-LAN.") In LAN switching, the feature that was brought about by the 802.1Q standard. It is feature on a LAN switch that makes selected ports behave as if they were attached to the same segment. Another good name for this feature would be V segment or virtual segment. Devices/users that exchange a large amount of information are usually placed within the same VLAN segment. This helps make the operation of the LAN switch more efficient, keeping traffic contained within specified ports. This allows other ports on separate VLANs to carry other nonrelated traffic simultaneously. Further, it also prevents devices in one VLAN from having access to devices on another. VLANs are configured by a network engineer, network analyst, or network administrator. When IP telephony is implemented over an Ethernet-switched network, the telephone devices connected to the network are best placed into their own VLANs. Most switches that are 802.1Q compatible can recognize more than 1,000 VLANs and up to 500 hosts per VLAN. Further, there are two kinds of VLANs: static and dynamic. Static VLANs are associated with switch ports, and dynamic VLANs are associated with the MAC addresses of devices attached to the switch. Dynamic VLANs allow users to move to another office that could have a data connection installed. The switch would recognize the MAC address of the device and automatically include its traffic in the same VLAN as the previously connected switch port. See also *Frame Tagging.*

Virtual Path Identifier (VPI)   The ATM transport method over a physical connection is divided into paths and channels. Paths and channels can be administrated by subscribers. Over one physical connection (i.e., OC-3 or DS3) are a possible 4096 VPIs. For each VPI is a possible 65,536 *VCIs (Virtual Channel Identifier).* For a diagram, see *VCI.*

Virtual PBX   A reference to IP telephony as an end user product, or Centron services from a PSTN services provider. IP telephony is replacing Centron services due to its cost effectiveness and ability to be remotely administered by an administrator or service provider. See also *IP Telephony* and *Centron/Centrex.*

Virtual Tributary (VT)   A virtual tributary is a communications channel or circuit that exists within another (larger) multiplexed communications channel. For example, a DS1 within a DS3 is a virtual tributary. Sometimes channels within an AT&T SLC96 carrier system (pair gain system for carrying 96 phone conversations on 8 pairs of wire) are referred to as *virtual pairs.*

Virtual Trunk Protocol (VTP)   In switched Ethernet LAN networks, a communications method that switches use to share VLAN information and which connections between switches are to carry traffic for specified VLANs. See also *802.1Q* and *Spanning Tree.*

Virtual Trunk Protocol Bomb (VTP Bomb)   (Slang) An event where an entire network policy area loses VTP trunking and VLAN configuration due to a switch being connected to the network that has a higher revision number in its VTP trunk revision table. This can cause a corporate network to go completely out of service. Switches update the version of VTP configuration every time a network change or VLAN change is made. They track these changes by giving them a VTP version number. The VTP revision number is used by LAN switches to determine which has the most up-to-date network information. Before a LAN switch is added to a network that has previously been commissioned in another network, its VTP and other VLAN information should always be configured to zero. This ensures it does not have a higher revision number from its old network that could override all VTP configurations in its new network surroundings. See also *802.1Q (VLAN)* and *Spanning Tree.*

VLAN (Virtual LAN)   (Pronounced "vee-LAN.") In LAN switching, the feature that was brought about by the 802.1Q standard. It is featured on a LAN switch that makes selected ports behave as if they were attached to the same segment. Another good name for this feature would be *V segment,* or *virtual segment.* Devices/users that exchange a large amount of information are usually placed within the same VLAN segment. This helps make the operation of the LAN switch more efficient, keeping traffic contained within specified ports. This allows other ports on separate VLANs to carry other nonrelated traffic simultaneously. VLANs are configured by a network engineer, network analyst, or network administrator. When IP telephony is implemented over an Ethernet-switched network, the telephone devices connected to the network are best placed into their own VLANs. Most switches that are 802.1Q compatible can recognize more than 1,000 VLANs and up to 500 hosts per VLAN. Further, there are two kinds of VLANs: static and dynamic. Static VLANs are associated with switch ports, and dynamic VLANs are associated with the MAC addresses of devices attached to the switch. Dynamic VLANs allow users to move to another office that could have a data connection installed. The switch would recognize the MAC address of the device and automatically include its traffic in the same VLAN as the previously connected switch port. See also *Frame Tagging.*

VLF (Very Low Frequency)   Radio frequencies in the range of 3 to 300 kHz.

VLSM (Variable-Length Subnet Mask)   A feature of a routing operating system to specify/identify a different subnet mask for the same network number on different subnets. VLSM can help optimize available address space. In order to use VLSM, a network administrator must use a routing protocol that supports it. For example, Cisco routers support VLSM with Open Shortest Path First (OSPF), Integrated Intermediate System-to-Intermediate System (Integrated IS-IS), Enhanced IGRP (EIGRP), and static routing.

Voice Band   A reference to the frequency range of a human voice that is transferred on a telephone line. The range of the human voice is about 200 Hz to 12 kHz. The range of the voice transmitted over a POTS phone line is flat (an even reproduction) from 500 Hz to 3,500 Hz (3 kHz in bandwidth).

Voice Enabled Router   A router that is QoS, MGCP, and H.323 enabled at minimum that would enable it to carry voice traffic in an IP network and/or translate it to the PSTN (Public Service Telephone Network). See also *MGCP, Gateway, H.323, WIC,* and *RTP.*

Voice Enabled Switch   A reference to an Ethernet switching device that is 802.1Q as well as 802.1p compliant that would enable the device to carry IP telephony traffic. IP telephony and VoIP are not compatible with token ring networks.

Voice eXtensible Markup Language (VXML)   VXML is similar to HTML. When a user calls a special phone number, the call is routed to a device called a Voice Response Unit (VRU). The VRU launches a web browser that finds and interprets a document written in VXML and then responds to the caller. Users can interact with the web either by voice or touch tones.

Voice Grade   A reference to a POTS (Plain Old Telephone Service) telephone line, like the ones that are subscribed to by residential telephone company customers. Local telephone companies do not guarantee any transfer rate of data over these lines. So, if you can't transmit data at 28.8 Kb/s with your new modem, the phone company will probably ask you to subscribe to a switched 56K data line (which is more costly than a POTS line).

Voice Mail   An answering machine system that integrates with a *PBX (Private Branch Exchange)* or key telephone system. Octel (now a part of Lucent Technologies) is a manufacturer of voice-mail systems that are used in business-office applications and in central-office applications for telephone companies to offer voice mail/voice messaging as a service to subscribers. Voice mail can also be purchased as a network interface card with software that runs on a PC. Figure V.6 shows an Octel/Lucent stand-alone voice-mail system.

Figure V.6 An Aspen Voice-Mail System

Voice-Mail Notification   A dial tone that blinks on and off when a telephone line initially goes into an off-hook state. A stutter dial tone is used to signal a user that a new voice message is in their voice mailbox.

Voice Over ATM   A method of transporting voice conversations over ATM. The ATM transport protocol has an interface specifically designed for real-time transmissions, such as voice and video. It is commonly known as the *AAL1 (ATM Adaptation Layer 1)*. See also *AAL1* and *AAL*.

Voice Over Frame Relay   A method of transporting time-sensitive data, such as digital voice over frame-relay connections. This can be accomplished by routing the voice through a *DLCI (Digital Link Connection Identifier)* that has appropriate classes of service on the end equipment, which is a router or a switch. See also *DLCI* and *Bandwidth Control Elements*.

Voice Over IP (VoIP)   A method of connecting voice conversations over controlled IP environments, such as private data networks. The private networks used are predominately comprised of frame relay, and ATM, which have committed information rates and *QOS (Quality of Service)* measures incorporated into them. Work is being done to make large-scale quality VoIP in public environments, such as the Internet, commonplace. See also *RTSP*.

VoIP   See *Voice Over IP.*

Voice Port   Also called a *cable voice port* or a *UNI (User Network Interface)*. In cable-TV networks, a device is located at the telephone subscriber's premises that modulates and demodulates the DS0 voice upstream and downstream channels (Fig. V.7). The modulated DS0 voice

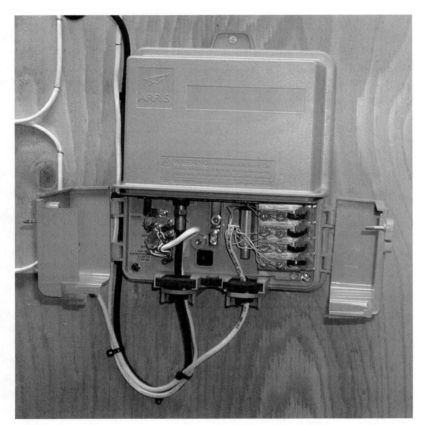

Figure V.7 Voice Port. In This Cable Voice Network Interface, the Cable TV Drop Connects at the Left and the Customer Premises' Inside Telephone Wire Connects at Right (This Voice Port Interfaces Four Telephone Lines)

signal is sent by the *HDT (Host Digital Terminal)* located at the cable-TV head end. The voice port provides connection to the customer premises telephone wiring. Channel control between the head end and the customer-site voice port can be maintained remotely, from the head end or other office. Because the voice service in cable telephony is provided via radio channels, the channels can be switched when a subscriber is experiencing static or radio interference on the line. This would be equivalent to a pair change in a twisted-pair-based telephone network. A pair change cannot be performed by remote control and requires a technician on site.

Voice Recognition   A reference to a computer or machine's ability to recognize a specific individual's voice, like a fingerprint. This technology is frequently confused with speech recognition, which recognizes words, not distinct voices.

Voice Response Unit (VRU)   Better known as *IVR (Interactive Voice Response)*. A telecommunications and data processing technology that interfaces a person to information held in a computer by using a phone line. If you have ever called your bank and entered your account number, a password, and a prompt so that a computerized voice can read back your bank-account balance, then you have used IVR.

Voice-Grade Private Line (VGPL)   There are two grades of private line service, voice grade and data grade. Both can be analog (via modem) or digital. A private line is a communications path from one point to another that is not switched, or dial-up, it is a physical pair of wires or a virtual connection through a transport network (such as SONET or T1 carrier).

Volt (V)   The basic unit of electric force, pressure, or *EMF (Electromotive Force)*. In Ohm's Law formulas, *EMF* is used as the designator for voltage, *E*. The two main components of electricity are current (amperage) and voltage. Voltage is also referred to as the *potential energy* between two points in a circuit.

Voltage   A reference to the amount of electric force, pressure, or *EMF (Electromotive Force)*. See *Volt*.

Voltage Drop   A reference to the amount of voltage difference from one point in a circuit to another. A voltage drop is usually in reference to one component and is useful to know when troubleshooting electronic circuitry.

Voltmeter   A meter used to verify the amount of voltage at a single point in a circuit (referenced to ground) or the amount of voltage potential between two different points in a circuit. Most meters for measuring voltage come in the form of a multimeter, which has a volt, ohm, and current (amp) meter incorporated into its design (Fig. V.8).

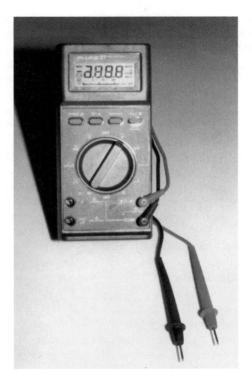

Figure V.8 Digital Voltmeter (DVM)

Volume Unit Meter (VU Meter)   See *VU Meter.*

VPI (Virtual Path Identifier)   The ATM transport method over a physical connection is divided into paths and channels. Paths and channels can be administrated by subscribers. Over one physical connection (i.e., OC-3 or DS3) are a possible 4096 VPIs. For each VPI is a possible 65,536 *VCIs (Virtual Channel Identifier).* For a diagram, see *VCI.*

VRC (Vertical Redundancy Check)   A part of *Longitudinal Redundancy Checking (LRC).* A method of checking for errors in communications/modem transmissions by combining vertical error checking and longitudinal error checking. A transmission device sends data in bytes, which are logically stacked on top of each other. The stack forms a block. The last bit of each line is used to form a check sequence. LRC is about

85% accurate in detecting and re-transmitting blocks containing errors. The newer method of error checking is *CRC (Cyclic Redundancy Checking)*. For a diagram, see *Vertical Redundancy Check*.

VT   1. Video terminal.   2. Virtual tributary. A virtual tributary is a communications channel or circuit that exists within another (larger) multiplexed communications channel. For example, a DS1 within a DS3 is a virtual tributary. Sometimes channels within an AT&T SLC96 carrier system (a pair-gain system for carrying 96 phone conversations on eight pairs of wire) are referred to as *virtual pairs*.

VT 100   A DEC device that consists of a monitor and a keyboard. It has no memory or programming, so it is called a *dumb terminal*. It is now possible to use a PC to act or communicate as a dumb I/O terminal. To use a PC in this application (such as to plug into a microwave link and boost its power), it must be equipped with terminal-emulation software. The microwave link device has its own microprocessor and only needs a device to communicate with; that device is usually a VT100 terminal. Terminal-emulation software allows your PC to "look" like a VT100 terminal to the microwave radio equipment.

VTA (Video Terminal Adapter)   A device that converts 4.320 (ISDN) video to 4.323 (Ethernet) video.

VTP (Virtual Trunk Protocol)   In switched Ethernet LAN networks, a communications method that switches use to share VLAN information and which connections between switches are to carry traffic for specified VLANs. See also *802.1Q* and *Spanning Tree*.

VTP Bomb (Virtual Trunk Protocol Bomb)   (Slang) An event where an entire network policy area loses VTP trunking and VLAN configuration due to a switch being connected to the network that has a higher revision number in its VTP trunk revision table. This can cause a corporate network to go completely out of service. Switches update the version of VTP configuration every time a network change or VLAN change is made. They track these changes by giving them a VTP version number. The VTP revision number is used by LAN switches to determine which has the most up-to-date network information. Before a LAN switch is added to a network that has previously been commissioned in another network, its VTP and other VLAN information should *always* be configured to zero. This ensures it does not have a higher revision number from its old network that could override all VTP configurations in its new network surroundings. See also *802.1Q (VLAN)* and *Spanning Tree*.

VU Meter (Volume Unit Meter)   The meter on tape recorders that moves while recordings are being made. The meter is used as a reference to make adjustments of the recording level (the strength of the signal that is sent to the tape). If the signal fed to the tape is strong enough to push the meter into the red, then the recording will be distorted. If the signal is not strong enough to push the meter, then the signal is not being recorded. Many cassette tape recorders today have LED VU meters. When recording messages for an *IVR (Integrated Voice Response)* system or other device, it is useful to watch the VU meter while recording.

W Band The band of frequencies designated by the IEEE between 75 GHz and 110 GHz. For a table, see *IEEE Radar Band Designation.*

W3C (World Wide Web Consortium) An international partnership of companies that are involved with Internet and the World Wide Web that was founded in 1994 by Tim Berners-Lee. The purpose of W3C is to oversee the development of open standards to evolve the Web in a uniform manner that is usable by everyone and to prevent competing factions from implementing proprietary browser formats. The W3C is the chief standards organization for *HTTP (Hypertext Transfer Protocol)* and *HTML (Hypertext Markup Language).*

Waffle Splice Closure A cable splice closure that is commonly used in air-pressure applications or shallow trenches where a strong casing is needed to protect the splice (Fig. W.1). See also *Air Pressure.*

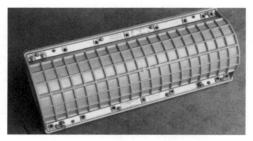

Figure W.1 Waffle Splice Closure

WAN (Wide-Area Network)   A network of computers or computing devices connected by telephone lines that extend beyond an area code's service area. An example of a WAN application is a computer that accesses another computer in another state to access information. Popular ways for computers to connect over long distances are by using a dial-up modem, a frame relay circuit, an ATM circuit, an ISDN circuit, or a 56K leased line. There are advantages and disadvantages to each of these services (which are offered by long-distance telephone companies). The faster and more reliable the service, the more expensive it is. Frame relay is rapidly becoming the most economical WAN protocol for applications that transfer data over long distances frequently (Fig. W.2).

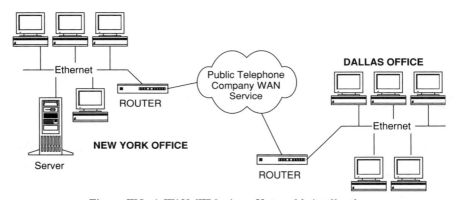

Figure W.2 A WAN (Wide-Area Network) Application

Watchdog Timer   1. A hardware or software mechanism that is used to trigger an event or an escape from a process unless the timer is periodically reset.   2. In NetWare, a timer that indicates the maximum period of time that a server will wait for a client to respond to a watchdog packet. If the timer expires, the server sends another watchdog packet (up to a set maximum).

WATS (Wide-Area Telephone Service)   A toll-free dialing service (800/888 lines) offered by telephone companies. In-WATS lines are priced and set up for incoming only calls, and sometimes calls from a certain area. You can also subscribe to Out-WATS service as well. WATS can be for interstate and intrastate long distance. If you call an 800 number, you are most likely calling an in-WATS service line that a company has set up for customers. The time to start checking into WATS service is when your long distance to or from a specific area exceeds $200.00 per month.

Watt (W)   The unit of electrical power, represented as P in Ohm's Law formulas. Wattage is calculated by $P = I \times E$, where P is power in watts, I is current in amps, and E is voltage in volts. For example, if you have a light bulb that draws 1 amp at 100 volts, it would be a 100-watt light bulb. The power consumed by the light bulb is radiated as heat and light. Just because a light bulb has a higher wattage rating does not mean that it is brighter, it could be hotter as well. Technically, one watt is equal to one joule per second. Another way to grasp the concept of a watt is to use the comparison that 746 watts is equal to one horsepower.

Waveguide   A device used to direct radio-frequency transmissions or light waves. Waveguides in radio transmitters look like high-tech plumbing. In lightwave applications, they are a small prism or optical fiber.

Wavelength   The wavelength of a radio signal in meters is equal to 300,000,000 m/s (the speed of light) divided by the frequency in hertz. For example, the wavelength for an FM radio station's signal if they are at 96.3 MHz on the radio dial is equal to (300,000,000 m/s ÷ 96,300,000 Hz = 3.115 meters). It is useful to know wavelength in the design of radio antennas, which are made to be the same length or a fraction of the length of a radio signal's wavelength.

Wavelength-Division Multiplexing (WDM)   A way of increasing a fiber optic's capacity by using multiple colors of light. Each color of light has its own wavelength (and its own frequency). The electronic equipment on each end of the fiber can distinguish the different signals by their color (frequency/wavelength). In most applications today, each fiber optic in a communications network carries one light signal that is one pure color. In the future, fiber optic will be wavelength-division multiplexed to carry many transmission signals.

WDCS (Wide-Band Digital Cross-Connect System)   Another name for *DCS (Digital Cross-Connect System)*.

WDM (Wavelength Division Multiplexing)   See *Wavelength Division Multiplexing*.

Web Browser   Also called a *browser* or an *Internet browser.* A computer program that allows users to download World Wide Web pages for viewing on their computers. Two popular browser programs are Netscape Navigator and Microsoft Internet Explorer. The first browser program was called *Mosaic,* and it was a text browser, as opposed to the newer graphical browsers.

WEP (Wired Equivalent Privacy)   In wireless 802.11 LAN networks, a DSSS encryption standard that can have different magnitudes of security. The most popular are 40 bit and 128 bit. These encryption standards provide security that is theoretically equal to a wired circuit that cannot be tapped, sniffed, or monitored in any other way.

Wet Circuit   A T1 circuit is a wet circuit when its 135V DC battery voltage is present. When the T1 has the battery voltage removed via a *CSU/DSU (Channel Service Unit/Digital Service Unit)* it is a dry T1 circuit or more commonly known as a *DS1 (Digital Service Level 1)*.

Wet T1   See *Wet Circuit*.

White Board   Also called a *mushroom board* or *peg board*. It is placed between termination blocks (such as 66M150 blocks) to provide a means of support for routing cross-connect wire. See *Fig. W.3*.

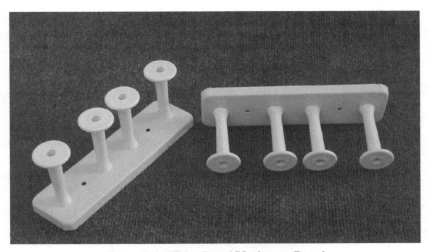

Figure W.3 White Board/Mushroom Board

White Noise   Random electrical noise, also called *ambient noise*. White noise is the sound you get from a TV or radio when it is not tuned to a station.

WIC (WAN Interface Card)   In Cisco Systems' router and gateway products, an add-on module, or card that enables connectivity to the link method of customer choice. WICs come in a variety of interface options, including T1, ISDN, and 56K digital. See also *VIC (Voice Interface Card)*.

Wide-Area Network (WAN)   See *WAN*.

Wide Band   Another name for *Broadband*. Incorporating more than one channel into a communications transmission. T1 is a broadband communications protocol because it carries 24 conversations over four wires. Cable-TV is also broadband because it carries many TV channels over one coax.

Wide-Band Division Multiplexing   The transmission of multiple SONET transmissions over a single fiber-optic pair. Each SONET transmission is sent via a different frequency (or color) of light. This method is used to obtain an OC-768 with four OC-192 transmissions of different frequencies.

Wildcard Mask   A 32-bit quantity used in conjunction with IP addresses to determine which bits in an IP address should be ignored when comparing that address with another IP address. A wildcard mask is specified when setting up access lists.

Wind Loading   In wireless communications, a reference to the amount of wind an antenna structure can withstand. For definitions of wind loading specifications for antennas and towers, it is best to check manufacturer specifications. For general engineering information, refer to the following specifications: TIA/EIA-195 (for antennas) or TIA/EIA-222 (for towers).

Wink   Another name for a hookflash that is sent by PBX systems to telephone company central-office switches that signals a request for dial tone or other services.

Wink Start   A reference to E&M signaling for analog voice circuits. E&M technology dates back to the time telegraphs were used and is an outdated service no longer offered as a service by most telephone companies, however, E&M trunking still has special uses because of its simplistic nature. E&M interfaces come in handy in the PBX environment when there is a need to connect to an analog audio device such as an overhead paging system, or tape recorder. There are 5 types of E&M interfaces that can have either two or six wires in the loop. The most common type of E&M signaling is the Four Wire Wink Start E&M, and the next most used is the Four Wire Immediate Start E&M. The Wink Start E&M operates as follows: The call originating switch goes off-hook and then waits for a "wink" from the terminating or destination switch. When the destination switch provides the 200ms off-hook "wink," then the originating switch sends dialed digits. After the dialed digits are received and a connection is made to the terminating loop by a handset being taken off hook, the same "off hook" condition is given over the E&M trunk connecting the terminating switch to the destination switch. When one switch goes "on-hook" or hangs up, the other does as well. The most simple E&M signaling method is the Immediate Start E&M, where the originating end goes "off-hook," or provides a 1000 ohm short on the line and sends digits without regard to the other end. The originating switch stays off-hook until the receiving switch goes off-hook and then back "on-hook," or the call originator goes back "on-hook" or hangs up. The E&M immediate start is the better choice for interfacing external audio devices to PBX systems, and is the less appropriate choice for PBX trunking because if the terminating switch does not answer the call, and the originating switch does not manually hang-up or go back "on-hook," then the loop is left connected. This problem with Immediate Start E&M is the reason that Wink Start E&M was brought about. Wink Start E&M is also called E&M with Answer Supervision.

Wink Start Signal   Another name for a hookflash that is sent by PBX systems to telephone company central-office switches that signals a request for dial tone or other services. Wink start signaling is slowly being replaced by T1 out-of-band signaling, which is much faster, offers many more services, is less expensive, and is, of course, digital.

Wire Center   A reference to a telephone company central office's geographical service area. The central office serves the area that its telephone wires (outside plant) reach to.

Wire Pair   A reference to two solid wires, twisted together, usually just called a *pair*. For more information, see *UTP*.

Wire Tap   A device used to monitor telephone lines. With newer technology, the telephone company is capable of tapping or monitoring a telephone line with a stroke of a few keys. They can even set the telephone line to be monitored by a different telephone line, anywhere they want. Telephone companies (especially the Bell Companies) have extremely strict security guidelines regarding the monitoring of telephone conversations. They will not set up a tap or a monitor service without legal procedures being followed according to the laws of the area they are operating in. Watch-dogs are in place to be sure that telephone company employees do not monitor telephone lines when they are not supposed to. Some "spy" shops sell telephone line-tapping and recording devices, but I would not recommend the use of them because of strict laws regarding telephone privacy.

Wireless LAN   1. A local-area network of computers and peripheral devices that communicates via radio signals or light waves (low-power laser beams). These systems are useful in situations where the cost of installing wiring between the devices is very expensive or for temporary/mobile applications.   2. A method of having PC or other network IP appliances communicate over a network without the use of wire or fiber optic. The 802.11 specification utilizes two methods of radio modulation: DSSS (Direct Sequence Spread Spectrum) and FHSS (Frequency Hopping Spread Spectrum). DSSS is the faster and longer range of the two, delivering up to 11 Mbps of throughput, including overhead.   Methods of operation vary among manufacturers, enabling different combinations of 802.11 features to be used.   Some of the features include WEP (Wire Equivalent Privacy), roaming, point-to-point MAN implementation, and load sharing (among multiple radios). Under 802.11, radios are allowed to transmit up to 50 mW, which can transmit up to 25 miles in perfect conditions at a 1 Mbps throughput, and up to 12 miles in perfect conditions at an 11 Mbps throughput. Typical indoor 30 mW radio deployments provide a range of 150 feet at throughput of 11 Mbps. See also *Free Space Path Loss, Link Budget, Rain Attenuation, Antenna Gain, 802.11b, DSSS,* and *FHSS.*

Wire-Wrap Termination   A type of twisted-pair wire termination used on DSX panels and other digital telecommunications services processing equipment (Fig. W.4). Wire-wrap termination is used because it requires very little space, compared to 110 blocks or 66M150 blocks.

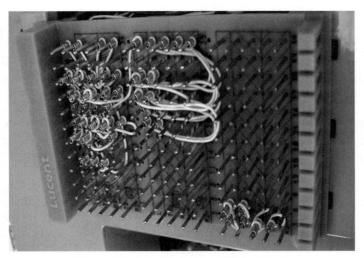

Figure W.4 Wire-Wrap Termination Block

WLAN   A reference to the wireless LAN, 802.11 Ethernet wireless LAN standard. Wireless radios manufactured under the description of this standard are intended for use in unlicensed LAN, point-to-point WAN, and point-to-multipoint WAN. This standard provides for standard technologies that provide relief in the bulging private line market. See also *RF, DSSS, Earth Bulge, Free Space Path Loss, Link Budget, Azimuth, Rain Attenuation, Co-Channel,* and *Antenna Gain.*

WLAN (Wireless LAN)   A method of having PC or other network IP appliances communicate over a network without the use of wire or fiber optic. The 802.11 specification utilizes two methods of radio modulation, DSSS (Direct Sequence Spread Spectrum) and FHSS (Frequency Hopping Spread Spectrum). DSSS is the faster and longer range of the two, delivering up to 11 Mbps of throughput including overhead. Methods of operation vary among manufacturers, enabling different combinations of 802.11 features to be used. Some of the features include WEP (Wire Equivalent Privacy), roaming, point-to-point MAN implementation, and load sharing (among multiple radios). Under 802.11, radios are allowed to transmit up to 50 mW, which can transmit up to 25 miles in perfect conditions at a 1 Mbps throughput, and up to 12 miles in perfect conditions at an 11 Mbps throughput. Typical indoor 30 mW radio deployments provide a range of 150 feet at throughput of 11 Mbps. See also *802.11, DSSS,* and *FHSS.*

Word   In computer memory, a word is 16 bits, which is one data unit processed by the bus. Newer computers and other processing systems are built with 32- and 64-bit busses, which gives them the ability to process double words (32 bits) and quad words (64 bits).

World Wide Web (WWW)   The *Graphical User Interface (GUI)* system that makes finding information on the Internet easier by organizing it into pages. WWW also provides hyperlinks, which, when "clicked on" with a mouse, take you to the corresponding page (it is actually another address) that contains the implied information.

World Wide Web Consortium (W3C)   An international partnership of companies that are involved with Internet and the World Wide Web that was founded in 1994 by Tim Berners-Lee. The purpose of W3C is to oversee the development of open standards to evolve the Web in a uniform manner that is usable by everyone and to prevent competing factions from implementing proprietary browser formats. The W3C is the chief standards organization for *HTTP (Hypertext Transfer Protocol)* and *HTML (Hypertext Markup Language).*

WWW (World Wide Web)   The *Graphical User Interface (GUI)* system that makes finding information on the Internet easier by organizing it into pages. WWW also provides hyperlinks, which, when "clicked on" with a mouse, take you to the corresponding page (it is actually another address) that contains the implied information.

X Band   The band of frequencies designated by the IEEE between 8 GHz and 12 GHz (3.75 cm to 2.5 cm). For a table, see *IEEE Radar Band Designation.*

X.3   A standard that defines a *PAD's (Packet Assembler/Disassembler)* operating parameters in an X.25 network. After X.25 became a standard, it was modified to be more efficient with the use of PAD devices. X.3 defines the way PAD overhead settings are communicated between the terminal equipment and the host. How the PAD gets the settings over the network is defined in X.28 and X.29. For X.3 specifics, see *PAD Parameters.*

X.20   The recommended standard by the *ITU (International Telecommunications Union)* for asynchronous communications between modems on public-switched telephone networks (dial-up lines).

X.21   X.25 without analog. An X.25 application for direct digital interface to ISDN or another nonanalog protocol.

X.25   A widely implemented packet service provided by telecommunications companies that runs at speeds up to 56/64 Kbps (Figs. X.1 and X.2). X.25 data-packet transfer services are named after the protocol that they are provided through. This service is usually billed by the byte (i.e., x dollars per million bytes transferred). This is good for short and bursty transmissions, such as those made by automated teller machines, credit-card transactions, terminal-to-host, or other similar traffic-producing applications. X.25 is also capable of reliably transporting TCP/IP and other protocols. Regarding operability, X.25 is a third-layer connection-oriented protocol. Error detection is performed at level two (LAPB) and three (X.25). This gives X.25 an advantage in areas where telephone facilities are of poor integrity. Because X.25 is responsible for detecting and discarding damaged packets, the customer only pays for packets received, rather than paying for a bandwidth, as in frame relay (Committed Information Rate). X.25 was developed from the *LAPB (Link-Access Procedure Balanced mode)* protocol. X.25 has remained a viable service for many because of modifications and additions to the protocol family, such as LAPBE, X.28, and X.75. Frame relay is commonly used as a higher-speed backbone enhancement to X.25 networks. Several illustrations of the X.25 frame structure are within this book under *X.25 Data Packet, X.25 Control Field,* and *X.25 Packet Control Header.*

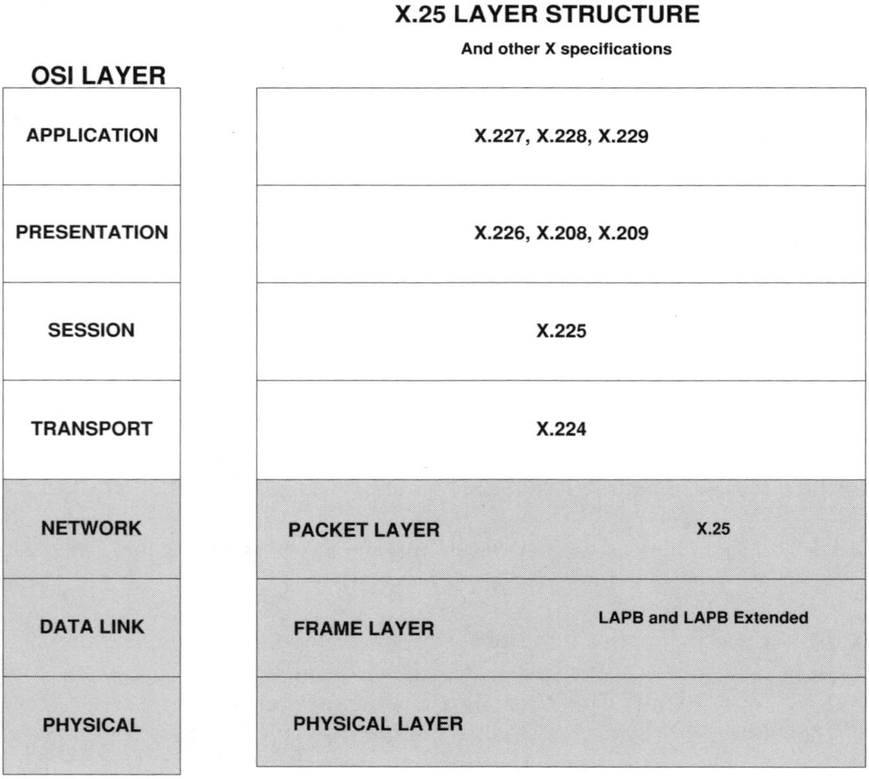

X.25 LAYER STRUCTURE
And other X specifications

OSI LAYER	
APPLICATION	X.227, X.228, X.229
PRESENTATION	X.226, X.208, X.209
SESSION	X.225
TRANSPORT	X.224
NETWORK	PACKET LAYER X.25
DATA LINK	FRAME LAYER LAPB and LAPB Extended
PHYSICAL	PHYSICAL LAYER

Figure X.1   X.25 Functional Layers

01111110	ADDRESS	CONTROL HEADER	PACKET/DATA	FRAME CHECK SEQUENCE	01111110
Flag					Flag

Figure X.2   X.25 Basic Frame Structure

X.25 Call-Request Packet   Also called a *call set-up packet.* This is an important Level-3 (data-link layer) control frame (Fig. X.3). This particular header, along with other headers that are sent in response to it, negotiate the rules that will be used to transfer data across the packet

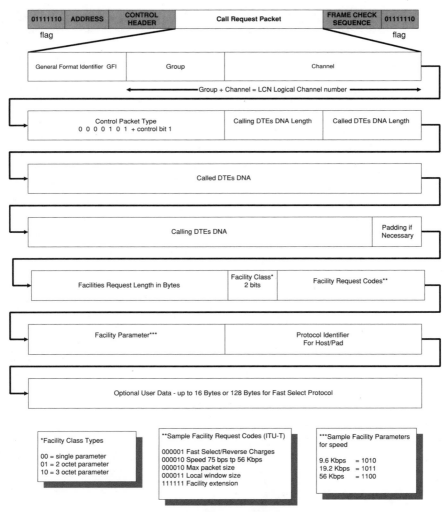

Figure X.3   X.25 Call-Request/Call Set-Up Packet

layer (data-link layer, in OSI terms). Parameter negotiation that is initiated by this packet include the *DNA (Data Network Address)* length, the window size (modulo 8 or 128), who will pay for the call, the *DNA (Data Network Address)* length, the maximum packet size, and also who will pay for the call. The previous parameters listed are referred to as *facilities* in X.25. Fourteen different packet-level control frames are defined by the ITU-T. They are listed under the *X.25 Packet-Control Header Structure* diagram.

X.25 Control Field or Control Byte   A reference to the 8-bit control field in an X.25 frame. At Level 2 of the X.25 protocol, the units of data are called *frames*. The frames are headers for packets and are, therefore, "seen" by network devices before Layer-3 packet information is "seen." The frame layer is Level 2 of the OSI and the control frames perform operating functions for Level-2 OSI operation. The different three types of control frames are: information, supervisory, and unnumbered. The explanations are as follows:

- *I Frame (Information Frame)*   If the first bit of a control field is a "0," then the data inside the Layer-3 packet is user information. The next three bits define which window number the particular frame is based on eight frames (if LAPB). The last three bits of the I-type control byte are for notification of the next window coming.

- *S Frame (Supervisory)*   If the first two bits of the control byte are "10," then the frame is a supervisory frame. The S frame is used to start and stop traffic.

- *U Frame*   The U frame is sometimes called a *control frame* in itself. The six different control frames are defined in X.25 (the maximum is 32 because five bits are used for control signaling). They are: DISC, DM, SARM, SABM, FRMR, and UA. They are used to negotiate connection parameters, request transfer/retransfer of packets, link maintenance, and link reset. All three types of control frames are diagrammed under their respective name. See *X.25 I Frame, X.25 S Frame,* and *X.25 U Frame.*

X.25 Data Packet   The Level-3 packet type that carries end-user information in X.25 connections. After all connection agreements are made by end devices, the header that is illustrated in Fig. X.4 comes into operation. The data packet header carries window information for Level 3, up to 256 bytes of user data, and other header information that is illustrated.

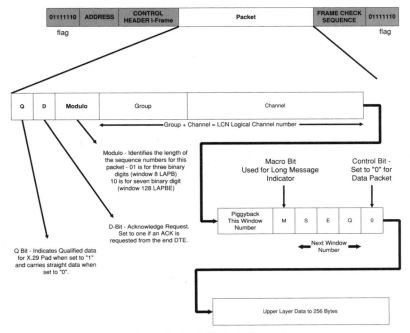

Figure X.4   X.25 Data Packet

X.25 I Frame   A type of frame used for transmission control in the frame layer of X.25 (*data-link layer*, in OSI terms). When the control header is an I frame, the frame layer is up and running, and data is able to be transferred between the packet layers (network layer, in OSI terms). The control header "tells" the receiver of data the window number of the packet it has received and the window number of the next packet (Fig. X.5). The window can be one of eight packets that are in transit for modulo 8, or one of 128 packets that are in transit for modulo 128 (also called *LAPB extended*). For more information regarding control bytes, see *X.25 Control Field*.

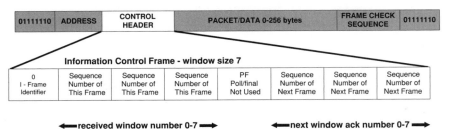

Figure X.5   X.25 I Frame (Information Frame) Types

X.25 Packet-Control Header Structure   This is the format for communicating Level-3 (data link layer) DTE and DCE call control for X.25 (Fig. X.6). In X.25, this is where the negotiation for data transfer begins. The end devices negotiate communications parameters. The initial packet-level control frame used to set up a call is the Call Request packet. This packet and the Level-3 header are illustrated under *X.25 Call Set-Up Packet*. When the call set up has been negotiated, the two ends begin sending data packets, which contain Level-4 information. A sample data packet and its header structure are shown under the definition *X.25 Data Packet.*

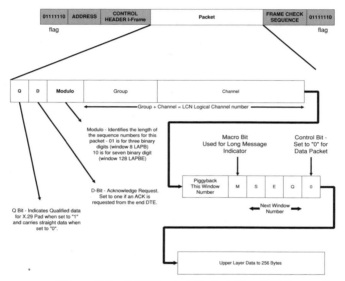

Figure X.6   X.25 Control Packet (for Call Request)

X.25 S Frame   For an explanation, see *X.25 Control Field* (Fig. X.7).

X.25 HDLC Supervisory Frame Structure: modulo 8

| 01111110 | ADDRESS | CONTROL HEADER | No Data Sent | FRAME CHECK SEQUENCE | 01111110 |

RR Receive Ready S Frame

| 1 S Frame Identifier | 0 S Frame Identifier | 0 S Frame-type Identifier Bit | 0 S Frame-type Identifier Bit | PF | Next Frame Requestor Bit | Next Frame Requestor Bit | Next Frame Requestor Bit |

REJ Reject S Frame

| 1 S Frame Identifier | 0 S Frame Identifier | 0 S Frame-type Identifier Bit | 1 S Frame-type Identifier Bit | PF | Next Frame Requestor Bit | Next Frame Requestor Bit | Next Frame Requestor Bit |

RNR Receive Not Ready S Frame

| 1 S Frame Identifier | 0 S Frame Identifier | 1 S Frame type Identifier Bit | 0 S Frame type Identifier Bit | PF | Next Frame Requestor Bit | Next Frame Requestor Bit | Next Frame Requestor Bit |

Figure X.7   X.25 S Frame Types

X.25 Triple X Protocol (XXX)   A term that refers to the three protocols used for X.25 transmission over a dial-up telephone line (Fig. X.8). Specifically, the protocols are X.25, X.28, and X.3.

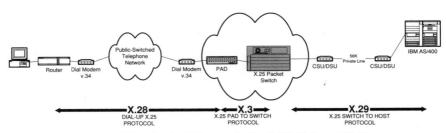

Figure X.8   Triple-X Protocols for X.25, X.3, and X.28

X.25 U Frame   For an explanation, see *X.25 Control Field* (Fig. X.9).

X.25 HDLC Unnumbered Frame Types - modulo 7

01111110	ADDRESS	CONTROL HEADER		No Data Sent		FRAME CHECK SEQUENCE	01111110
Flag		8 BITS					Flag

DISC Disconnect U Frame   Used for going offline

1 U Frame identifier	1 U Frame identifier	0 U Frame type identifier bit	0 U Frame type identifier bit	Poll-Final 1 = responce requested	0 U Frame identifier	1 U Frame identifier	0 U Frame identifier

SARM Set Asynchronous Response Mode U Frame   Used for back in service announcement LAP

1 U Frame identifier	1 U Frame identifier	1 U Frame type identifier bit	1 U Frame type identifier bit	Poll-Final 1 = responce requested	0 U Frame identifier	0 U Frame identifier	0 U Frame identifier

SABM Set Asynchronous Balanced Mode U Frame   Resets communication link and all buffers LAPB

1 U Frame identifier	1 U Frame identifier	1 U Frame type identifier bit	1 U Frame type identifier bit	Poll-Final 1 = responce requested	1 U Frame identifier	0 U Frame identifier	0 U Frame identifier

DM Disconnect Mode U Frame   Back in service announcement without resetting link - send SABM

1 U Frame identifier	1 U Frame identifier	1 U Frame type identifier bit	1 U Frame type identifier bit	Poll-Final 1 = responce requested	0 U Frame identifier	0 U Frame identifier	0 U Frame identifier

FRMR Frame Reject U Frame   Request for SABM to reset link and retransmit

1 U Frame identifier	1 U Frame identifier	1 U Frame type identifier bit	0 U Frame type identifier bit	Poll-Final 1 = responce requested	0 U Frame identifier	0 U Frame identifier	1 U Frame identifier

UA Unnumbered Acknowledgment U Frame   Used for U Frame receipt acknowledgment

1 U Frame identifier	1 U Frame identifier	0 U Frame type identifier bit	0 U Frame type identifier bit	Poll-Final 1 = responce requested	1 U Frame identifier	1 U Frame identifier	0 U Frame identifier

Figure X.9   X.25 U Frame Types

X.28   After the X.25 protocol standard came out, it was modified by making it able to connect over dial-up telephone lines and to manage *STDM (Statistical Time-Division Multiplexing)*. The rules and methods that define how this is done are defined within the X.28 and the X.29 standard. X.29 specifically allows the host to remotely change parameters of a PAD's communication parameters with a terminal.

X.29   A protocol that defines the connection of an X.25 network between a packet switch and a host computer.

X.61   The recommended standard by the *ITU (International Telecommunications Union)* for the data-user section of Signaling System 7 (SS7).

X.75   A modified version of the X.25 protocol that is used for call transition over international boundaries. It has an additional facilities field in the call-request packet called the *utilities field*. The utilities field records the DNICs used in data transfers.

X.121   The international addressing scheme for X.75.

X.130   The recommended standard by the *ITU (International Telecommunications Union)* for circuit set-up and tear-down times.

X.200   The recommended standard by the *ITU (International Telecommunications Union)* for the *OSI (Open Systems Interconnection)* model, and the functions and protocols of each layer.

X.400   An X-series standard published by the ITU-T that defines a universal e-mail protocol that can transmit messages between virtually all e-mail applications. In X.400, the sender's name is not included with the address.

X.500   A newer revision of the X.400 e-mail standard published by the ITU-T that extends the data-interchange protocol to include address formats, including the sender's.

X. Standards   *ITU/T (International Telecommunications Union/ Telegraphy)*, formerly known as *CCITT (Consultative Committee International Telephony and Telegraphy)* standards for digital communications.

X Windows   A protocol that connects *GUI (Graphical User Interface)* workstations with application programs/servers using TCP/IP.

Xaga   A type of "hot," buried splice enclosure for telephone cable that requires heating a special wrap that is placed around the splice (Fig. X.10). When the wrapping is heated, it shrinks into a sturdy casing. If it is installed properly, it is waterproof. *Xaga* is a registered trademark of Raychem.

Figure X.10   XAGA Splice Apparatus

XC   Cross Connect.

xDSL (x Digital Subscriber Loop)   A collective term used to refer to ADSL, HDSL, SDSL, IDSL, and VDSL. DSL is a family of telecommunications services that can provide over 50 Mbps of downstream bandwidth and 2.3 Mbps of upstream bandwidth (Fig. X.11). Furthermore, the xDSL formats can be transmitted over a copper pair that is providing POTs dial tone. DSL technology incorporates *DMT (Discrete Multi Tone)* line coding. This enables telephone companies to provide simultaneous high-speed Internet access, video on demand, telephone, and videophone on the same line. xDSL is an advancement of *ISDN (Integrated Services*

Digital Network), which incorporates a *QAM (Quadrature Amplitude Modulation)* as a line-coding technique. QAM is the building block for DMT. Originally, ISDN was intended as a flexible and feature-rich service. The ISDN transmission technique has now been incorporated into xDSL as a workhorse to carry other existing and future services, rather than be a service in itself. This improvement, along with various compression techniques, provides the large increase in speed. For specific information regarding DSL protocols, see *ADSL, RADSL, SDSL, IDSL,* and *VDSL.* See the chart for a comparison among them.

TYPE	Symmetry	Xmission range	# of pairs	Downstream rate	Upstream rate
ADSL	asymmetrical	18,000 feet	1	9 Mbps	1 Mbps
SDSL	symmetrical	24,000 feet	1	1.544 Mbps	1.544 Mbps
HDSL	symmetrical	20,000 feet	2 or 3	1.544 Mbps	1.544 Mbps
IDSL	symmetrical	18,000 feet	1	64 to 144 Kbps	64 to 144 Kbps
RADSL	asymmetrical	18,000 feet	1	9 Mbps	1 Mbps
VDSL	asymmetrical	4500 feet	1	52 Mbps	2.3 Mbps

Figure X.11   DSL Family Comparison

Xmit   Abbreviation for *transmit.*

XML (Extensible Markup Language)   A database format that is intended to enhance the capabilities of content services and delivery over the Internet. XML is more flexible than HTML because it allows for database formats to be self-defining. What makes XML able to work this way is that it does not have a fixed tag set or fixed semantics, which means that the semantics and tag sets are defined by the application programs that utilize the XML. XML could be thought of as a translation book where users can place a word, whatever word it may be in any language with any pronunciation or sound, and then place a picture or meaning next to the word, and further, order the pages of the book in any way they like. However, some connection to another book and page needs to be made for reference (this would be one of the other databases, such as a banking record). When the book is completed, it is, of course, a database in itself. XML also features a simpler method of set-up than HTML, and is widely available. It is commonplace for XML to be used to access a portion of data from an Internet web page and transfer it to another display method, such as a stock ticker display, a pager or cellular phone display, or telephone display.

XXX (Triple X Protocol)   A term that refers to the three protocols used for X.25 transmission over a dial-up telephone line. Specifically, the protocols are X.25, X.28, and X.3.

Y

Yagi Antenna A directional transmitting antenna (Fig. Y.1). The Yagi antenna utilizes one or more director elements to "focus" a radiated signal in one direction or plane, and a reflector element that is placed at a wavelength distance 180 degrees from the driven element to null the transmitted signal in the backward direction.

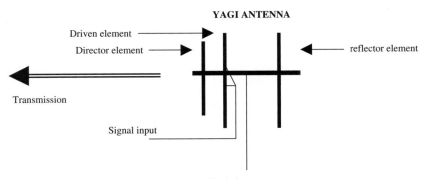

Figure Y.1 Yagi Antenna

Z

Zener Diode   A diode that operates in the reversed biased/avalanche mode. When a zener diode is in this mode, it has a constant voltage from its anode to cathode (Fig. Z.1). Zener diodes are used to regulate voltage, and they are available in many voltage values, such as 5 V, 12 V, 24 V, etc.

Figure Z.1 Zener Diode Schematic Symbol

Zero-Byte Time-Slot Interchange   A method of line coding, where the quantity of zeros transmitted is reduced. The reduction is accomplished by recognizing bytes that are all zeros (eight consecutive bits) and replacing them with an alternate byte (or flag) that does not contain all zeros.

Zero Suppression   To omit zeros from a number field.

Zulu Time   Formerly known as *Greenwich Mean Time.* Time that is kept on an atomic 24-hour clock in France. The Zulu time standard is now more accurate by innovation of the cesium timing reference standard, which is the time-keeping element in "bits clocks" (timing devices used in central office nodes to synchronize SONET equipment). Zulu time is slang for *Universal Time Coordinated.*

Vertical Horizontal Coordinates Table for United States Cities

City	V coordinate	H coordinate
Aberdeen, SD	5992	5308
Akron, OH	5637	2472
Albany, NY	4639	1629
Albuquerque, NM	8549	5887
Allentown, PA	5166	1585
Altoona, PA	5460	1972
Amarillo, TX	8266	5076
Anaheim, CA	9250	7810
Appleton, WI	5589	3776
Asheville, NC	6749	2001
Atlanta, GA	7260	2083
Atlantic City, NJ	5284	1284
Augusta, GA	7089	1674
Augusta, ME	3961	1870
Austin, TX	9005	3996
Bakersfield, CA	8947	8060
Baltimore, MD	5510	1575
Baton Rouge, LA	8476	2874
Bellingham, WA	6087	8933
Billings, MT	6391	6790
Biloxi, MS	8296	2481
Binghamton, NY	4943	1837

City	V coordinate	H coordinate
Birmingham, AL	7518	2304
Bismarck, ND	5840	5736
Blacksburg, VA	6247	1687
Bloomington, IN	6417	2984
Boise, ID	7096	7869
Boston, MA	4422	1249
Bridgeport, CT	4841	1360
Buffalo, NY	5076	2326
Burlington, IA	6417	2984
Burlington, VT	4270	1808
Camden, NJ	5249	1453
Canton, OH	5676	2419
Carson City, NV	8139	8306
Casper, WY	6918	6297
Cedar Rapids, IA	6261	4021
Charleston, SC	7021	1281
Charlotte, NC	6657	1698
Chattanooga, TN	7098	2366
Cheyenne, WY	7203	5958
Chicago, IL	5986	3426
Cincinnati, OH	6263	2679
Clarksburg, WV	5865	2095
Clearwater, FL	8203	1206
Cleveland, OH	5574	2543
Columbia, SC	6901	1589
Columbus, OH	5872	2555
Concord, NH	4326	1426
Corpus Christi, TX	9475	3739
Dallas, TX	8436	4034
Danville, KY	6558	2561
Dayton, OH	6113	2705
Daytona Beach, FL	7791	1052
Denver, CO	7501	5899
Des Moines, IA	6471	4275
Detroit, MI	5536	2828
Dodge City, KS	7640	4958
Dubuque, IA	6088	3925
Duluth, MN	5352	4530
Eau Claire, WI	5698	4261
El Paso, TX	9231	5655
Fargo, ND	5615	5182
Farmingham, MA	4472	1249

City	V coordinate	H coordinate
Fayetteville, AR	7600	3872
Fayetteville, NC	6501	1385
Flagstaff, AZ	8746	2367
Flint, MI	5461	2993
Fort Collins, CO	7331	5965
Fort Lauderdale, FL	8282	1557
Fort Wayne, IN	5942	2982
Fort Worth, TX	8479	4122
Frankfort, KY	6462	2634
Fresno, CA	8669	8239
Grand Forks, ND	5420	5300
Grand Island, NE	6901	4936
Grand Junction, CO	7804	6438
Grand Rapids, MI	5628	3261
Greeley, CO	7345	5895
Green Bay, WI	5512	3747
Hackensack, NJ	4976	1432
Harrisburg, PA	5363	1733
Hartford, CT	4687	1373
Helena, MT	6336	7348
Hot Springs, AR	7827	3554
Houston, TX	8938	3563
Huntsville, AL	7267	2535
Huron, SD	6201	5183
Indianapolis, IN	6272	2992
Iowa City, IA	6313	3972
Jackson, MS	8035	2880
Jacksonville, FL	7649	1276
Johnson City, TN	6595	2050
Joliet, IL	6088	3454
Joplin, MO	7421	4015
Kalamazoo, MI	5749	3177
Kansas City, MO	7027	4203
Kennewick, WA	6595	8391
Knoxville, TN	6801	2251
La Crosse, WI	5874	4133
Lansing, MI	5584	3081
Laredo, TX	9681	4099
Las Cruces, NM	9132	5742
Las Vegas, NV	8665	7411
Lawton, OK	8178	4451
Leesburg, VA	5634	1685

City	V coordinate	H coordinate
Lewiston, ME	4042	1391
Logan, UT	7367	7102
Long Beach, CA	9217	7856
Los Angeles, CA	9213	7878
Lubbock, TX	8596	4962
Lynchburg, VA	6093	1703
Macon, GA	7364	1865
Madison, WI	5887	3796
Madisonville, KY	6845	2942
Manchester, NH	4354	1388
Medford, OR	7503	8892
Memphis, TN	7471	3125
Meridian, MS	7899	2639
Miami, FL	8351	1527
Milwaukee, WI	5788	3589
Minneapolis, MN	5777	4513
Missoula, MT	6336	7650
Mobile, AL	8167	2367
Montgomery, AL	7692	2247
Morgantown, WV	5764	2083
Morristown, NJ	5035	1478
Muncie, IN	6130	2925
Nashua, NH	4394	1356
Nashville, TN	7010	2710
Nassau, NY	4961	1355
New Brunswick, NJ	5085	1434
New Haven, CT	4792	1342
New London, CT	4700	1242
New Orleans, LA	8483	2638
New York City, NY	4977	1406
Newark, NJ	5015	1430
Norfolk, VA	5918	1223
North Bend, WA	6354	8815
Oakland, CA	8486	8695
Ogden, UT	7480	7100
Oklahoma City, OK	7947	4373
Omaha, NE	6687	4595
Orlando, FL	7954	1031
Paduca, KY	6982	3088
Pendleton, OR	6707	8326
Peoria, IL	6362	3592
Philadelphia, PA	5257	1501

City	V coordinate	H coordinate
Phoenix, AZ	9135	6748
Pine Bluff, AR	7803	3358
Pittsburgh, PA	5621	2185
Pocatello, ID	7146	7250
Portland, ME	4121	1384
Portland, OR	6799	8914
Poughkeepsie, NY	4821	1526
Providence, RI	4550	1219
Provo, UT	7680	7006
Pueblo, CO	7787	5742
Racine, WI	5837	3535
Raleigh, NC	6344	1436
Reading, PA	5258	1612
Redwood City, CA	8556	8682
Reno, NV	8064	8323
Richmond, VA	5906	1472
Roanoke, VA	6196	1801
Rochester, NY	4913	2195
Rock Island, Il	6276	3816
Sacramento, CA	8304	8580
Salt Lake City, UT	7576	7065
San Antonio, TX	9225	4062
San Bernardino, CA	9172	7710
San Diego, CA	9468	7629
San Francisco, CA	8492	8719
San Jose, CA	8583	8619
Santa Fe, NM	8389	5804
Santa Monica, CA	9227	7920
Santa Rosa, CA	8354	8787
Savannah, GA	7266	1379
Scranton, PA	5042	1715
Seattle, WA	6336	8896
Shreveport, LA	8272	3495
Sioux City, IA	6468	4768
Sioux Falls, SD	6279	4900
South Bend, IN	5918	3206
Spartanburg, SC	6811	1833
Spokane, WA	6247	8180
Springfield, IL	6539	3518
Springfield, MA	4620	1408
Springfield, MO	7310	3836
St. Joseph, MO	6913	4301

City	V coordinate	H coordinate
St. Paul, MN	5776	4498
Stamford, CT	4897	1388
Sunnyvale, CA	8576	8643
Syracuse, NY	4798	1990
Tallahassee, FL	7877	1716
Tampa, FL	8173	1147
Terre Haute, IN	6428	3145
Toledo, OH	5704	2820
Topeka, KS	7110	4369
Trenton, NJ	5164	1440
Troy, NY	4616	1633
Tulsa, OK	7707	4173
Tucson, AZ	9345	6485
Van Nuys, CA	9197	7919
Washington, DC	5622	1583
Westchester, NY	4912	1330
Wheeling, WV	5755	2241
Wichita, KS	7489	4520
Wilmington, DE	5326	1485
Winchester, KY	6441	2509
Winston-Salem, NC	6440	1710
Worchester, MA	4513	1330
Yakima, WA	6533	8607
Yuma, AZ	9385	7171

Appendix

B

Calling Countries from the United States: Country and City Codes

Country	Code	City	Code
ALGERIA	213	city code not used	
AMERICAN SAMOA	684	city code not used	
ANDORRA	33	ALL CITIES	628
ARGENTINA	54	BUENOS ARIES	1
ARGENTINA	54	CORDOBA	1
ARGENTINA	54	LA PLATA	21
ARGENTINA	54	ROSARIO	41
ARUBA	297	ALL CITIES	8
ASCENSION ISL.	247	city code not used	
AUSTRALIA	61	ADELAIDE	8
AUSTRALIA	61	BRISBANE	7
AUSTRALIA	61	MELBOURNE	3
AUSTRALIA	61	SYDNEY	2
AUSTRIA	43	GRAZ	316
AUSTRIA	43	LINZ DONAU	732
AUSTRIA	43	VIENNA	1
BAHRAIN	973	city code not used	
BANGLEDESH	880	BARISAL	431
BANGLEDESH	880	CHITTAGONG	31
BANGLEDESH	880	DHAKA	2
BANGLEDESH	880	KHULNA	41
BELGIUM	32	ANTWERP	3

Country	Code	City	Code
BELGIUM	32	BRUSSELS	2
BELGIUM	32	GHENT	91
BELGIUM	32	LIEGE	41
BELIZE	501	BELIZE CITY	direct
BELIZE	501	COROZAL TOWN	04
BELIZE	501	PUNTA GORDA	07
BENIN	229	city code not used	
BOLIVIA	591	COCHABAMBA	42
BOLIVIA	591	LA PAZ	2
BOLIVIA	591	SANTA CRUZ	33
BRAZIL	55	BELE HORIZONTE	31
BRAZIL	55	RIO DE JANEIRO	21
BRAZIL	55	SAO PAULO	11
BRUNEI	673	BANDER SERI BEGAWAN	2
BRUNEI	673	KUALA BELAIT	3
BRUNEI	673	TUTONG	4
BULGARIA	359	PLOVDIV	32
BULGARIA	359	ROUSSE	82
BULGARIA	359	SOFIA	2
BULGARIA	359	VARNA	52
CAMEROON	237	city code not used	
CHILE	56	CONCEPCION	41
CHILE	56	SANTIAGO	2
CHILE	56	VALPARAISO	32
CHINA	86	BEJIENG (PEKING)	1
CHINA	86	FUZHOU	591
CHINA	86	GHUANGZHOU (CANTON)	20
CHINA	86	SHANGHAI	21
COLUMBIA	57	BARRANQUILLA	5
COLUMBIA	57	BOGATA	1
COLUMBIA	57	CALI	23
COLUMBIA	57	MEDELLIN	4
COSTA RICA	506	city code not used	
CYPRUS	357	LIMASSOL	51
CYPRUS	357	NICOSIA	2
CYPRUS	357	PAPHOS	61
CZECHOSLOVAKIA	42	BRATISLAVA	7
CZECHOSLOVAKIA	42	BRNO	5
CZECHOSLOVAKIA	42	HAVIROV	6994
CZECHOSLOVAKIA	42	OSTRAVA	69
CZECHOSLOVAKIA	42	PRAGUE	2
DENMARK	45	AALBORG	8

Country	Code	City	Code
DENMARK	45	AARHUS	6
DENMARK	45	COPENHAGEN	1
DENMARK	45	ODDENSE	7
ECUADOR	593	AMBATO	2
ECUADOR	593	CUENCA	7
ECUADOR	593	GUAYAQUIL	4
ECUADOR	593	QUITO	2
EGYPT	20	ALEXANDRIA	3
EGYPT	20	ASWAN	97
EGYPT	20	ASYUT	88
EGYPT	20	BENHA	13
EGYPT	20	CAIRO	2
EL SALVADOR	503	city code not used	
ETHIOPIA	251	ADDIS ABABA	1
ETHIOPIA	251	AKAKI	1
ETHIOPIA	251	ASMARA	4
ETHIOPIA	251	ASSAB	3
ETHIOPIA	251	AWASSA	6
FAEROE ISLANDS	298	city code not used	
FIJI ISLANDS	679	city code not used	
FINLAND	358	EPPO EBBO	15
FINLAND	358	HELSINKI	0
FINLAND	358	TAMMEFORS-TAMPERE	31
FINLAND	358	TURKU	21
FRANCE	33	LYON	7
FRANCE	358	MARSEILLE	91
FRANCE	358	NICE	93
FRANCE	358	PARIS	1
FRENCH ANTILLES	596	city code not used	
FRENCH GUIANA	594	city code not used	
FRENCH POLYNESIA	689	city code not used	
GABON	241	city code not used	
GAMBIA	220	city code not used	
GERMAN DEM. REP.	37	BERLIN	2
GERMAN DEM. REP.	37	DRESDEN	51
GERMAN DEM. REP.	37	LEIPZIG	41
GERMAN DEM. REP.	37	MAGDEBURG	91
GERMANY FED. REP	49	BERLIN	30
GERMANY FED. REP	49	BONN	228
GERMANY FED. REP	49	FRANKFURT	69
GERMANY FED. REP	49	MUNICH	89
GIBRALTAR	350	city code not used	

Country	Code	City	Code
GREECE	30	ATHENS	1
GREECE	30	IRAKLION	81
GREECE	30	LARISSA	41
GREECE	30	PIRAEUS PIREEFS	1
GREENLAND	299	GODTHAB	2
GREENLAND	299	SONDRE STROMFJORD	11
GREENLAND	299	THULE	50
GUADELOUPE	590	city code not used	
GUAM	671	city code not used	
GUANTANAMO BAY USN	5399	city code not used	
GUATEMALA	502	ANTIGUA	9
GUATEMALA	502	GUATEMALA CITY	2
GUATEMALA	502	QUEZALTENANGO	9
GUYANA	592	BARTICA	5
GUYANA	592	GEORGETOWN	2
GUYANA	592	NEW AMSTERDAM	3
HAITI	509	CAP-HAITIEN	3
HAITI	509	CAYES	5
HAITI	509	GONAIVE	2
HAITI	509	PORT AU PRINCE	1
HONDURAS	504	city code not used	
HONG KONG	852	HONG KONG	5
HONG KONG	852	KOWLOON	3
HONG KONG	852	NEW TERRITORIES	0
HUNGARY	36	BUDAPEST	1
HUNGARY	36	DERBRECEN	52
HUNGARY	36	GYOR	96
HUNGARY	36	MISKOLC	46
ICELAND	354	AKUREYRI	6
ICELAND	354	KEFLAVIC	2
ICELAND	354	REYKJAVIK	1
INDIA	91	BOMBAY	22
INDIA	91	CALCUTTA	33
INDIA	91	MADRAS	44
INDIA	91	NEW DELHI	11
INDONESIA	62	JAKARTA	21
INDONESIA	62	MEDAN	61
INDONESIA	62	SEMARANG	24
IRAN	98	ESFAHAN	31
IRAN	98	MASHAD	51
IRAN	98	TABRIZ	41
IRAN	98	TEHRAN	21

Country	Code	City	Code
IRELAND	353	CORK	21
IRELAND	353	DUBLIN	1
IRELAND	353	GALWAY	91
IRELAND	353	LIMERICK	61
ISRAEL	972	HAIFA	4
ISRAEL	972	JERUSALEM	2
ISRAEL	972	RAMAT GAN	3
ISRAEL	972	TEL AVIV	3
ITALY	39	FLORENCE	55
ITALY	39	GENOA	10
ITALY	39	MILAN	2
ITALY	39	NAPLES	81
ITALY	39	ROME	6
IVORY COAST	255	city code not used	
JAPAN	81	KYOTO	75
JAPAN	81	OSAKA	6
JAPAN	81	SAPORRO	11
JAPAN	81	TOKYO	3
JAPAN	81	YOKOHAMA	45
JORDAN	962	AMMAN	6
JORDAN	962	IRBID	2
JORDAN	962	JERASH	4
JORDAN	962	KARAK	3
JORDAN	962	MA'AN	3
KENYA	254	KISUMU	35
KENYA	254	MOMBASA	11
KENYA	254	NAIROBI	2
KENYA	254	NAKURU	37
KOREA	82	INCHEON	32
KOREA	82	PUSAN	51
KOREA	82	SEOUL	2
KOREA	82	TAEGU	53
KUWAIT	965	city code not used	
LESOTHO	266	city code not used	
LIBERIA	231	city code not used	
LIBYA	218	BENGHAZI	61
LIBYA	218	MISURATHA	51
LIBYA	218	TRIPOLI	21
LIBYA	218	ZAWAI	23
LEICHTENSTIEN	41	ALL CITIES	75
LUXEMBOURG	352	city code not used	
MACAO	853	city code not used	

Country	Code	City	Code
MALAWI	265	DOMASI	531
MALAWI	265	MAKWASA	474
MALAWI	265	ZOMBA	50
MALAYSIA	60	IPOH	5
MALAYSIA	60	JOHOR BAHRU	7
MALAYSIA	60	KAJANG	3
MALAYSIA	60	KUALA LUMPUR	3
MALTA	356	city code not used	
MARSHALL ISLANDS	692	EBEYE	871
MARSHALL ISLANDS	692	MAJURO	9
MICRONESIA	691	KOSREA	851
MICRONESIA	691	PONAPE	9
MICRONESIA	691	TRUK	8319
MICRONESIA	691	YAP	841
MIQUELON	508	city code not used	
MONACO	33	ALL CITIES	93
MOROCCO	212	AGADIR	8
MOROCCO	212	BENI-MELLAL	48
MOROCCO	212	CASABLANCA	direct
MOROCCO	212	EL JADIDA	34
NAMBIA	264	GROOTFONTEIN	673
NAMBIA	264	KEETMANSHOOP	631
NAMBIA	264	MARIENTAL	661
NETHERLANDS	31	AMSTERDAM	20
NETHERLANDS	31	ROTTERDAM	10
NETHERLANDS	31	THE HAGUE	70
NETHERLANDS ANTILLES	599	BONAIRE	7
NETHERLANDS ANTILLES	599	CURACAO	9
NETHERLANDS ANTILLES	599	ST. EUSTATIUS	3
NETHERLANDS ANTILLES	599	ST. MAARTEN	5
NEW CALEDONIA	687	city code not used	
NEW ZEALAND	64	AUKLAND	9
NEW ZEALAND	64	CHRISTCHURCH	3
NEW ZEALAND	64	DUNEDIN	24
NEW ZEALAND	64	HAMILTON	71
NICARAGUA	505	CHINANDEGA	341
NICARAGUA	505	DIRIAMBA	42
NICARAGUA	505	LEON	311
NICARAGUA	505	MANAGUA	2
NIGERIA	234	LAGOS	1
NORWAY	47	BERGEN	5
NORWAY	47	OSLO	2

Country	Code	City	Code
NORWAY	47	STAVANGER	4
NORWAY	47	TRONDHEIM	7
OMAN	968	city code not used	
PAKISTAN	92	ISLAMABAD	51
PAKISTAN	92	KARICHI	21
PAKISTAN	92	LAHORE	42
PANAMA	507	city code not used	
PAPUA NEW GUINEA	675	city code not used	
PARAGUAY	595	ASUNCION	21
PARAGUAY	595	CONCEPCION	31
PERU	51	AREQUIPA	54
PERU	51	CALLAO	14
PERU	51	LIMA	14
PERU	51	TRUJILLO	44
PHILLIPINES	63	CEBU	32
PHILLIPINES	63	DAVAO	35
PHILLIPINES	63	MANILLA	2
POLAND	48	CRAKOW	12
POLAND	48	GDANSK	58
POLAND	48	WARSAW	22
PORTUGAL	351	COIMBRA	39
PORTUGAL	351	LISBON	1
PORTUGAL	351	PORTO	2
PORTUGAL	351	SETUBAL	65
QATAR	974	city code not used	
ROMANIA	40	BUCHAREST	0
ROMANIA	40	CLUJ-NAPOCA	51
ROMANIA	40	CONSTANTA	16
SAIPAN	670	ROTA ISL.	532
SAIPAN	670	SUSUPE CITY	234
SAIPAN	670	TINIAN ISL.	433
SAN MARINO	39	ALL CITIES	541
SAUDI ARABIA	966	HOFUF	3
SAUDI ARABIA	966	JEDDAH	2
SAUDI ARABIA	966	MAKKAH	2
SAUDI ARABIA	966	RIYADH	1
SENEGAL	221	city code not used	
SINGAPORE	65	city code not used	
SOUTH AFRICA	27	CAPE TOWN	21
SOUTH AFRICA	27	DURBAN	31
SOUTH AFRICA	27	JOHANNESBURG	11
SPAIN	34	BARCELONA	3

Country	Code	City	Code
SPAIN	34	MADRID	1
SPAIN	34	SEVILLE	54
SPAIN	34	VALENCIA	6
SRI LANKA	94	COLOMBO CENTRAL	1
SRI LANKA	94	KANDY	8
SRI LANKA	94	KOTTE	1
ST. PIERRE	508	city code not used	
SURINAME	597	city code not used	
SWAIZLAND	268	city code not used	
SWEDEN	46	GOTEBORG	31
SWEDEN	46	MALMO	40
SWEDEN	46	STOCKHOLM	8
SWEDEN	46	VASTERAS	21
SWITZERLAND	41	BASEL	61
SWITZERLAND	41	BERNE	31
SWITZERLAND	41	GENEVA	22
SWITZERLAND	41	ZURICH	1
TAIWAN	886	KAOHSIUNG	7
TAIWAN	886	TAINAN	6
TAIWAN	886	TAIPEI	2
TANZANIA	255	DAR ES SALAAM	51
TANZANIA	255	DODOMA	61
TANZANIA	255	MWANZA	68
TANZANIA	255	TANGA	53
THAILAND	66	BANGKOK	2
THAILAND	66	BURIRUM	44
THAILAND	66	CHANTHABURI	39
TOGO	228	city code not used	
TUNSIA	216	BIZERTE	2
TUNSIA	216	KAIROUAN	7
TUNSIA	216	MSEL BOURGUIBA	2
TUNSIA	216	TUNIS	1
TURKEY	90	ADANA	711
TURKEY	90	ANKARA	41
TURKEY	90	ISTANBUL	1
TURKEY	90	IZMIR	51
UGANDA	256	ENTEBBE	42
UGANDA	256	JINJA	43
UGANDA	256	KAMPALA	41
UGANDA	256	KYAMBOGO	41
UNITED ARAB EMIRATES	971	ABU DHABI	2
UNITED ARAB EMIRATES	971	AJMAN	6

Country	Code	City	Code
UNITED ARAB EMIRATES	971	AL AIN	3
UNITED ARAB EMIRATES	971	DUBAI	4
UNITED ARAB EMIRATES	971	SHARJAH	6
UNITED KINGDOM	44	BELFAST	232
UNITED KINGDOM	44	BIRMINGHAM	21
UNITED KINGDOM	44	GLASGOW	41
UNITED KINGDOM	44	LONDON	1
URAGUAY	598	CANELONES	332
URAGUAY	598	MERCEDES	532
URAGUAY	598	MONTEVIDEO	2
VATICAN CITY	39	ALL OF VATICAN CITY	6
VENEZUELA	58	BARQUISIMETO	51
VENEZUELA	58	CARACAS	2
VENEZUELA	58	MARACAIBO	61
VENEZUELA	58	VALENCIA	41
YEMEN ARAB REPUBLIC	967	AMRAN	2
YEMEN ARAB REPUBLIC	967	SANAA	2
YEMEN ARAB REPUBLIC	967	TAIZ	4
YEMEN ARAB REPUBLIC	967	YARIM	4
YEMEN ARAB REPUBLIC	967	ZABID	3
YUGOSLAVIA	38	BELGRADE	11
YUGOSLAVIA	38	SARAJEVO	71
YUGOSLAVIA	38	ZAGREB	41
ZAIRE	243	KINSHASA	12
ZAIRE	243	LUBUMBASHI	222
ZAMBIA	260	CHINGOLA	2
ZAMBIA	260	KITWE	2
ZAMBIA	260	LUANSHYA	2
ZAMBIA	260	LUSAKA	1
ZAMBIA	260	NDOLA	26
ZIMBABWE	263	BULAWAYO	9
ZIMBABWE	263	HARARE	4
ZIMBABWE	263	MUTARE	20

Area Codes of the NANP Listed by Location

State/Province/Country	City	Area Code
ALABAMA	AUBURN	334
ALABAMA	BIRMINGHAM	205
ALABAMA	DECATUR	256
ALABAMA	GADSDEN	256
ALABAMA	HUNTSVILLE	256
ALABAMA	MOBILE	334
ALABAMA	MONTGOMERY	334
ALABAMA	SELMA	334
ALABAMA	TUSCALOOSA	205
ALASKA	ALL LOCATIONS	907
ALBERTA	CALGARY	403
ALBERTA	EDMONTON	780
ALBERTA	GRANDE PRAIRIE	780
ANGUILLA	ALL LOCATIONS	264
ANTIGUA/BARBUDA	ALL LOCATIONS	268
ARIZONA	FLAGSTAFF	520
ARIZONA	PHOENIX	602
ARIZONA	PRESCOTT	520
ARIZONA	SUBURBAN EAST PHOENIX	480
ARIZONA	SUBURBAN WEST PHOENIX	623

State/Province/Country	City	Area Code
ARIZONA	TUCSON	520
ARIZONA	YUMA	520
ARKANSAS	ALL LOCATIONS	501
ARKANSAS	JONESBORO	870
ARKANSAS	PINE BLUFF	870
ARKANSAS	TEXARKANA	870
BAHAMAS	ALL LOCATIONS	242
BARBADOS	ALL LOCATIONS	246
BERMUDA	ALL LOCATIONS	441
BRITISH COLUMBIA	KAMLOOPS	250
BRITISH COLUMBIA	KELOWNA	250
BRITISH COLUMBIA	VANCOUVER	604
BRITISH COLUMBIA	VICTORIA	250
BRITISH VIRGIN ISLANDS	ALL LOCATIONS	284
CALIFORNIA	ALHAMBRA	626
CALIFORNIA	ANAHEIM	657
CALIFORNIA	ANAHEIM	714
CALIFORNIA	BAKERSFIELD	661
CALIFORNIA	BAKERSFIELD	805
CALIFORNIA	BEVERLY HILLS	310
CALIFORNIA	BEVERLY HILLS	424
CALIFORNIA	BISHOP	760
CALIFORNIA	BURBANK	818
CALIFORNIA	CHICO	530
CALIFORNIA	CHULA VISTA	935
CALIFORNIA	CORONA	951
CALIFORNIA	COSTA MESA	949
CALIFORNIA	EL CAJON	935
CALIFORNIA	FRESNO	559
CALIFORNIA	GARDENA	310
CALIFORNIA	GARDENA	424
CALIFORNIA	IRVINE	949
CALIFORNIA	LA MESA	935
CALIFORNIA	LA PUENTE	626
CALIFORNIA	LANCASTER	661
CALIFORNIA	LONG BEACH	562
CALIFORNIA	LOS ANGELES	323
CALIFORNIA	METRO LOS ANGELES	213
CALIFORNIA	MODESTO	209
CALIFORNIA	MONTEREY	831
CALIFORNIA	MOUNTAIN VIEW	650

State/Province/Country	City	Area Code
CALIFORNIA	MOUNTAIN VIEW	764
CALIFORNIA	NAPA	707
CALIFORNIA	NEEDLES	760
CALIFORNIA	NEWPORT BEACH	949
CALIFORNIA	OAKLAND	341
CALIFORNIA	OAKLAND	510
CALIFORNIA	ONTARIO	752
CALIFORNIA	ONTARIO	909
CALIFORNIA	PALM SPRINGS	760
CALIFORNIA	PALMDALE	661
CALIFORNIA	PASADENA	818
CALIFORNIA	PASADENA	626
CALIFORNIA	PETALUMA	707
CALIFORNIA	PLEASANTON	925
CALIFORNIA	POMONA	752
CALIFORNIA	POMONA	909
CALIFORNIA	REDDING	530
CALIFORNIA	RIVERSIDE	951
CALIFORNIA	SACRAMENTO	916
CALIFORNIA	SALINAS	831
CALIFORNIA	SAN DIEGO	619
CALIFORNIA	SAN FRANCISCO	628
CALIFORNIA	SAN FRANCISCO	415
CALIFORNIA	SAN JOSE	408
CALIFORNIA	SAN JOSE	669
CALIFORNIA	SAN MATEO	764
CALIFORNIA	SAN MATEO	650
CALIFORNIA	SANTA BARBARA	805
CALIFORNIA	SANTA MONICA	562
CALIFORNIA	STOCKTON	209
CALIFORNIA	UPLAND	752
CALIFORNIA	UPLAND	909
CALIFORNIA	VAN NUYS	818
CALIFORNIA	VISALIA	559
CALIFORNIA	WALNUT CREEK	925
CALIFORNIA	WHITTIER	562
CALIFORNIA	YREKA	530
CALIFORNIA	SAN DIEGO	858
CANADA (SERVICES)		600
CAYMAN ISLANDS	ALL LOCATIONS	345
CNMI	ALL LOCATIONS	670

State/Province/Country	City	Area Code
COLORADO	ASPEN	970
COLORADO	COLORADO SPRINGS	719
COLORADO	DENVER	303
COLORADO	DENVER	720
COLORADO	PUEBLO	719
COLORADO	VAIL	970
CONNECTICUT	BLOOMFIELD	860
CONNECTICUT	CANAAN	860
CONNECTICUT	FAIRFIELD	203
CONNECTICUT	FAIRFIELD	475
CONNECTICUT	HARTFORD	860
CONNECTICUT	NEW HAVEN	475
CONNECTICUT	NEW HAVEN	203
CONNECTICUT	NORWICH	860
CONNECTICUT	SOUTHERN COUNTIES	203
CONNECTICUT	BLOOMFIELD	959
CONNECTICUT	CANAAN	959
CONNECTICUT	HARTFORD	959
CONNECTICUT	NORWICH	959
DELAWARE	ALL LOCATIONS	302
DISTRICT OF COLUMBIA	ALL LOCATIONS	202
DOMINICA	ALL LOCATIONS	767
DOMINICAN REPUBLIC	ALL LOCATIONS	809
FLORIDA	ARCADIA	863
FLORIDA	BOCA RATON	561
FLORIDA	BRADENTON	941
FLORIDA	BROWARD COUNTY	954
FLORIDA	CAPE CANAVERAL	321
FLORIDA	CLEARWATER	727
FLORIDA	CLEARWATER	813
FLORIDA	DAYTONA BEACH	904
FLORIDA	FT. LAUDERDALE	954
FLORIDA	FT. MYERS	941
FLORIDA	GAINESVILLE	352
FLORIDA	JACKSONVILLE	904
FLORIDA	KEY WEST	305
FLORIDA	KISSIMMEE	407
FLORIDA	LAKELAND	863
FLORIDA	MIAMI	305
FLORIDA	MIAMI	786
FLORIDA	NAPLES	941

State/Province/Country	City	Area Code
FLORIDA	NORTH DADE	786
FLORIDA	ORLANDO	407
FLORIDA	SARASOTA	941
FLORIDA	SEBRING	863
FLORIDA	ST. PETERSBURG	727
FLORIDA	ST. PETERSBURG	813
FLORIDA	STUART	561
FLORIDA	TAMPA	813
FLORIDA	WEST PALM BEACH	561
FLORIDA	TALLAHASSEE	850
GEORGIA	ATLANTA	404
GEORGIA	ATLANTA AND SUBURBAN AREAS	678
GEORGIA	AUGUSTA	706
GEORGIA	COLUMBUS	706
GEORGIA	MACON	478
GEORGIA	SAVANNAH	912
GEORGIA	SUBURBAN ATLANTA (THE AREA OUTSIDE THE INTERSTATE 285 BELTWAY)	770
GEORGIA	ALBANY	229
GRENADA	ALL LOCATIONS	473
GUAM	ALL LOCATIONS	671
HAWAII	ALL LOCATIONS	808
IDAHO	ALL LOCATIONS	208
ILLINOIS	BLOOMINGTON	309
ILLINOIS	CHAMPAIGN	217
ILLINOIS	CHICAGO	312
ILLINOIS	CHICAGO	773
ILLINOIS	CHICAGO	872
ILLINOIS	COLLINSVILLE	618
ILLINOIS	DES PLAINES	708
ILLINOIS	GRANITE CITY	618
ILLINOIS	JOLIET	815
ILLINOIS	PALATINE	708
ILLINOIS	PEORIA	309
ILLINOIS	ROCKFORD	815
ILLINOIS	SPRINGFIELD	217
ILLINOIS	SUBURBAN CHICAGO	331
ILLINOIS	SUBURBAN CHICAGO	464

State/Province/Country	City	Area Code
ILLINOIS	SUBURBAN CHICAGO	630
ILLINOIS	SUBURBAN CHICAGO	224
ILLINOIS	WAUKEGAN	708
ILLINOIS	ELGIN	847
INDIANA	BLOOMINGTON	812
INDIANA	EVANSVILLE	812
INDIANA	GARY	219
INDIANA	HAMMOND	219
INDIANA	INDIANAPOLIS	317
INDIANA	KOKOMO	765
INDIANA	LAFAYETTE	765
INDIANA	RICHMOND	765
INDIANA	SOUTH BEND	219
INDIANA	TERRA HAUTE	812
INDIANA	WARSAW	219
IOWA	COUNCIL BLUFFS	712
IOWA	DES MOINES	515
IOWA	DUBUQUE	319
IOWA	SIOUX CITY	712
JAMAICA	ALL LOCATIONS	876
KANSAS	DODGE CITY	316
KANSAS	JUNCTION	785
KANSAS	KANSAS CITY	913
KANSAS	LAWRENCE	785
KANSAS	TOPEKA	785
KANSAS	TOPEKA	913
KANSAS	WICHITA	316
KENTUCKY	ASHLAND	606
KENTUCKY	BOWLING GREEN	270
KENTUCKY	LOUISVILLE	502
KENTUCKY	PADUCAH	270
LOUISIANA	ALEXANDRIA	318
LOUISIANA	BATON ROUGE	225
LOUISIANA	LAFAYETTE	337
LOUISIANA	LAKE CHARLES	337
LOUISIANA	MONROE	318
LOUISIANA	NEW ORLEANS	504
LOUISIANA	PLAQUEMINE	225
MAINE	ALL LOCATIONS	207
MANITOBA	ALL LOCATIONS	204
MARYLAND	BALTIMORE	410

State/Province/Country	City	Area Code
MARYLAND	BALTIMORE	443
MARYLAND	BETHESDA	240
MARYLAND	BETHESDA	301
MARYLAND	ROCKVILLE	240
MARYLAND	ROCKVILLE	301
MARYLAND	SILVER SPRING	240
MARYLAND	SILVER SPRING	301
MARYLAND	TOWSON	410
MARYLAND	TOWSON	443
MASSACHUSETTS	ACTON	978
MASSACHUSETTS	BOSTON	617
MASSACHUSETTS	BURLINGTON	781
MASSACHUSETTS	FRAMINGHAM	508
MASSACHUSETTS	LOWELL	978
MASSACHUSETTS	NEW BEDFORD	508
MASSACHUSETTS	NORTH READING	978
MASSACHUSETTS	PLYMOUTH	508
MASSACHUSETTS	READING	781
MASSACHUSETTS	SPRINGFIELD	413
MASSACHUSETTS	WALTHAM	781
MEXICO ROAMING-TEMP		521
MEXICO ROAMING-TEMP		523
MEXICO ROAMING-TEMP		524
MEXICO ROAMING-TEMP		525
MEXICO ROAMING-TEMP		526
MEXICO ROAMING-TEMP		527
MEXICO ROAMING-TEMP		528
MEXICO ROAMING-TEMP		529
MICHIGAN	ANN ARBOR	278
MICHIGAN	ANN ARBOR	313
MICHIGAN	ANN ARBOR	734
MICHIGAN	BATTLE CREEK	616
MICHIGAN	BIRMINGHAM	586
MICHIGAN	BIRMINGHAM	810
MICHIGAN	DETROIT	313
MICHIGAN	FLINT	586
MICHIGAN	FLINT	810
MICHIGAN	GRAND RAPIDS	616
MICHIGAN	KALAMAZOO	616
MICHIGAN	LANSING	517
MICHIGAN	LIVONIA	278

State/Province/Country	City	Area Code
MICHIGAN	LIVONIA	734
MICHIGAN	MARIE	906
MICHIGAN	MARQUETTE	906
MICHIGAN	MUSKEGON	231
MICHIGAN	OAKLAND COUNTY	248
MICHIGAN	OAKLAND COUNTY	947
MICHIGAN	PONTIAC	586
MICHIGAN	PONTIAC	810
MICHIGAN	SAULT STE. MARIE	906
MICHIGAN	TRAVERSE CITY	231
MICHIGAN	YPSILANTI	278
MICHIGAN	YPSILANTI	734
MICHIGAN	ALPENA	989
MICHIGAN	DETROIT	679
MICHIGAN	MIDLAND	989
MICHIGAN	SAGINAW	989
MINNESOTA	DULUTH	218
MINNESOTA	MINNEAPOLIS	952
MINNESOTA	MINNEAPOLIS	612
MINNESOTA	ROCHESTER	507
MINNESOTA	ST. CLOUD	320
MINNESOTA	ST. PAUL	651
MINNESOTA	WILLMAR	320
MISSISSIPPI	BILOXI	228
MISSISSIPPI	GREENVILLE	662
MISSISSIPPI	GULFPORT	228
MISSISSIPPI	HATTIESBURG	601
MISSISSIPPI	JACKSON	601
MISSISSIPPI	OXFORD	662
MISSISSIPPI	PASCAGOULA	228
MISSISSIPPI	TUPELO	662
MISSOURI	CHESTERFIELD	636
MISSOURI	GRAY SUMMIT	636
MISSOURI	JEFFERSON CITY	573
MISSOURI	JOPLIN	417
MISSOURI	KANSAS CITY	816
MISSOURI	KIRKSVILLE	660
MISSOURI	MARYVILLE	660
MISSOURI	O'FALLON	636
MISSOURI	OUTSIDE ST. LOUIS	573
MISSOURI	SPRINGFIELD	417

State/Province/Country	City	Area Code
MISSOURI	ST. LOUIS	314
MISSOURI	TRENTON	660
MONTANA	ALL LOCATIONS	406
MONTSERRAT	ALL LOCATIONS	664
NEBRASKA	LINCOLN	402
NEBRASKA	NORTH PLATTE	308
NEBRASKA	OMAHA	402
NEVADA	CARSON CITY	775
NEVADA	LAS VEGAS	702
NEVADA	RENO	775
NEW BRUNSWICK	ALL LOCATIONS	506
NEW HAMPSHIRE	ALL LOCATIONS	603
NEW JERSEY	ASBURY PARK	908
NEW JERSEY	ATLANTIC CITY	609
NEW JERSEY	ELIZABETH	908
NEW JERSEY	ESSEX COUNTIES	973
NEW JERSEY	HACKENSACK	201
NEW JERSEY	JERSEY CITY	201
NEW JERSEY	MIDDLESEX	732
NEW JERSEY	MONMOUTH	732
NEW JERSEY	MORRIS	973
NEW JERSEY	MORRISTOWN	201
NEW JERSEY	NEW BRUNSWICK	908
NEW JERSEY	NEWARK	201
NEW JERSEY	OCEAN COUNTIES	732
NEW JERSEY	PASSAIC	973
NEW JERSEY	PRINCETON	609
NEW JERSEY	TRENTON	609
NEW JERSEY	WESTFIELD	908
NEW JERSEY	CHERRY HILL	856
NEW MEXICO	ALL LOCATIONS	505
NEW YORK	ALBANY	518
NEW YORK	BINGHAMTON	607
NEW YORK	BRONX	347
NEW YORK	BROOKLYN	347
NEW YORK	BROOKLYN	718
NEW YORK	BUFFALO	716
NEW YORK	ELMIRA	607
NEW YORK	MANHATTAN	646
NEW YORK	NASSAU COUNTY	516
NEW YORK	NEW YORK CITY	212

State/Province/Country	City	Area Code
NEW YORK	NEW YORK CITY	917
NEW YORK	NEWBURGH	914
NEW YORK	NIAGARA FALLS	716
NEW YORK	OSSINING	914
NEW YORK	POUGHKEEPSIE	914
NEW YORK	QUEENS	347
NEW YORK	QUEENS	718
NEW YORK	QUEENS	917
NEW YORK	ROCHESTER	716
NEW YORK	SCHENECTADY	518
NEW YORK	STATEN ISLAND	347
NEW YORK	STATEN ISLAND	718
NEW YORK	STATEN ISLAND	917
NEW YORK	SUFFOLK COUNTY	631
NEW YORK	SYRACUSE	315
NEWFOUNDLAND	ALL LOCATIONS	709
NORTH CAROLINA	ASHEBORO	336
NORTH CAROLINA	ASHEVILLE	828
NORTH CAROLINA	BURLINGTON	336
NORTH CAROLINA	CHARLOTTE	704
NORTH CAROLINA	DURHAM	919
NORTH CAROLINA	FAYETTEVILLE	910
NORTH CAROLINA	GREENSBORO	336
NORTH CAROLINA	GREENVILLE	252
NORTH CAROLINA	HENDERSONVILLE	828
NORTH CAROLINA	MOREHEAD CITY	252
NORTH CAROLINA	RALEIGH	919
NORTH CAROLINA	ROCKY MOUNT	252
NORTH CAROLINA	WINSTON-SALEM	910
NORTH CAROLINA	CHARLOTTE	980
NORTH DAKOTA	ALL LOCATIONS	701
NOVA SCOTIA	ALL LOCATIONS	902
OHIO	AKRON	330
OHIO	ASHTABULA	440
OHIO	ATHENS	740
OHIO	CINCINNATI	513
OHIO	CLEVELAND	216
OHIO	COLUMBUS	614
OHIO	DAYTON	937
OHIO	HAMILTON	937
OHIO	LANCASTER	740

State/Province/Country	City	Area Code
OHIO	LORAIN	440
OHIO	MARIETTA	740
OHIO	TOLEDO	419
OHIO	WESTLAKE	440
OHIO	YOUNGSTOWN	330
OKLAHOMA	ARDMORE	580
OKLAHOMA	BROKEN ARROW	918
OKLAHOMA	ENID	580
OKLAHOMA	LAWTON	580
OKLAHOMA	MUSKOGEE	918
OKLAHOMA	NORMAN	405
OKLAHOMA	TULSA	918
ONTARIO	HAMILTON	905
ONTARIO	LONDON	519
ONTARIO	OTTAWA	613
ONTARIO	SAULT STE. MARIE	705
ONTARIO	THUNDER BAY	807
ONTARIO	TORONTO	416
ONTARIO	TORONTO	647
ONTARIO	UNIONVILLE	905
ONTARIO	WINDSOR	519
ONTARIO	WOODBRIDGE	905
OREGON	EUGENE	541
OREGON	MEDFORD	541
OREGON	PORTLAND	503
OREGON	PORTLAND	971
OREGON	SALEM	971
PENNSYLVANIA	ALLENTOWN	610
PENNSYLVANIA	ALTOONA	814
PENNSYLVANIA	BETHLEHEM	610
PENNSYLVANIA	ERIE	814
PENNSYLVANIA	HARRISBURG	717
PENNSYLVANIA	HERSHEY	717
PENNSYLVANIA	LATROBE	724
PENNSYLVANIA	NEW CASTLE	724
PENNSYLVANIA	PHILADELPHIA	215
PENNSYLVANIA	PHILADELPHIA	267
PENNSYLVANIA	PITTSBURGH	412
PENNSYLVANIA	READING	610
PENNSYLVANIA	SCRANTON	570
PENNSYLVANIA	WEST CHESTER	610

State/Province/Country	City	Area Code
PENNSYLVANIA	WILKES-BARRE	570
PENNSYLVANIA	WILLIAMSPORT	570
PENNSYLVANIA	YORK	717
PENNSYLVANIA	PHILADELPHIA	484
PUERTO RICO	ALL LOCATIONS	787
QUEBEC	AREA OUTSIDE MONTREAL	450
QUEBEC	MONTREAL	514
QUEBEC	QUEBEC	418
QUEBEC	SHERBROOKE	819
QUEBEC	TROIS RIVIERES	819
RHODE ISLAND	ALL LOCATIONS	401
SASKATCHEWAN	ALL LOCATIONS	306
SOUTH CAROLINA	CHARLESTON	843
SOUTH CAROLINA	COLUMBIA	803
SOUTH CAROLINA	GREENVILLE	864
SOUTH CAROLINA	LEXINGTON	843
SOUTH CAROLINA	SPARTANBURG	864
SOUTH CAROLINA	WINNSBORO	803
SOUTH DAKOTA	ALL LOCATIONS	605
ST. KITTS & NEVIS	ALL LOCATIONS	869
ST. LUCIA	ALL LOCATIONS	758
ST. VINCENT & GRENADINES	ALL LOCATIONS	784
TENNESSEE	CHATTANOOGA	423
TENNESSEE	CLARKSVILLE	931
TENNESSEE	KNOXVILLE	865
TENNESSEE	MEMPHIS	901
TENNESSEE	NASHVILLE	615
TEXAS	ABILENE	915
TEXAS	AMARILLO	806
TEXAS	ARLINGTON	817
TEXAS	AUSTIN	512
TEXAS	BROWNSVILLE	956
TEXAS	CONROE	936
TEXAS	CORPUS CHRISTI	361
TEXAS	DALLAS	469
TEXAS	DALLAS	972
TEXAS	DALLAS	214
TEXAS	DENTON	940
TEXAS	EL PASO	915
TEXAS	FORT WORTH	817
TEXAS	FREDERICKSBURG	830

State/Province/Country	City	Area Code
TEXAS	GALVESTON	409
TEXAS	HOUSTON	281
TEXAS	HOUSTON	832
TEXAS	HOUSTON	713
TEXAS	LAREDO	956
TEXAS	LONGVIEW	903
TEXAS	LUBBOCK	806
TEXAS	MCALLEN	956
TEXAS	NACOGDOCHES	936
TEXAS	NEW BRAUNFELS	830
TEXAS	SAN ANGELO	915
TEXAS	SAN ANTONIO	210
TEXAS	TEXARKANA	903
TEXAS	TYLER	903
TEXAS	UVALDE	830
TEXAS	WACO	254
TEXAS	WICHITA FALLS	940
TEXAS	BRYAN-COLLEGE STATION	979
TEXAS	FREEPORT	979
TRINIDAD AND TOBAGO	ALL LOCATIONS	868
TURKS & CAICOS ISLANDS	ALL LOCATIONS	649
US VIRGIN ISLANDS	ALL LOCATIONS	340
UTAH	CEDAR CITY	435
UTAH	OGDEN	801
UTAH	PROVO	801
UTAH	SALT LAKE CITY	801
UTAH	ST. GEORGE	435
VERMONT	ALL LOCATIONS	802
VIRGINIA	ALEXANDRIA	703
VIRGINIA	ARLINGTON	571
VIRGINIA	ARLINGTON	703
VIRGINIA	COVINGTON	540
VIRGINIA	FAIRFAX	571
VIRGINIA	FAIRFAX	703
VIRGINIA	FALLS CHURCH	703
VIRGINIA	NEWPORT NEWS	757
VIRGINIA	NORFOLK	757
VIRGINIA	NORFOLK	804
VIRGINIA	RICHMOND	804
VIRGINIA	ROANOKE	540
VIRGINIA	VIENNA	571

State/Province/Country	City	Area Code
VIRGINIA	VIRGINIA BEACH	757
VIRGINIA	WINCHESTER	540
WASHINGTON	AUBURN	253
WASHINGTON	BAINBRIDGE ISLAND	206
WASHINGTON	BELLINGHAM	360
WASHINGTON	BELLINGHAM	564
WASHINGTON	EVERETT	425
WASHINGTON	GIG HARBOR	253
WASHINGTON	KENT	425
WASHINGTON	OLYMPIA	360
WASHINGTON	OLYMPIA	564
WASHINGTON	SEATTLE	206
WASHINGTON	SPOKANE	509
WASHINGTON	TACOMA	253
WASHINGTON	VANCOUVER	360
WASHINGTON	VANCOUVER	564
WEST VIRGINIA	ALL LOCATIONS	304
WISCONSIN	APPLETON	920
WISCONSIN	EAU CLAIRE	715
WISCONSIN	GREEN BAY	920
WISCONSIN	KENOSHA	262
WISCONSIN	MADISON	608
WISCONSIN	MILWAUKEE	414
WISCONSIN	RACINE	262
WISCONSIN	SHEBOYGAN	920
WISCONSIN	WAUKESHA	262
WISCONSIN	WAUSAU	715
WYOMING	ALL LOCATIONS	307
YUKON & NW TERRITORIES	ALL LOCATIONS	867

Area Codes of the NANP Listed by Number

State/Province/Country	City	Area Code
EASILY RECOGNIZABLE CODE		200
NEW JERSEY	HACKENSACK	201
NEW JERSEY	MORRISTOWN	201
NEW JERSEY	NEWARK	201
NEW JERSEY	JERSEY CITY	201
DISTRICT OF COLUMBIA	ALL LOCATIONS	202
CONNECTICUT	FAIRFIELD	203
CONNECTICUT	NEW HAVEN	203
CONNECTICUT	SOUTHERN COUNTIES	203
MANITOBA	ALL LOCATIONS	204
ALABAMA	BIRMINGHAM	205
ALABAMA	TUSCALOOSA	205
WASHINGTON	SEATTLE	206
WASHINGTON	BAINBRIDGE ISLAND	206
MAINE	ALL LOCATIONS	207
IDAHO	ALL LOCATIONS	208
CALIFORNIA	MODESTO	209
CALIFORNIA	STOCKTON	209
TEXAS	SAN ANTONIO	210
NOT AVAILABLE		211
NEW YORK	NEW YORK CITY	212
CALIFORNIA	METRO LOS ANGELES	213

State/Province/Country	City	Area Code
TEXAS	DALLAS	214
PENNSYLVANIA	PHILADELPHIA	215
OHIO	CLEVELAND	216
ILLINOIS	CHAMPAIGN	217
ILLINOIS	SPRINGFIELD	217
MINNESOTA	DULUTH	218
INDIANA	GARY	219
INDIANA	HAMMOND	219
INDIANA	SOUTH BEND	219
INDIANA	WARSAW	219
GENERAL PURPOSE CODE		220
GEOGRAPHIC RELIEF CODE		221
EASILY RECOGNIZABLE CODE		222
GEOGRAPHIC RELIEF CODE		223
ILLINOIS	SUBURBAN CHICAGO	224
LOUISIANA	BATON ROUGE	225
LOUISIANA	PLAQUEMINE	225
GEOGRAPHIC RELIEF CODE		226
GEOGRAPHIC RELIEF CODE		227
MISSISSIPPI	BILOXI	228
MISSISSIPPI	GULFPORT	228
MISSISSIPPI	PASCAGOULA	228
GEORGIA	ALBANY	229
GEOGRAPHIC RELIEF CODE		230
MICHIGAN	MUSKEGON	231
MICHIGAN	TRAVERSE CITY	231
GEOGRAPHIC RELIEF CODE		232
EASILY RECOGNIZABLE CODE		233
GEOGRAPHIC RELIEF CODE		234
GEOGRAPHIC RELIEF CODE		235
GENERAL PURPOSE CODE		236
GEOGRAPHIC RELIEF CODE		237
GENERAL PURPOSE CODE		238
GEOGRAPHIC RELIEF CODE		239
MARYLAND	ROCKVILLE	240
MARYLAND	SILVER SPRING	240
MARYLAND	BETHESDA	240
GENERAL PURPOSE CODE		241
BAHAMAS	ALL LOCATIONS	242
GEOGRAPHIC RELIEF CODE		243
EASILY RECOGNIZABLE CODE		244

State/Province/Country	City	Area Code
GENERAL PURPOSE CODE		245
BARBADOS	ALL LOCATIONS	246
GEOGRAPHIC RELIEF CODE		247
MICHIGAN	OAKLAND COUNTY	248
GENERAL PURPOSE CODE		249
BRITISH COLUMBIA	KAMLOOPS	250
BRITISH COLUMBIA	KELOWNA	250
BRITISH COLUMBIA	VICTORIA	250
GEOGRAPHIC RELIEF CODE		251
NORTH CAROLINA	GREENVILLE	252
NORTH CAROLINA	MOREHEAD CITY	252
NORTH CAROLINA	ROCKY MOUNT	252
WASHINGTON	AUBURN	253
WASHINGTON	GIG HARBOR	253
WASHINGTON	TACOMA	253
TEXAS	WACO	254
EASILY RECOGNIZABLE CODE		255
ALABAMA	DECATUR	256
ALABAMA	GADSDEN	256
ALABAMA	HUNTSVILLE	256
GENERAL PURPOSE CODE		257
GEOGRAPHIC RELIEF CODE		258
GENERAL PURPOSE CODE		259
GEOGRAPHIC RELIEF CODE		260
GEOGRAPHIC RELIEF CODE		261
WISCONSIN	KENOSHA	262
WISCONSIN	RACINE	262
WISCONSIN	WAUKESHA	262
GENERAL PURPOSE CODE		263
ANGUILLA	ALL LOCATIONS	264
GENERAL PURPOSE CODE		265
EASILY RECOGNIZABLE CODE		266
PENNSYLVANIA	PHILADELPHIA	267
ANTIGUA/BARBUDA	ALL LOCATIONS	268
GENERAL PURPOSE CODE		269
KENTUCKY	BOWLING GREEN	270
KENTUCKY	PADUCAH	270
GEOGRAPHIC RELIEF CODE		271
GEOGRAPHIC RELIEF CODE		272
GEOGRAPHIC RELIEF CODE		273
GENERAL PURPOSE CODE		274

State/Province/Country	City	Area Code
GEOGRAPHIC RELIEF CODE		275
GEOGRAPHIC RELIEF CODE		276
EASILY RECOGNIZABLE CODE		277
MICHIGAN	ANN ARBOR	278
MICHIGAN	LIVONIA	278
MICHIGAN	YPSILANTI	278
GENERAL PURPOSE CODE		279
GEOGRAPHIC RELIEF CODE		280
TEXAS	HOUSTON	281
GEOGRAPHIC RELIEF CODE		282
GENERAL PURPOSE CODE		283
BRITISH VIRGIN ISLANDS	ALL LOCATIONS	284
GENERAL PURPOSE CODE		285
GEOGRAPHIC RELIEF CODE		286
GEOGRAPHIC RELIEF CODE		287
EASILY RECOGNIZABLE CODE		288
GEOGRAPHIC RELIEF CODE		289
EXPANSION CODE		290
EXPANSION CODE		291
EXPANSION CODE		292
EXPANSION CODE		293
EXPANSION CODE		294
EXPANSION CODE		295
EXPANSION CODE		296
EXPANSION CODE		297
EXPANSION CODE		298
EXPANSION CODE		299
EASILY RECOGNIZABLE CODE		300
MARYLAND	ROCKVILLE	301
MARYLAND	SILVER SPRING	301
MARYLAND	BETHESDA	301
DELAWARE	ALL LOCATIONS	302
COLORADO	DENVER	303
WEST VIRGINIA	ALL LOCATIONS	304
FLORIDA	MIAMI	305
FLORIDA	KEY WEST	305
SASKATCHEWAN	ALL LOCATIONS	306
WYOMING	ALL LOCATIONS	307
NEBRASKA	NORTH PLATTE	308
ILLINOIS	BLOOMINGTON	309
ILLINOIS	PEORIA	309

State/Province/Country	City	Area Code
CALIFORNIA	BEVERLY HILLS	310
CALIFORNIA	GARDENA	310
NON-EMERGENCY ACCESS		311
ILLINOIS	CHICAGO	312
MICHIGAN	ANN ARBOR	313
MICHIGAN	DETROIT	313
MISSOURI	ST. LOUIS	314
NEW YORK	SYRACUSE	315
KANSAS	DODGE CITY	316
KANSAS	WICHITA	316
INDIANA	INDIANAPOLIS	317
LOUISIANA	ALEXANDRIA	318
LOUISIANA	MONROE	318
IOWA	DUBUQUE	319
MINNESOTA	ST. CLOUD	320
MINNESOTA	WILLMAR	320
FLORIDA	CAPE CANAVERAL	321
EASILY RECOGNIZABLE CODE		322
CALIFORNIA	LOS ANGELES	323
GEOGRAPHIC RELIEF CODE		324
GEOGRAPHIC RELIEF CODE		325
GEOGRAPHIC RELIEF CODE		326
GEOGRAPHIC RELIEF CODE		327
GEOGRAPHIC RELIEF CODE		328
GENERAL PURPOSE CODE		329
OHIO	AKRON	330
OHIO	YOUNGSTOWN	330
ILLINOIS	SUBURBAN CHICAGO	331
GEOGRAPHIC RELIEF CODE		332
EASILY RECOGNIZABLE CODE		333
ALABAMA	AUBURN	334
ALABAMA	MOBILE	334
ALABAMA	MONTGOMERY	334
ALABAMA	SELMA	334
GEOGRAPHIC RELIEF CODE		335
NORTH CAROLINA	ASHEBORO	336
NORTH CAROLINA	BURLINGTON	336
NORTH CAROLINA	GREENSBORO	336
LOUISIANA	LAFAYETTE	337
LOUISIANA	LAKE CHARLES	337
GEOGRAPHIC RELIEF CODE		338

State/Province/Country	City	Area Code
GENERAL PURPOSE CODE		339
US VIRGIN ISLANDS	ALL LOCATIONS	340
CALIFORNIA	OAKLAND	341
GENERAL PURPOSE CODE		342
GEOGRAPHIC RELIEF CODE		343
EASILY RECOGNIZABLE CODE		344
CAYMAN ISLANDS	ALL LOCATIONS	345
GENERAL PURPOSE CODE		346
NEW YORK	BRONX	347
NEW YORK	BROOKLYN	347
NEW YORK	QUEENS	347
NEW YORK	STATEN ISLAND	347
GEOGRAPHIC RELIEF CODE		348
GENERAL PURPOSE CODE		349
GEOGRAPHIC RELIEF CODE		350
GEOGRAPHIC RELIEF CODE		351
FLORIDA	GAINESVILLE	352
GEOGRAPHIC RELIEF CODE		353
GENERAL PURPOSE CODE		354
EASILY RECOGNIZABLE CODE		355
GEOGRAPHIC RELIEF CODE		356
GENERAL PURPOSE CODE		357
GEOGRAPHIC RELIEF CODE		358
GENERAL PURPOSE CODE		359
WASHINGTON	BELLINGHAM	360
WASHINGTON	OLYMPIA	360
WASHINGTON	VANCOUVER	360
TEXAS	CORPUS CHRISTI	361
GENERAL PURPOSE CODE		362
GENERAL PURPOSE CODE		363
GEOGRAPHIC RELIEF CODE		364
GEOGRAPHIC RELIEF CODE		365
EASILY RECOGNIZABLE CODE		366
GENERAL PURPOSE CODE		367
GENERAL PURPOSE CODE		368
GEOGRAPHIC RELIEF CODE		369
RESERVED		370
RESERVED		371
RESERVED		372
RESERVED		373
RESERVED		374

State/Province/Country	City	Area Code
RESERVED		375
RESERVED		376
RESERVED		377
RESERVED		378
RESERVED		379
GEOGRAPHIC RELIEF CODE		380
GEOGRAPHIC RELIEF CODE		381
GENERAL PURPOSE CODE		382
GEOGRAPHIC RELIEF CODE		383
GEOGRAPHIC RELIEF CODE		384
GEOGRAPHIC RELIEF CODE		385
GEOGRAPHIC RELIEF CODE		386
GEOGRAPHIC RELIEF CODE		387
EASILY RECOGNIZABLE CODE		388
GEOGRAPHIC RELIEF CODE		389
EXPANSION CODE		390
EXPANSION CODE		391
EXPANSION CODE		392
EXPANSION CODE		393
EXPANSION CODE		394
EXPANSION CODE		395
EXPANSION CODE		396
EXPANSION CODE		397
EXPANSION CODE		398
EXPANSION CODE		399
EASILY RECOGNIZABLE CODE		400
RHODE ISLAND	ALL LOCATIONS	401
NEBRASKA	LINCOLN	402
NEBRASKA	OMAHA	402
ALBERTA	CALGARY	403
GEORGIA	ATLANTA	404
OKLAHOMA	NORMAN	405
MONTANA	ALL LOCATIONS	406
FLORIDA	ORLANDO	407
FLORIDA	KISSIMMEE	407
CALIFORNIA	SAN JOSE	408
TEXAS	GALVESTON	409
MARYLAND	BALTIMORE	410
MARYLAND	TOWSON	410
LOCAL DIRECTORY ASSISTANCE		411
PENNSYLVANIA	PITTSBURGH	412

State/Province/Country	City	Area Code
MASSACHUSETTS	SPRINGFIELD	413
WISCONSIN	MILWAUKEE	414
CALIFORNIA	SAN FRANCISCO	415
ONTARIO	TORONTO	416
MISSOURI	JOPLIN	417
MISSOURI	SPRINGFIELD	417
QUEBEC	QUEBEC	418
OHIO	TOLEDO	419
GEOGRAPHIC RELIEF CODE		420
GEOGRAPHIC RELIEF CODE		421
EASILY RECOGNIZABLE CODE		422
TENNESSEE	CHATTANOOGA	423
EXPANSION CODE	BEVERLY HILLS	424
CALIFORNIA	GARDENA	424
WASHINGTON	EVERETT	425
WASHINGTON	KENT	425
GEOGRAPHIC RELIEF CODE		426
GENERAL PURPOSE CODE		427
GENERAL PURPOSE CODE		428
GENERAL PURPOSE CODE		429
GEOGRAPHIC RELIEF CODE		430
GEOGRAPHIC RELIEF CODE		431
GEOGRAPHIC RELIEF CODE		432
EASILY RECOGNIZABLE CODE		433
GENERAL PURPOSE CODE		434
UTAH	CEDAR CITY	435
UTAH	ST. GEORGE	435
GEOGRAPHIC RELIEF CODE		436
GEOGRAPHIC RELIEF CODE		437
GENERAL PURPOSE CODE		438
GEOGRAPHIC RELIEF CODE		439
OHIO	ASHTABULA	440
OHIO	LORAIN	440
OHIO	WESTLAKE	440
BERMUDA	ALL LOCATIONS	441
CALIFORNIA		442
MARYLAND	BALTIMORE	443
MARYLAND	TOWSON	443
EASILY RECOGNIZABLE CODE		444
GEOGRAPHIC RELIEF CODE		445
GEOGRAPHIC RELIEF CODE		446

State/Province/Country	City	Area Code
GEOGRAPHIC RELIEF CODE		447
GENERAL PURPOSE CODE		448
GEOGRAPHIC RELIEF CODE		449
QUEBEC	AREA OUTSIDE MONTREAL	450
GENERAL PURPOSE CODE		451
GENERAL PURPOSE CODE		452
GENERAL PURPOSE CODE		453
GENERAL PURPOSE CODE		454
EASILY RECOGNIZABLE CODE		455
INBOUND INTERNATIONAL		456
GEOGRAPHIC RELIEF CODE		457
GEOGRAPHIC RELIEF CODE		458
GEOGRAPHIC RELIEF CODE		459
GEOGRAPHIC RELIEF CODE		460
GEOGRAPHIC RELIEF CODE		461
GENERAL PURPOSE CODE		462
GEOGRAPHIC RELIEF CODE		463
ILLINOIS	SUBURBAN CHICAGO	464
GEOGRAPHIC RELIEF CODE		465
EASILY RECOGNIZABLE CODE		466
GENERAL PURPOSE CODE		467
GEOGRAPHIC RELIEF CODE		468
GEOGRAPHIC RELIEF CODE		469
GEOGRAPHIC RELIEF CODE		470
GENERAL PURPOSE CODE		471
GEOGRAPHIC RELIEF CODE		472
GRENADA	ALL LOCATIONS	473
GEOGRAPHIC RELIEF CODE		474
CONNECTICUT	FAIRFIELD	475
CONNECTICUT	NEW HAVEN	475
GENERAL PURPOSE CODE		476
EASILY RECOGNIZABLE CODE		477
GEORGIA	MACON	478
GEOGRAPHIC RELIEF CODE		479
ARIZONA	SUBURBAN EAST PHOENIX	480
GENERAL PURPOSE CODE		481
GEOGRAPHIC RELIEF CODE		482
GEOGRAPHIC RELIEF CODE		483
PENNSYLVANIA	PHILADELPHIA	484
GENERAL PURPOSE CODE		485
GEOGRAPHIC RELIEF CODE		486

State/Province/Country	City	Area Code
GENERAL PURPOSE CODE		487
EASILY RECOGNIZABLE CODE		488
GENERAL PURPOSE CODE		489
EXPANSION CODE		490
EXPANSION CODE		491
EXPANSION CODE		492
EXPANSION CODE		493
EXPANSION CODE		494
EXPANSION CODE		495
EXPANSION CODE		496
EXPANSION CODE		497
EXPANSION CODE		498
EXPANSION CODE		499
PCS		500
ARKANSAS	ALL LOCATIONS	501
KENTUCKY	LOUISVILLE	502
OREGON	PORTLAND	503
LOUISIANA	NEW ORLEANS	504
NEW MEXICO	ALL LOCATIONS	505
NEW BRUNSWICK	ALL LOCATIONS	506
MINNESOTA	ROCHESTER	507
MASSACHUSETTS	FRAMINGHAM	508
MASSACHUSETTS	NEW BEDFORD	508
MASSACHUSETTS	PLYMOUTH	508
WASHINGTON	SPOKANE	509
CALIFORNIA	OAKLAND	510
NOT AVAILABLE		511
TEXAS	AUSTIN	512
OHIO	CINCINNATI	513
QUEBEC	MONTREAL	514
IOWA	DES MOINES	515
NEW YORK	NASSAU COUNTY	516
MICHIGAN	LANSING	517
NEW YORK	ALBANY	518
NEW YORK	SCHENECTADY	518
ONTARIO	LONDON	519
ONTARIO	WINDSOR	519
ARIZONA	TUCSON	520
ARIZONA	FLAGSTAFF	520
ARIZONA	PRESCOTT	520
ARIZONA	YUMA	520

State/Province/Country	City	Area Code
MEXICO ROAMING-TEMP		521
FUTURE PCS		522
MEXICO ROAMING-TEMP		523
MEXICO ROAMING-TEMP		524
MEXICO ROAMING-TEMP		525
MEXICO ROAMING-TEMP		526
MEXICO ROAMING-TEMP		527
MEXICO ROAMING-TEMP		528
MEXICO ROAMING-TEMP		529
CALIFORNIA	CHICO	530
CALIFORNIA	REDDING	530
CALIFORNIA	YREKA	530
GEOGRAPHIC RELIEF CODE		531
GEOGRAPHIC RELIEF CODE		532
FUTURE PCS		533
GENERAL PURPOSE CODE		534
GEOGRAPHIC RELIEF CODE		535
GEOGRAPHIC RELIEF CODE		536
GENERAL PURPOSE CODE		537
GEOGRAPHIC RELIEF CODE		538
GEOGRAPHIC RELIEF CODE		539
VIRGINIA	COVINGTON	540
VIRGINIA	ROANOKE	540
VIRGINIA	WINCHESTER	540
OREGON	EUGENE	541
OREGON	MEDFORD	541
GEOGRAPHIC RELIEF CODE		542
GENERAL PURPOSE CODE		543
FUTURE PCS		544
GEOGRAPHIC RELIEF CODE		545
GENERAL PURPOSE CODE		546
GEOGRAPHIC RELIEF CODE		547
GENERAL PURPOSE CODE		548
GENERAL PURPOSE CODE		549
GEOGRAPHIC RELIEF CODE		550
GEOGRAPHIC RELIEF CODE		551
GEOGRAPHIC RELIEF CODE		552
GEOGRAPHIC RELIEF CODE		553
GEOGRAPHIC RELIEF CODE		554
NOT AVAILABLE		555
GEOGRAPHIC RELIEF CODE		556

State/Province/Country	City	Area Code
GEOGRAPHIC RELIEF CODE		557
GEOGRAPHIC RELIEF CODE		558
CALIFORNIA	FRESNO	559
CALIFORNIA	VISALIA	559
GEOGRAPHIC RELIEF CODE		560
FLORIDA	BOCA RATON	561
FLORIDA	STUART	561
FLORIDA	WEST PALM BEACH	561
CALIFORNIA	SANTA MONICA	562
CALIFORNIA	LONG BEACH	562
CALIFORNIA	WHITTIER	562
GEOGRAPHIC RELIEF CODE		563
WASHINGTON	BELLINGHAM	564
WASHINGTON	OLYMPIA	564
WASHINGTON	VANCOUVER	564
GEOGRAPHIC RELIEF CODE		565
FUTURE PCS		566
GEOGRAPHIC RELIEF CODE		567
GENERAL PURPOSE CODE		568
GENERAL PURPOSE CODE		569
PENNSYLVANIA	SCRANTON	570
PENNSYLVANIA	WILKES-BARRE	570
PENNSYLVANIA	WILLIAMSPORT	570
VIRGINIA	ARLINGTON	571
VIRGINIA	FAIRFAX	571
VIRGINIA	VIENNA	571
GEOGRAPHIC RELIEF CODE		572
MISSOURI	JEFFERSON CITY	573
MISSOURI	OUTSIDE ST. LOUIS	573
GEOGRAPHIC RELIEF CODE		574
GENERAL PURPOSE CODE		575
GENERAL PURPOSE CODE		576
FUTURE PCS		577
GEOGRAPHIC RELIEF CODE		578
GEOGRAPHIC RELIEF CODE		579
OKLAHOMA	ARDMORE	580
OKLAHOMA	ENID	580
OKLAHOMA	LAWTON	580
GENERAL PURPOSE CODE		581
GENERAL PURPOSE CODE		582
GEOGRAPHIC RELIEF CODE		583

State/Province/Country	City	Area Code
GEOGRAPHIC RELIEF CODE		584
GEOGRAPHIC RELIEF CODE		585
MICHIGAN	BIRMINGHAM	586
MICHIGAN	FLINT	586
MICHIGAN	PONTIAC	586
GEOGRAPHIC RELIEF CODE		587
FUTURE PCS		588
GEOGRAPHIC RELIEF CODE		589
EXPANSION CODE		590
EXPANSION CODE		591
EXPANSION CODE		592
EXPANSION CODE		593
EXPANSION CODE		594
EXPANSION CODE		595
EXPANSION CODE		596
EXPANSION CODE		597
EXPANSION CODE		598
EXPANSION CODE		599
CANADA (SERVICES)		600
MISSISSIPPI	HATTIESBURG	601
MISSISSIPPI	JACKSON	601
ARIZONA	PHOENIX	602
NEW HAMPSHIRE	ALL LOCATIONS	603
BRITISH COLUMBIA	VANCOUVER	604
SOUTH DAKOTA	ALL LOCATIONS	605
KENTUCKY	ASHLAND	606
NEW YORK	BINGHAMTON	607
NEW YORK	ELMIRA	607
WISCONSIN	MADISON	608
NEW JERSEY	ATLANTIC CITY	609
NEW JERSEY	PRINCETON	609
NEW JERSEY	TRENTON	609
PENNSYLVANIA	ALLENTOWN	610
PENNSYLVANIA	BETHLEHEM	610
PENNSYLVANIA	READING	610
PENNSYLVANIA	WEST CHESTER	610
REPAIR SERVICE		611
MINNESOTA	MINNEAPOLIS	612
ONTARIO	OTTAWA	613
OHIO	COLUMBUS	614
TENNESSEE	NASHVILLE	615

State/Province/Country	City	Area Code
MICHIGAN	BATTLE CREEK	616
MICHIGAN	GRAND RAPIDS	616
MICHIGAN	KALAMAZOO	616
MASSACHUSETTS	BOSTON	617
ILLINOIS	COLLINSVILLE	618
ILLINOIS	GRANITE CITY	618
CALIFORNIA	SAN DIEGO	619
GEOGRAPHIC RELIEF CODE		620
GEOGRAPHIC RELIEF CODE		621
EASILY RECOGNIZABLE CODE		622
ARIZONA	SUBURBAN WEST PHOENIX	623
GEOGRAPHIC RELIEF CODE		624
GENERAL PURPOSE CODE		625
CALIFORNIA	ALHAMBRA	626
CALIFORNIA	LA PUENTE	626
CALIFORNIA	PASADENA	626
GENERAL PURPOSE CODE		627
CALIFORNIA	SAN FRANCISCO	628
GENERAL PURPOSE CODE		629
ILLINOIS	SUBURBAN CHICAGO	630
NEW YORK	SUFFOLK COUNTY	631
GENERAL PURPOSE CODE		632
EASILY RECOGNIZABLE CODE		633
GENERAL PURPOSE CODE		634
GENERAL PURPOSE CODE		635
MISSOURI	CHESTERFIELD	636
MISSOURI	O'FALLON	636
MISSOURI	GRAY SUMMIT	636
GEOGRAPHIC RELIEF CODE		637
GENERAL PURPOSE CODE		638
GENERAL PURPOSE CODE		639
GEOGRAPHIC RELIEF CODE		640
GEOGRAPHIC RELIEF CODE		641
GEOGRAPHIC RELIEF CODE		642
GENERAL PURPOSE CODE		643
EASILY RECOGNIZABLE CODE		644
GENERAL PURPOSE CODE		645
NEW YORK	MANHATTAN	646
ONTARIO	TORONTO	647
GEOGRAPHIC RELIEF CODE		648
TURKS & CAICOS ISLANDS	ALL LOCATIONS	649

State/Province/Country	City	Area Code
CALIFORNIA	MOUNTAIN VIEW	650
CALIFORNIA	SAN MATEO	650
MINNESOTA	ST. PAUL	651
GENERAL PURPOSE CODE		652
GENERAL PURPOSE CODE		653
GEOGRAPHIC RELIEF CODE		654
EASILY RECOGNIZABLE CODE		655
GEOGRAPHIC RELIEF CODE		656
CALIFORNIA	ANAHEIM	657
GENERAL PURPOSE CODE		658
GENERAL PURPOSE CODE		659
MISSOURI	KIRKSVILLE	660
MISSOURI	MARYVILLE	660
MISSOURI	TRENTON	660
CALIFORNIA	BAKERSFIELD	661
CALIFORNIA	LANCASTER	661
CALIFORNIA	PALMDALE	661
MISSISSIPPI	GREENVILLE	662
MISSISSIPPI	OXFORD	662
MISSISSIPPI	TUPELO	662
GEOGRAPHIC RELIEF CODE		663
MONTSERRAT	ALL LOCATIONS	664
GENERAL PURPOSE CODE		665
EASILY RECOGNIZABLE CODE		666
GENERAL PURPOSE CODE		667
GEOGRAPHIC RELIEF CODE		668
CALIFORNIA	SAN JOSE	669
CNMI	ALL LOCATIONS	670
GUAM	ALL LOCATIONS	671
GENERAL PURPOSE CODE		672
GEOGRAPHIC RELIEF CODE		673
GENERAL PURPOSE CODE		674
GENERAL PURPOSE CODE		675
GEOGRAPHIC RELIEF CODE		676
EASILY RECOGNIZABLE CODE		677
GEORGIA	ATLANTA AND SUBURBAN AREAS	678
MICHIGAN	DETROIT	679
GEOGRAPHIC RELIEF CODE		680
GEOGRAPHIC RELIEF CODE		681
GEOGRAPHIC RELIEF CODE		682

State/Province/Country	City	Area Code
GEOGRAPHIC RELIEF CODE		683
RESERVED FOR NANP COUNTRY		684
GENERAL PURPOSE CODE		685
GEOGRAPHIC RELIEF CODE		686
GENERAL PURPOSE CODE		687
EASILY RECOGNIZABLE CODE		688
GEOGRAPHIC RELIEF CODE		689
EXPANSION CODE		690
EXPANSION CODE		691
EXPANSION CODE		692
EXPANSION CODE		693
EXPANSION CODE		694
EXPANSION CODE		695
EXPANSION CODE		696
EXPANSION CODE		697
EXPANSION CODE		698
EXPANSION CODE		699
IC SERVICES	ALL LOCATIONS	700
NORTH DAKOTA	ALL LOCATIONS	701
NEVADA	LAS VEGAS	702
VIRGINIA	ARLINGTON	703
VIRGINIA	ALEXANDRIA	703
VIRGINIA	FAIRFAX	703
VIRGINIA	FALLS CHURCH	703
NORTH CAROLINA	CHARLOTTE	704
ONTARIO	SAULT STE. MARIE	705
GEORGIA	AUGUSTA	706
GEORGIA	COLUMBUS	706
CALIFORNIA	PETALUMA	707
CALIFORNIA	NAPA	707
ILLINOIS	DES PLAINES	708
ILLINOIS	PALATINE	708
ILLINOIS	WAUKEGAN	708
NEWFOUNDLAND	ALL LOCATIONS	709
U.S. GOVERNMENT		710
TRS ACCESS		711
IOWA	COUNCIL BLUFFS	712
IOWA	SIOUX CITY	712
TEXAS	HOUSTON	713
CALIFORNIA	ANAHEIM	714
WISCONSIN	EAU CLAIRE	715

State/Province/Country	City	Area Code
WISCONSIN	WAUSAU	715
NEW YORK	BUFFALO	716
NEW YORK	NIAGARA FALLS	716
NEW YORK	ROCHESTER	716
PENNSYLVANIA	HARRISBURG	717
PENNSYLVANIA	HERSHEY	717
PENNSYLVANIA	YORK	717
NEW YORK	BROOKLYN	718
NEW YORK	QUEENS	718
NEW YORK	STATEN ISLAND	718
COLORADO	COLORADO SPRINGS	719
COLORADO	PUEBLO	719
COLORADO	DENVER	720
GEOGRAPHIC RELIEF CODE		721
EASILY RECOGNIZABLE CODE		722
GENERAL PURPOSE CODE		723
PENNSYLVANIA	LATROBE	724
PENNSYLVANIA	NEW CASTLE	724
GENERAL PURPOSE CODE		725
GEOGRAPHIC RELIEF CODE		726
FLORIDA	CLEARWATER	727
FLORIDA	ST. PETERSBURG	727
GENERAL PURPOSE CODE		728
GENERAL PURPOSE CODE		729
GEOGRAPHIC RELIEF CODE		730
GEOGRAPHIC RELIEF CODE		731
NEW JERSEY	MIDDLESEX	732
NEW JERSEY	MONMOUTH	732
NEW JERSEY	OCEAN COUNTIES	732
EASILY RECOGNIZABLE CODE		733
MICHIGAN	ANN ARBOR	734
MICHIGAN	LIVONIA	734
MICHIGAN	YPSILANTI	734
GEOGRAPHIC RELIEF CODE		735
GEOGRAPHIC RELIEF CODE		736
GEOGRAPHIC RELIEF CODE		737
GENERAL PURPOSE CODE		738
GENERAL PURPOSE CODE		739
OHIO	ATHENS	740
OHIO	LANCASTER	740
OHIO	MARIETTA	740

State/Province/Country	City	Area Code
GEOGRAPHIC RELIEF CODE		741
GENERAL PURPOSE CODE		742
GENERAL PURPOSE CODE		743
EASILY RECOGNIZABLE CODE		744
GEOGRAPHIC RELIEF CODE		745
GEOGRAPHIC RELIEF CODE		746
GEOGRAPHIC RELIEF CODE		747
GENERAL PURPOSE CODE		748
GEOGRAPHIC RELIEF CODE		749
GEOGRAPHIC RELIEF CODE		750
GEOGRAPHIC RELIEF CODE		751
CALIFORNIA	ONTARIO	752
CALIFORNIA	POMONA	752
CALIFORNIA	UPLAND	752
GEOGRAPHIC RELIEF CODE		753
GENERAL PURPOSE CODE		754
EASILY RECOGNIZABLE CODE		755
GENERAL PURPOSE CODE		756
VIRGINIA	NEWPORT NEWS	757
VIRGINIA	NORFOLK	757
VIRGINIA	VIRGINIA BEACH	757
ST. LUCIA	ALL LOCATIONS	758
GENERAL PURPOSE CODE		759
CALIFORNIA	BISHOP	760
CALIFORNIA	NEEDLES	760
CALIFORNIA	PALM SPRINGS	760
GEOGRAPHIC RELIEF CODE		761
GENERAL PURPOSE CODE		762
MINNESOTA	EXPANSION	763
CALIFORNIA	MOUNTAIN VIEW	764
CALIFORNIA	SAN MATEO	764
INDIANA	KOKOMO	765
INDIANA	LAFAYETTE	765
INDIANA	RICHMOND	765
EASILY RECOGNIZABLE CODE		766
DOMINICA	ALL LOCATIONS	767
GEOGRAPHIC RELIEF CODE		768
GENERAL PURPOSE CODE		769
GEORGIA	SUBURBAN ATLANTA (THE AREA OUTSIDE THE INTERSTATE 285 BELTWAY)	770

State/Province/Country	City	Area Code
GENERAL PURPOSE CODE		771
GENERAL PURPOSE CODE		772
ILLINOIS	CHICAGO	773
GEOGRAPHIC RELIEF CODE		774
NEVADA	CARSON CITY	775
NEVADA	RENO	775
GEOGRAPHIC RELIEF CODE		776
EASILY RECOGNIZABLE CODE		777
GEOGRAPHIC RELIEF CODE		778
GEOGRAPHIC RELIEF CODE		779
ALBERTA	EDMONTON	780
ALBERTA	GRANDE PRAIRIE	780
MASSACHUSETTS	BURLINGTON	781
MASSACHUSETTS	READING	781
MASSACHUSETTS	WALTHAM	781
GEOGRAPHIC RELIEF CODE		782
GENERAL PURPOSE CODE		783
ST. VINCENT & GRENADINES	ALL LOCATIONS	784
KANSAS	JUNCTION	785
KANSAS	LAWRENCE	785
KANSAS	TOPEKA	785
FLORIDA	MIAMI	786
FLORIDA	NORTH DADE	786
PUERTO RICO	ALL LOCATIONS	787
EASILY RECOGNIZABLE CODE		788
GEOGRAPHIC RELIEF CODE		789
EXPANSION CODE		790
EXPANSION CODE		791
EXPANSION CODE		792
EXPANSION CODE		793
EXPANSION CODE		794
EXPANSION CODE		795
EXPANSION CODE		796
EXPANSION CODE		797
EXPANSION CODE		798
EXPANSION CODE		799
SERVICE		800
UTAH	SALT LAKE CITY	801
UTAH	OGDEN	801
UTAH	PROVO	801
VERMONT	ALL LOCATIONS	802

State/Province/Country	City	Area Code
SOUTH CAROLINA	COLUMBIA	803
SOUTH CAROLINA	WINNSBORO	803
VIRGINIA	NORFOLK	804
VIRGINIA	RICHMOND	804
CALIFORNIA	BAKERSFIELD	805
CALIFORNIA	SANTA BARBARA	805
TEXAS	AMARILLO	806
TEXAS	LUBBOCK	806
ONTARIO	THUNDER BAY	807
HAWAII	ALL LOCATIONS	808
DOMINICAN REPUBLIC		809
MICHIGAN	BIRMINGHAM	810
MICHIGAN	FLINT	810
MICHIGAN	PONTIAC	810
BUSINESS OFFICE		811
INDIANA	BLOOMINGTON	812
INDIANA	EVANSVILLE	812
INDIANA	TERRA HAUTE	812
FLORIDA	CLEARWATER	813
FLORIDA	ST. PETERSBURG	813
FLORIDA	TAMPA	813
PENNSYLVANIA	ALTOONA	814
PENNSYLVANIA	ERIE	814
ILLINOIS	JOLIET	815
ILLINOIS	ROCKFORD	815
MISSOURI	KANSAS CITY	816
TEXAS	ARLINGTON	817
TEXAS	FORT WORTH	817
CALIFORNIA	BURBANK	818
CALIFORNIA	PASADENA	818
CALIFORNIA	VAN NUYS	818
QUEBEC	SHERBROOKE	819
QUEBEC	TROIS RIVIERES	819
GEOGRAPHIC RELIEF CODE		820
GEOGRAPHIC RELIEF CODE		821
FUTURE 800 SERVICE		822
GEOGRAPHIC RELIEF CODE		823
GENERAL PURPOSE CODE		824
GEOGRAPHIC RELIEF CODE		825
GEOGRAPHIC RELIEF CODE		826
GENERAL PURPOSE CODE		827

State/Province/Country	City	Area Code
NORTH CAROLINA	ASHEVILLE	828
NORTH CAROLINA	HENDERSONVILLE	828
GENERAL PURPOSE CODE		829
TEXAS	FREDERICKSBURG	830
TEXAS	NEW BRAUNFELS	830
TEXAS	UVALDE	830
CALIFORNIA	MONTEREY	831
CALIFORNIA	SALINAS	831
TEXAS	HOUSTON	832
FUTURE 800 SERVICE		833
GENERAL PURPOSE CODE		834
GEOGRAPHIC RELIEF CODE		835
GENERAL PURPOSE CODE		836
GENERAL PURPOSE CODE		837
GEOGRAPHIC RELIEF CODE		838
GEOGRAPHIC RELIEF CODE		839
GEOGRAPHIC RELIEF CODE		840
GENERAL PURPOSE CODE		841
GEOGRAPHIC RELIEF CODE		842
SOUTH CAROLINA	CHARLESTON	843
SOUTH CAROLINA	LEXINGTON	843
FUTURE 800 SERVICE		844
GEOGRAPHIC RELIEF CODE		845
GEOGRAPHIC RELIEF CODE		846
ILLINOIS		847
GEOGRAPHIC RELIEF CODE		848
GEOGRAPHIC RELIEF CODE		849
FLORIDA	TALLAHASSEE	850
GEOGRAPHIC RELIEF CODE		851
GENERAL PURPOSE CODE		852
GEOGRAPHIC RELIEF CODE		853
GENERAL PURPOSE CODE		854
FUTURE 800 SERVICE		855
NEW JERSEY	CHERRY HILL	856
GEOGRAPHIC RELIEF CODE		857
CALIFORNIA		858
EXPANSION CODE		859
CONNECTICUT	BLOOMFIELD	860
CONNECTICUT	CANAAN	860
CONNECTICUT	HARTFORD	860
CONNECTICUT	NORWICH	860

State/Province/Country	City	Area Code
GEOGRAPHIC RELIEF CODE		861
GEOGRAPHIC RELIEF CODE		862
FLORIDA	ARCADIA	863
FLORIDA	LAKELAND	863
FLORIDA	SEBRING	863
SOUTH CAROLINA	GREENVILLE	864
SOUTH CAROLINA	SPARTANBURG	864
TENNESSEE	KNOXVILLE	865
FUTURE 800 SERVICE		866
YUKON & NW TERRITORIES	ALL LOCATIONS	867
TRINIDAD AND TOBAGO	ALL LOCATIONS	868
ST. KITTS & NEVIS	ALL LOCATIONS	869
ARKANSAS	JONESBORO	870
ARKANSAS	PINE BLUFF	870
ARKANSAS	TEXARKANA	870
GEOGRAPHIC RELIEF CODE		871
ILLINOIS	CHICAGO	872
GENERAL PURPOSE CODE		873
GENERAL PURPOSE CODE		874
GEOGRAPHIC RELIEF CODE		875
JAMAICA	ALL LOCATIONS	876
TOLL FREE		877
GEOGRAPHIC RELIEF CODE		878
GEOGRAPHIC RELIEF CODE		879
PAID-800 SERVICE		880
PAID-888 SERVICE		881
PAID-877 SERVICE		882
GEOGRAPHIC RELIEF CODE		883
GEOGRAPHIC RELIEF CODE		884
GEOGRAPHIC RELIEF CODE		885
NDTP CODE (TEMPORARY)		886
GEOGRAPHIC RELIEF CODE		887
TOLL FREE		888
TOLL FREE		800
NDTP CODE (TEMPORARY)		889
EXPANSION CODE		890
EXPANSION CODE		891
EXPANSION CODE		892
EXPANSION CODE		893
EXPANSION CODE		894
EXPANSION CODE		895

State/Province/Country	City	Area Code
EXPANSION CODE		896
EXPANSION CODE		897
EXPANSION CODE		898
EXPANSION CODE		899
SERVICE		900
TENNESSEE	MEMPHIS	901
NOVA SCOTIA	ALL LOCATIONS	902
TEXAS	LONGVIEW	903
TEXAS	TEXARKANA	903
TEXAS	TYLER	903
FLORIDA	DAYTONA BEACH	904
FLORIDA	JACKSONVILLE	904
ONTARIO	HAMILTON	905
ONTARIO	UNIONVILLE	905
ONTARIO	WOODBRIDGE	905
MICHIGAN	MARQUETTE	906
MICHIGAN	SAULT STE. MARIE	906
ALASKA	ALL LOCATIONS	907
NEW JERSEY	ASBURY PARK	908
NEW JERSEY	ELIZABETH	908
NEW JERSEY	NEW BRUNSWICK	908
NEW JERSEY	WESTFIELD	908
CALIFORNIA	ONTARIO	909
CALIFORNIA	POMONA	909
CALIFORNIA	UPLAND	909
NORTH CAROLINA	FAYETTEVILLE	910
NORTH CAROLINA	WINSTON-SALEM	910
EMERGENCY		911
GEORGIA	SAVANNAH	912
KANSAS	KANSAS CITY	913
KANSAS	TOPEKA	913
NEW YORK	NEWBURGH	914
NEW YORK	OSSINING	914
NEW YORK	POUGHKEEPSIE	914
TEXAS	ABILENE	915
TEXAS	EL PASO	915
TEXAS	SAN ANGELO	915
CALIFORNIA	SACRAMENTO	916
NEW YORK	NEW YORK CITY	917
NEW YORK	QUEENS	917
NEW YORK	STATEN ISLAND	917

State/Province/Country	City	Area Code
OKLAHOMA	BROKEN ARROW	918
OKLAHOMA	MUSKOGEE	918
OKLAHOMA	TULSA	918
NORTH CAROLINA	DURHAM	919
NORTH CAROLINA	RALEIGH	919
WISCONSIN	APPLETON	920
WISCONSIN	GREEN BAY	920
WISCONSIN	SHEBOYGAN	920
GEOGRAPHIC RELIEF CODE		921
EASILY RECOGNIZABLE CODE		922
GENERAL PURPOSE CODE		923
GEOGRAPHIC RELIEF CODE		924
CALIFORNIA	PLEASANTON	925
CALIFORNIA	WALNUT CREEK	925
GEOGRAPHIC RELIEF CODE		926
GEOGRAPHIC RELIEF CODE		927
GEOGRAPHIC RELIEF CODE		928
GEOGRAPHIC RELIEF CODE		929
GEOGRAPHIC RELIEF CODE		930
TENNESSEE	CLARKSVILLE	931
GEOGRAPHIC RELIEF CODE		932
EASILY RECOGNIZABLE CODE		933
GEOGRAPHIC RELIEF CODE		934
CALIFORNIA	CHULA VISTA	935
CALIFORNIA	EL CAJON	935
CALIFORNIA	LA MESA	935
TEXAS	CONROE	936
TEXAS	NACOGDOCHES	936
OHIO	DAYTON	937
OHIO	HAMILTON	937
GEOGRAPHIC RELIEF CODE		938
GEOGRAPHIC RELIEF CODE		939
TEXAS	DENTON	940
TEXAS	WICHITA FALLS	940
FLORIDA	BRADENTON	941
FLORIDA	FT. MYERS	941
FLORIDA	NAPLES	941
FLORIDA	SARASOTA	941
GEOGRAPHIC RELIEF CODE		942
GEOGRAPHIC RELIEF CODE		943
EASILY RECOGNIZABLE CODE		944

State/Province/Country	City	Area Code
GEOGRAPHIC RELIEF CODE		945
GEOGRAPHIC RELIEF CODE		946
MICHIGAN	OAKLAND COUNTY	947
GEOGRAPHIC RELIEF CODE		948
CALIFORNIA	COSTA MESA	949
CALIFORNIA	IRVINE	949
CALIFORNIA	NEWPORT BEACH	949
NOT AVAILABLE		950
CALIFORNIA	CORONA	951
CALIFORNIA	RIVERSIDE	951
MINNESOTA	MINNEAPOLIS	952
GEOGRAPHIC RELIEF CODE		953
FLORIDA	FT. LAUDERDALE	954
FLORIDA	BROWARD COUNTY	954
EASILY RECOGNIZABLE CODE		955
TEXAS	BROWNSVILLE	956
TEXAS	LAREDO	956
TEXAS	MCALLEN	956
GEOGRAPHIC RELIEF CODE		957
GEOGRAPHIC RELIEF CODE		958
CONNECTICUT	BLOOMFIELD	959
CONNECTICUT	CANAAN	959
CONNECTICUT	HARTFORD	959
CONNECTICUT	NORWICH	959
RESERVED		960
RESERVED		961
RESERVED		962
RESERVED		963
RESERVED		964
RESERVED		965
RESERVED		966
RESERVED		967
RESERVED		968
RESERVED		969
COLORADO	ASPEN	970
COLORADO	VAIL	970
OREGON	PORTLAND	971
OREGON	SALEM	971
TEXAS	DALLAS	972
NEW JERSEY	MORRIS	973
NEW JERSEY	PASSAIC	973

State/Province/Country	City	Area Code
NEW JERSEY	ESSEX COUNTIES	973
GEOGRAPHIC RELIEF CODE		974
GEOGRAPHIC RELIEF CODE		975
GENERAL PURPOSE CODE		976
EASILY RECOGNIZABLE CODE		977
MASSACHUSETTS	ACTON	978
MASSACHUSETTS	LOWELL	978
MASSACHUSETTS	NORTH READING	978
TEXAS	BRYAN-COLLEGE STATION	979
TEXAS	FREEPORT	979
NORTH CAROLINA	CHARLOTTE	980
GEOGRAPHIC RELIEF CODE		981
GEOGRAPHIC RELIEF CODE		982
GEOGRAPHIC RELIEF CODE		983
GEOGRAPHIC RELIEF CODE		984
GEOGRAPHIC RELIEF CODE		985
GEOGRAPHIC RELIEF CODE		986
GENERAL PURPOSE CODE		987
EASILY RECOGNIZABLE CODE		988
MICHIGAN	ALPENA	989
MICHIGAN	MIDLAND	989
MICHIGAN	SAGINAW	989
EXPANSION CODE		990
EXPANSION CODE		991
EXPANSION CODE		992
EXPANSION CODE		993
EXPANSION CODE		994
EXPANSION CODE		995
EXPANSION CODE		996
EXPANSION CODE		997
EXPANSION CODE		998
EXPANSION CODE		999

Binary, Decimal, and Hexadecimal Conversions

b2 = binary, b10 = decimal, b16 = hexadecimal

b10	b2	b16
0	0	0
1	1	1
2	10	2
3	11	3
4	100	4
5	101	5
6	110	6
7	111	7
8	1000	8
9	1001	9
10	1010	A
11	1011	B
12	1100	C
13	1101	D
14	1110	E
15	1111	F
16	10000	10
17	10001	11
18	10010	12
19	10011	13

b10	b2	b16
20	10100	14
21	10101	15
22	10110	16
23	10111	17
24	11000	18
25	11001	19
26	11010	1A
27	11011	1B
28	11100	1C
29	11101	1D
30	11110	1E
31	11111	1F
32	100000	20
33	100001	21
34	100010	22
35	100011	23
36	100100	24
37	100101	25
38	100110	26
39	100111	27
40	101000	28
41	101001	29
42	101010	2A
43	101011	2B
44	101100	2C
45	101101	2D
46	101110	2E
47	101111	2F
48	110000	30
49	110001	31
50	110010	32
51	110011	33
52	110100	34
53	110101	35
54	110110	36
55	110111	37
56	111000	38
57	111001	39
58	111010	3A
59	111011	3B
60	111100	3C
61	111101	3D

b10	b2	b16
62	111110	3E
63	111111	3F
64	1000000	40
65	1000001	41
66	1000010	42
67	1000011	43
68	1000100	44
69	1000101	45
70	1000110	46
71	1000111	47
72	1001000	48
73	1001001	49
74	1001010	4A
75	1001011	4B
76	1001100	4C
77	1001101	4D
78	1001110	4E
79	1001111	4F
80	1010000	50
81	1010001	51
82	1010010	52
83	1010011	53
84	1010100	54
85	1010101	55
86	1010110	56
87	1010111	57
88	1011000	58
89	1011001	59
90	1011010	5A
91	1011011	5B
92	1011100	5C
93	1011101	5D
94	1011110	5E
95	1011111	5F
96	1100000	60
97	1100001	61
98	1100010	62
99	1100011	63
100	1100100	64
101	1100101	65
102	1100110	66
103	1100111	67

b10	b2	b16
104	1101000	68
105	1101001	69
106	1101010	6A
107	1101011	6B
108	1101100	6C
109	1101101	6D
110	1101110	6E
111	1101111	6F
112	1110000	70
113	1110001	71
114	1110010	72
115	1110011	73
116	1110100	74
117	1110101	75
118	1110110	76
119	1110111	77
120	1111000	78
121	1111001	79
122	1111010	7A
123	1111011	7B
124	1111100	7C
125	1111101	7D
126	1111110	7E
127	1111111	7F
128	10000000	80
129	10000001	81
130	10000010	82
131	10000011	83
132	10000100	84
133	10000101	85
134	10000110	86
135	10000111	87
136	10001000	88
137	10001001	89
138	10001010	8A
139	10001011	8B
140	10001100	8C
141	10001101	8D
142	10001110	8E
143	10001111	8F
144	10010000	90
145	10010001	91

b10	b2	b16
146	10010010	92
147	10010011	93
148	10010100	94
149	10010101	95
150	10010110	96
151	10010111	97
152	10011000	98
153	10011001	99
154	10011010	9A
155	10011011	9B
156	10011100	9C
157	10011101	9D
158	10011110	9E
159	10011111	9F
160	10100000	A0
161	10100001	A1
162	10100010	A2
163	10100011	A3
164	10100100	A4
165	10100101	A5
166	10100110	A6
167	10100111	A7
168	10101000	A8
169	10101001	A9
170	10101010	AA
171	10101011	AB
172	10101100	AC
173	10101101	AD
174	10101110	AE
175	10101111	AF
176	10110000	B0
177	10110001	B1
178	10110010	B2
179	10110011	B3
180	10110100	B4
181	10110101	B5
182	10110110	B6
183	10110111	B7
184	10111000	B8
185	10111001	B9
186	10111010	BA
187	10111011	BB

b10	b2	b16
188	10111100	BC
189	10111101	BD
190	10111110	BE
191	10111111	BF
192	11000000	C0
193	11000001	C1
194	11000010	C2
195	11000011	C3
196	11000100	C4
197	11000101	C5
198	11000110	C6
199	11000111	C7
200	11001000	C8
201	11001001	C9
202	11001010	CA
203	11001011	CB
204	11001100	CC
205	11001101	CD
206	11001110	CE
207	11001111	CF
208	11010000	D0
209	11010001	D1
210	11010010	D2
211	11010011	D3
212	11010100	D4
213	11010101	D5
214	11010110	D6
215	11010111	D7
216	11011000	D8
217	11011001	D9
218	11011010	DA
219	11011011	DB
220	11011100	DC
221	11011101	DD
222	11011110	DE
223	11011111	DF
224	11100000	E0
225	11100001	E1
226	11100010	E2
227	11100011	E3
228	11100100	E4
229	11100101	E5

b10	b2	b16
230	11100110	E6
231	11100111	E7
232	11101000	E8
233	11101001	E9
234	11101010	EA
235	11101011	EB
236	11101100	EC
237	11101101	ED
238	11101110	EE
239	11101111	EF
240	11110000	F0
241	11110001	F1
242	11110010	F2
243	11110011	F3
244	11110100	F4
245	11110101	F5
246	11110110	F6
247	11110111	F7
248	11111000	F8
249	11111001	F9
250	11111010	FA
251	11111011	FB
252	11111100	FC
253	11111101	FD
254	11111110	FE
255	11111111	FF

Color Codes

The twisted-pair cable color code

This system has tip colors, ring colors, group colors, and binder colors. The twisted-pair color code is used to identify the number of a pair within a cable. For example, a pair in the Orange-Yellow binder group that is Yellow-Green is pair 418 (O-Y+Y-O). The twisted-pair color code in our telephone network is:

Pair codes for pairs 1 to 25 within binders.

Pair #	Tip color	Ring color	Abbreviation
1	white	blue	W-BL
2	white	orange	W-O
3	white	green	W-G
4	white	brown	W-BR
5	white	slate	W-S
6	red	blue	R-BL
7	red	orange	R-O
8	red	green	R-G
9	red	brown	R-BR
10	red	slate	R-S
11	black	blue	BK-BL
12	black	orange	BK-O
13	black	green	BK-G
14	black	brown	BK-BR
15	black	slate	BK-S
16	yellow	blue	Y-BL
17	yellow	orange	Y-O
18	yellow	green	Y-G
19	yellow	brown	Y-BR
20	yellow	slate	Y-S
21	violet	blue	V-BL

22	violet	orange	V-O
23	violet	green	V-G
24	violet	brown	V-BR
25	violet	slate	V-SL

A *binder* is a plastic ribbon wrapped around 25 pairs of wire or 600 pairs of wire. Binder codes are for identifying groups of 25 or 600. The binder code is the same as the pair code, only the group colors and the pair colors are reversed (e.g., white-blue becomes blue-white), so when the abbreviated colors are put together, they are easier to understand. Binder codes are good for 600 pairs of wire. After that, each group of 600 pairs gets an additional plastic binder.

Binder color code

Binder#	Binder color	Abbreviation	Pairs in binder
1	blue white	BL-W	1–25
2	orange white	O-W	26–50
3	green white	G-W	51–75
4	brown white	BR-W	76–100
5	slate white	S-W	101–125
6	blue red	BL-R	126–150
7	orange red	O-R	151–175
8	green red	G-R	176–200
9	brown red	BR-R	201–225
10	slate red	S-R	226–250
11	blue black	BL-BK	251–275
12	orange black	O-BK	276–300
13	green black	G-BK	301–325
14	brown black	BR-BK	326–350
15	slate black	S-BK	351–375
16	blue yellow	BL-Y	376–400
17	orange yellow	O-Y	401–425
18	green yellow	G-Y	426–450
19	brown yellow	BR-Y	451–475
20	slate yellow	S-Y	476–500
21	blue violet	BL-V	501–525
22	orange violet	O-V	526–550
23	green violet	G-V	551–575
24	brown violet	BR-V	576–600

Large cable binders:

Pair 1 to 600, surrounded by a white binder
Pair 601 to 1200, surrounded by a red binder
Pair 1201 to 1800, surrounded by a white binder

Fiber-optic color code

Groups of 12 fibers are placed in loose tube buffers:

Fiber Optic Color Code Reference

Fiber #	Blue Tube	Orange Tube	Green Tube	Brown Tube	Slate Tube	White Tube	Red Tube	Black Tube	Yellow Tube	Violet Tube	Rose Tube
Blue Strand	1	13	25	37	49	61	73	85	97	109	121
Orange Strand	2	14	26	38	50	62	74	86	98	110	122
Green Strand	3	15	27	39	51	63	75	87	99	111	123
Brown Strand	4	16	28	40	52	64	76	88	100	112	124
Slate Strand	5	17	29	41	53	65	77	89	101	113	125
White Strand	6	18	30	42	54	66	78	90	102	114	126
Red Strand	7	19	31	43	55	67	79	91	103	115	127
Black Strand	8	20	32	44	56	68	80	92	104	116	128
Yellow Strand	9	21	33	45	57	69	81	93	105	117	129
Violet Strand	10	22	34	46	58	70	82	94	106	118	130
Rose Strand	11	23	35	47	59	71	83	95	107	119	131
Aqua Strand	12	24	36	48	60	72	84	96	108	120	132

The resistor color code

Resistors have four color bands on them. They are regarded to as the first, second, third, and fourth bands. The first band is the closest to one side of the resistor and the following bands count to the inside. The first band indicates the first integer of the value of resistance. The second band indicates the second integer of the value of resistance. The third band indicates a multiplier or number of zeros to be placed after the first two band numbers. A diagram is the easiest way to demonstrate this code.

Color	band 1	band 2	×	band 3
Black	0	0	×	1
Brown	1	1	×	10
Red	2	2	×	100
Orange	3	3	×	1000
Yellow	4	4	×	10,000
Green	5	5	×	100,000
Blue	6	6	×	1,000,000
Violet	7	7	×	10,000,000
Gray	8	8	×	100,000,000
White	9	9		none

Wiring Standards

Arcnet 2   Used with Arcnetworks.

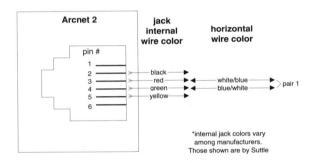

Ethernet 10/100 BaseT   Used in 10BaseT-100BaseT applications.

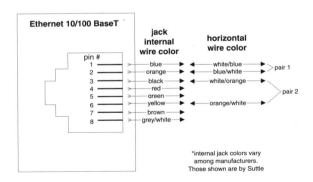

RS-232C   Used in serial data applications.

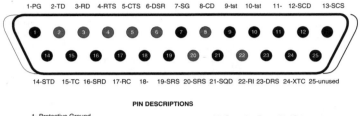

PIN DESCRIPTIONS

1- Protective Ground
2- Transmitted Data
3-Received data
4-Request To Send
5- Clear To Send
6 - Data Set ready
7 - Signal Ground
8 - Received Line Signal Detector
9 - Reserved for Data Set Testing
10 - Reserved for Data Set Testing
11 - Unassigned
12 - Secondary Recieved Line Signal Detector
13 - Secondary Clear to Send

14 - Secondary Transmitted Data
15 - Transmit Signal Element Timing
16 - Secondary Received Data
17 - Reciever Signal Element Timing
18 - Unassigned
19 - Secondary Request to Send
20 - Data Terminal ready
21 - Signal Quality Detector
22 - Ring Indicator
23 - Data Signal Rate Selector
24 - Transmit Signal Element Timing
25 - Unassigned

T568A   Used in Ethernet wiring applications.

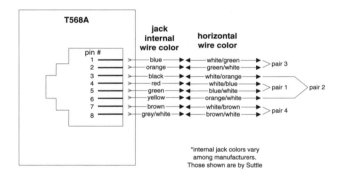

T568B   Used in Ethernet wiring applications.

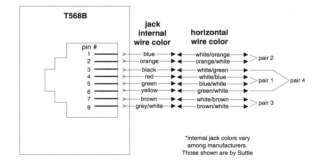

Token Ring   Used in token-ring networks.

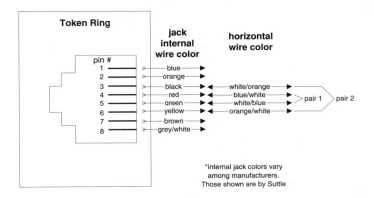

RJ11C USOC POTs   Used for plain old telephone service. Single-pair dial tone service.

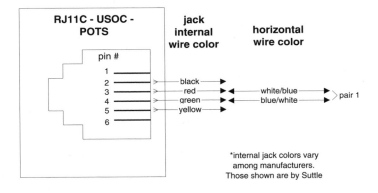

RJ61 USOC   Used for dial-modem and private-line modem connections.

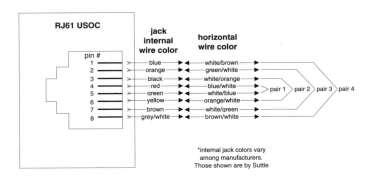

RJ45S USOC   Used for dial-modem and private-line modem connections.

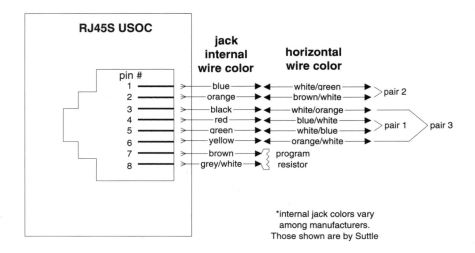

RJ48S USOC   Used for dial-modem and private-line modem connections.

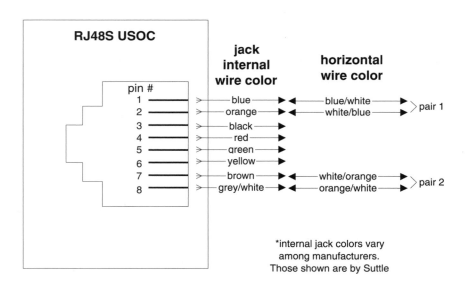

TP-PMD   Used for dial-modem and private-line modem connections.

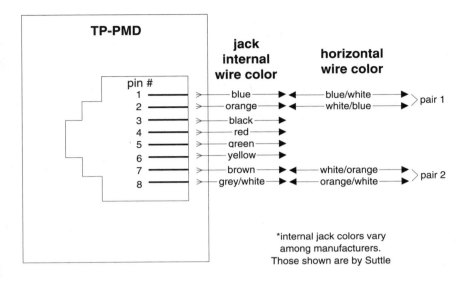

TP-PMD

jack internal wire color

horizontal wire color

pin #
1 ——— >— blue ——►◄——— blue/white ——►⟩ pair 1
2 ——— >— orange ——►◄——— white/blue ——►
3 ——— >— black ——►
4 ——— >— red ——►
5 ——— >— green ——►
6 ——— >— yellow ——►
7 ——— >— brown ——►◄——— white/orange ——►⟩ pair 2
8 ——— >— grey/white ——►◄——— orange/white ——►

*internal jack colors vary
among manufacturers.
Those shown are by Suttle

Index

Note: **Boldface** numbers indicate illustrations.

Q